T5-ADP-963

# IRON AND COPPER PROTEINS

# ADVANCES IN EXPERIMENTAL MEDICINE AND BIOLOGY



## Recent Volumes in this Series

Volume 67
ATHEROSCLEROSIS DRUG DISCOVERY
Edited by Charles E. Day • 1976

Volume 68
CURRENT TRENDS IN SPHINGOLIPIDOSES AND ALLIED DISORDERS
Edited by Bruno W. Volk and Larry Schneck • 1976

Volume 69
TRANSPORT PHENOMENA IN THE NERVOUS SYSTEM: Physiological and Pathological Aspects
Edited by Giulio Levi, Leontino Battistin, and Abel Lajtha • 1976

Volume 70
KININS: Pharmacodynamics and Biological Roles
Edited by F. Sicuteri, Nathan Back, and G. L. Haberland • 1976

Volume 71
GANGLIOSIDE FUNCTION: Biochemical and Pharmacological Implications
Edited by Giuseppe Porcellati, Bruno Ceccarelli, and Guido Tettamanti • 1976

Volume 72
FUNCTIONS AND METABOLISM OF PHOSPHOLIPIDS IN THE CENTRAL AND PERIPHERAL NERVOUS SYSTEMS
Edited by Giuseppe Porcellati, Luigi Amaducci, and Claudio Galli • 1976

Volume 73A
THE RETICULOENDOTHELIAL SYSTEM IN HEALTH AND DISEASE: Functions and Characteristics
Edited by Sherwood M. Reichard, Mario R. Escobar, and Herman Friedman • 1976

Volume 73B
THE RETICULOENDOTHELIAL SYSTEM IN HEALTH AND DISEASE: Immunologic and Pathologic Aspects
Edited by Herman Friedman, Mario R. Escobar, and Sherwood M. Reichard • 1976

Volume 74
IRON AND COPPER PROTEINS
Edited by Kerry T. Yasunobu, Howard F. Mower, and Osamu Hayaishi • 1976

# IRON AND COPPER PROTEINS

Edited by
Kerry T. Yasunobu and Howard F. Mower
University of Hawaii School of Medicine
Honolulu, Hawaii

and
Osamu Hayaishi
Kyoto University Faculty of Medicine
Kyoto, Japan

PLENUM PRESS • NEW YORK AND LONDON

Library of Congress Cataloging in Publication Data

Main entry under title:

Iron and copper proteins.

(Advances in experimental medicine and biology; v. 74)
"Proceedings of a symposium . . . held at the East West Center, University of Hawaii, December 15-17, 1975 and sponsored by the United States–Japan Cooperative Science Program."
Includes index.
1. Iron proteins–Congresses. 2. Copper proteins–Congresses. I. Yasunobu, Kerry Tsuyoshi, 1925- II. Mower, Howard F. III. Hayaishi, Osamu, 1920-
IV. United States–Japan Cooperative Science Program. V. Series.
QP552.I67I76 574.1'9245 76-20631
ISBN 0-306-39074-4

Proceedings of a symposium on Iron and Copper Proteins held at the East West Center-University of Hawaii, December 15-17, 1975 and sponsored by the United States–Japan Cooperative Science Program


A Division of Plenum Publishing Corporation
227 West 17th Street, New York, N.Y. 10011



Printed in the United States of America

# Preface

An Fe- and Cu-Protein Symposium was held on December 15-18, 1975 at the East West Center-University of Hawaii and was sponsored by the United States-Japan Cooperative Science Program under the auspices of the National Science Foundation and the Japan Society for the Promotion of Science.

It was recognized by the organizers of the symposium that metalloproteins are very important in the field of health science and a subject worthy of discussion by experts from the United States, Japan and Europe. The meeting was restricted to Fe- and Cu-proteins but this is still a very broad subject matter and therefore, selected topics of current interest in this field were chosen. This book contains the collected papers from most of the symposium participants.

The subject matter covered in this book is divided into four parts. These are: 1) the iron-sulfur proteins (which are not a part of the mitochondrial electron transport system); 2) the iron-sulfur proteins and heme proteins of the mitochondrial electron transport system; 3) other heme and nonheme iron proteins; and 5) selected copper proteins.

The organizers of the symposium wish to express their gratitude to the participants, the session chairmen, and Drs. I.C. Gunsalus and E. Frieden who assisted in the organization of the symposium. Also, we wish to express our gratitude to Dr. H. McKaughan, Director of Research and Dean of the Graduate School, University of Hawaii, Dr. George Cypher of the International Copper Association, and Damon Foundation for their special assistance in connection with the operation of the meeting and in the preparation of the book. Dr. O'Connell of the National Science Foundation and Mr. J. McMahon of the East West Center are congratulated for the excellent arrangement and operation of the meeting facilities. Finally, we wish to thank the staff of Plenum Publishing Co., especially Linda Coleman, for the efficient professional assistance received throughout the preparation of this book.

Kerry T. Yasunobu
University of Hawaii

Howard Mower
University of Hawaii

Osamu Hayaishi
Kyoto University

March 27, 1976

## Contents

Part I
IRON-SULFUR PROTEINS
(Non-mitochondrial electron transport proteins)


1. Structure and Function of Chloroplast-Type Ferredoxins . . . . . . . . . . . . . . . . . . . . . . . . 1
H. Matsubara, K. Wada, and R. Masaki

2. Structural Investigations of the Environment of the Iron-Sulfur Cluster of the 2-Iron Ferredoxins . . . . . . 16
J. Drum, K.T. Yasunobu, and R.E. Cramer

3. Studies on Bovine Adrenal Ferredoxin . . . . . . . . . . 36
S. Takemori, K. Suhara, and M. Katagiri

4. Some Insight into Intramolecular Electron Transfer of an Adrenodoxin Molecule . . . . . . . . . . . . . . . . 45
T. Kimura and T. Taniguchi

5. Structural Studies of Electron Transport Proteins from Sulfate Reducing Bacteria: The Amino Acid Sequence of Two Rubredoxins Isolated from *Desulfovibrio vulgaris* and *Desulfovibrio gigas* . . . . . 57
M. Bruschi and J. Le Gall

6. The Iron-Sulfur Centers and the Function of Hydrogenase from *Clostridium pasteurianum* . . . . . . . . 68
J.-S. Chen, L.E. Mortenson, and G. Palmer

7. Sequence Investigation of the *Clostridium pasteurianum* Nitrogenase: The Partial Amino Acid Sequence of Azoferredoxin . . . . . . . . . . . . . . . . . . 83
M. Tanaka, M. Haniu, K.T. Yasunobu, and L.E. Mortenson

8. On the Nature of an Intermediate that is Formed During the Enzymatic Conversion of Phenylalanine to Tyrosine . . . . . . . . . . . . . . . . . . . . . 91
S. Kaufman

9. Some Properties of Bovine Pineal Tryptophan Hydroxylase . . . . . . . . . . . . . . . . . . . . . 103
A. Ichiyama, H. Hasegawa, C. Tohyama, C. Dohmoto, and T. Kataoka

10. Evidence for Participation of NADH-Dependent Reductase in the Reaction of Benzoate 1,2-Dioxygenase (Benzoate Hydroxylase) . . . . . . . . . 118
H. Fujisawa, M. Yamaguchi, and T. Yamauchi

11. Subunit Structure of Nonheme Iron-Containing Dioxygenases . . . . . . . . . . . . . . . . . . . . . 127
M. Nozaki, R. Yoshida, C. Nakai, M. Iwaki, Y. Saeki, and H. Kagamiyama

Part II

IRON-SULFUR PROTEINS AND HEME PROTEINS OF THE MITOCHONDRIAL ELECTRON TRANSPORT SYSTEM

12. Iron-Sulfur Porteins, the Most Numerous and Most Diversified Components of the Mitochondrial Electron Transfer System . . . . . . . . . . . . . . . . . . . . . 137
H. Beinert

13. Composition and Enzymatic Properties of the Mitochondrial NADH- and NADPH-Ubiquinone Reductase (Complex I) . . . . . . . . . . . . . . . . . . . . 150
Y. Hatefi

14. Factors Controlling the Turnover Number of Succinate Dehydrogenase: A New Look at an Old Problem . . . . . . . . . . . . . . . . . . . . . . . . 161
B.A.C. Ackrell, E.B. Kearney, P. Mowery, T.P. Singer, H. Beinert, A.D. Vinogradov, and G.A. White

15. Biochemical and EPR Probes for Structure-Function Studies of Iron Sulfur Centers of Succinate Dehydrogenase . . . . . . . . . . . . . . . . . 182
T.E. King, T. Ohnishi, D.B. Winter, and J.T. Wu

16. Heme a and Copper Environments in Cytochrome Oxidase . . . . . . . . . . . . . . . . . . . . . . . . . 228
Y. Orii, S. Yoshida, and T. Iizuka

17. Heme Interactions in Pseudomonas Cytochrome Oxidase . . . 240
D.C. Wharton, K. Hill, and Q.H. Gibson

Part III
OTHER HEME AND NON-HEME Fe-PROTEINS

18. Equilibrium States and Dynamic Reactions of Iron in the Camphor Monoxygenase System . . . . . . . . . 254
I.C. Gunsalus and S.G. Sligar

19. Current Status of the Sequence Studies of the Pseudomonas putida Camphor Hydroxylase System . . . . . . 263
M. Tanaka, S. Zeitlin, K.T. Yasunobu, and I.C. Gunsalus

20. Highly Purified Cytochrome P-450 from Liver Microsomal Membranes: Recent Studies on the Mechanism of Catalysis . . . . . . . . . . . . . . . . . . 270
M.J. Coon, D.P. Ballou, F.P. Guengerich, G.D. Nordblom, and R.E. White

21. Characterization of Purified Cytochrome P-$450_{scc}$ and P-$450_{11\beta}$ from Bovine Adrenocortical Mitochondria . . . . . . . . . . . . . . . . . . . . . . 281
M. Katagiri, S. Takemori, E. Itagaki, K. Suhara, T. Gomi, and H. Sato

22. Purification and Properties of Cytochrome P-450 from Adrenocortical Mitochondria and Its Interaction with Adrenodoxin . . . . . . . . . . . . . . . . . . . . 290
T. Sugiyama, R. Miura, and T. Yamano

23. The Role of Cytochrome P-450 in the Regulation of Steroid Biosynthesis . . . . . . . . . . . . . . . . . . 303
P.F. Hall

24. On the Participation of Cytochrome P-450 in the Mechanism of Prevention of Hepatic Carcinogenesis . . . . 314
A.N. Saprin

25. Model System Studies of Axial Ligation in the Oxidized Reaction States of Cytochrome P-450 Enzymes . . . . . . . . . . . . . . . . . . . . . . . . 321
R.H. Holm, S.C. Tang, S. Koch, G.C. Papefthymiou, S. Foner, R.B. Frankel, and J.A. Ibers

26. A New Assay Procedure for Indoleamine 2,3-Dioxygenase . . 335
O. Hayaishi, F. Hirata, T. Ohnishi, R. Yoshida, and T. Shimizu

27. Is Indoleamine 2,3-Dioxygenase Another Heme and Copper Containing Enzyme? . . . . . . . . . . . . . . . . . 343
F.O.Brady and A. Udom

28. Copper Content of Indoleamine 2,3-Dioxygenase . . . . . . 354
F. Hirata, T. Shimizu, R. Yoshida, T. Ohnishi, M. Fujiwara, and O. Hayaishi

29. Pseudomonad and Hepatic L-Tryptophan 2,3-Dioxygenase . . 358
P. Feigelson

30. On the Prosthetic Groups of L-Tryptophan 2,3-Dioxygenase from Pseudomonas: Evidence for Noninvolvement of Copper in the Reaction . . . . . . . . . . . . . . . . . 363
Y. Ishimura and O. Hayaishi

31. The Search for Copper in L-Tryptophan 2,3-Dioxygenases. . 374
F.O. Brady

32. Comparison of Function of the Distal Base Between Myoglobin and Peroxidase . . . . . . . . . . . . . . . . 382
I. Yamazaki, Y. Hayashi, R. Makino, and H. Yamada

33. X-Ray Absorption Spectroscopy: Probing the Chemical and Electronic Structure of Metalloproteins . . . . . . . 389
W.E. Blumberg, P. Eisenberg, J. Peisach, and R.G. Shulman

34. Carbonate: Key to Transferrin Chemistry . . . . . . . . 400
G.W. Bates and G. Graham

Part IV
COPPER-PROTEINS

35. Oxidation and Reduction of Copper Ions in Catalytic Reactions of RHUS Laccase . . . . . . . . . . . . . . . . 408
T. Nakamura

36. Recent Studies on Copper Containing Oxidases . . . . . . 424
B. Mondoví, L. Morpurgo, G. Rotilio, and A. Finazzi-Agró

37. Collagen Cross-Linking: The Substrate Specificity of Lysyl Oxidase . . . . . . . . . . . . . . . . . . . . . . 438
R.C. Siegel, J.C.C. Fu and Y. Chang

38. Purification and Properties of Lung Lysyl Oxidase, a Copper-Enzyme . . . . . . . . . . . . . . . . . . . . . 447
J.J. Shieh and K.T. Yasunobu

39. Binuclear Copper Clusters as Active Sites for Oxidases . . . . . . . . . . . . . . . . . . . . . 464
H.S. Mason

40. An Interesting Reaction of Cupric Ions with Ferricyanide and Ferrocyanide . . . . . . . . . . . . 470
P. McMahill, N. Blackburn, and H.S. Mason

41. Formal Catalytic Mechanism of Ascorbate Oxidase . . . . 472
S.R. Burstein, B. Gerwin, H. Taylor, and J. Westley

42. The Involvement of Superoxide and Trivalent Copper in the Galatose Oxidase Reaction . . . . . . . . . . . 489
G.A. Hamilton, G.R. Dyrkacz, and R.D. Libby

43. The Biological Role of Ceruloplasmin and Its Oxidase Activity . . . . . . . . . . . . . . . . . . . . 505
E. Frieden and H.S. Hsieh

44. Superoxide Dismutases: Studies of Structure and Mechanism . . . . . . . . . . . . . . . . . . . . 530
I. Fridovich

45. Iron- and Manganese-Containing Superoxide Dismutases: Structure, Distribution, and Evolutionary Relationships . . . . . . . . . . . . . . . . . . . . . 540
J.M. McCord

46. Superoxide Dismutase in Photosynthetic Organisms . . . . 551
K. Asada, S. Kanematsu, M. Takahashi, and Y. Kona

47. Chemistry and Biology of Copper-Chelatin . . . . . . . . 565
K.V. Rajagopalan, D.R. Winge, and R. Premakumar

48. Circular Dichroism Study of Bovine Plasma Amine Oxidase, a Cu-Amine Oxidase . . . . . . . . . . . . . 575
H. Ishizaki and K.T. Yasunobu

Index . . . . . . . . . . . . . . . . . . . . . . . . . 589

# STRUCTURE AND FUNCTION OF CHLOROPLAST-TYPE FERREDOXINS[1]

Hiroshi Matsubara, Keishiro Wada and Ryuichi Masaki

Department of Biology, Faculty of Science, Osaka University, Toyonaka, Osaka (560), Japan

Ferredoxin is a general name given to a group of proteins containing equal amounts of non-heme iron and inorganic sulfur with low oxidation-reduction potentials and characteristic EPR signals. They are biologically active as electron carriers in diverse systems as reviewed in many occasions (Buchanan and Arnon, 1972; Hall et al., 1974; Lovenberg, 1973; Orme-Johnson, 1973; Matsubara, 1972). The amino acid sequences are exemplified by those of _Clostridium_ pasteurianum ferredoxin (Tanaka et al., 1966), spinach ferredoxin (Matsubara et al., 1967) and adrenodoxin (Tanaka et al., 1973).

Several unique papers on the structure-function relationships of the bacterial ferredoxins have recently been published. (Adman et al., 1973; Orme-Johnson, 1973; Lode et al., 1974a,b; Hong and Rabinowitz, 1970). However, the research on the ferredoxins of the chloroplast type are very limited and therefore, several experiments have been undertaken in our laboratory to elucidate the functional roles of amino acid residues in the ferredoxins isolated from the chloroplast.

## COMPARISON OF AMINO ACID SEQUENCE OF VARIOUS CHLOROPLAST-TYPE FERREDOXINS

So far only eight complete sequences of ferredoxins are known. They are ferredoxins from L. glauca (Benson and Yasunobu,1969), spinach (Matsubara et al., 1967), alfalfa (Keresztes-Nagy et al., 1969), taro (Rao and Matsubara, 1970), _Scenedesmus_ _quadricauda_[2] (Sugeno and Matsubara, 1969), _Aphanothece_ _sacrum_ (major) (Wada et al., 1975a; Hase et al., 1976), _Spirulina_ _maxima_ (Tanaka et al., 1975) and _Spirulina_ _platensis_ (Wada et al., 1975b). The partial amino acid sequences of ferredoxins from Equisetum telmateia (major) (Aggawal et al., 1971; Kagamiyama et al., 1975), Aphanothece sacrum (minor)[3], Zea mays[3], and Spirogyra sp.[3] have been published. Fig. 1 summarizes the distribution of various amino acid residues in the sequences of chloroplast-type ferredoxins.

Although tyrosine residues were once suggested to be the possible residues which function directly in the electron transfer reaction in clostridial-type ferredoxins (Adman et al., 1973), the sequence of Clostridium M-E ferredoxin (Tanaka et al., 1974) and semisynthetic ferredoxin derivatives of Clostridium acidi-urici ($Leu^2$)-ferredoxin (Lode et al., 1974a) and Clostridium M-E ($Leu^2$)-ferredoxin (Lode et al., 1974b) have revealed that the tyrosine residues have no direct functional role in the electron transfer reaction. Histidine and tryptophan are lacking in many clostridial-type ferredoxins and therefore, these aromatic residues are not involved in electron transfer.

Chloroplast-type ferredoxins have many tyrosine residues, but none of them are found in common positions in the sequence. X at position 39 is replaced by an unknown amino acid in Aphanothece ferredoxin (minor) and X at position 82 by an unknown residue in Equisetum ferredoxin. Tryptophan is not present in all of the ferredoxin isolated and the Spirogyra ferredoxin has no histidine residue. Thus, we conclude that these residues are not involved directly in electron transfer reactions of chloroplast-type ferredoxin.

The most important residues in all ferredoxin molecules are the cysteine residues which chelate the non-heme iron atoms which are bridged by sulfide atoms to produce the iron-sulfur cluster. Chloroplast-type ferredoxins have been reported to have four to six cysteine residues depending on their origins. Although no three-dimensional structural data is available yet, only four cysteine residues are thought to be indispensable for the construction of the iron-sulfur cluster according to physicochemical experiments and studies of model compounds (Orme-Johnson, 1973; Sands and Dunham, 1975; Hall, D.O. et al., 1975). This suggestion was supported by chemical studies. The first clue came from the amino acid composition of the cotton ferredoxin which contained only four cysteine residues (Newman et al., 1969). However, the location of these four cysteine residues in ferredoxin were not determined. The partial sequence studies of horsetail ferredoxin, which also contains only four cysteine residues, established that the four residues which probably chelate the iron atoms are residues 39, 44, 47, and 77 present in spinach ferredoxin (Aggawal et al., 1971; Kagamiyama et al., 1975). These residues are circled in Fig. 1. The sequence studies of ferredoxins from Aphanothece sacrum (major) (Wada et al., 1975a; Hase et al., 1976), Aphanothece sacrum (minor) (Hase and Matsubara[3]) and

Zea mays (Matsubara et al.[3]) have confirmed the above mentioned comments about the cysteine residues.

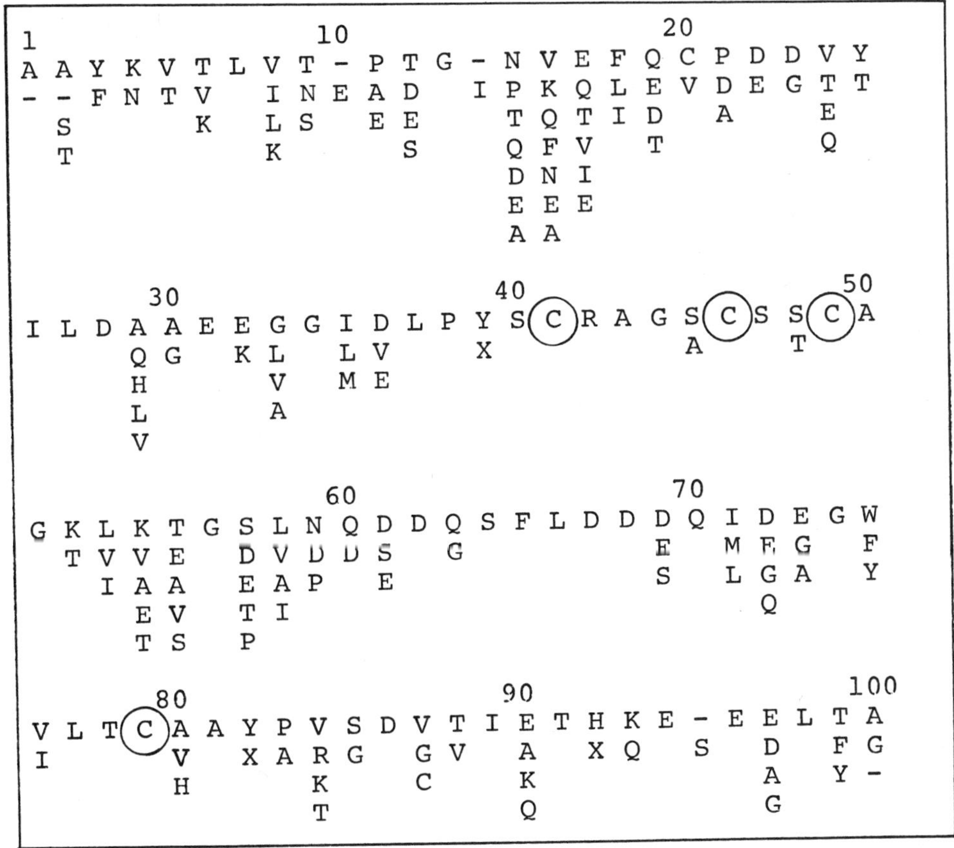


Fig. 1. Distribution of various amino acids in the chloroplast-type ferredoxin sequences. Spinach ferredoxin sequence is used as a standard (upper row with three gaps). These gaps are inserted to obtain maximum homology of the various ferredoxin sequences. Various residues occupying the corresponding loci in other ferredoxins are listed below the standard sequence. X and - represent unknown amino acids and deletions respectively. Circles show the probable essential cysteine residues which are chelated to the iron atoms in the molecule. See the text for the origins of ferredoxins. The residue 9 in Spirulina platensis ferredoxin has been revised to be asparagine instead of aspartic acid (Wada et al., 1975b).

TABLE I

The amino acid composition of pyridylethyl-ferredoxin of _Spirulina platensis_ obtained after the reaction with vinylpyridine in the presence of guanidine-HCl

| Amino acid | Lysine | Histidine | Pyridylethyl- | Arginine |
|---|---|---|---|---|
| Moles/mole of protein | 2.00(2) | 0.97 (1) | 1.83 (2) | 1.00 (1) |

_Spirulina platensis_ ferredoxin dissolved in 0.1 M Tris-HCl buffer solution containing 0.7 M NaCl, pH 7.5, was reacted with 4-vinylpyridine (8 fold molar excess) in the presence of 5 volumes of 6 M guanidine-HCl, pH 7.5, for 60 min. The reaction mixture was passed through a Sephadex column and an aliquot of apoferredoxin solution was analyzed for its amino acid composition. Only the basic residues are shown in this table.

_Spirulina platensis_ ferredoxin has six cysteine residues (Wada et al. 1975b) and none of them reacted with an excess amount of 4-vinylpyridine in the native state. However, two out of six cysteine residues reacted with this reagent in the presence of 6 M guanidine-HCl. The experimental conditions and the analytical data are shown in Table I. This product which contained two residues of S-pyridylethyl cysteine could be reconstituted with 2-mercaptoethanol, $Na_2S$ and $FeSO_4$ by the method of Hong and Rabinowitz (1967) to produce an active ferredoxin derivative. The absorption spectrum of this reconstituted material was quite similar to that of the unmodified ferredoxin and therefore,strongly supports the concept that four cysteine residues are needed for the construction of the iron-sulfur cluster in the chloroplast-type ferredoxin. The result is in agreement with the observation that one of five cysteine residues in spinach ferredoxin was easily carboxymethylated under the right conditions without perturbation (Petering and Palmer, 1970) of the ferredoxin.

## SUSCEPTIBILITY OF CARBOXYL-TERMINAL RESIDUES OF SPINACH FERREDOXIN TO CARBOXYPEPTIDASE A HYDROLYSIS AND THE EFFECT OF THESE RESIDUES ON THE RECONSTITUTION BY IRON AND SULFIDE

The carboxyl-terminal sequence of spinach ferredoxin is -Leu-Thr-Ala (Matsubara et al., 1967) and the dependence of structural integrity of the iron-sulfur cluster on this terminal backbone was examined.

Spinach ferredoxin was incubated with carboxypeptidase A at pH

8 in the presence of ascorbic acid which stabilized ferredoxin from both denaturation during storage and proteolysis (Matsubara and Shin[3]).

TABLE II

Susceptibility of carboxyl-terminal residues of spinach ferredoxin to carboxypeptidase A

| | | Number of residues released by carboxypeptidase A degestion | |
|---|---|---|---|
| | | untreated ferredoxin | trichloroacetic acid-treated ferredoxin |
| Leucine | | 0.11 | 0.94 |
| Threonine | | 0.27 | 0.99 |
| Alanine | | 1.00 | 1.00 |
| Absorbance at 420 nm | before digestion | 0.59 | |
| | after digestion | 0.58 | |
| $A_{420}/A_{270}$ | before digestion | 0.47 | |
| | after digestion | 0.46 | |

Spinach ferredoxin ($A_{420}/A_{270}$=0.47) dissolved in 0.1 M Tris-HCl buffer solution, pH 8.0, was digested with carboxypeptidase A (1/25 w/w) for 20 hr at room temperature in the presence of ascorbic acid (4 mg/ml). After digestion, the reaction mixture was passed through a Sephadex column and the absorption spectrum of the digested ferredoxin was recorded. An aliquot was analyzed for the amino acid composition. As a control experiment, the trichloroacetic acid treated apoferredoxin was digested by the enzyme under the same conditions.

The carboxyl-terminal alanine was released completely in the presence of ascorbate without any significant change of the absorbance in the visible wavelength spectrum and in the absorbance ratio between the visible and ultraviolet region (Table II). On the contrary, carboxypeptidase A released quantitatively the three carboxyl-terminal residues from the apoferredoxin prepared by trichloroacetic acid treatment. These results are consistent with the structural features of other chloroplasts-type ferredoxins such as those from Scenedesmus and Aphanothece which lack one residue corresponding to the carboxyl-terminal alanine in spinach ferredoxin. A similar observation was also previously reported for a bacterial ferredoxin (Hong and Rabinowitz, 1970).

After removal of the three residues from carboxyl-terminal region of the spinach apoferredoxin, we tried to reconstitute the ferredoxin lacking these residues by the addition of $FeSO_4$ and $Na_2S$ in the presence of 6 M guanidine-HCl and 2-mercaptoethanol under nitrogen. The mixture was dialyzed against 0.01 M Tris-HCl buffer solution containing 0.07 M NaCl, pH 7.6, in the presence of ascorbic acid. The solution did not show any color typical for the chloroplast-type ferredoxin. On the other hand, the apoferredoxin without proteolysis could be reconstituted (about 30-35% recovery) to ferredoxin after the same treatment as mentioned above. The reconstituted ferredoxin was resistant to endopeptidases in the presence of ascorbic acid as observed for the native ferredoxin (Matsubara and Shin[3]).

These results suggest that the carboxyl-terminal alanine of spinach ferredoxin is exposed and susceptible to hydrolysis by carboxypeptidase A and is not important at all for conformational stability of the protein. However, the neighboring threonine of leucine residues or both residues are conformationally important.

### ACETYLATION OF AMINO GROUPS IN SPINACH FERREDOXIN AND SALT EFFECT ON ACETYLATED FERREDOXIN

As shown in Fig. 1, there are no lysine residues which occupy identical positions when the chloroplast-type ferredoxin sequences are compared. Therefore, the lysine residues are probably not involved directly in the electron transfer reaction, but it may be involved in producing a stable conformation of ferredoxin and a reactive site against the flavin enzyme, ferredoxin-NADP reductase (Shin et al. 1963).

Spinach ferredoxin has four lysine residues and one terminal amino group. Of these five residues, about four residues were acetylated after the reaction with a 50-fold molar excess of acetic anhydride. The reaction product was rather stable when it was kept in a phosphate buffer solution, pH 7.5, instead of Tris-HCl buffer solution usually used for the other experiments. The absorption spectrum of this product was slightly different from that of the native ferredoxin, particularly from 420 to 470 nm as shown in Fig.2. This spectrum is, however, reversed to that of the native ferredoxin by the addition of certain amounts of NaCl, $MgCl_2$ or $MgSO_4$. In the discussion to follow, the modified form is designated the D-form and the reversed form is the N-form (compared with the native or N-form ferredoxin). The D-form showed shifts of the absorption maxima to the shorter wavelengths, decreased absorbances in the ultraviolet region, increased absorbances in the visible region, and a disappearance of the maximum near 465 nm.

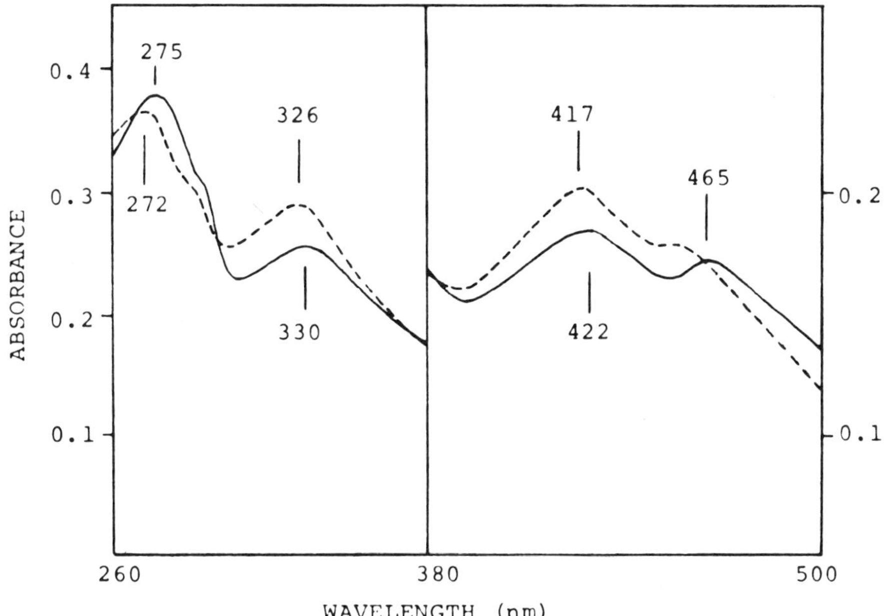


Fig. 2. Absorption spectra of native ferredoxin, N-form of ferredoxin and D-form of ferredoxin. Spinach ferredoxin was acetylated with a 50 fold molar excess of acetic anhydride. About four amino groups were modified according to the trinitrophenylation method (Habeeb, 1966). The spectrum of the modified ferredoxin (D-form) was recorded in 5 mM phosphate buffer solution, pH 7.5, and is shown by the dotted line. The spectrum of the reversed form (N-form) was recorded in 5 mM phosphate buffer solutin, pH 7.5, containing 1 M NaCl and is shown by the solid line. This spectrum is indistinguishable from that of the native ferredoxin.

The absorption spectrum of D-form resembled closely that of ferredoxin treated with denaturing agents (Petering and Palmer, 1970; Padmanabhan and Kimura, 1970) or that of model iron-sulfur compounds (Mayerle et al. 1973). The reversibility of the spectra between D-form and N-form of the modified ferredoxin was dependent on the concentration and the kind of salt used. There were isosbestic points in the absorption spectra at 266, 292, 370 and 465 nm.

Since the absorbance change of D-form upon the addition of ions was most remarkable at around 420-470 nm, the absorbance ratio, R= $A_{420}/A_{465}$, was plotted against the various concentrations of $Na^+$ or $Mg^{++}$ added to the solution of D-form of ferredoxin. The results are shown in Fig. 3.

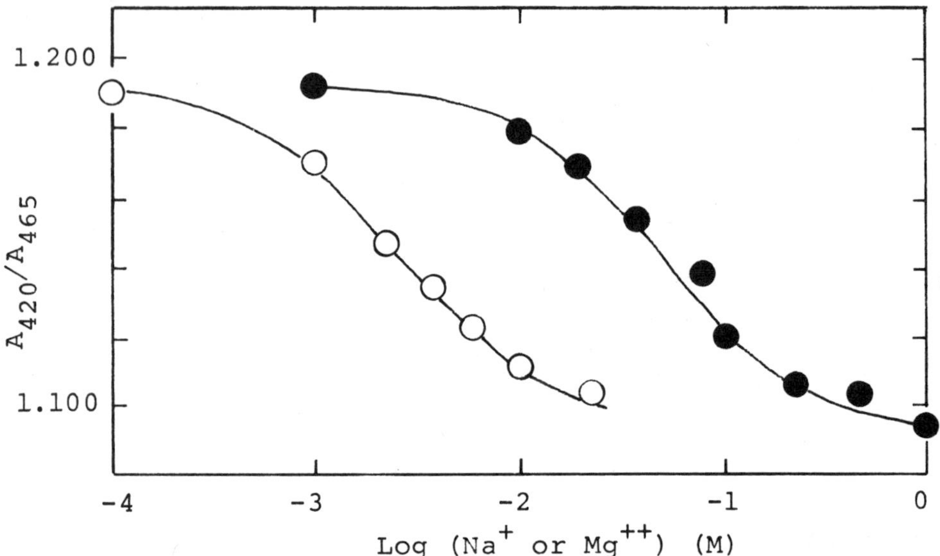


Fig. 3. Effects of Na+ and Mg++ ions on the absorbance ratio, $A_{420}/A_{465}$. The absorbance change of D-form of ferredoxin (prepared as mentioned in Fig. 2) was recorded after the addition of various concentrations of Na+ or Mg++ in 5 mM phosphate buffer solution, pH 7.5. The curve with open circles is for the Mg++ ion and that with filled circles for the Na+ ion.

The highest value of R, 1.191, was obtained at 5 mM Na+ ion and the lowest value, 1.094, for 1 M Na+ ion. The inflection point of the curve was observed at about $6.5 \times 10^{-2}$ M for Na+. The inflection point for Mg++ ion was about $3 \times 10^{-3}$ M. A similar phenomenon was previously reported for spinach ferredoxin in 5 M urea and 1 M NaCl under the anaerobic conditions (Petering and Palmer, 1970). They studied the salt and $O_2$ effect on the conformation and iron-sulfur structure of ferredoxin by absorption and CD spectral methods. The denaturing effect of urea was reversed by the addition of salt and it was suggested that urea and high ionic strength affected independently the protein conformation without altering the structure of the Fe-S cluster.

Padmanabhan and Kimura (1969,1970) suggested from the measurements of the optical activity of ferredoxins with two irons and two sulfurs that the iron-sulfur chromophore of these proteins was optically inactive and that there was an asymetric interaction of this center with the polypeptide chain.

We have studied the salt effect on the acetylated spinach ferredoxin by CD spectral measurements (Masaki et al[3]). The CD spectrum of the acetylated ferredoxin was essentially the same as the spectrum of the ferredoxin treated with denaturing agents (Petering and Palmer, 1970; Padmanabhan and Kimura, 1970) and the CD spectrum was completely reversed to that of the native ferredoxin by the addition of salts.

These observations together with the results obtained from the Hill plots for cation binding calculated from the curve in Fig. 3, have lead us to suggest that when the amino groups in spinach ferredoxin were acetylated, a salt bridge which normally exists between an amino group and an acidic group was broken. This change probably lead to a repulsion between the acidic group and a neighboring acidic group which also loosened the interaction between the iron-sulfur center and polypeptide backbone. However, the iron-sulfur structure itself was left intact. This effect was eliminated by adding ions which bind to at least one of the acidic group involved in the interaction.

The acetylated ferredoxin lost the ability to form a complex with NADP-ferredoxin reductase, but by adding ions the binding was restored to the same extent as observed for native ferredoxin (Masaki et al.[3]). Therefore, the amino group themselves have no practical role for complex formation with NADP-ferredoxin reductase.

## STATUS OF TRYPTOPHAN AND ARGININE RESIDUE IN CHLOROPLAST-TYPE FERREDOXINS

The tryptophan residue at position 73 in spinach ferredoxin is replaced by other aromatic residues, namely phenylalanine or tyrosine in the other ferredoxins which are shown in Fig. 1. The sole tryptophan residue in spinach ferredoxin was used as a structural probe by monitoring its fluorescence.

When spinach ferredoxin was excited at 295 nm, only a very weak emission spectrum appeared at around 345 nm. When the acetylated ferredoxin prepared as mentioned in the previous section was excited at 295 nm, a larger emission spectrum was observed as shown in Fig. 4. This spectrum was very similar to that recorded immediately after the treatment of ferredoxin with 4 M guanidine-

HCl at pH 7.5. The emission spectrum of this ferredoxin approached that of free tryptophan if left standing for over 2 hrs until the ferredoxin was completely denatured. The emission spectrum of acetyl-ferredoxin was reversed to the spectrum of the native ferredoxin at high ionic strengths which produce the N-form of ferredoxin.

Fig. 5 shows the effect of Na+ concentration on the fluorescence of the tryptophan residue in acetyl ferredoxin. The inflection point occured at about $5 \times 10^{-2}$ M Ma+ concentration which is very close to the value obtained from the spectral absorption changes in the visible wavelength regions. Table III gives the quantum yields of the tryptophan residues obtained under various conditions.

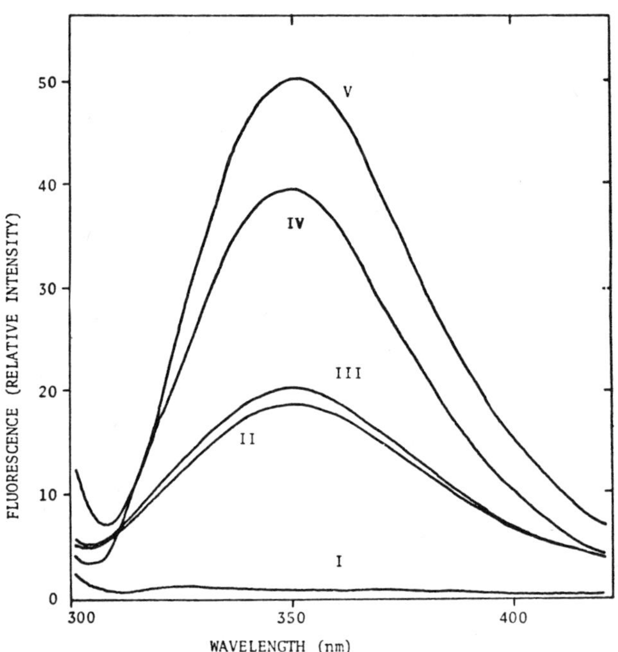


Fig. 4. Fluorescence spectra of spinach ferredoxin under the various conditions. The excitation wavelength was 295 nm and the emission spectrum were recorded. I, Native ferredoxin dissolved in 5 mM phosphate buffer solution, pH 7.5; III, the acetylferredoxin dissolved in the same buffer solution; II, the ferredoxin in 4 M guanidine-HCl (recorded immediately after addition of the denaturant); and IV, the ferredoxin in 4 M guanidine-HCl (recorded long after addition of denaturant ; and V, free tryptophan.

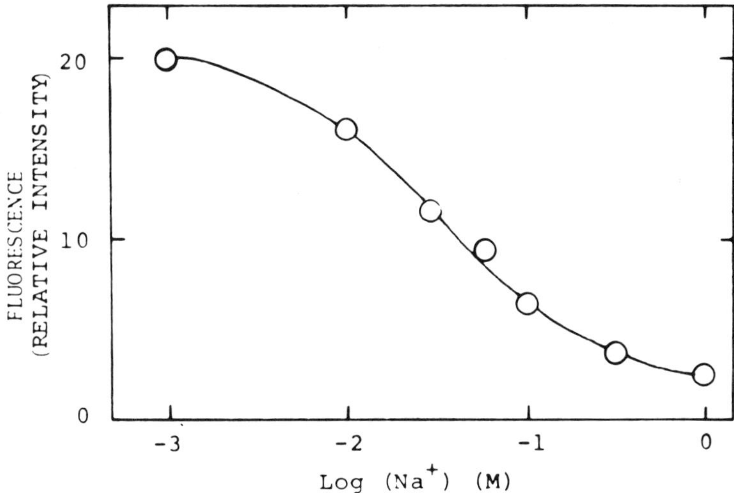


Fig. 5. Effects of Na+ concentrations on the flourescence of the tryptophan residue in the acetylated spinach ferredoxin. The acetylferredoxin in 5 mM phosphate buffer solution, pH 7.5 was excited at 295 nm in the presence of various concentrations of Na+ ion and the intensities of the flourescence at 350 nm were plotted.

TABLE III

Quantum yields of tryptophan residue of spinach ferredoxin under the different conditions

| Ferredoxin | Condition | Quantum yield at 350 nm |
|---|---|---|
| Native ferredoxin | in low ionic concentrations | 0.008 |
| | in high ionic concentrations | 0.004 |
| | in 4 M guanidine-HCl | |
| | immediately after addition | 0.074 |
| | over 2 hr after addition | 0.159 |
| Acetylated ferredoxin | at low ionic strength | 0.082 |
| | at high ionic strength | 0.011 |
| Free tryptophan | - | 0.20 |

The native spinach ferredoxin, acetylated ferredoxin and free tryptophan in 5 mM phosphate buffer solution, pH 7.5, were excited at 295 nm and the quantum yields were calculated from the intensities of fluorescence at 350 nm.

These results suggest that the fluorescence changes of the tryptophan residue is closely related to changes in the gross structure near the chromophore iron-sulfur center and that probably the tryptophan residue occupies a position near the center. Incidently the tryptophan residue 73 in spinach ferredoxin is located close to one of the four essential chromophore-forming cysteine residues, namely at position 77.

The common arginine residue present in all chloroplast-type ferredoxin is found at position 42 as shown in Fig. 1. The accessibility of this residue to a modifying reagent was first examined with *Spirulina platensis* ferredoxin. The native ferredoxin and the completely denatured ferredoxin were treated with 0.05 M cyclohexanedione in 0.2 M sodium borate, pH 9.0 at $37^{\circ}$ for about 1 hr and the reactions were terminated with acetic acid. The extent of modification of arginine residue was determined by amino acid analysis after HCl hydrolysis (Patthy and Smith, 1975). The arginine residue in the denatured ferredoxin was completely modified under these conditions by the reagent, but in the native ferredoxin the modification of the arginine residue was less than 50%. Therefore, this arginine residue in ferredoxin is probably in a state which is not completely accessible to this reagent.

## SUMMARY

Comparison of various chloroplast-type ferredoxin sequences, chemical and enzymic modifications, reconstitution experiments, and fluorescence measurement of chloroplast-type ferredoxins have led to the following conclusions.

1. Tyrosine, histidine, and tryptophan residues are not directly involved in the oxidation-reduction mechanism of ferredoxins. The four indispensible cysteine residues in spinach ferredoxin which constitutes a part of the iron-sulfur cluster are located at residues 39, 44, 47 and 77. Two out of six cysteine residues in *Spirulina* ferredoxin could be easily modified with vinylpyridine without the loss of reconstitutive ability i.e. the apoferredoxin could be converted to the holoform by the addition of iron and sulfide.

2. Spinach ferredoxin was digested with carboxypeptidase A and the terminal alanine could be removed without loss of the spectral properties of native ferredoxin. However, the removal of the terminal three residues gave rise to the loss of reconstitutive ability.

3. The amino groups of spinach ferredoxin were modified by acetic anhydride and four residues were acetylated. The acetylated preparation of ferredoxin had an unique spectrum. Upon the addition of high concentration of ions the spectrum of this derivative resembled the spectrum of native ferredoxin. Acetylferredoxin did not combine with ferredoxin-NADP reductase, but upon the addition of moderate concentrations of cations, it did bind to this enzyme.

We believe that a salt bridge exists between an amino group and an acidic group in native ferredoxin which affects the Fe-S cluster in ferredoxin.

4. Flourescence measurements of the sole tryptophan residue in spinach ferredoxin was made under various conditions and we have concluded that tryptophan 73 in spinach ferredoxin was located near the iron-sulfur center.

The common arginine residue which occurs at position 42 in *Spirulina* ferredoxin, was treated with cyclohexanedione and it was found that this residue in native ferredoxin was rather resistant to this reagent. This arginine residue is probably in a position which is readily accessible to the reagent.

## ACKNOWLEDGEMENTS

We express our thanks to Dr. M. Shin, Mr. T. Hase, Mrs. H. Tokunaga, Mr. M. Ohmiya and Mr. H. Koike for their cooperation and discussions in various parts of the experiments. We also thank Dr. D.I. Arnon and Mr. R.K. Chain for the gift of NADP-ferredoxin reductase and Drs. D.O. Hall and K.K. Rao for the permission of citing *Zea mays* ferredoxin work.

## FOOTNOTES

1. This work was supported in part by a grant from the Japanese Ministry of Education.

2. The green alga previously reported as an unknown species *Scenedesmus* sp., has recently been identified by Dr. Mark A. Ragan, Dept. of Biology, Dalhousie University, Halifax, Nova Scotia, Canada, as variety of *Scenedesmus quadricauda*.

3. Unpublished results.

## REFERENCES

Adman, E.T., Sieker, L.C., and Jensen, L.H. 1973. The Structure of a Bacterial Ferredoxin. *J. Biol. Chem.* 248: 3987-3996.

Aggawal, S.J., Rao, K.K. and Matsubara, H. 1971. Horsetail Ferredoxin: Isolated and Some Chemical Studies. *J. Biol. Chem.* 69: 601-603.

Benson, A.M. and Yasunobu, K.T. 1969. Non-Heme Iron Proteins X. The Amino Acid Sequences of Ferredoxin from Leucaena glauca. *J. Biol. Chem.* 244: 955-963.

Buchanan, B.B. and Arnon, D.I. 1970. Ferredoxins: Chemistry and Function in Photosynthesis, Nitrogen Fixation, and Fermentative Metabolism. *Adv. in Enzymol.* 33: 119-176.

Habeeb, A.F.S.A. 1966. Determination of the Free Amino Groups in Proteins by Trinitrobenzenesulfonic Acid. *Anal. Biochem.* 14: 328-336.

Hall, D. O., Cammack, R. and Rao, K. K. 1974. The Iron-Sulfur Proteins: Evolution of a Ubiquitous Protein from Model Systems to Higher Organisms. Origins of Life 5: 363-386.

Hall D. O., Rao, K. K. and Cammack, R. 1975. The Iron-Sulfur Proteins: Structure, Function and Evolution of a Ubiquitous Group of Proteins. Sci. Prog., Oxford. 62: 285-317.

Hase, T., Wada, K. and Matsubara, H. 1976. The Amino Acid Sequence of the Major Component of Aphanothece sacrum Ferredoxin. J. Biochem. 79: in press.

Hong, J.-S. and Rabinowitz, J. C. 1967. Preparation and Properties of Clostridial Apoferredoxins. Biochem. Biophys. Res. Commun. 29: 246- 252.

Hong, J.-S. and Rabinowitz, J. C. 1970. The Effects of Chemical Modifications on the Reconstitution, Activity, and Stability of Clostridial Ferredoxin. J. Biol. Chem. 245: 4988-4994.

Kagamiyama, H., Rao, K. K., Hall, D. O., Cammack, R. and Matsubara, H. 1975. Equisetum (Horsetail) Ferredoxin: Characterization of the Active Center and Positions of the Four Cysteine Residues in this 2Fe-2S Protein. Biochem. J. 145: 121-123.

Keresztes-Nagy, S., Perini, F. and Margoliash, E. 1969. Primary Structure of Alfalfa Ferredoxin. J. Biol. Chem. 244: 981-955.

Lode, E. T., Murray, C. L., Sweeney, W. V. and Rabinowitz, J. C. 1974a. Synthesis and Properties of C. acidi-urici [Leu]-Ferredoxin: A function of the Peptide Chain and Evidence Against the Direct Role of the Aromatic Residue in Electron Transfer. Proc. Natl. Acad. Sci., USA 71: 1361-1365.

Lode, E. T., Murray, C. L. and Rabinowitz, J. C. 1974b. Semisynthetic synthesis and Biological Activity of a Clostridial-type Ferredoxin Free of Aromatic Amino Acid Residues. Biochem. Biophys. Res. Commun. 61: 163-169.

Lovenberg, W. (ed.) 1973. Iron-Sulfur Proteins, I and II. Acad. Press, New York and London.

Matsubara, H. 1972. Chemistry and Evolution of Ferredoxin Molecules. in Aspects of Cellular and Molecular Physiology (K. Hamaguchi, ed.) pp. 31-55, University of Tokyo Press, Tokyo.

Matsubara, H., Sasaki, R. M. and Chain, R. K. 1967. The Amino Acid Sequence of Spinach Ferredoxin. Proc. Natl. Acad. Sci., USA 57: 439-445.

Mayerle, J. J., Frankel, R. B., Holm, R. H., Ibers, J. A., Phillips, W. D. and Weiher, J. F. 1973. Synthetic Analogs of the Active Sites of Iron-Sulfur Proteins. Structure and Properties of Bis [o-xylyldithiolato-$\mu_2$-Sulfidoferrate(III)], an Analog of the 2Fe-2S Proteins. Proc. Natl. Acad. Sci., USA 70: 2429-2433.

Newman, D. J., Ihle, J. N. and Dure, III, L. 1969. Chemical Composition of a Ferredoxin Isolated from Cotton. Biochem. Biophys. Res. Commun. 36: 947-950.

Orme-Johnson, W. H. 1973. Iron-Sulfur Proteins: Structure and Function. Ann. Rev. Biochem. 42: 159-204.

Orme-Johnson, W. H. 1973. Tryptic Cleavage of Clostridium acidi-urici Apoferredoxin, and Reconstitution of the Separated Fragments. Biochem. Soc. Trans. 1: 4-5.

Padmanabhan, R. and Kimura, T. 1969. Extrinsic Properties of Optical Activity in Adrenal Iron-Sulfur Protein (Adrenodoxin). Biochem. Biophys. Res. Commun. 37: 363-368.

Padmanabhan, R. and Kimura, T. 1970. Studies on Adrenal Steroid Hydroxylases. Extrinsic Properties of the Optical Activity in Adrenal Iron-Sulfur Protein (Adrenodoxin). J. Biol. Chem. 245: 2469-2475.

Patthy, L. and Smith, E. L. 1975. Identification of Functional Arginine Residues in Ribonuclease A and Lysozyme. J. Biol. Chem. 250: 565-569.

Petering, D. H. and Palmer, G. 1970. Properties of Spinach Ferredoxin in Anaerobic Urea Solution: A Comparison with the Native Protein. Arch. Biochem. Biophys. 141: 456-464.

Rao, K. K. and Matsubara, H. 1970. The Amino Acid Sequence of Taro Ferredoxin. Biochem. Biophys. Res. Commun. 38: 500-506.

Sands, R. H. and Dunham, W. R. 1975. Spectroscopic Studies on Two-Iron Ferredoxin. Quarterly Reviews of Biophysics 7: 443-504.

Shin, M., Tagawa, K. and Arnon, D. I. 1963. Crystallization of Ferredoxin-TPN Reductase and its Role in the Photosynthetic Apparatus of Chloroplasts. Biochem. Z. 338: 84-96.

Sugeno, K. and Matsubara, H. 1969. The Amino Acid Sequence of Scenedesmus Ferredoxin. J. Biol. Chem. 244: 2979-2989.

Tanaka, M., Haniu, M., Yasunobu, K.T., Jones, J. B. and Stadtman, T. C. 1974. Amino Acid Sequence Determination of the Clostridium M-E Ferredoxin and a Comment on the Role of the Aromatic Residues in the Clostridial Ferredoxin. Biochemistry 13: 5284-5289.

Tanaka, M., Haniu, M., Yasunobu, K.T. and Kimura, T. 1973. The Amino Acid Sequence of Bovine Adrenodoxin. J. Biol. Chem. 248: 1141-1157.

Tanaka, M., Haniu, M., Zeitlin, S. and Yasunobu, K.T. 1975. Amino Acid Sequence of the Spirulina maxima Ferredoxin: A Ferredoxin from a Procaryote. Biochem. Biophys. Res. Commun. 64: 399-407.

Tanaka, M., Nakashima, T., Benson, A., Mower, H. and Yasunobu, K.T. 1965. The Amino Acid Sequence of Clostridium pasteurianum Ferredoxin. Biochemistry 5: 1666-1681.

Wada, K., Hase, T. and Matsubara, H. 1975a. Evolutionary Information Involved in Primary Structure of Chloroplast-type Ferredoxin. J. Biochem. 78: 637-639.

Wada, K., Hase, T., Tokunaga, H. and Matsubara, H. 1975b. Amino Acid Sequence of Spirulina platensis Ferredoxin: A Far Divergency of Blue-Green Algal Ferredoxins. FEBS Lett. 55: 102-104.

# STRUCTURAL INVESTIGATIONS OF THE ENVIRONMENT OF THE IRON-SULFUR CLUSTER OF THE 2-IRON FERREDOXINS

J. Drum, K.T. Yasunobu and R.E. Cramer*

Dept. of Biochemistry and Dept. of Chemistry*

University of Hawaii, Honolulu, Hawaii 96822

Ferredoxins are a class of proteins involved in electron transfer reactions. Their distinguishing characteristics is the presence of sulfur-bound ("non-heme") iron atoms through which electrons are carried. These proteins include the rubredoxins (1), which contain ∿50 residues and a single iron, coordinated by 4 cysteines, as shown in Fig. 1A; the "clostridial" ferredoxins (2), possess ∿60 residues and 2 active sites of the type shown in Fig. 1B; and the photosynthetic plant and algal ferredoxins (3) which have ∿100 residues and an active site containing 2 irons, 2 labile sulfurs, and 4 cysteines. Adrenodoxin (4) and putidaredoxin (5) have similar iron-sulfur centers, though their sequences are quite dissimilar. The sequences of 7 of the 2-iron ferredoxins have been determined (6-12) and are illustrated in Fig. 2. Difference matrices, shown in Fig. 3, reveal a high degree of invariance and conservatism in these sequences. Horsetail (13) and Suringer (14) ferredoxins have also been partially sequenced, and both lack cysteine 18, indicating that cysteines 39, 44, 47, and 77 coordinate the iron-sulfur center.

The structure of this center in the 2-iron ferredoxins is not known with certainty, since their x-ray structure has not been determined. However, various forms of evidence (15,16) suggest that it is a bridged structure, as depicted in Fig. 1C or 1D. In addition synthetic analogs of the iron-sulfur cluster of the 2-iron ferredoxins have been made (17), and their iron coordination is tetrahedral, as in Fig. 1D. Anacystis nidulans ferredoxin (18) has been successfully crystallized, and x-ray work on it is now in progress; however, no sequence work on this ferredoxin has been reported.

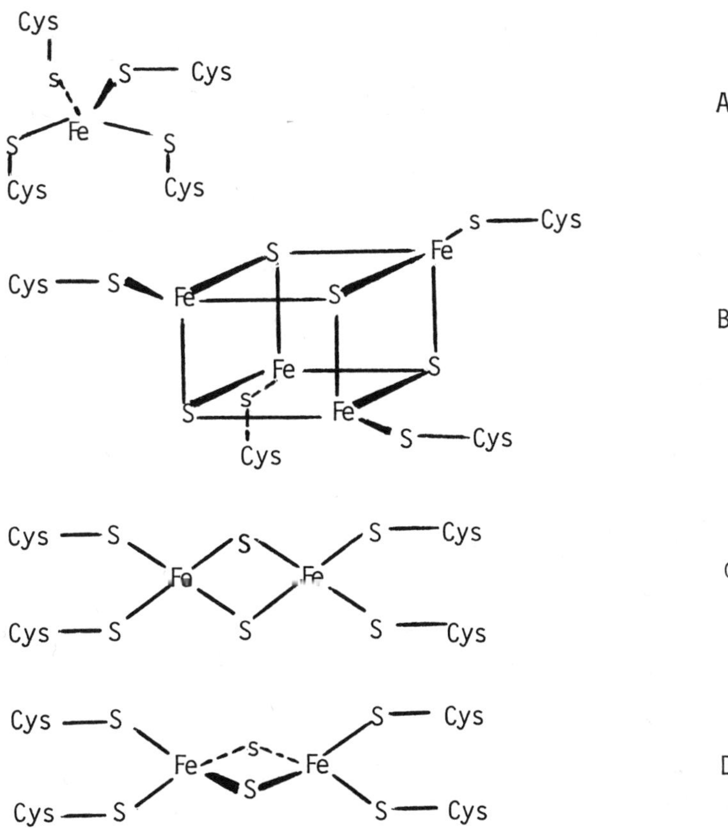


Fig. 1. Iron-sulfur clusters of the various ferredoxins: A, rubredoxins; B, clostridial ferredoxins; C, proposed structure for plant and algal ferredoxins with iron in square planar coordination; D, similar proposed structure with iron in tetrahedral coordination.

| | 1 | | | | 5 | | | | | | | | | | | | 15 |
|---|---|---|---|---|---|---|---|---|---|---|---|---|---|---|---|---|---|
| 1. Taro | A | T | Y | K | V | K | L | V | T | - | P | S | G | - | Q | Q | E |
| 2. Spinach | A | A | Y | K | V | T | L | V | T | - | P | T | G | - | N | V | E |
| 3. Alfalfa | A | S | Y | K | V | K | L | V | T | - | P | E | G | - | T | Q | E |
| 4. L. glauca | - | A | F | K | V | K | L | L | T | - | P | D | G | - | P | K | E |
| 5. Scenedesmus | A | T | Y | K | V | T | L | K | T | - | P | S | G | - | D | Q | T |
| 6. S. platensis | A | T | Y | K | V | T | L | I | N | E | A | E | G | I | N | E | T |
| 7. S. maxima | A | T | Y | K | V | T | L | I | S | E | A | E | G | I | N | E | T |

| | | | | | | | | | | 25 | | | | | | | | | | | | | | | 40 |
|---|---|---|---|---|---|---|---|---|---|---|---|---|---|---|---|---|---|---|---|---|---|---|---|---|---|
| 1. | F | Q | C | P | D | D | V | Y | I | L | D | Q | A | E | E | V | G | I | D | L | P | Y | S | C | R |
| 2. | F | Q | C | P | D | D | V | Y | I | L | D | A | A | E | E | E | G | I | D | L | P | Y | S | C | R |
| 3. | F | E | C | P | D | D | V | Y | I | L | D | H | A | E | E | E | G | I | V | L | P | Y | S | C | R |
| 4. | F | E | C | P | D | D | V | Y | I | L | D | Q | A | E | E | L | G | I | D | L | P | Y | S | C | R |
| 5. | I | E | C | P | D | D | T | Y | I | L | D | A | A | E | E | A | G | L | D | L | P | Y | S | C | R |
| 6. | I | D | C | D | D | D | T | Y | I | L | D | A | A | E | E | A | G | L | D | L | P | Y | S | C | R |
| 7. | I | D | C | E | E | E | T | Y | I | L | D | A | A | E | E | A | G | L | D | L | P | Y | S | C | R |

| | | | | | | | | | | 50 | | | | | | | | | | | | | | | 65 |
|---|---|---|---|---|---|---|---|---|---|---|---|---|---|---|---|---|---|---|---|---|---|---|---|---|---|
| 1. | A | G | S | C | S | S | C | A | G | K | V | K | V | G | D | V | D | Q | S | D | G | S | F | L | D |
| 2. | A | G | S | C | S | S | C | A | G | K | L | K | T | G | S | L | N | Q | S | D | Q | S | F | L | D |
| 3. | A | G | S | C | S | S | C | A | G | K | V | A | A | G | E | V | N | Q | S | D | G | S | F | L | D |
| 4. | A | G | S | C | S | S | C | A | G | K | L | V | E | G | D | L | D | Q | S | D | Q | S | F | L | D |
| 5. | A | G | A | C | S | S | C | A | G | K | V | E | A | G | T | V | D | Q | S | D | Q | S | F | L | D |
| 6. | A | G | A | C | S | T | C | A | G | T | I | T | S | G | T | I | D | Q | S | D | Q | S | F | L | D |
| 7. | A | G | A | C | S | T | C | A | G | K | I | T | S | G | S | I | D | Q | S | D | Q | S | F | L | D |

| | | | | | | | | | | 75 | | | | | | | | | | | | | | | 90 |
|---|---|---|---|---|---|---|---|---|---|---|---|---|---|---|---|---|---|---|---|---|---|---|---|---|---|
| 1. | D | E | Q | I | G | E | G | W | V | L | T | C | V | A | Y | P | V | S | D | G | T | I | E | T | H |
| 2. | D | D | Q | I | D | Q | G | W | V | L | T | C | A | A | Y | P | V | S | D | V | T | I | E | T | H |
| 3. | D | D | Q | I | E | E | G | W | V | L | T | C | V | A | Y | A | K | S | D | V | T | I | E | T | H |
| 4. | D | E | Q | I | E | E | G | W | V | L | T | C | A | A | Y | P | R | S | D | V | V | I | E | T | H |
| 5. | D | S | Q | M | D | G | G | F | V | L | T | C | V | A | Y | P | T | S | D | C | T | I | A | T | H |
| 6. | D | D | Q | I | E | A | G | Y | V | L | T | C | V | A | Y | P | T | S | D | C | T | I | L | T | H |
| 7. | D | D | Q | I | E | A | G | Y | V | L | T | C | V | A | Y | P | T | S | D | C | T | I | Q | T | H |

| | | | | | | | 97 |
|---|---|---|---|---|---|---|---|
| 1. | K | E | E | E | L | T | A |
| 2. | K | E | E | E | L | T | A |
| 3. | K | E | E | E | L | T | A |
| 4. | K | E | E | E | L | T | A |
| 5. | K | E | E | D | L | F | - |
| 6. | Q | E | E | G | L | Y | - |
| 7. | Q | E | E | G | L | Y | - |

Fig. 2. Aligned sequences of plant and algal ferredoxins.

| | Taro | Sp. | Alf. | L.g. | Sc. | S.p. | S.m. |
|---|---|---|---|---|---|---|---|
| Taro | | 18 | 16 | 19 | 26 | 37 | 35 |
| Spinach | 18 | | 19 | 18 | 30 | 35 | 33 |
| Alfalfa | 16 | 19 | | 22 | 29 | 39 | 35 |
| L. glauca | 19 | 18 | 22 | | 33 | 40 | 37 |
| Scenedesmus | 26 | 30 | 29 | 33 | | 24 | 23 |
| S. plat. | 37 | 35 | 39 | 40 | 24 | | 7 |
| S. maxima | 35 | 33 | 35 | 37 | 23 | 7 | |
| Total | 151 | 153 | 160 | 169 | 165 | 182 | 170 |

Fig. 3A. Amino acid difference matrix of sequences shown in Fig. 1.

| | Taro | Sp. | Alf. | L.g. | Sc. | S.p. | S.m. |
|---|---|---|---|---|---|---|---|
| Taro | | 24 | 20 | 23 | 35 | 50 | 52 |
| Spinach | 24 | | 26 | 27 | 37 | 44 | 49 |
| Alfalfa | 20 | 26 | | 24 | 38 | 46 | 52 |
| L. glauca | 23 | 27 | 24 | | 44 | 54 | 55 |
| Scenedesmus | 35 | 37 | 38 | 44 | | 30 | 36 |
| S. plat. | 50 | 44 | 46 | 54 | 30 | | 11 |
| S. maxima | 52 | 49 | 52 | 55 | 36 | 11 | |
| Total | 204 | 207 | 206 | 227 | 220 | 235 | 255 |

Fig. 3B. MBDC difference matrix of the same sequences.

A most unusual property of the ferredoxins is their highly negative E°' (∿-400 MV (2)). The possible cause of this may be extreme hydrophobicity of the environment of the iron-sulfur cluster. If this is the case, the residues forming the "cage" should be in the interior of the protein and inaccessible to proteolytic attack. In addition, a degree of paramagnetic shifting should be observed by NMR upon reduction. The main objectives of this investigation, then, were to ascertain which residues comprise the environment of the iron-sulfur cluster and which, if any, are involved in electron transfer.

## METHODS

Taro ferredoxin was prepared by a procedure based on that of Rao (19) and incorporating features of Benson (9) and Keresztes-Nagy and Margoliash (20); the preparation is summarized in Fig. 4. All steps were performed at 4°C or less and at pH 7.7, except for the initial homogenization, which was begun at pH 8.2. In steps using acetone, dry ice was the coolant. The purified, concentrated ferredoxin could be stored indefinitely under nitrogen in 0.05 M Tris, pH 7.7, containing 2 M NaCl.

For $^{13}C$ NMR experiments, ammonium phosphate was substituted for the usual Tris in the final elution; the Tris was removed prior to elution by washing with 0.05 M ammonium phosphate-0.1 M NaCl, then eluted with 0.05 M ammonium phosphate-0.5 M NaCl. No adverse effects of ammonium phosphate on the ferredoxin were observed. The ferredoxin could also be lyophilized in Tris or ammonium phosphate at pH 7.7 without deterioration.

*Spirulina maxima* ferredoxin was used for the NMR spectra illustrated in Fig. 8. This ferredoxin was treated with nucleases, purified on Sephadex and DEAE, and concentration to ∿6 ml in the same manner as the taro ferredoxin. It was then lyophilized to 0.8 ml prior to transfer to the NMR tube.

Ferredoxin reductase (EC 1.6.7.1) was prepared by a modified procedure of Shin (2), interfaced with the existing ferredoxin preparation, as shown in Fig. 4. Purity of the reductase was estimated by $A_{458}/A_{275}$ ratio. Iron content of the ferredoxin was determined by atomic absorption, and purity was checked by $A_{422}/A_{277}$ ratio and by amino acid composition.

Activity of the ferredoxin was measured by both the chloroplast assay and the cytochrome *c* reduction assay (21). The chloroplast assay was run in a Perkin-Elmer 356 spectrophotometer, with actinic illumination at 670 nm. Taro chloroplasts were prepared by a modified procedure of Whatley and Arnon (22), at pH 8, using 0.04 M Tris-0.35 M NaCl as the grinding medium and 0.01 M Tris-0.035 M NaCl as the lysing medium. Chlorophyll was determined according to the method of Arnon (23). The reaction mixture contained 0.05 M Tris, pH 8, 2.5 mM NADP, 0.25 mM ascorbate, 1 μM BSA, and a chloroplast suspension containing 20 μM chlorophyll.

For the cytochrome *c* reduction assay, the reaction mixture contained 0.05 M Tris, pH 8, 1 mM $MgCl_2$, 10 μM reductase, 0.5 mM NADPH, and 50 μM cytochrome *c*. From one experiment to another, the concentration of ferredoxin varied between 1 and 10 μM. All results are corrected for the marginal activity of the reductase in the absence of ferredoxin. In the absence of reductase, the ferredoxin is completely inactive.

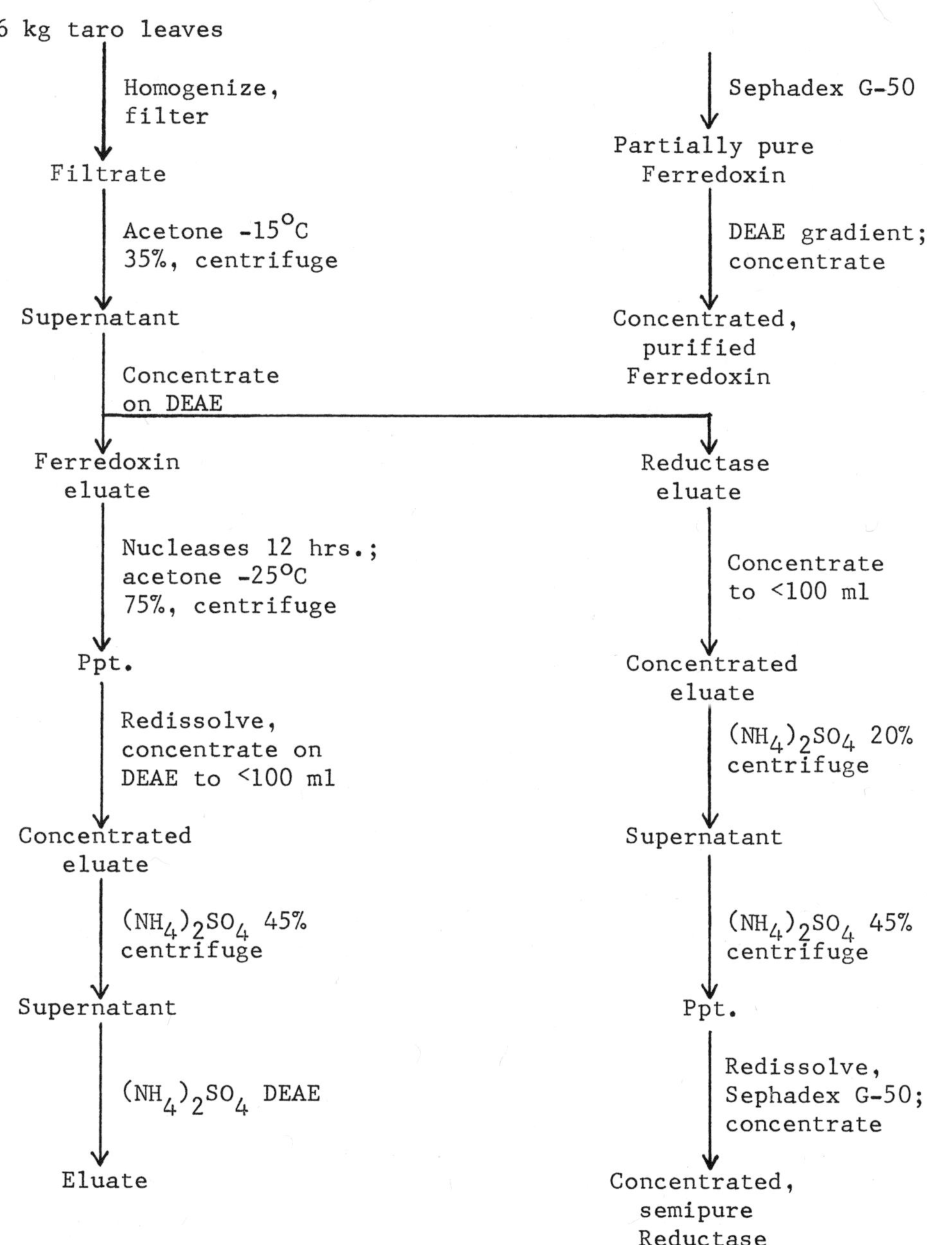


Fig. 4. Preparation of taro ferredoxin and reductase.

Proteolytic digests for the determination of cleavage site were carried out in 0.2 M N-ethylmorpholine acetate buffer at pH 8.5. Aliquots were quenched at the indicated times, dried, redissolved in the same buffer, and readjusted to pH 8.5 (which, in practice, was seldom necessary). Half of each aliquot was then treated with leucine aminopeptidase M, the other half with carboxypeptidase A (CPA) (except for the tryptic digests, in which carboxypeptidase B was used). After incubation with these proteases for ~8 hours, the samples were redried, redissolved in citrate buffer at pH 3, and their free amino acids determined on a Beckman 120 amino acid analyzer.

Further studies to determine the effects of chymotrypsin on activity were performed using 0.05 M Tris, pH 8, containing 1 mM $CaCl_2$. The integrity (fraction of the ferredoxin with its iron-sulfur cluster intact) was read at various times, and simultaneously 2 aliquots were withdrawn. One was assayed by the cytochrome c method; the other was treated with benzylsulfonyl fluoride (BSF) to quench the chymotrypsin. The BSF-treated aliquots were then treated with CPA for 6 hours, and the released tyrosine was determined on the analyzer.

Finally, natural-abundance $^{13}C$ NMR spectra were taken by the Fourier-transform method on a Varian XL-100-15 NMR spectrometer with a Digilab computer system and self-contained proton decoupling. A 12-mm NMR tube was used, and the sample volume was 3.2 ml. It contained 75% $D_2O$ and 2.5 mM ferredoxin. During the runs, anaerobicity was maintained, and the temperature was kept between 0 and 4°C. The pulse width was 200 msec. followed by a delay of 500 msec.

For the second run, the ferredoxin was reduced with dithionite in the tube under nitrogen, resealed, and run again. Each spectrum was run for 3 days and required over 300,000 scans. Interpretation of the spectra was based largely on the works of Allerhand (24), Gurd et al. (26).

## RESULTS

### Preparation

When the preparation was carried out as described, 90-100 mg of ferredoxin was obtained, with the expected amino acid composition and an $A_{422}/A_{277}$ ratio of 0.46-0.48, and containing 1.8-2 irons/mole. About 135 mg of reductase could also be claimed; and from spectral measurements, its purity was estimated at 50%.

## Preliminary Chemical Studies

Incubation of the ferredoxin with chaotropic agents or detergents has no effect on the rate of denaturation. Similarly, EDTA or $CN^-$ has no effect even in 100-fold excess. Organics of low dielectric constant in sufficient concentration cause denaturation; but in sub-denaturing concentrations, they do not assist EDTA or $CN^-$. This implies that the iron-sulfur cluster is inaccessible to the solvent and that before it can be directly attacked, a conformational change or unfolding is necessary to expose it.

TABLE I

Effects of Proteases on Integrity of Ferredoxin

| | % Integrity | | | | | |
|---|---|---|---|---|---|---|
| Time, hrs | Control | Trypsin | Chymotrypsin | Thermolysin | Subtilisin | Papain |
| 0 | 95 | 92 | 95 | 95 | 92 | 95 |
| 1 | 90 | | 78 | | 65 | 35 |
| 2 | 88 | 87 | 65 | 65 | 20 | 0 |
| 3 | 85 | | | | 0 | |
| 4 | 82 | | 50 | 25 | | |
| 5 | 80 | 80 | 40 | | | |
| 6 | 77 | | | 8 | | |
| 7 | 77 | | 35 | | | |
| 9 | 75 | 70 | | | | |
| 11 | | 68 | 20 | | | |

## Cleavage Studies

The effects of proteases on the integrity of the ferredoxin are summarized in Table I. The ferredoxin is resistant to trypsin, as Table II demonstrates. However, when chymotrypsin is used, a new N-terminal lysine and a new C-terminal tyrosine appear after only 2 hours, as can be seen in Table III. The amounts of these amino acids are increased at 4 hours; but after this, many others appear indicating that the denaturation exposes new cleavage sites, and thus rendering further analysis unfruitful.

TABLE II

New N- and C-Terminals Resulting from Cleavage by Trypsin

| Amino Acid | 2 hrs | | 5 hrs | | 9 hrs | |
|---|---|---|---|---|---|---|
| | C | N | C | N | C | N |
| D | 0 | 0 | 0 | 0 | 0 | 0 |
| T | 0.19 | 0.25 | 0.25 | 0.15 | 0.20 | 0.20 |
| S | 0.04 | 0.13 | 0.08 | 0.12 | 0.11 | 0.07 |
| E | 0.08 | 0 | + | 0 | 0 | 0 |
| P | 0 | 0 | 0 | 0 | 0 | 0 |
| G | 0.08 | 0.17 | 0.23 | 0.40 | 0.37 | 0.18 |
| A | 0.53 | 0.60 | 0.73 | 0.55 | 0.62 | 0.51 |
| V | 0 | 0 | 0 | 0 | 0.10 | 0 |
| I | 0 | + | 0 | 0 | + | + |
| L | 0 | + | + | + | 0.85 | 0.20 |
| Y | 0 | 0.10 | 0 | 0.10 | 0 | 0.10 |
| F | 0 | + | 0 | 0 | 0 | + |
| W | 0 | 0 | 0 | 0 | 0 | 0 |
| K | + | + | + | + | + | 0.18 |
| H | 0 | 0 | 0 | 0 | 0 | 0 |
| R | 0 | 0 | 0 | + | + | 0 |

+, trace

TABLE III

New N- and C-Terminals Resulting from Cleavage by Chymotrypsin

| Amino Acid | 2 hrs | | 4 hrs | | 6 hrs | |
|---|---|---|---|---|---|---|
| | C | N | C | N | C | N |
| D | 0 | 0.21 | 0 | 0.25 | 0 | * |
| T | 0.20 | 0.34 | 0.38 | 0.48 | 0.39 | 0.83 |
| S | 0 | 0.18 | 0 | 0.54 | 0 | 0.77 |
| E | 0 | 0 | 0 | + | 0 | + |
| P | 0 | 0 | 0 | 0 | 0 | 0 |
| G | 0.20 | 0.20 | 0.20 | 0.31 | 0.20 | 0.50 |
| A | 0.70 | 0.67 | 0.75 | 0.78 | 1.35 | 1.82 |
| V | 0 | + | 0 | 0.19 | + | 0.65 |
| I | 0.20 | 0.14 | 0.20 | 0.51 | 0.45 | 0.89 |
| L | 0.15 | 0.62 | 0.62 | 0.42 | 1.60 | 1.66 |
| Y | 0.51 | 0.59 | 0.72 | 0.63 | 1.15 | 1.71 |
| Γ | 0 | 0 | 0 | 0 | + | 0 |
| W | 0 | 0 | 0 | 0 | * | 0 |
| K | 0.10 | 0.43 | 0.28 | 0.68 | 0.72 | 1.50 |
| H | 0 | 0 | 0 | 0 | 0 | 0 |
| R | 0 | + | 0 | 0 | 0 | 0 |

+, trace
*, obscured by anomalous peak

The exposure of non-native cleavage sites by initial cleavage, as well as the appearance of anomalous peaks (those in a position not corresponding to any amino acid), frustrated studies with other proteases. However, the results obtained demonstrate that chymotrypsin cleaves between a tyrosine and a lysine; the only such combination is tyrosine 3 and lysine 4.

## Activity Studies

Treatment of the ferredoxin with chymotrypsin causes loss of activity in both the photosynthetic and cytochrome c reduction assays, as displayed in Table IV. Further investigation reveals that the loss of activity is greater than can be accounted for by destruction of the iron-sulfur cluster. This data is summarized in Table V and illustrated in Fig. 5. In addition, as Fig. 6 shows, the loss of activity correlates well with the amount of tyrosine liberated by CPA treatment. (This is partially obscured in the case of the *S. maxima* ferredoxin, since its C-terminal is a tyrosine.) As the earlier cleavage studies revealed, the first point of attack by chymotrypsin is after tyrosine 3. From this data, then, it can be gleaned that tyrosine 3 is required for activity, presumably due to involvement in electron transfer.

TABLE IV

Effect of Chymotrypsin on Ferredoxin Activity

| Sample | % Int. | Chloroplast Assay | | Cytochrome *c* Assay | |
|---|---|---|---|---|---|
| | | Activity $A_{340}$/min x $10^3$ | % of Control | Activity $A_{550}$/min x $10^3$ | % of Control |
| Taro | | | | | |
| Control | 82 | 51 | 100 | 75 | 100 |
| Incub. Blank | 67 | 37 | 73 | 54 | 72 |
| Chymo-trypsin | 40 | 16 | 33 | 30 | 40 |
| *S. maxima* | | | | | |
| Control | 78 | 38 | 100 | 66 | 100 |
| Incub. Blank | 62 | 23 | 59 | 45 | 58 |
| Chymo-trypsin | 45 | 9 | 22 | 20 | 30 |

TABLE V

Integrity and Activity of Ferredoxins as a Function of Cleavage

A. Taro Ferredoxin

| Time, hrs | % Int. | % Activity | % Loss of Activity | Tyrosines Evolved* |
|---|---|---|---|---|
| 0 | 92 | 100 | 0 | 0 |
| 0.5 | 92 | 81 | 19 | 0.15 |
| 1.5 | 85 | 54 | 46 | 0.26 |
| 3 | 60 | 34 | 66 | 0.59 |
| 6 | 50 | 27 | 73 | 0.65 |
| 12 | 35 | 22 | 78 | 1.49 |

B. *S. maxima* Ferredoxin

| Time, hrs | % Int. | % Activity | % Loss of Activity | Tyrosines Evolved* |
|---|---|---|---|---|
| 0 | 100 | 100 | 0 | 0 |
| 0.5 | 97 | 84 | 16 | 0.28 |
| 1.5 | 97 | 75 | 25 | 0.64 |
| 3 | 85 | (86) | (14) | (0.50) |
| 6 | 69 | 50 | 50 | 1.28 |
| 12 | 47 | 40 | 60 | 3.78 |

* Corrected in *S. maxima* for C-terminal tyrosine.

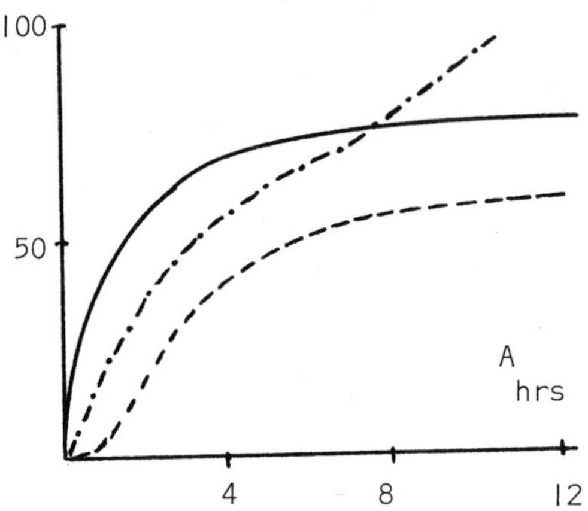


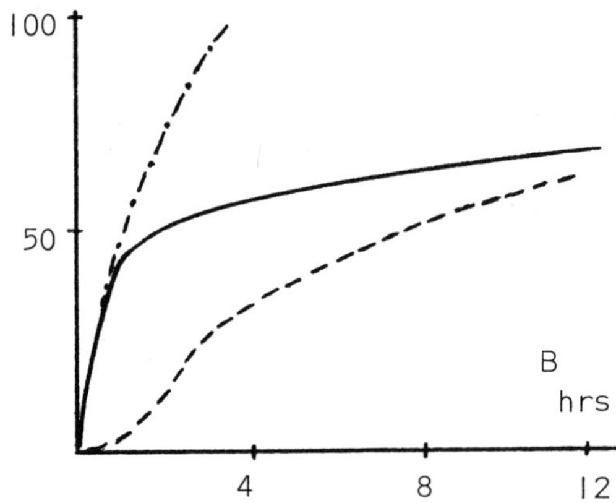


Figure 5. Effects of Chymotrypsin on Ferredoxins

A, taro; B, S. maxima; ———— , % loss of activity; – – – –, % loss of integrity; —·—·—·—, Tyrosines Liberated, % of ferredoxin concentration.

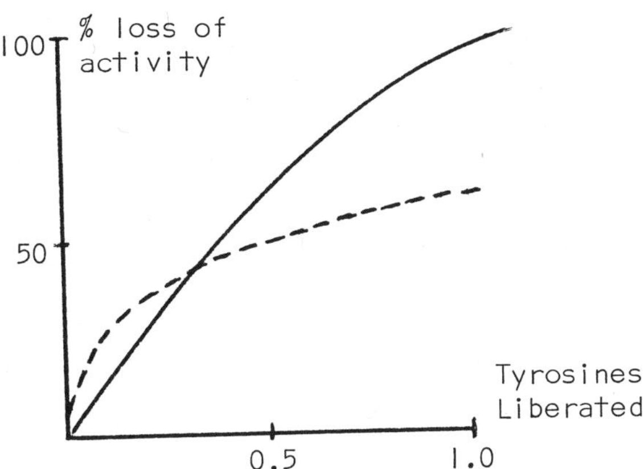


Figure 6. Activity of Ferredoxins as a Function of Tyrosine Liberated
——taro; – – – – S. maxima

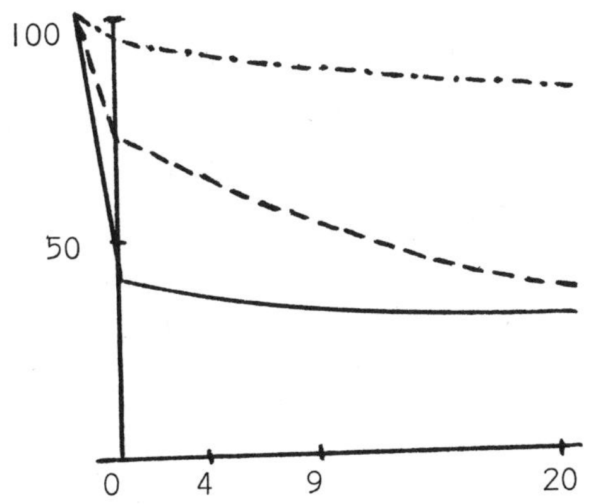


Figure 7. Deterioration of Ferredoxin After Limited Cleavage
– – – –% integrity of N-des-3 ferredoxin.
—— % activity of N-des-3 ferredoxin.
–·–·–% activity/integrity of control.

Also, either tyrosine 3 or the N-terminal tripeptide (N3) is required for stability, as is shown in Table VI A and displayed in Fig. 7. If the chymotryptic digestion is terminated by BSF, there is only a miniscule loss of activity from then on. However, the loss of integrity proceeds more rapidly than that of the native ferredoxin, until the remaining integrity falls to the value expected on the basis of the remaining activity. This implies that the N-des-3 ferredoxin is much less stable than the native ferredoxin and is denaturing more rapidly.

The accelerated denaturation of the N-des-3 ferredoxin might be explained by assuming that the removal of N3 exposes the iron-sulfur cluster. However, if this were the case, the iron-sulfur cluster of the N-des-3 ferredoxin would be destroyed very quickly by iron-binding agents. As Table VI B discloses, EDTA has virtually no effect, and the effect of $CN^-$ is within experimental error. Thus removal of N3 does not directly expose the iron-sulfur cluster, and it is more likely that destabilization results from the disruption of a hydrophobic interaction of tyrosine 3, probably with another aromatic residue between it and the iron-sulfur cluster, and/or a salt linkage between the N-terminal amino group and one of the carboxyl side chains.

TABLE VI A

Integrity and Activity of Ferredoxin
After Limited Cleavage

| Time, hrs | Control | | Chymotrypsin 2 hrs | |
|---|---|---|---|---|
| | % Int. | % Act. | % Int. | % Act. |
| Before Treatment | 98 | 100 | 98 | 100 |
| 0 | 95 | 100 | 72 | 37 |
| 4 | 90 | 90 | 60 | 34 |
| 9 | 85 | 85 | 50 | 34 |
| 20 | 82 | 80 | 32 | 29 |

TABLE VI B

Effect of Iron-Binding Agents on Ferredoxin After Limited Cleavage

| Time, hrs | % Activity | % Integrity | | |
|---|---|---|---|---|
| | | Blank | EDTA | $CN^-$ |
| 0 | 45 | 80 | 80 | 80 |
| 2 | 45 | 72 | 72 | 70 |
| 5 | 40 | 60 | 60 | 58 |
| 8 | 40 | 55 | 55 | 52 |
| 11 | 37-40 | 50 | 50 | 45 |
| 22 | 35 | 40 | 38 | 38 |

## $^{13}C$ NMR Studies

The natural abundance spectrum of oxidized S. maxima ferredoxin is displayed in Fig. 8A. Below 60 ppm, a number of high-field peaks can be seen. The peak nearest the standard can be assigned to the γ-carbons of the valine residues. The two following peaks correspond, in order, to the δ- and the γ-carbons of both the isoleucines and the leucines. The next peak arises from the β-carbons of asparagine, glutamine, aspartic and glutamic acids, and valine. The β-cysteine peak appears next, and the proximate envelope of high-field peaks consists largely of α-carbons; it probably also contains a number of β-carbons and some others. In the low-field end of the spectrum, the two peaks between 100 and 130 ppm are those of the aromatic carbons of the (6) tyrosines and the single phenylalanine and the #4 and #5 carbons of the single histidine. Finally, the cluster of peaks at 170-185 ppm is composed mainly of backbone carbonyl carbons and also includes the amido and carboxyl carbons of asparagine, glutamine, and aspartic and glutamic acids, which comprise ∿20% of the carbons in this region.

A modicum of change can be seen in the spectrum of the reduced ferredoxin, in Fig. 8B. The γ-valine peak is markedly diminished, and the isoleucine and leucine peaks are slightly altered. The new peak appearing between these two could be a shifted γ-valine peak

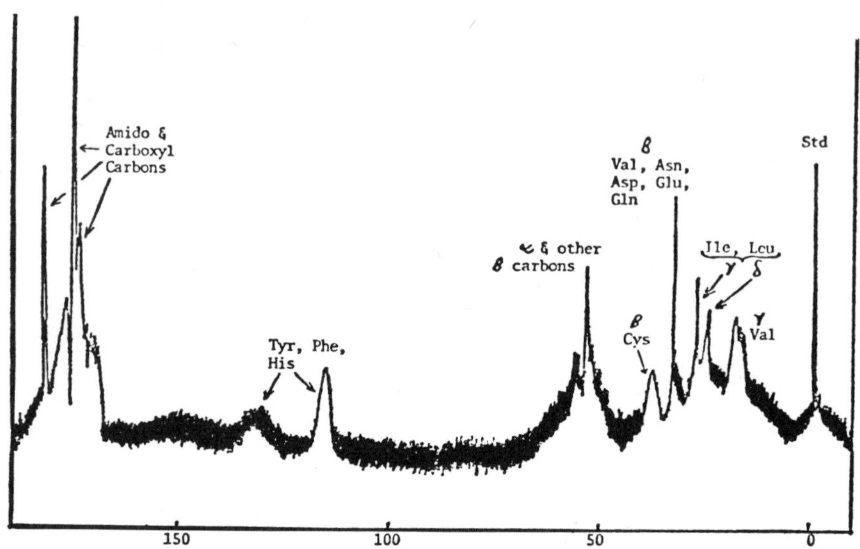


Fig. 8A. Natural abundance $^{13}C$ NMR spectrum of oxidized S. maxima ferredoxin. Concentration = 2.5 mM in ∿ 0.1 M ammonium phosphate, pH 7.7, 1 M NaCl, containing 75% $D_2O$. Chemical shift scale is relative to 1-trimethylsilylpropane-3-sulfonate standard. ∿ 300,000 scans.

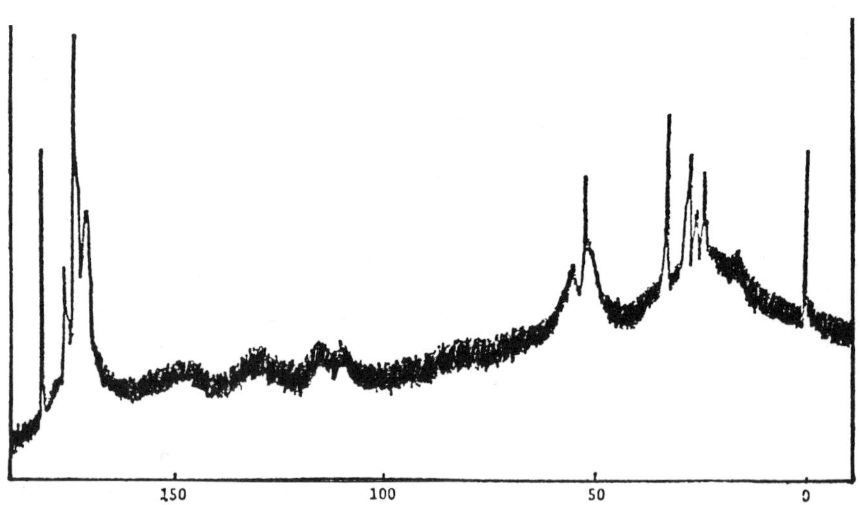


Fig. 8B. Natural abundance $^{13}C$ NMR spectrum of S. maxima ferredoxin reduced with dithionite. Concentrations and pH are same as oxidized form. ∿ 375,000 scans.

or one or more shifted peaks from the (8) isoleucines and/or the (7) leucines. The β-cysteine peak has all but vanished, and the α-carbon envelope is somewhat diminished but not significantly altered. The higher-field of the two aromatic peaks is substantially altered, but due to the excessive noise level, it is not possible to discern whether the peak is split or broadened or some of its constituent signals have simply vanished due to contact shift. There is no observable change in the lower-field aromatic peak, but the noise obscures detail. Finally, some of the carbonyl carbons are shifted 1 or 2 ppm, though most of them are unaffected.

Broadening was not obvious in the reduced spectrum, as might be expected from bulk paramagnetism; Hv the large intrinsic widths of these peaks may well mask any broadening. (These peaks can hardly be referred to as "lines," since each is composed of a number of lines; and each line may be relatively broad (24).)

From these observations, it is evident that the β-cysteinyl carbons are contact-shifted and/or -broadened beyond detectability. It is also entirely possible that one or more of the aromatic residues are paramagnetically shifted and are thus in close proximity to (though not necessarily in contact with) the iron-sulfur cluster. High-field shifts implicate valines as well as isoleucines and/or leucines in constituting the environment of the iron-sulfur cluster, although it is not possible to ascertain how many or which of these residues are involved.

## DISCUSSION AND CONCLUSIONS

Iron-sulfur proteins have been under investigation for some time, for many reasons and by a variety of techniques. $^{13}C$ NMR has been used to detect electron density overlap between both tyrosines and the iron-sulfur cluster in C. acidi-urici ferredoxin (26). However, the $^{13}C$ content of these tyrosines was selectively enriched (to ∿55%); and in natural abundance, neither the selectivity nor this concentration of $^{13}C$ could be obtained.

Nevertheless, it can be seen that several non-aromatic hydrophobic residues could be in close proximity to the iron-sulfur cluster, serving to protect it from the solvent and providing the low dielectric constant needed for the highly negative $E^{o\prime}$. More important, proximity and/or contact of aromatic residues with the iron-sulfur cluster is quite likely. This is indirectly confirmed by the chymotryptic studies. If tyrosine 3 is required for activity, it is almost certainly in close proximity to the iron-sulfur cluster. However, since removal of N3 does not expose the iron-sulfur cluster, tyrosine 3 cannot pass electrons directly to the cluster; rather, they must be passed through another intervening aromatic residue.

Pinpointing of this residue, as well as the hydrophobic residues shielding the iron-sulfur cluster, may provide a great deal of insight into the conformation of the 2-iron ferredoxins. Phenylalanine 16 and the C-terminal aromatic residue cannot be involved, since these are replaced by non-aromatic residues in some species. There are 5 remaining invariant aromatic positions, and three of these are very near cysteines 39 and 77 (in the sequence), inviting much speculation. In any case, the x-ray work on the *A. nidulans* ferredoxin has been proceeding for some time; and thus the solution of the structure of this protein so vital to nature is within measurable distance of its end!

The author gratefully acknowledges:

Supporting grants from the National Institutes of Health, the American Heart Association, and the Hawaii Heart Association.

The generous gift of the *S. maxima* ferredoxin from Dr. T.E. Meyer of the University of California at San Diego.

Technical assistance from Jim Loo, Phil Dahlstrom, and Dr. Harry Yamamoto in the operation of various equipment used in this project.

REFERENCES

1. Sobel, B.E. & Lovenberg, W. Proc. Nat. Acad. Sci., 54, 1 (1965), 193.

2. Lovenberg, W., Buchanan, B.B. & Rabinowitz, J.C. J. Biol. Chem., 238, 13 (1963), 3899.

3. San Pietro, A. & Lang, H.M. J. Biol. Chem., 231, 1 (1958), 211.

4. Kimura, T. & Ohno, H. J. Biochem. (Japan), 63, 3 (1968), 716.

5. Eaton, W., Palmer, G., Fee, J.A., Kimura, T. & Lovenberg, W. Proc. Nat. Acad. Sci., 68, 12 (1971), 3015.

6. Rao, K.K. & Matsubara, H. Biochem. Biophys. Res. Commun., 38, 3 (1970), 500.

7. Matsubara, H., Sasaki, R.M. & Chain, R.K. Proc. Nat. Acad. Sci., 57, 2 (1967), 439.

8. Keresztes-Nagy, S. & Margoliash, E. J. Biol. Chem., 244, 4 (1969), 981.

9. Benson, A.M. & Yasunobu, K.T. J. Biol. Chem., 244, 4 (1969), 955.

10. Sugeno, K. & Matsubara, H. J. Biol. Chem., 244, 11 (1969), 2979.

11. Tanaka, M., Haniu, M., Zeitlin, S., Yasunobu, K.T., Evans, M.C.W., Rao, K.K. & Hall, D.O. Biochem. Biophys. Res. Commun., 64, 1 (1975), 399.

12. Tanaka, M., Haniu, M., Yasunobu, K.T., Rao, K.K. & Hall, D.O. Unpublished Manuscript.

13. Kagamiyama, H., Rao, K.K., Hall, D.O., Cammack, R. & Matsubara, H. Biochem. J., 145, 1 (1975), 121.

14. Wada, K., Kamagiyama, H., Shin, M. & Matsubara, H. J. Biochem. (Japan), 76, 6 (1974), 1217.

15. Palmer, G., Sands, R.H. J. Biol. Chem., 241, 1 (1966), 253.

16. Tang, S.P.W., Spiro, T.G., Mukai, K. & Kimura, T. Biochem. Biophys. Res. Commun., 53, 3 (1973), 869.

17. Mayberle, J.J., Frankel, R.B., Holm, R.H., Ibers, J.A., Phillips, W.D. & Weiher, J.F. Proc. Nat. Acad. Sci., 70, 8 (1973), 2429.

18. Yamanaka, T., Takenami, S., Wada, K. & Okunuki, K. Biochem. Biophys. Acta, 180 (1969), 196.

19. Rao, K.K. Phytochemistry, 8 (1969), 1379.

20. Keresztes-Nagy, S. & Margoliash, E. J. Biol. Chem., 241, 24 (1966), 5955.

21. Shin, M. Methods Enzymol., 23 (1971), 440.

22. Whatley, F.R. & Arnon, D.I. Methods Enzymol., 6 (1963), 308.

23. Arnon, D.I. Plant Physiol., 24 (1949), 1.

24. Allerhand, A., Cochran, D.W. & Doddrell, D. Proc. Nat. Acad. Sci., 67, 3 (1970), 1093.

25. Gurd, F.R.N., Lawson, P.J., Cochran, D.W. & Wenker, E. J. Biol. Chem., 246, 11 (1971), 3725.

26. Packer, E.L., Sternleicht, H. & Rabinowitz, J.C. Proc. Nat. Acad. Sci., 69, 11 (1972), 3278.

STUDIES ON BOVINE ADRENAL FERREDOXIN

S. Takemori*, K. Suhara and M. Katagiri

Department of Chemistry, Faculty of Science

Kanazawa University, Ishikawa 920, Japan

SUMMARY:Native and reconstituted adrenal ferredoxins have been obtained in crystalline form. The apoprotein was prepared by treatment of the native protein with trichloroacetic acid. When the apoprotein was incubated with ferrous ion, sulfide and 2-mercaptoethanol, the recovery of the reconstituted protein was considerably low. The reconstitution was greatly enhanced by the presence of 8 M urea. The reconstituted protein was indistinguishable from the native protein with respect to enzymic activity, spectral properties and iron content. The apparent contents of labile sulfide in both native and reconstituted proteins were about one mole per mole of protein when analyzed according to the original methylene blue method without alkaline zinc incubation. Extension of the alkaline zinc incubation over 2 hr resulted in release of two moles of labile sulfide per mole of protein.

---

The steroid hydroxylation system of adrenal mitochondria requires three components:a flavoprotein, NADPH-adrenal ferredoxin reductase, an iron-sulfur protein, adrenal ferredoxin and a hemoprotein, cytochrome P-450(1), all of which have recently been obtained in highly purified form(2-7). Due to its excellent solubility, adrenal ferredoxin can be easily extracted into the sucrose solution during the isolation of mitochondria from bovine adrenal cortexes and has been obtained in crystalline form(3). The values of iron and sulfide content and the absorbance ratio of 414 nm/276 nm were greater than those of any preparation yet reported. The iron and sulfide of adrenal ferredoxin could be separated from the apoprotein moiety by the treatment with trichloroacetic acid. The apoprotein thus pre-

---

* Present address: Department of Environmental Sciences, Faculty of Integrated Arts and Sciences, Hiroshima University, Hiroshima.

pared could recombine with ferrous ion and sulfide to form the reconstituted protein. Both native and reconstituted proteins have been crystallized in quite stable forms (3,8), and which promise to give materials suitable for chemical and physical studies of the iron-sulfur protein. The results of these investigations are described here

## PREPARATION OF CRYSTALLINE NATIVE AND RECONSTITUTED ADRENAL FERREDOXINS

The frozen adrenal cortexes were thawed and homogenized with 250 mM sucrose solution. The homogenate was centrifuged at 700 x g for 10 min. The cloudy supernatant was decanted off and saved. The precipitate was homogenized again and centrifuged as above. The supernatants of the first and second centrifugations were combined and centrifuged at 9,000 x g for 30 min. The supernatant ($S_1$) was saved. The sedimented pellets were suspended in the sucrose solution and the differential centrifugation (700 x g for 10 min and 9,000 x g for 30 min) was repeated. The supernatant ($S_2$) was saved. Most of adrenal ferredoxin were extracted in the $S_1$ and $S_2$ fractions. Mitochondrial pellets thus obtained were used for isolation of NADPH-adrenal ferredoxin reductase and cytochrome P-450. The $S_1$ and $S_2$ fractions containing adrenal ferredoxin were absorbed on a large column (5 x 25 cm) of DEAE cellulose. The eluted preparation was then chromatographed on DEAE cellulose columns (2.8 x 30 cm) twice, and was followed by gel filtration on Sephadex G-100 (2.8 x 70 cm). The final preparation was then crystallized. Ammonium sulfate was used to obtain the brown-red crystals shown in Fig. 1. The yield of twice-recrystallized adrenal ferredoxin was about 60 mg from 1 kg of adrenal cortex. A summary of the results of the preparation of adrenal ferredoxin from the $S_1$ and $S_2$ fractions is illustrated in Table I (3).

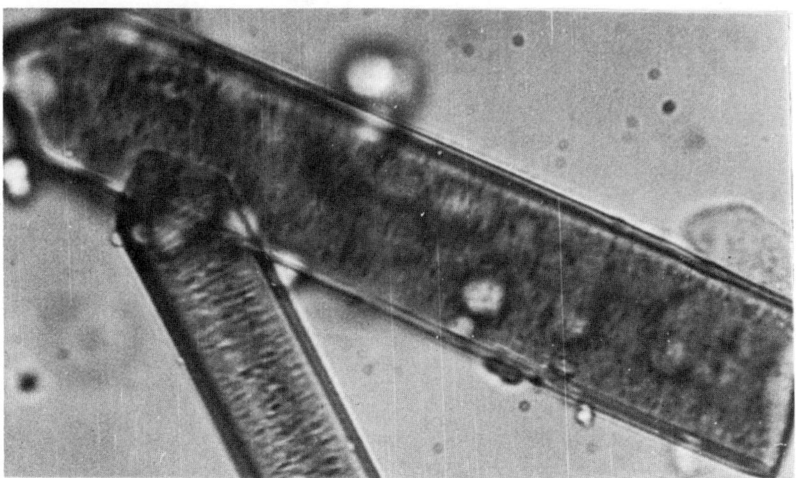

Fig. 1. Crystals of adrenal ferredoxin.

TABLE I The Purification of Adrenal Ferredoxin from the Sucrose Extract($S_1$ and $S_2$) of Adrenal Cortex(1 kg)

| Preparation | Protein(mg) | Purity index ($A_{414nm}/A_{276nm}$) |
|---|---|---|
| First DEAE cellulose | 2 600 | -* |
| Third DEAE cellulose | 230 | 0.46 |
| Sephadex G-100 | 123 | 0.73 |
| First crystals | 80 | 0.82 |
| Second crystals | 60 | 0.86 |

* Omitted because the preparation in this stage contained a large amount of colored impurities.

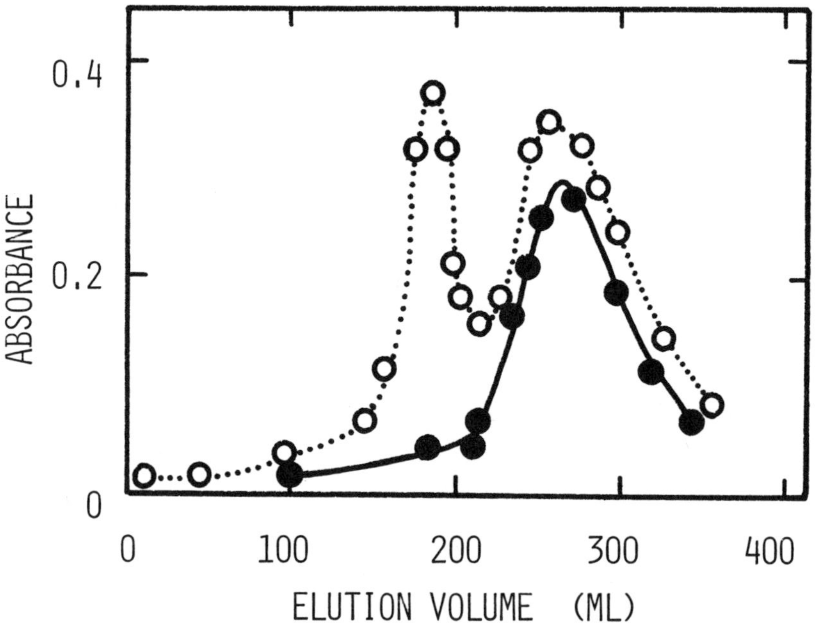


Fig. 2. Chromatography of the reconstituted adrenal ferredoxin on a DEAE cellulose column(2.8 x 26 cm) equilibrated with 10 mM potassium phosphate buffer(pH 7.4) containing 170 mM KCl. The column was washed with about 400 ml of 10 mM potassium phosphate buffer(pH 7.4) containing 170 mM KCl and adrenal ferredoxin was then eluted with the same buffer containing 300 mM KCl. ------- 276-nm absorbance; ______ 414-nm absorbance.

Apo-adrenodoxin was prepared by treating the native protein with trichloroacetic acid. When adrenal ferredoxin was treated with trichloroacetic acid, the color of the protein was bleached, and the apoprotein formed was free of both iron and sulfide. This protein was converted to ferredoxin by the addition of ferrous ion, sulfide and 2-mercaptoethanol. However, the recovery of the reconstituted protein was considerably low(20%). The recombination reaction of the apoprotein with ferrous ion and sulfide required keeping the apoprotein in 8 M urea. The yield could be as high as 75%, under these conditions. The excess reagents were removed by dialysis and most of the unreacted apoprotein was separated by DEAE cellulose chromatography as shown in Fig. 2. The reconstituted adrenal ferredoxin was finally crystallized.

## COMPARISON OF CRYSTALLINE NATIVE AND RECONSTITUTED ADRENAL FERREDOXINS

Spectral properties:The absorption spectrum of native adrenal ferredoxin is shown in Fig. 3. The spectrum shows peaks in the visible region at 414nm(ε:4960) and 455nm(ε:4210), a near-ultraviolet absorption peak at 320nm(ε:6410) and a broad protein band at 276nm (ε:5790). The specific extinction coefficients[K expressed in (g protein/ml)$^{-1}$.cm$^{-1}$] at 276 and 414 nm were 675 and 578, respectively. The absorbance ratio, 414nm/276nm, in the oxidized form was 0.86 at pH 7.4. The spectrum of apoprotein exhibited a maximum in the ultraviolet region, but no absorption in the visible region. The reconstituted protein had spectral properties identical with the native protein except that the absorbance ratio, 414nm/276nm, was 0.78.

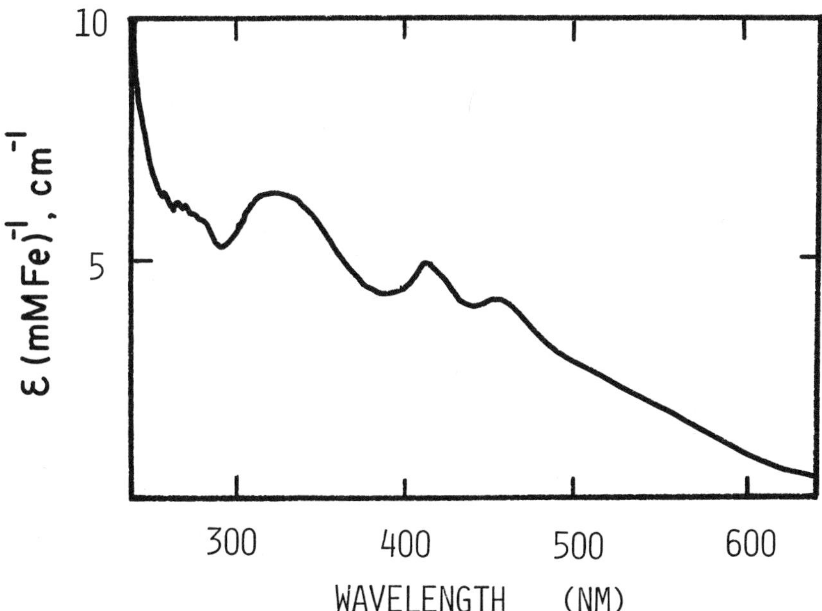


Fig. 3. Absorption spectrum of native adrenal ferredoxin.

Homogeneity: The native and reconstituted adrenal ferredoxins were tested for homogeneity by electrophoresis in polyacrylamide gel. Both preparations were found to move as a single, well-defined band. When the apoprotein was mixed with either native or reconstituted protein, the electrophoretic pattern of the mixture gave a single sharp band. The electrophoretic mobilities were essentially the same for the apo and holoproteins under specified condition employed. The ultracentrifuge patterns of native and reconstituted proteins showed only one symmetrical peak. The values of $s_{20,w}$ for native and reconstituted proteins were 1.65 and 1.83 S, respectively. These values were with that of the apoprotein (1.48S).

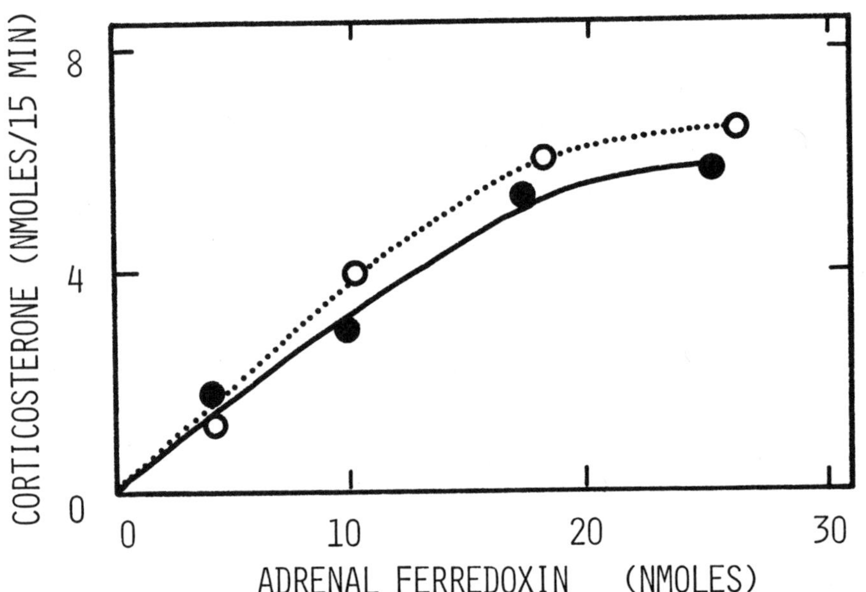


Fig. 4. Activity of adrenal ferredoxin in the 11β-hydroxylation of deoxycorticosterone. The reactions were carried out in 0.5 ml incubation mixtures containing 25 μmoles of potassium phosphate buffer (pH 7.4), 100 nmoles of deoxycorticosterone, 370 pmoles of cytochrome P-450, 500 pmoles of adrenal ferredoxin reductase, 8 μmoles of KCl, 8 μmoles of $MgCl_2$, 580 nmoles of NADP+, 4.5 μmoles of glucose 6-phosphate, 1.5 units of glucose-6-phosphate dehydrogenase and appropriate amounts of adrenal ferredoxin. The amount of corticosterone formed after 15 min incubation was determined fluorometrically(9). (------- reconstituted adrenal ferredoxin; ______ native adrenal ferredoxin).

Catalytic properties: The activity of adrenal ferredoxin in the reconstituted steroid 11β-hydroxylase system was estimated in the presence of a NADPH generating system, NADPH-adrenal ferredoxin reductase, cytochrome P-450 and deoxycorticosterone. As shown in Fig. 4, the native and reconstituted proteins functioned equally well in the hydroxylation of deoxycorticosterone.

The function of the both proteins was also examined by measuring the stimulation of the rate of adrenal ferredoxin-dependent cytochrome c reduction in the presence of NADPH-adrenal ferredoxin reductase and NADPH. As shown in Table II, the activity of the reconstituted protein was identical with that of the native one. The apoprotein did not replace the holoprotein either in cytochrome c reduction or in 11β-hydroxylase system. When adrenal ferredoxin was reduced with NADPH and adrenal ferredoxin reductase, the superoxide anion was generated as judged by accumulation of adrenochrome in the presence of epinephrine and $O_2$(Table III). No $O_2^{-1}$ generation was observed in the apoprotein. The conversion of epinephrine to adrenochrome was greatly inhibited by superoxide dismutase. The concentration of superoxide dismutase at 50 % inhibition was estimated to be 5 nM for the native and reconstituted adrenodoxin.

TABLE II Activity of Adrenal Ferredoxin in Cytochrome c Reduction in the Presence of NADPH-Adrenal Ferredoxin Reductase

| Adrenal ferredoxin | Cytochrome c reductase activity (moles of cytochrome c reduced/ min/mole of protein) |
|---|---|
| Native | 620 |
| Apo | 3 |
| Reconstituted | 640 |

The reactions were carried out in 1 ml incubation mixtures containing 100 μmoles of potassium phosphate buffer(pH 7.4), 50 nmoles of NADPH, 400 pmoles of adrenal ferredoxin reductase, 20 nmoles of cytochrome c and 10-50 pmoles of adrenal ferredoxin. Reaction rates were estimated at 25°C by measuring the increase of absorbance at 550 nm.

TABLE III The Adrenal Ferredoxin Dependent Formation of Superoxide Anion

| Adrenal ferredoxin | Superoxide anion formation (moles of $O_2^{-1}$/ min / mole of protein) |
|---|---|
| Native | 44 |
| Apo | 1 |
| Reconstituted | 48 |

The adrenal ferredoxin dependent formation of superoxide anion was assayed with epinephrine by measuring increase in absorbance at 480 nm as described by Misra and Fridovich(10). The reactions were carried out in 1 ml incubation mixtures containing 50 μmoles of potassium phosphate buffer(pH 7.4), 100 nmoles of EDTA, 200 nmoles of epinephrine, 40 nmoles of NADPH, 400 pmoles of adrenal ferredoxin reductase and 10-50 pmoles of adrenal ferredoxin. The rate of generation of superoxide anion was calculated by assuming that each adrenochrome corresponded to 1.39 $O_2^{-1}$(10).

Iron content: The iron content of the native protein was 120 ngram atoms of iron per mg of protein on biuret basis. The apoprotein contained no significant amounts of iron(less than 5 ngram atoms of iron per mg of protein). The iron content of the reconstituted protein was estimated to be 118 ngram atoms of iron per mg of protein. When a correction factor(0.81) was used for conversion into the dry weight of protein, the data indicate the presence of two gram atoms of iron per mole of adrenal ferredoxin, assuming a molecular weight of 13,094(11).

Estimation of labile sulfide: The labile sulfide was measured by a colorimetric assay based on methylene blue formation(12,13). The sample was directly treated with a mixture of zinc acetate and sodium hydroxide, and then coupled with N,N'-dimethyl-p-phenylenediami in the presence of ferric chloride to give methylene blue, which was then determined spectrophotometrically. The application of this metho to the sulfide analysis indicated that the apparent content of labile sulfide in the native protein was about one mole per mole of protein. Thus the molar ratio of iron to sulfur was two since the protein contained two gram atoms of iron per mole of protein. As shown in Table IV, extension of the incubation time to over 2 hr with the alkaline zinc reagent resulted in additional release of labile sulfide and the molar ratio of iron to sulfide approached one. A similar phenomenon was also observed with the reconstituted protein: After 2 hr of incubation with the alkaline zinc reagent, the amount of detectable labile sulfide was equivalent to the amount of iron present. The freshly prepared apoprotein yielded essentially no detectable sulfide. Labile sulfide analysis of Spirulina platensis ferredoxin, with two iron atoms (14), was performed in collaboration with Drs. K. Wada and H. Matsubara of Osaka University and provided further evidence that the reaction of the protein-bound sulfide with the alklaine zinc reagent was time dependent. As shown in Table IV, the value of one mole of labile sulfide per mole of protein was obtained under the original assay conditions, while extension of the alkaline zinc incubation to over 2 hr resulted in release of two moles of labile sulfide per mole of protein.

TABLE IV Effect of Incubation with the Alkaline Zinc Reagent upon Labile Sulfide Analysis

| Preincubation time (min) | Native adrenal ferredoxin | | Reconstituted adrenal ferredoxin | | *Spirulina platensis* ferredoxin | |
|---|---|---|---|---|---|---|
| | a | b | a | b | a | b |
| 0 | 0.98 | 2.03 | 0.82 | 2.43 | 0.98 | 1.94 |
| 20 | 1.29 | 1.55 | 1.48 | 1.35 | 1.35 | 1.41 |
| 40 | 1.44 | 1.39 | 1.68 | 1.19 | 1.72 | 1.10 |
| 60 | 1.76 | 1.14 | 1.84 | 1.09 | 1.79 | 1.07 |
| 120 | 2.00 | 1.00 | 1.99 | 1.00 | 1.95 | 0.98 |
| 180 | 1.98 | 1.01 | 2.02 | 0.99 | 2.05 | 0.93 |

a: moles of detectable labile sulfide per mole of protein, and b: the molar ratio of iron to labile sulfide.

In contrast to the case of the ferredoxins, all labile sulfide in both spinach and parsley chloroplast ferredoxins was rapidly released when added to the alkaline zinc reagent(8,15). It was considered that the additional detectable sulfide liberated from the prolonged incubation might be due to $H_2S$ derived directly from cysteine residues by a β-elimination reaction, because $H_2S$ could be liberated from cysteine residues in protein under basic conditions(16). If this were the case, the cysteine residue should be consumed in amounts equivalent to the additional detectable sulfide. When the cysteine residues of the protein were determined as cysteic acid after oxidation by performic acid and hydrolysis, there was no effect on the amount of cysteic acid recovered after a 2 hr incubation. Thus sulfide detectable under these conditions is not from the cysteine residue. It seems likely that the protein-bound sulfide of adrenal ferredoxin is relatively more stable toward the alkaline zinc reagent when compared with the chloroplast ferredoxins.

## ACKNOWLEDGEMENTS

We gratefully acknowledge the help of Dr. R. Sato, Institute for Protein Research, Osaka University, for supplying the adrenal glands. We also thank Drs. K. Wada and H. Matsubara, Faculty of Science, Osaka University, for performing sulfide analysis of Spirulina platensis ferredoxin. This investigation was supported by the Scientific Research Fund B-947051 from the Ministry of Education of Japan.

## REFERENCES

1. Omura, T., Sanders, E., Estabrook, R. W., Cooper, D. Y. and Rosenthal, O. (1966) Arch. Biochem. Biophys. 117, 660.
2. Kimura, T. and Suzuki, K. (1967) J. Biol. Chem. 242, 485.
3. Suhara, K., Takemori, S., and Katagiri, M. (1972) Biochim. Biophys. Acta 263, 272.
4. Suhara, K., Ikeda, Y., Takemori, S. and Katagiri, M. (1972) FEBS Lett. 28, 45.
5. Takemori, S., Suhara, K., Hashimoto, S., Hashimoto, M., Sato, H., Gomi, T. and Katagiri, M. (1975) Biochem. Biophys. Res. Commun. 63, 588.
6. Takemori, S., Sato, H., Gomi, T., Suhara, K. and Katagiri, M. Biochem. Biophys. Res. Commun. In press.
7. Sugiyama, T. and Yamano, T. (1975) FEBS Lett. 52, 145.
8. Suhara, K., Kanayama, K., Takemori, S. and Katagiri, M. (1974) Biochim. Biophys. Acta 336, 309.
9. Rosenthal, O. and Narasimhulu, S. (1969) in Methods in Enzymology (Colowick, S. P. and Kaplan, N. O., eds), Vol. XV, pp. 604, Academic Press, New York.
10. Misra, H. P. and Fridovich, I. (1971) J. Biol. Chem. 246, 6886.
11. Tanaka, M., Haniu, M. and Yasunobu, K. T. (1970) Biochem. Biophys. Res. Commun. 39, 1182.
12. Fogo, J. K. and Popowsky, M. (1949) Anal. Chem. 21, 732.
13. Lovenberg, W., Buchanan, B. B. and Rabinowitz, J. C. (1963) J. Biol. Chem. 238, 3899.
14. Wada, K. and Matsubara, H., Unpublished work.
15. Suhara, K., Takemori, S., Katagiri, M., Wada, K., Kobayashi, H. and Matsubara, H. Anal. Biochem. In press.
16. Greenstein, J. P. and Winits, M. (1961) in Chemistry of the Amino Acids, Vol. III, pp. 1899, John Wiley and Sons, New York.

# SOME INSIGHT INTO INTRAMOLECULAR ELECTRON TRANSFER OF AN ADRENODOXIN MOLECULE

Tokuji Kimura and Taketoshi Taniguchi

Department of Chemistry, Wayne State University

Detroit, Michigan 48202

The question of whether or not an aromatic amino acid residue participates in the electron transfer reaction from the surface of protein molecule to the iron-sulfur cluster has been a subject of considerable interest and debate. In particular, heme and non-heme iron proteins have been extensively investigated by utilizing such varied approaches as chemical modification, comparison of the oxidized and reduced structures by means of X-ray crystallography, and kinetic pulse radiolysis studies. In bacterial ferredoxins, both 13-C nuclear magnetic resonance (1,2) and X-ray crystallography studies have revealed unequivocally the proximity of a tyrosine residue to the 4Fe-4S cluster. An elegant study by Rabinowitz's group indicated that a modified ferredoxin from *Clostridum M-E* free of any aromatic residue is as active as the native ferredoxin from *Clostridium acidi urici* (5) in the phosphoroclastic reaction. They concluded that the aromatic residues do not participate in the intramolecular electron transfer reaction in the bacterial ferredoxin.

In the adrenal iron-sulfur protein (adrenodoxin), there is only one tyrosine residue at position 82 of the polypeptide chain. In addition, it contain 4 phenylalanine but no tryptophan residues per molecule (6). The tyrosine residue gives rise to an anomalous fluorescence at 331 nm, whereas the apoprotein, free from the iron-sulfur chromophore, exhibits the normal emission at 305 nm. This observation has suggested that the tyrosine residue might be under the weak influence if the iron-sulfur chromophore (7). Furthermore, the adrenal iron-sulfur protein accepts one electron per 2Fe-2S center upon reduction by either dithionite of flavoprotein reductase and NADPH (8). One may, therefore, speculate

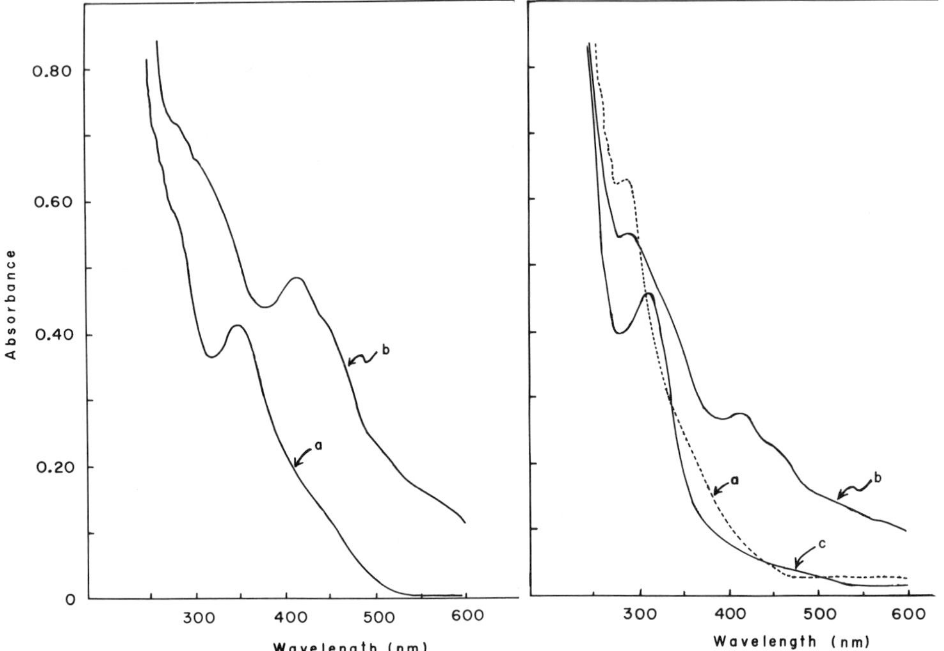


Figure 1 (left) Optical absorption spectra of reconstituted nitro-adrenodoxin and nitro-apoadrenodoxin.

(a): nitro-apoadrenodoxin ($11.2 \times 10^{-5}$ M)
(b): reconstituted nitroadrenodoxin ($5.6 \times 10^{-6}$ M)

Figure 2 (right) Optical absorption spectra of reconstituted aminoadrenodoxin

(a) amino-apoadrenodoxin ($15.1 \times 10^{-5}$ M)
(b) oxidized aminoadrenodoxin ($3.9 \times 10^{-5}$ M)
(c) reduced aminoadrenodoxin ($3.9 \times 10^{-5}$ M)

that another protein ligand besides the sulfur atoms is close to one of the two iron atoms in a manner required by the one-electron accepting mechanism. In fact, a recent observation by Gray's laboratory suggests this possiblity , based on their interpretation of the low-temperature near infrared d-d transition (9).

In this context, we decided to examine whether or not the tyrosine residue does influence the iron-sulfur chromophore, and to see if this residue plays any role in the intramolecular electron transfer of this protein. We have prepared the nitro-, and amino-tyrosyl derivatives of adrenal iron-sulfur protein by the reaction with tetranitromethane. As will be reported elsewhere (10), the nitration of the tyrosine residue of the apoprotein occurs satisfactorily and specifically under certain conditions. The amino-derivative was prepared by the reduction with dithionite. The reconstitution of the iron-sulfur chromophore in the modified proteins was carried out by a similar method to that for the native protein (11).

In this report, we wish to describe: first, the properties of the iron-sulfur chromophore in the modified derivatives; and second, the enzymatic activity of the modified proteins as compared with the native protein. Detailed experimental procedures will be found in publications from this laboratory (10,12).

## RESULTS

### Conditions for nitration

The best condition for the nitration of the apo-protein by tetranitromethane was obtained with an 11.7 molar excess of tetranitromethane, using 3.2 mg protein per ml for 3 hours at 22$^{o}$ under anaerobic conditions. The yield was estimated to be nearly 100% on the basis of the absorbance values at the isosbestic point of 382 nm in the presence of 4M urea. The amino acid composition of the nitro-derivative was found to be same as that of the native protein, except for tyrosine. A slight decrease in the half cystine content was observed, probably due to the oxidation of sulfhydryl group to sulfinic acid. From the results of SDS-gel electrophoresis of the modified samples, there was no indication of the formation of intramolecular cross linkages between the protein molecules caused by tetranitromethane treatment.

The dissociation constant for the phenolic group in the nitro-derivative was determined by observing the absorbance changes at 428 nm at various pHs. A pK of 7.4 was determined experimentally.

The amino-derivative was prepared by the reduction of the nitro-derivative by dithionite, following the method of Sokolvsky et al.

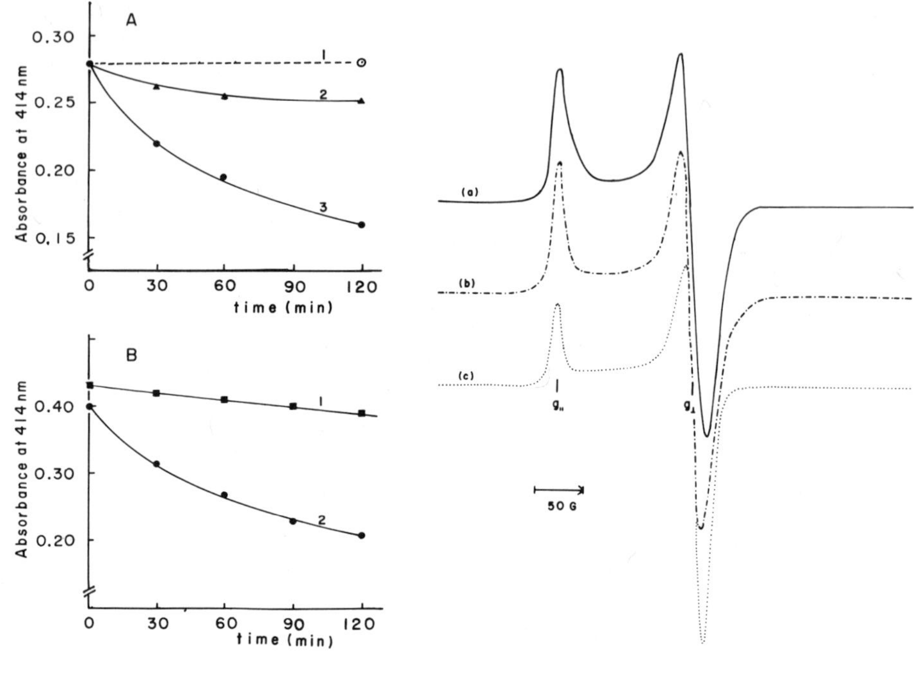


Figure 3 Figure 4

Figure 3. Stability of modified adrenodoxin (left)

A, (1): native adrenodoxin at 25°
(2): nitro-adrenodoxin at 0°
(3): nitro-adrenodoxin at 25°
B. (1): amino-adrenodoxin at 25° in the presence of 20% glycerol and dithiothreitol
(2): amino-adrenodoxin at 25°

Figure 4. Electron paramagnetic resonance apsctra of reduced modified proteins (right)

(a) nitroadrenodoxin (0.56 mM)
(b) aminoadrenodoxin (0.18 mM)
(c) native adrenodoxin (0.26 mM)

Conditions: microwave power, 10 mW; frequency, 9.169 GHz; modulation amplitude, 10 gauss; 77° K; gain, 160 (a), 320 (b), and 100 (c).

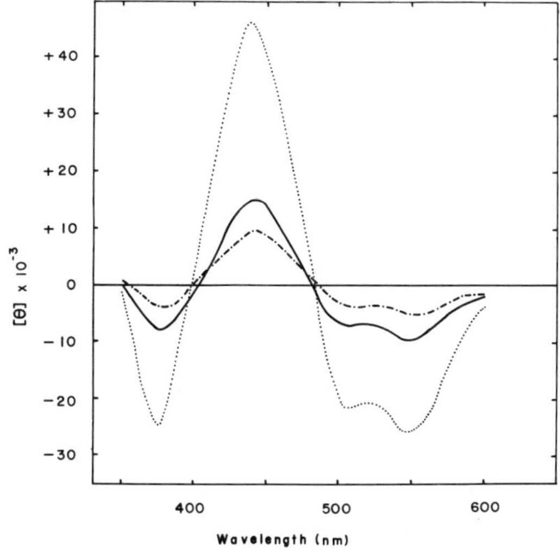


Figure 5. Circular dichroism of modified adrenodoxin

---------: native adrenodoxin

————————: nitro-adrenodoxin

———o———: amino-adrenodoxin

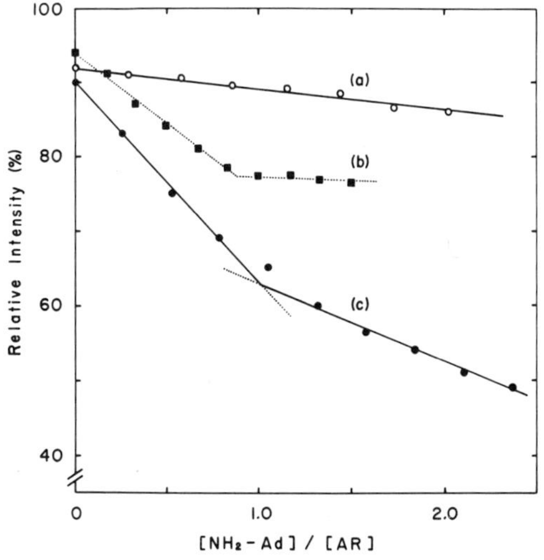


Figure 6. Complex formation of modified adrenodoxin with adrenodoxin reductase

(a) amino apoadrenodoxin
(b) native adrenodoxin
(c) amino-adrenodoxin

The tyrosine fluorescence emission intensity of adrenodoxin reductase at 335 nm (excitation wavelength of 290 nm) upon titration with the modified proteins.

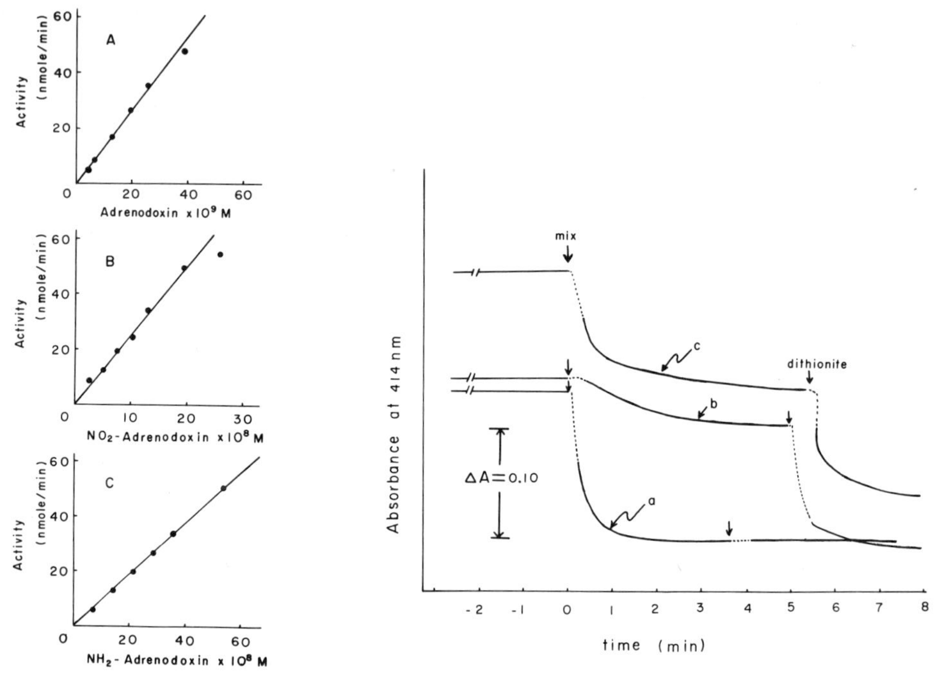


Figure 7. The activity of modified adrenodoxin in the cytochrome c reduction (left) assay.

The reaction mixture contained 10 mM phosphate buffer (pH 7.4), 2.7 x $10^{-5}$ M cytochrome c, 4.0 x $10^{-5}$ M NADPH, adrenodoxin reductase, and various amounts of either native of modified adrenodoxin.

Figure 8. The enzymatic reduction of modified adrenodoxin (right).

The reaction mixture contained, 10 mM phosphate buffer (pH 7.4), 20% glycerol, 1 mM dithiothreitol, adrenodoxin reductase, 1.2 μM NADPH, and 0.4 μM adrenodoxin (a), 0.65 μM aminoadrenodoxin (b) or 0.16 μM adrenodoxin plus 0.52 μM amino-adrenodoxin (c).

(13). A quantitative reduction of the nitrated apoprotein was obtained.

Reconstitution of iron-sulfur center

The reconstitution of the modified protein was carried out by adding dithiothreitol, $Na_2S$, and $FeCl_3$. After 30 minutes at 22°, the reaction mixture was subjected to a DEAE-cellulose column chromatography. The resulting brown protein was used for further experiments. The iron and labile sulfur contents of the modified and reconstituted samples were found to have a molar ratio of labile sulfur to iron near unity.

Optical absorption spectra

The reconstituted nitro-protein exhibited an absorption maximum at 414 nm and a slight shoulder at 450 nm. The extinction coefficient at 414 nm was $3.4 \times 10^3$ $cm^{-1}$ per gram atom of iron, after subtracting the amount of absorbance contribution by the nitro-apoprotein (Figure 1). The spectra of the reconstituted amino-protein has a peak at 410 nm with an extinction coefficient of $3.8 \times 10^3$ $cm^{-1}$ per gram atom of iron, and a shoulder at 450 nm. (Figure 2).

Stability of iron-sulfur chromophore in the modified protein

The stability of the reconstituted sample was examined by measuring the change in absorbance at 414 nm, due to the iron-sulfur chromophore. The native protein was stable enough at 25° for 2 hours at pH 7.4. However, a distinct decrease in the absorbance of the iron-sulfur center was observed in both the nitro-, and amino derivatives under same conditions. At 0°, or in the presence of dithiothreitol and 20% glycerol, the decomposition of the chromophore was largely protected (Figure 3).

Electron paramagnetic resonance

Upon reduction of the modified samples, distinct electron paramagnetic resonance signals were observed at $g = 1.94$ and $g = 2.01$. The g-values and the line shape of the paramagnetic centers of these modified proteins were exactly identical with that of the native proteins (Figure 4). From these results, the instability induced by the modification (Figure 3) is not due to the instability of the iron-sulfur chromophore *per se*, but appears to be due to alterations in the polypeptide conformation.

Circular dichroism of the modified proteins

The CD spectra of the native and modified proteins were determined with the native and modified proteins. All samples gave rise to identical extrema at 375, 440, 510, 525, and 550 nm (Figure 5). These results again support our conclusion that the iron-sulfur chromophore of the modified proteins are identical before and after the chemical **modification** of the tyrosine residue. However, it is to be noted here that the values of the molar ellipticities of the modified proteins at 440 nm are about 1/4 that of the native protein. Since the chromophores in both the native and modified proteins are identical, this difference must then be explained by the changes in the asymmetric polypeptide environment adjacent to the chromophore: the decrease in the total optical activity is due to extrinsic effects rather than the intrinsic effects of the chromophore. These results support our previous finding obtained from the deterioration studies of the chromophore by protein denaturants, that large Cotton effects in the visible wavelength region are mainly due to the contributions from the polypeptide chain near the chromophore, rather than due to the iron-sulfur center itself (14).

## Complex formation with flavoprotein reductase

Chu and Kimura (15) have reported that the adrenal iron-sulfur protein forms a 1 : 1 molar complex with the flavoprotein reductase at low ionic strengths. The complex was found to be crucially important for cytochrome c reduction, in the presence of flavoprotein and NADPH. Since this complex dissociates into its components as the ionic strength increases, the interacting forces between the two proteins appear to be largely ionic.

Complex formation of the modified proteins with the reductase was studied by measuring the fluorescence quenching of the tryptophan residues of the reductase molecule upon binding (Figure 6). In agreement with the previous results, the native protein binds to the reductase at a molar ratio of 1 : 1. The aminoapoprotein does not form the complex. In addition, the amino-derivative forms a similar one to one complex. The tryptophan residues of the free reductase exhibit the emission maximum at 335 nm, indicating that these residues are buried in the interior of the protein molecule. Both the native and modified complexes show an identical maximum at 335 nm. Therefore, upon binding, a gross movement of the tryptophan residues to a hydrophilic environment is very unlikely. However, it is noted that the extent of quenching upon complexing with the modified protein is significantly larger than the quenching by the native protein. This indicates that the bound form of the reductase with the modified protein has small differences in structure from that of the native protein.

## Enzymatic activities of modified proteins

The initial reaction rate of cytochrome c reduction in the presence of excess amounts of NADPH and the flavoprotein reductase with limited amounts of the native or modified proteins was determined (Figure 7). Under these conditions, the molecular activities of the native, nitro-, and amino-derivatives were calculated to be 434, 80.0 and 30.6 moles of cytochrome c reduced per min per mole of protein respectively. Therefore, these modified proteins had a much lower activity than the native protein. Next, we examined the direct reduction of the modified iron-sulfur proteins by the reductase and NADPH under anaerobic conditions. As illustrated in Figure 8, the enzymatic reduction of the native protein can be completed within one minute, while the amino-derivative shows a very slow rate of reduction. When both samples are mixed, double kinetics of reduction are clearly observed. We can therefore, conclude that the amino derivative has 10% of the activity of the native protein. Furthermore, the rate of reduction of cytochrome c by the amino-derivative is comparable to the rate of the direct reduction of this derivative by the flavoprotein reductase. This suggests that the interaction between the iron-sulfur protein and cytochrome c does not change with the chemical modification.

<u>Dissociation constants of the modified complexes</u>

We have reported in detail (15,16) that the presence of the iron-sulfur protein sharply changes the NADPH-dichlorophenolino-phenol reductase activity. By utilizing this property we have determined the dissociation constants of the complexes between the flavoprotein and the modified iron-sulfur proteins.

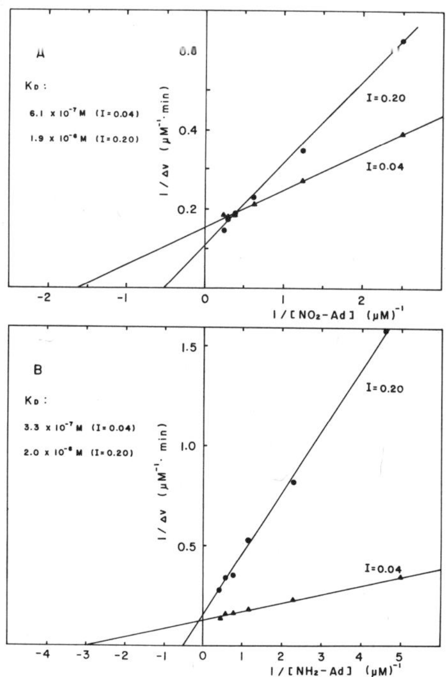


Figure 9. Determination of dissociation constants of modified complexes.

The reaction mixture contained 0.01 M phosphate buffer, pH 7.4, 0.12 μM NADPH; 0.17 μM dichlorophenolindophenol; $6.6 \times 10^{-8}$ M adrenodoxin reductase, and nitroadrenodoxin (A) or aminoadrenodoxin (B) The ionic strength was adjusted by the addition of NaCl.

The results obtained at diferent ionic strengths for the nitro-, and amino-derivatives are shown in Figure 9. The dissociation constants were estimated from the intercept on the abscissa. The values at $\mu$ = 0.04 were 6.1 x $10^{-7}$ M and 3.3 x $10^{-7}$ M for the nitro-, and amino-derivatives, respectively. When compared with the value of $10^{-9}$ M for the native protein, these values are about 100 times greater than the value for the native protein. It is clear that, the chemical modification of the tyrosine residue at position 82 changes the affinity of the iron-sulfur protein for the flavoprotein.

## DISCUSSION

From the results presented here, it is concluded that upon chemical modification of the tyrosine residue at position 82, the iron-sulfur protein chromophore *per se* does not change its corrdination structure and retains its basic physical properties. However, the enzymatic activities of the modified proteins are drastically decreased as a result of changes in the dissociation constant of the respective complexes with the flavoprotein reductase.

Chemical analyses and the absorption spectrum showed that the iron-sulfur chromophore contained 2 iron and 2 labile sulfur atoms, as observed in the native protein. The paramagnetic resonance and circular dichroism data of both the native and modified proteins safely exclude the participation of the tyrosine residue as an additional ligand to the iron-sulfur chromophore. The data on the stability of the chromophore and the circular dichroism ellipticity strongly imply that the modified proteins has a different conformation adjacent to the chromophore from that of the native protein.

Considering the fact that the electronically different nitro-, and amino-derivatives give approximately 10-20% of the enzymatic activities relative to that of the native protein, the involvement of the tyrosine residue as an intermediate in the intramolecular electron transfer reaction is unlikely. In addition to this, we found that the direct chemical reduction of the modified protein by excess dithionite occured at a comparable rate to that of the native protein. Our preliminary results on the direct reduction of the native protein by pulse radiolysis done at the Weizman Institute in collaboration with Dr. Pecht have shown that its reduction by solvated electrons is surprisingly slow. This was in the order of milliseconds, while the kinetics followed was first order. This slow rate might be due to the repulsion of electrons by the negative charges present in this strong acidic protein. Consequently, the loss of the enzyme activity induced by the modification can reasonably be explained by a decrease in their affinity towards the flavoprotein. In this relation, when the total amino acid sequence of the adrenal iron-sulfur protein (6) is compared with that of Pseudomonas protein (17), 37% of the sequences are genetically homologous. Yet, the corresponding position of the tyrosine 82 in the Pseudomonas protein is not an aromatic amino acid but is alanine. Our results do not exclude the possibility of the involvement of phenylalanine residues. In this regard, we can only point out the slight resemblence in the sequences of adrenodoxin and putidaredoxin with respect to the phenylalanine residue at position 46 in the former and at position 56 in the Pseudomonas protein.

The detailed mechanistic role of the tyrosine residue 82 in the complex formation with the flavoprotein is not clear from the studies to date. However, at least two possibilities can be pointed out: first, the tyrosine residue present in the reductase and second, the chemical modification of the tyrosine residue rearranges the structure of the binding site so as to decrease its affinity towards the flavoprotein molecule. Our present evidence for concerning the location of the tyrosine residue in this molecule at the following data: the moderately high pK value (about 10.6) of the phenolic hydroxyl group; the sluggish reactivity of the group to tetranitromethane; and the influence of the buried iron-sulfur chromophore on the fluorescence properties of the protein. It is, thus, likely that the tyrosine residue is not exposed on the protein surface but is buried.

With regards to the structure of the iron-sulfúr-flavin complex, we have previously observed that the optical absorbance of the reductase in the flavin region increases upon the formation of the complex with the iron-sulfur protein. This result together with our present observations on the quenching of the tryptophan fluorescence suggest that the adrenodoxin molecule may bind to the reductase in the vicinity of the tryptophan residues and thereby influences the FAD moiety of this protein. Alternatively, the structure of the flavoprotein may change minutely upon binding, as has been demonstrated *per se* by the changes in catalytic activities of the reductase (18). In addition to this the observed large quenching of the modified protein over that of the native protein makes it somewhat difficult to accept a simple interpretation of this problem at the present time.

Investigations which can show the exact location of the tyrosine residue in the iron-sulfur protein as well as the tryptophan residues in the reductase molecule will provide further knowledge concerning the mechanism of the electron-transfer in this intriguing iron-sulfur-flavin complex.

Acknowledgements

This study was supported by a Research Grant from the National Institutes of Health (AM-12713-07).

## REFERENCES


1. Packer, E.L., Sterlicht, H., and Rabinowitz, J.C. (1972) Proc. Natl. Acad. Sci. U.S., 69, 3278-82
2. Packer, E.L., Sterlicht, H., and Rabinowitz, J.C. (1975) J. Biol. Chem. 250, 2062-72.
3. Adman, E.T., Sieker, L.C., and Jensen, L.H., (1973) J. Biol. Chem. 248, 3987-96.
4. Carter, C.W., Jr., Kraut, J., Freer, S.T., Zuong, N.H., Alden, R.A., and Bartsch, R.G. (1974) J. Biol. Chem. 249, 4212-25.
5. Lode, E.T., Murray, C.L., and Rabinowitz, J.C. (1974) Biochem. Biophys. Res. Commun. 61, 163-9.
6. Tanaka, M., Haniu, M., Yasunobu, K.T. and Kimura, T. (1973) J. Biol. Chem. 248, 1141-57.
7. Kimura, T., Ting, J.J., and Huang, J.J. (1972) J. Biol. Chem. 247, 4476-79.
8. Huang, J.J., and Kimura, T. (1973) Biochemistry 12, 406-9.
9. Rawlings, J., Siiman, O., and Gray, H.B. (1974) Proc. Natl. Acad. Sci. U.S. 71, 125-7.
10. Taniguchi, T., and Kimura, T. (1975) Biochemistry in press.
11. Mukai, K., Huang, J.J. and Kimura, T. (1974) Biochim. Biophys. Acta. 336, 427-36.
12. Taniguchi, T., and Kimura, T., to be submitted.
13. Sokolovsky, R., Riorden, J.F. and Vallee, B.L. (1966) Biochemistry 5, 3582-9.
14. Padmanabhan, R., and Kimura, T. (1970) J. Biol. Chem. 245, 2469-75.
15. Chu, J-W., and Kimura, T. (1973) J. Biol. Chem. 248, 5183-7.
16. Nakamura, S., and Kimura, T. (1971) J. Biol. Chem. 246, 6235-41.
17. Tanaka, M., Haniu, M., Yasunobu, K.T., Dus, K., and Gunsalus, I.C. (1974) J. Biol. Chem. 249, 3689-3701.
18. Kimura, T., Chu, J-W., and Parcells, J. (1975) in "5th International Symposium on Flavins and Flavoproteins", edited by Thomas P. Singer, in press.

# STRUCTURAL STUDIES OF ELECTRON TRANSFER PROTEINS FROM SULFATE REDUCING BACTERIA: THE AMINO ACID SEQUENCE OF TWO RUBREDOXINS ISOLATED FROM DESULFOVIBRIO VULGARIS AND DESULFOVIBRIO GIGAS

Mireille Bruschi and Jean Le Gall

Laboratoire de Chimie Bacterienne, C.N.R.S.

13274 Marseille Cedex 2, France

## INTRODUCTION

The sulfate reducing bacteria are still able to perform today a very ancient process: the dissimilatory reduction of sulfates. Ault and Kulp (1) have shown that a biological reduction of sulfates did occur on this planet as far back as 1 to 2.5 billion years, that is to say before the appearance of the atmospheric oxygen. Sulfate reducing bacteria were for a long time thought to contain a rather truncated electron transfer chain, but recent work in this field brought to light the fact that the 8 electron pairs necessary to reduce sulfate into hydrogen sulfide necessitate the presence of a rather sophisticated set of electron carriers.

## GENERAL CONSIDERATION ON SOME ELECTRON CARRIER PROTEINS FROM *DESULFOVIBRIO*

Some electron transfer proteins from *Desulfovibrio* which have been isolated and studied are presented in Table I. These electron transfer proteins to be used for structural studies must answer to some criteria:

- be useful to understand the mechanisms of a class of protein in which they represent simple models.
- establish phylogenetical relationships between bacterial groups.

Table I

Some Properties of Electron Carrier Proteins from *Desulfovibrio* Species

| | Redox Potential | Active Center | Molecular Weight | Primary Structure | Tertiary Structure |
|---|---|---|---|---|---|
| Cytochromes | | | | | |
| $c_{553}$ | 0 to - 100 mV | 1 mesoheme | 9,000 | completed | under way |
| $c_3$ (MW 13,000) | - 210 mV | 4 mesohemes | 13,000 | completed | under way |
| $c_3$ (MW 26,000) | negative | 8 mesohemes | 26,000 | not determined | under way |
| Flavodoxin | - 150 ; - 440 mV | FMN | 15,000 | completed | completed 2 Å resolution |
| Non-heme iron proteins | | | | | |
| Ferredoxin | - 350 mV | 1 cluster 4 Fe - 4S | 6,000 | completed | under way |
| Rubredoxin | 0 to - 100 mV | 1 Fe | 6,000 | completed | under way |

- understand the electron transfer mechanisms which are a special property of the group of bacteria to which they belong.

- in addition, they must be relatively easy to purify and be present in reasonable amounts in the cells.

Cytochrome $c_{553}$ has been purified from the strain Hildenborough of Desulfovibrio vulgaris (2). It has a molecular weight of 9,000 and its single heme is bound to the apoprotein through a typical Cys-2 a.a.-Cys-His cluster indicating that histidine is probably the 5th ligand position of the heme iron. The presence of an absorption band at 695 nm shows that methionine sulfur is the 6th ligand. Its amino-acid sequence (3) indicates a relationship with pseudomonads cytochrome $c_{551}$.

Cytochrome $c_3$ (MW 13,000) are present in all desulfovibriones. It has been shown to contain four hemes per molecule (4). The four hemes are attached to the protein by two types of sites, two with the typical arrangement Cys-2 a.a.-Cys-His and two others with the sequence Cys-4 a.a.-Cys-His. The 5th and 6th coordination positions of the four heme iron are occupied by histidinyl nitrogens.

The sequence of cytochromes $c_3$ (MW 13,000) are extremely variable from one strain to another. Only 25% of the total residues are kept in the same positions indicating their importance for the function of the protein. This has been used, together with EPR and NMR data, to propose a three dimensional picture of the molecule (5). Crystals suitable for X-ray crystallography have recently been obtained from the Norway strain of Desulfovibrio and are presently under study. Another cytochrome $c_3$ (MW 26,000) of 26,000 molecular weight has been purified. It contains the required number of cysteines necessary to bind eight hemes. Crystals of this cytochrome are under investigation.

Ferredoxins from desulfovibriones are different from other bacterial ferredoxins in that they contain only one 4 Fe-4 S cluster (6).

The sequence of D. vulgaris flavodoxin has been elucidated (7) and the tertiary structure of the oxidized form is known at 2 Å resolution (8).

The amino acid composition of several rubredoxins from desulfovibriones show that the molecule is related to clostridial rubredoxins. They all contain one atom of iron for a molecular weight of 6,000 daltons and four cysteine residues necessary to bind the iron. The amino acid sequence of two rubredoxins isolated from D. vulgaris and D. gigas has been elucidated.

## AMINO ACID SEQUENCES OF D. VULGARIS AND D. GIGAS RUBREDOXINS

*D. vulgaris* and *D. gigas* rubredoxins were prepared as previously described (9,10). The material used showed an $A_{280}:A_{490}$ ratio of 2.35. Conventional methods were used in which tryptic and chymotryptic peptides were prepared. The sequence of the first 30 residues of the *D. vulgaris* rubredoxin was determined using an automatic sequencer, after removal of the N-terminal methionine by cyanogen bromide. The complete amino acid sequence reported in Fig. 1 was deduced only after cleavage of the Asn-Gly bond (residues 22-23) by hydroxylamine (11). The total sequence of *D. gigas* rubredoxin reported in Fig. 2 was elucidated by degradation using a Protein Sequencer, which established the first 42 residues from the $NH_2$ terminal end, and by degradation of the peptide (38-53) obtained from specific cleavage by BNPS-Skatole (12) at the only tryptophan residue of the molecule. However the preparation of the tryptic peptides of the protein was necessary for characterization of some residues which could not be identified in the whole protein.

If the two sequences are aligned for maximum homology (Fig. 3), it is evident that they are similar, but many more differences occur than would be expected for homologous proteins isolated from two sulfate reducing bacteria. Thirty-seven amino acid residues are found in identical positions in *D. vulgaris* and *D. gigas* rubredoxins. *D. vulgaris* and *D. gigas* rubredoxins have an N-terminal methionine, which is blocked in the *D. gigas* rubredoxin but not in the *D. vulgaris* rubredoxin. The N-terminal part of these two proteins is identical between the residues 4 and 16.

In considering the functional relationships between the rubredoxins of the sulfate reducing bacteria, it is to be noted that an NADH + $H^+$ rubredoxin oxydo-reductase has been described in *D. gigas* (9). This enzyme is ten times more reactive with *D. gigas* rubredoxin than with *D. vulgaris* rubredoxin and shows no activity towards *Clostridium pasteurianum* and *Pseudomonas oleovorans* rubredoxins (H.D. Peck Jr. and M. Odum, preliminary results). In *D. vulgaris* which does not contain an NADH·$H^+$ rubredoxin oxydo-reductase, the function of the rubredoxin has not yet been elucidated. This functional difference observed between the *Desulfovibrio* rubredoxins is reflected in their amino acid sequences.

Alignment of the amino acid sequences of the three other rubredoxins so far studied (*C. pasteurianum* (13), *Peptostreptococcus elsdenii* (14) and *Micrococcus aerogenes* (15)) (Fig. 4) shows 18 amino acid residues in identical positions. These are of interest concerning the structural requirements for the active site regions. No differences are found in the position of the two clusters Cys-2 a.a.-Cys-Gly which are involved in the binding of the iron.

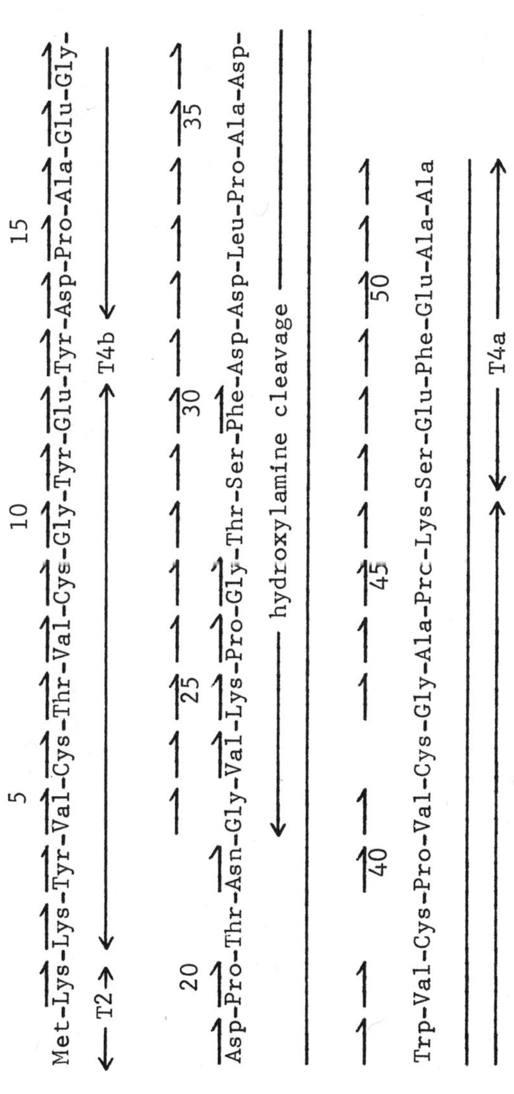


Figure 1. Amino Acid Sequence of Desulfovibrio vulgaris Rubredoxin

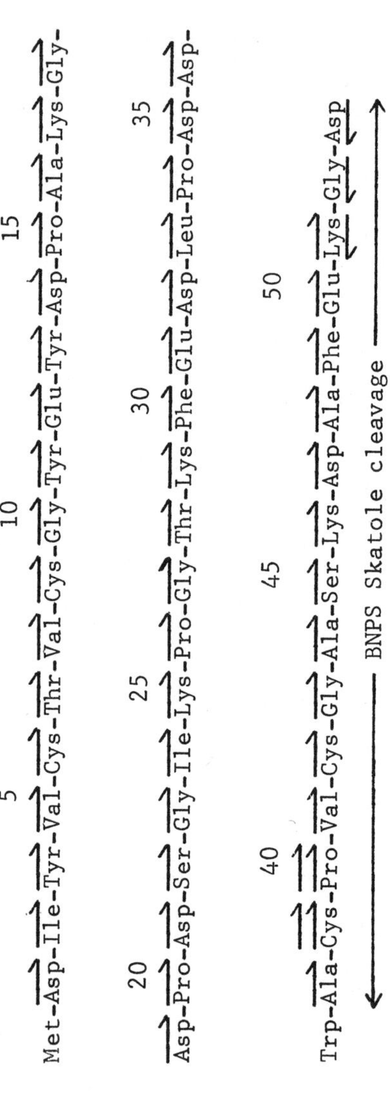


Figure 2. Amino Acid Sequence of <u>Desulfovibrio gigas</u> Rubredoxin

```
                  5                  10                  15
D.V.  Met-Lys-Lys-Tyr-Val-Cys-Thr-Val-Cys-Gly-Tyr-Glu-Tyr-Asp-Pro-Ala-Glu-
D.G.  Met-Asp-Ile-Tyr-Val-Cys-Thr-Val-Cys-Gly-Tyr-Glu-Tyr-Asp-Pro-Ala-Lys-

              20                  25                  30
      Gly-Asp-Pro-Thr-Asn-Gly-Val-Lys-Pro-Gly-Thr-Ser-Phe-Asp-Asp-Leu-Pro-
      Gly-Asp-Pro-Asp-Ser-Gly-Ile-Lys-Pro-Gly-Thr-Lys-Phe-Glu-Asp-Leu-Pro-

      35                  40                  45                  50
      Ala-Asp-Trp-Val-Cys-Pro-Val-Cys-Gly-Ala-Pro-Lys-Ser-Glu-Phe-Glu-Ala-Ala
      Asp-Asp-Trp-Ala-Cys-Pro-Val-Cys-Gly-Ala-Ser-Lys-Asp-Ala-Phe-Glu-Lys-Gly-Asp
```

Figure 3. Comparison of the Amino Acid Sequences of Rubredoxin from *Desulfovibrio vulgaris* and *Desulfovibrio gigas*. Invariant residues are underlined.

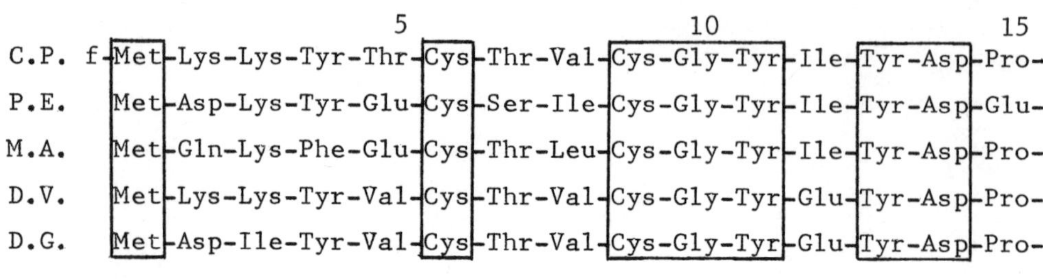


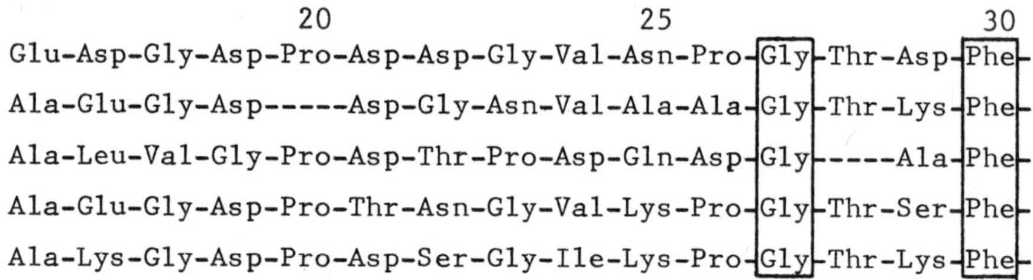


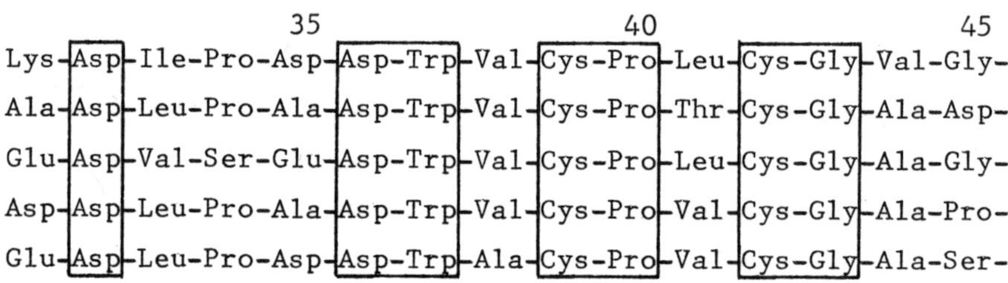


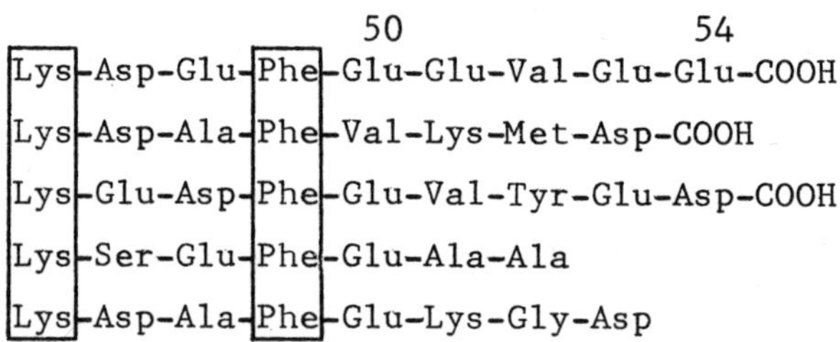


Figure 4. Primary Structure of the Clostridial Type of Rubredoxin Proteins. In the figure, C.P., P.E., M.A., D.V. and D.G. are abbreviations for Clostridium pasteurianum, Peptostreptococcus elsdenii, Micrococcus aerogenes, Desulfovibrio vulgaris and Desulfovibrio gigas.

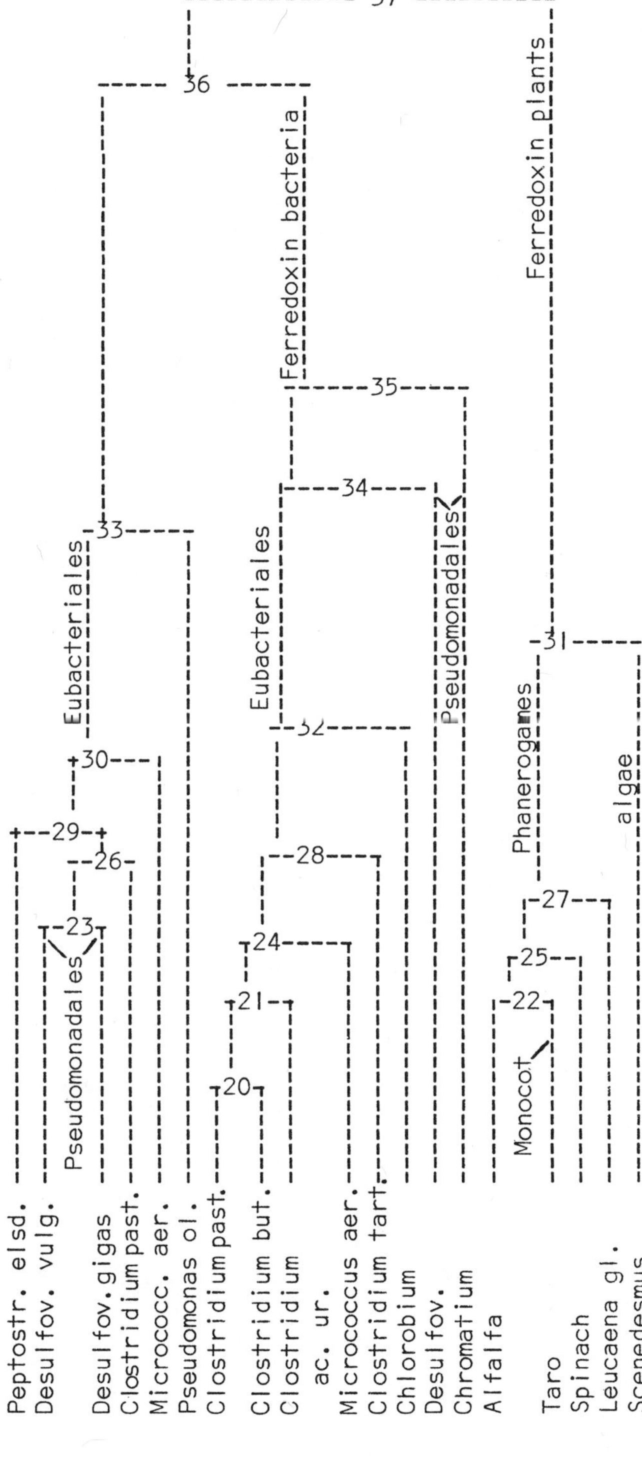


Figure 5 - Tentative phylogenetical tree of plant ferredoxin, bacterial ferredoxin and bacterial rubredoxin (kindly provided to us by H. Vogel, Centre de Recherches de Biochimie Macromoleculaire, C.M.R.S., Montpellier, France.)

The sequence Asp-Trp-Val-Cys-Pro (residues 36-40) is conserved in all the rubredoxins, as are the residues of lysine (46), tyrosine (11-13) and phenylalanine (30-40). The proline residues are not in identical positions, but they are located in the middle part of the proteins.

*Pseudomonas oleovorans* rubredoxin, found in an aerobic bacteria, is much larger than the clostridial type of rubredoxin. There is evidence for gene duplication and for homology with the rubredoxins from anaerobic bacteria. This had led to the hypothesis that the various rubredoxins have all arisen from the same common ancestor (16).

In view of the functional interchange between the bacterial ferredoxins and rubredoxins, homology might be expected between these proteins. However Weinstein (17) found no direct relationship between ferredoxin and rubredoxin isolated from *Micrococcus aerogenes*. The amino acid sequence of ferredoxin from *D. gigas* has been elucidated (6) and a comparison with the sequence of *D. gigas* rubredoxin shows no direct homology. When selected regions of spinach ferredoxins are compared with *M. aerogenes* rubredoxin, a related number of homologies can be found, and they are possibly related to the plant ferredoxins. A phylogenetic tree is presented in Fig. 5 similar to that proposed by Benson et al. (16). *D. vulgaris* and *D. gigas* rubredoxins have been crystallized and an X-ray study of the two proteins is in progress. This will allow a comparison with the tertiary structure of *C. pasteurianum* rubredoxin (18) and give further valuable information on this class of non-heme iron protein.

## REFERENCES


1. AULT, W.U. and KULP, J.L., Geochim. Cosmochim. Acta, 16 (1959) 201-235.

2. LE GALL, J. and BRUSCHI-HERIAUD, M., 1968, in "Structure and Function of Cytochromes" (Okunuki, K., Kamen, M.D. and Sekusu, I., Eds.) p. 467, Univ. of Tokyo Press and University Park Press.

3. BRUSCHI, M. and LE GALL, J., Biochim. Biophys. Acta, 271 (1972) 48-60.

4. DERVARTANIAN, D.V. and LE GALL, J., Biochim. Biophys. Acta, 346 (1974) 79-99.

5. DOBSON, C.M., HOYLE, N.J., GERALDES, C.F., BRUSCHI, M., LE GALL, J., WRIGHT, P.E. and WILLIAMS, R.J.P., Nature, 249 (1974) 425.

6. TRAVIS, J., NEWMAN, D.J., LE GALL, J. and PECK, H.D. Jr., Biochem. Biophys. Res. Commun., 45 (1971) 452-458.

7. DUBOURDIEU, M., LE GALL, J. and FOX, J.L., Biochem. Biophys. Res. Commun., 54 (1973) 1418-1425.

8. WATENPAUGH, K.D., SIEKER, L.C. and JENSEN, L.H., Proc. Natl. Acad. Sci. U.S., 70 (1973) 3857-3860.

9. BRUSCHI, M. and LE GALL, J., Biochim. Biophys. Acta, 263 (1972) 279-282.

10. LE GALL, J., Ann. Inst. Pasteur, 114 (1968) 109-115.

11. BORNSTEIN, P., Biochem. Biophys. Res. Commun., 36 (1969) 957-964.

12. OMENN, G.S., FONTANA, A. and ANFINSEN, C.B., J. Biol. Chem., 245 (1970) 1895.

13. Mc CARTHY, K.F. (1972) Ph.D. dissertation George Washington University.

14. BACHMAYER, H., YASUNOBU, K.T., PEEL, J.L. and MAYHEW, S., J. Biol. Chem., 243 (1968) 1022-1030.

15. BACHMAYER, H., BENSON, A.H., YASUNOBU, K.T., GARRARD, W.T. and WHITELEY, H.R., Biochem., 7 (1968) 986-996.

16. BENSON, A., TOMODA, K., CHANG, J., MATSUEDA, G., LODE, E.T., COON, M.J. and YASUNOBU, K.T., Biochem. Biophys. Res. Commun., 42 (1971) 640-646.

17. WEINSTEIN, B., Biochem. Biophys. Res. Commun., 35 (1969) 109-114.

18. HERRIOTT, J.R., SIEKER, L.C., JENSEN, L.H. and LOVENBERG, W., J. Mol. Biol., 50 (1970) 391-406.

# THE IRON-SULFUR CENTERS AND THE FUNCTION OF HYDROGENASE FROM CLOSTRIDIUM PASTEURIANUM

Jiann-Shin Chen, Leonard E. Mortenson and Graham Palmer

Department of Biological Sciences, Purdue University
West Lafayette, IN 47907 and Department of Biochemistry
Rice University, Houston, Texas 77001

## INTRODUCTION

Hydrogenase has widespread occurrence among bacteria and algae. Its function allows the organisms to grow under some stringent conditions. In essence, activation of molecular hydrogen by hydrogenase (e.g., $H_2$:$NAD^+$ oxidoreductase EC 1.12.1.2; $H_2$:ferricytochrome $c_3$ oxidoreductase EC 1.12.2.1) enables the organisms to use $H_2$ as the primary source of energy and/or reductant. Alternatively, production of $H_2$ through hydrogenase (e.g., $H_2$:ferredoxin oxidoreductase EC 1.12.7.1) permits the organism to use protons, perhaps the most available oxidant under anaerobic conditions, as the terminal electron acceptor to dispose of the excess electrons produced in oxidative metabolism. The latter is particularly important for obligate anaerobes such as *Clostridium pasteurianum*. A more detailed discussion on the physiological role of hydrogenase among microorganisms appeared recently (Mortenson and Chen, 1974).

The extensive purification of hydrogenase from several different organisms (Chen and Mortenson, 1974; Erbes, Burris and Orme-Johnson, 1975; Gitlitz and Krasna, 1975; LeGall et al., 1971; Nakos and Mortenson, 1971; Yagi, 1970) has established that hydrogenase is an iron-sulfur protein. Recent studies on purified hydrogenase from *C. pasteurianum* (Mortenson and Chen, 1975) further suggested the involvement of the iron-sulfur centers in the oxidation-reduction reactions carried out by hydrogenase. From quantitative electron paramagnetic resonance (EPR) measurement and extrusion studies, we have obtained important information on the properties of the iron-sulfur centers of hydrogenase.

In this communication, we will summarize the physical and chemical properties of hydrogenase from C. pasteurianum, and present a mechanism based on the present understanding of the iron-sulfur centers and the catalytic properties of this enzyme.

## METHODS

Hydrogenase was purified from $N_2$-fixing cells of C. pasteurianum as described by Chen and Mortenson (1974). A second hydroxyapatite column has since been added as the final step to improve the resolution of hydrogenase from a leading and trailing contaminant, and the yield of pure hydrogenase was thus increased.

Hydrogenase activity was routinely measured by the $H_2$ production assay with 1 mM methyl viologen as the electron carrier and 15 mM $Na_2S_2O_4$ as the reductant. The reaction was carried out in 0.05 M Tris·HCl buffer at pH 8 in a Warburg apparatus at 30° C under $H_2$. Bovine serum albumin (1 mg/ml) was included in the reaction mixture when pure hydrogenase was assayed. The concentration of enzyme used in an assay must be less than 80 nM (5 μg/ml when 2 ml of reaction mixture is used in a 15 ml Warburg manometric flask) in order to keep the reaction progress curve linear for several minutes. When the activity of hydrogenase was reported to be 330 μmoles $H_2$ produced per minute per mg protein (Chen and Mortenson, 1974), the assay was carried out with 155 nM of hydrogenase and no correction was made for the nonlinear (declining) reaction progress curve. The corrected activity was 500 μmoles $H_2$ produced per minute per mg protein when the initial velocity was obtained from the nonlinear reaction progress curve (by the procedure of Alberty and Koerber (1957)).

The analytical procedures for protein, iron and acid-labile sulfide were referenced previously (Chen and Mortenson, 1974). The accuracy of the o-phenanthroline method for iron determination was established by adding to the hydrogenase samples fixed amounts of iron as an internal standard. The additional iron was always recovered to within 91-101% of that added.

Circular dichroism spectra of hydrogenase were obtained at 22° C with a recording Cary 60 spectropolarimeter equipped with a 6001 CD attachment. The molar ellipticity was calculated using an average residue weight of 113 for hydrogenase. The α-helical content of hydrogenase was estimated with synthetic poly-L-glutamic acid as a reference (Wu and Yang, 1970).

Electron paramagnetic resonance (EPR) spectra were measured with a Varian E-6 X-band spectrometer equipped with a low temperature attachment. The spin intensity was integrated using an on-line minicomputer with the use of spinach ferredoxin and copper EDTA as standards. Potentiometric titration of hydrogenase was monitored by EPR spectroscopy and the detailed procedure will be reported elsewhere.

Extrusion studies on hydrogenase were carried out as described by Que, Holm and Mortenson (1975).

## RESULTS AND DISCUSSION

### Physical and Chemical Properties of Hydrogenase

Hydrogenase from *C. pasteurianum* is an oxygen-sensitive, iron-sulfur protein with a mol. wt. of 60,500. Table I summarizes some physical and chemical properties of hydrogenase. Purified hydrogenase showed a specific activity of 550 µmoles $H_2$ produced per minute per mg hydrogenase when assayed under conditions as described in Methods. Hydrogenase at this stage of purity showed a single, sharp protein band when electrophoresed on a polyacrylamide gel column containing sodium dodecyl sulfate (SDS).

Iron analyses of hydrogenase samples yielded values ranging from 9.1 to 12.3 iron atoms per molecule of hydrogenase. None of the iron atoms react with added o-phenanthroline under anaerobic conditions.

The method of Fogo and Popowsky (1949) as modified by Brumby et al. (1965) was unsuitable for the determination of acid-labile sulfide in hydrogenase (Chen and Mortenson, 1974). Titration of hydrogenase with p-hydroxymercuribenzoate at pH 4.6 and in the presence of SDS yielded 30-33 sulfhydryl equivalents per mole of protein. Since the amino acid analysis showed 12 half-cystine residues per molecule of hydrogenase, these results suggested an equimolar ratio of Fe:$S^=$:-SH in hydrogenase even though the $S^=$ was lower than 12 (Table I).

### Catalytic Properties of Hydrogenase

The oxidation of $H_2$ by purified hydrogenase can be coupled with the reduction of oxidized ferredoxin (Fd), benzyl viologen (BV), methyl viologen (MV) or methylene blue. The reduced form of the first three compounds can also donate electrons for reduction of $H^+$

TABLE I

A SUMMRAY OF SOME PHYSICAL AND CHEMICAL PROPERTIES OF HYDROGENASE FROM *CLOSTRIDIUM PASTEURIANUM*

| | |
|---|---|
| Molecular Weight | 60,500 |
| Polypeptide Chain | 1 |
| Iron (atoms/mole) | 9.1-12.3 |
| Acid-Labile Sulfide (a) (moles/mole) | 9-10.5 |
| Half-Cystine (moles/mole) | 12 |
| Extinction Coefficient ($cm^{-1}$ $mM^{-1}$ at 400 nm with $H_2$-reduced hydrogenase at pH 7.2) | 25.3 |
| Turnover Number ($H_2$ production assay) | $2.40 \times 10^5$ (b)<br>$3.36 \times 10^5$ (c) |

(a) Total titratable $S^=$ and SH was determined by titration with p-hydroxymercuribenzoate at pH 4.6 and in the presence of sodium dodecyl sulfate. From this 12 cysteines were substracted.

(b) Methyl viologen was the electron carrier. Turnover number was based on a $V_{max}$ (MV) of 4000 μmoles $H_2$ produced/min/mg hydrogenase.

(c) Ferredoxin (plus 1 mM methyl viologen) was the electron carrier. Turnover number was based on a $V_{max}$ of 5600 μmoles $H_2$ produced/min/mg hydrogenase.

to produce $H_2$. With dithionite as the reductant, the apparent $K_m$ values for reduced Fd and MV in the $H_2$ production reaction are 0.051 and 6.25 mM, respectively.

Purified hydrogenase produces $H_2$ from dithionite in the absence of an added electron carrier but only at $10^{-3}$ times the rate of the turnover number obtained with ferredoxin plus 1 mM MV. The function of the electron carrier, therefore, is to efficiently transfer electron from the reductant, in this case dithionite, to hydrogenase. In a steady state reaction mixture of hydrogenase and dithionite, the predominant form of hydrogenase is a form incapable of producing $H_2$. The implication of this will be further discussed in following sections.

Hydrogenase activity is inhibited by carbon monoxide and cyanide two well known ligands for transition metals. Strong inhibition of hydrogenase by CO is accompanied by distinct changes in the EPR spectra of hydrogenase which strongly suggests the involvement of iron-sulfur centers in hydrogenase function. The inhibition by CO is reversed by physical removal of the inhibitor (Thauer et al., 1974; Chen and Mortenson, unpublished). On the other hand, hydrogenase is both inhibited and gradually inactivated by cyanide even in the presence of dithionite and $H_2$ (Chen and Mortenson, unpublished). This result is in contrast with earlier results (Fisher et al., 1954; Hoberman and Rittenberg, 1943) that indicated only "oxygenated" or "oxidized" hydrogenase was inhibited by cyanide.

## Spectral Porperties of Hydrogenase

Optical absorption spectra. The UV-VIS absorption spectrum of $H_2$-reduced hydrogenase (Fig. 5 of Chen and Mortenson, 1974) resembles that of iron-sulfur proteins with tetrameric iron-sulfur clusters. Hydrogenase shows increasing absorption with decreasing wavelength in the range of 600 nm to 280 nm, and has a shoulder in the 400 nm region. At 400 nm, an extinction coefficient of 25.3 $cm^{-1}$ $mM^{-1}$ was found for $H_2$-reduced hydrogenase. Controlled oxygen-oxidation increases the absorption in the 400 nm region by a factor of 1.3 but excess oxygen eventually bleaches and inactivates the protein.

CD spectra. In contrast to a previous result (Multani and Mortenson, 1972) the highly purified $H_2$ase (0.2 mg/ml) does not show any CD absorption between 300 nm and 600 nm. This constitutes a substantial difference between hydrogenase and two-iron and eight-iron ferredoxins (Palmer et al., 1967; Atherton et al., 1966). CD bands attributable to the polypeptide chain of hydrogenase (below 250 nm) show an α-helical content of about 10% with $H_2$-reduced hydrogenase.

EPR spectra. The EPR spectrum A of hydrogenase undergoes a series of changes with varying redox potentials. Fig. 1 illustrates two distinct spectral forms (tracing A and C) and a transient form (tracing B). Spectrum A is observed with hydrogenase at redox potentials ($E_h$) more negative than -400 mV, i.e., in the presence of dithionite or under 1 atm $H_2$ at pH 7, with or without catalytic amounts of Fd or MV added. When hydrogenase is oxidatively titrated with ferricyanide, spectrum A changes first into a complex family of weak signals (represented by spectrum B) which appear at potential around -360 mV, and which in turn changes into a distinct rhombic type spectrum (spectrum C) when the $E_h$ is made more positive than -330 mV. In so far as all g-values are greater than 2.0, this latter resonance can be considered of the Hipip type. Even though the lineshape is not typical, a well-defined set of 3 g-values is present. These changes are reversible when the oxidized hydrogenase is reduced with either $H_2$ or dithionite in the absence of catalytic amounts of Fd or MV.

The features of the complex spectra observed at -360 mV (Fig. 1B) can be assigned. In addition to contributions from the "reduced" species (Fig. 1A) and the "oxidized" species (Fig. 1C) one can identify lines due to a second rhombic species with approximate g-values 2.077, ~1.98, 1.908. This appears to be conventional single-center ferredoxin resonance and presumably arises from those protein molecules in which only one cluster is present as the trianion.

## Iron-Sulfur Clusters of Hydrogenase

The EPR spectrum (Fig. 1A) exhibited by hydrogenase at potential more negative than -400 mV is much more complex than those seen in simple s = 1/2 paramagnetic centers and bears a striking resemblance to the EPR spectrum of the bacterial ferredoxin from Micrococcus lactilyticus (Mathews et al., 1974). For this latter protein there is good evidence for two S = 1/2 centers each arising from a tetrameric iron-sulfur cluster and for coupling, most probably dipolar in origin, between these two centers that produces the additional structure present in the spectrum. Integration of spin intensity when using spinach Fd and Cu EDTA as standards gave values of 1.66 - 1.80 spins per molecule of hydrogenase ($H_2$-or dithionite-reduced), suggesting at least two iron-sulfur centers per hydrogenase molecule. Table II lists the specific activity, iron content and the spin content of several different hydrogenase preparations. These values are significantly higher than those (0.7 - 1.0 spin/mole) recently reported on extensively purified hydrogenase from the same organism (Erbes, Burris and Orme-Johnson, 1975). Differences in specific activity and iron content between preparations undoubtedly cause this discrepency.

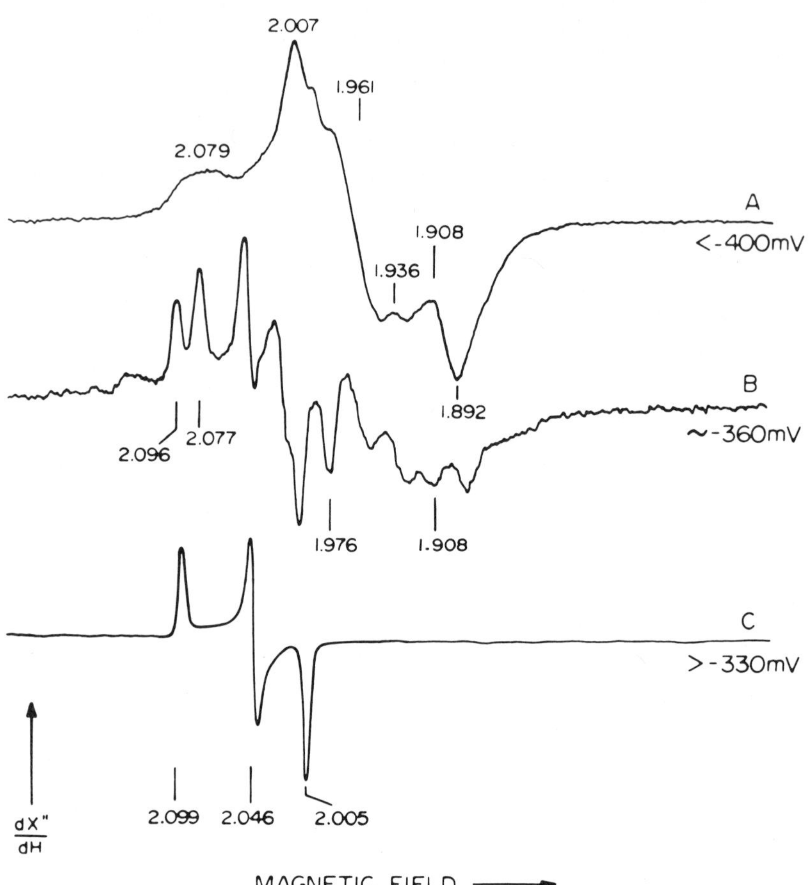


Fig. 1. Electron paramagnetic resonance spectra of hydrogenase at different redox states. Hydrogenase was in 0.1 M potassium phosphate buffer at pH 7.2. Spectrum A was recorded under 1 atm $H_2$; spectra B and C were under $N_2$. The redox potential in each case was estimated by comparing the spectrum with those obtained in other redox titrations in which a mediator mixture was used and the redox potential was monitored. These spectra were recorded with a Varian E-6 x-band spectrometer under the following conditions: temperature, 14.8$^{\circ}$ K; frequency, 9.228 GHz; time constant, 0.3 s. Other conditions were: (A) Hydrogenase, 35 μM; microwave power (MP), 1 mW; modulation amplitude (MA), 10 gauss; gain, 3200. (B) Hydrogenase 11 μM; MP, 1 mW; MA, 20 gauss; gain, 10,000. (C) Hydrogenase, 50 μM; MP, 16 mW; MA, 10 gauss; gain, 400.

TABLE II

SPECIFIC ACTIVITY, IRON CONTENT AND SPIN CONTENT OF HYDROGENASE PREPARATIONS

| Preparation | Specific Activity (units/mg) | Iron Content (atom/mole) | Spin** Content (spin/mole) |
|---|---|---|---|
| 1 | 378 | 9.1 | 1.35 |
| 2 | 562 | not determined | 1.66-1.72 |
| 3 | 521 | 12.2 | 1.56-1.66 |
| 4 | 552 | 12.3 | 1.80 |
| 3a | 521 | 12.2 | 1.64 |

* Hydrogenase activity was assayed as described in Methods. One unit is defined as one μmole $H_2$ produced per minute.

** Spin content was determined from EPR signals of hydrogenase at -430 mV (1 atm $H_2$, pH 7.2) except for 3a which was at -540 mV (1 atm $H_2$, pH 9.0).

Between the $E_h$ range of -430 mV to -540 mV, the spin intensity was essentially unchanged and the EPR spectrum (Fig. 1A) remained the same except that the small signal at g = 1.936 decreased at potentials more negative than -430 mV.

Integration of the rhombic spectrum observed at the most positive potentials (Fig. 1C) gave a spin content of 0.69 spin/mole for hydrogenase at an $E_h$ of -154 mV. This suggests that only one of the hydrogenase centers, possibly a center that is silent at -400 mV is oxidized to this EPR positive state.

The optical absorption spectrum of hydrogenase is very similar to that found with classical bacterial ferredoxins and suggests tetrameric iron-sulfur center(s). Extrusion of the iron-sulfur centers of hydrogenase with thiophenol in 80% hexamethylphosphoramide:20% 0.02 M Tris·HCl at pH 8.0 confirmed the presence of $Fe_4S_4(SR)_4$ clusters (Erbes, Burris and Orme-Johnson, 1975; Mortenson, Chen, Gillam and Holm, unpublished). Quantitative extrusion studies, with clostridial Fd as the standard, again pointed to the presence of at least two and possibly three tetrameric iron-sulfur clusters in hydrogenase (details of the experiments will be reported elsewhere). If there are three iron-sulfur tetramers in hydrogenase and only two are detected under reduced conditions by EPR, then a third would exist in an EPR silent form, perhaps as a (-2) tetramer. The role of these centers in hydrogenase activity is not clear at present although possible functions are suggested in Table III and in the following section.

## Mechanistic Consideration

Attempts were made to correlate the EPR-distinguishable states of hydrogenase with those involved in the physiological reactions. Data presented in the previous section suggested that EPR spectrum A resulted from two tetrameric Fe·S centers in the (-3) state, $[Fe_4S_4(SR)_4]^{-3}$. EPR spectrum C probably results from the presence of one $[Fe_4S_4(SR)_4]^{-1}$ center. Table III shows the postulated valence states of the Fe·S centers of hydrogenase at specified redox states. At least three electrons seem to be involved when hydrogenase is oxidized from state A to state C. In addition, potentiometric titration indicated that there were two midpoint potentials between state A and state C. Because of the two midpoint potentials between states A and C and because states A and C are separated by an EPR silent state, it seems that the Fe·S center seen in C does not have to arise from those seen in A. For example, a third tetrameric Fe·S center may be in the (-2) state when spectrum A is seen and it gets oxidized to the (-1) state to

TABLE III

PREDICTED PROPERTIES OF HYDROGENASE AT THE PROPOSED REDOX STATES*

| Redox State | H | | A | | B | | C |
|---|---|---|---|---|---|---|---|
| $E_h$ | <-400 mV | | ~-400 mV | | ~-360 mV | | ≥-330 mV |
| Valence of the $Fe_4S_4$ centers | $[Fe_4S_4(SR)_4]^{-3}$<br>$[Fe_4S_4(SR)_4]^{-3}$<br>$X \cdot 2e^-$ | $H_2$ ⇌ $H_2$ or reduced $e^-$ carrier | $[Fe_4S_4(SR)_4]^{-3}$<br>$[Fe_4S_4(SR)_4]^{-3}$<br>X | $O_2$ or oxidized $e^-$ carrier ⇌ $H_2$ $S_2O_4^=$ or reduced $e^-$ carrier | { $[Fe_4S_4(SR)_4]^{-3}$<br>$[Fe_4S_4(SR)_4]^{-2}$<br>X (?) , $[Fe_4S_4(SR)_4]^{-2}$<br>$[Fe_4S_4(SR)_4]^{-2}$<br>X (?) } | $O_2$ or oxidized $e^-$ carrier ⇌ $H_2$, $S_2O_4^=$ or reduced $e^-$ carrier | $[Fe_4S_4(SR)_4]^{-2}$<br>$[Fe_4S_4(SR)_4]^{-1}$<br>X (?) |
| EPR Signal | Fig. 1A | | Fig. 1A | | Fig. 1B | | Fig. 1C |
| Spin/mole | 2 | | 2 | | (1 or 0) | | 1 |

* Site X could be a third iron-sulfur center that is postulated to be the active center ($H_2$-activating center) of hydrogenase. The other centers probably are involved in electron transfer to and/or from the active center Fe·S cluster (see illustration in Fig. 2). Hence the (-1) tetramer of C may arise instead from X that is EPR silent at -400 mV.

show the spectrum C. The above arguments tend to eliminate state B as the $H_2$ producing state and suggest state A as the probable $H_2$ producing state. However, the facts outlined in the following paragraph suggest that state A is the oxidized form of the active center of hydrogenase and not the reduced form.

It was observed that the rate of $H_2$ production from reduced MV always dropped drastically when the $E_h$ reached around -400 mV at pH 7.0 and at this potential only spectrum A was seen. Possible inhibition of $H_2$ evolution by the oxidized form of MV at this $E_h$ might complicate the situation and hence strong conclusions cannot be drawn yet from this observation. On the other hand, like the situation in the presence of catalytic amounts of Fd and MV, EPR spectrum A at full intensity was obtained when hydrogenase was incubated with excess dithionite where the $H_2$ production is only 0.1% of the maximum obtained in the presence of the carrier. Because substrate levels of electron carriers (Fd or MV) would mask the EPR spectrum of hydrogenase, the EPR spectrum of hydrogenase at its full activity ($H_2$ production from substrate level MV at -550 mV) was not measured. However, the EPR spectrum and spin intensity of hydrogenase at -540 mV (achieved by adjusting the $H_2$ partial pressure and pH) is the same as that at -430 mV. Because of the reversible nature of the hydrogenase reaction with $H_2$, one could assume that hydrogenase at its full $H_2$-producing activity also shows EPR spectrum A. Therefore, it suggests that EPR analysis cannot detect the redox state(s) which exist at potentials more negative than -400 mV at pH 7.0. State H (Table III), which is more reduced than state A, is postulated to be the $H_2$-producing state of hydrogenase. The designation of state H as being more reduced than state A is compatible with 1) the kinetic data which show that the reduced electron carriers function as electron donors to form A in the $H_2$ production reaction and 2) the fact that when $H_2$ evolution ceases at -400 mV at pH 7.0, form A is still fully present.

Because the EPR spectrum A and spin intensity remained the same between -540 mV and -430 mV, it does not seem likely that the two tetrameric Fe·$S_4$ centers are reduced to the -4 state at -540 mV since $[Fe_4S_4(SR)_4]^{-4}$ would be EPR silent. A significant concentration of $[Fe_4S_4(SR)_4H]^{-3}$ probably does not exist under the latter conditions because a change in the EPR spectrum A would be expected. Site (X) was thus proposed (Fig. 2 and Table III) to be the $H_2$-activating site. In its oxidized form, site (X) oxidizes $H_2$ to $H^+$ and transfers the electrons first to the tetrameric Fe·S centers if they are in the (-2) state and then to other oxidized electron carriers. Alternatively, it might accept electrons from the tetrameric Fe·S centers to reduce $H^+$ to produce $H_2$. Reduced electron carriers at low $E_h$ such as fully reduced Fd are postulated to efficiently donate electrons to site (X) possibly through the tetrameric Fe·S centers (Fig. 2).

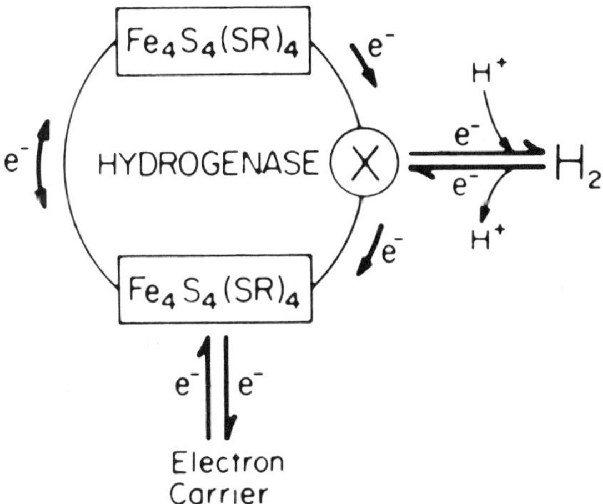


Fig. 2. A model of hydrogenase. Site X is the proposed $H_2$-activating site. Arrows show the proposed path of electron flow. It is possible that electrons from external carriers such as reduced ferredoxin also can be transferred directly to X.

This mechanism suggests that state H is responsible for the H/D exchange activity of hydrogenase. It also allows $H_2$ oxidation (uptake to occur at a wide range of $E_h$. In fact, $H_2$ oxidation was observed between -330 mV and -540 mV for hydrogenase from C. pasteurianum.

## SUMMARY

Hydrogenase from C. pasteurianum is an iron-sulfur protein containing at least two tetrameric iron-sulfur centers. Information on the structure of the remaining iron atoms must await future investigation.

Although the EPR spectra of dithionite-reduced hydrogenase and eight-iron Fd showed some similarity, the CD spectra clearly indicated a difference. The tetrameric iron-sulfur centers of hydrogenase were shown to undergo redox changes when hydrogenase was oxidized or reduced. However, no evidence is now available to support a role for the tetrameric Fe·S centers, responsible for the EPR spectrum A, as the primary site for $H_2$ binding and activation. Because we have found that the $[Fe_4S_4(SR)_4]$-containing ferredoxins do not have hydrogenase activity, it is conceivable that the additional iron atoms and/or certain amino acid residues of hydrogenase also contribute to the unique catalytic properties of this enzyme.

Chemical synthesis of Fe·S clusters with different peptide environments and with hydrogenase function would lead to the identification of these functional groups. X-ray diffraction studies on hydrogenase will certainly complement the other approaches.

Knowledge of the structure of the active site of hydrogenase will certainly accelerate research into: (1) the synthesis of a stable catalyst to replace hydrogenase in systems designed to produce $H_2$ by coupling this catalyst to a photoreducing system; and (2) the elucidation of the active sites of more complicated iron-sulfur enzyme such as nitrogenase.

## ACKNOWLEDGMENTS

This study was supported by grants to L.E.M. from the National Science Foundation (GB-22629) and from National Institutes of Health (AI 04865) and to G.P. from the National Institutes of Health (GM 21337).

## REFERENCES

Alberty, R. A. and B. M. Koerber. 1957. J. Am. Chem. Soc. 79:6379 6382.

Atherton, N. M., K. Garbett, R. D. Gillard, R. Mason, S. J. Mayhew, J. L. Peel and J. E. Stangroom. 1966. Nature 212:590-593.

Brumby, P. E., R. W. Miller and V. Massey. 1965. J. Biol. Chem. 240:2222-2228.

Chen, J. S. and L. E. Mortenson. 1974. Biochim. Biophys. Acta 371:283-298.

Erbes, D. L., R. H. Burris and W. H. Orme-Johnson. 1975. Proc. Nat. Acad. Sci. U.S.A. In Press.

Fisher, H. F., A. I. Krasna and D. Rittenberg. 1954. J. Biol. Chem. 209:569-578.

Gitliz, P. H. and Krasna, A. I. 1975. Biochemistry 14:2561-2568.

Fogo, J. K. and M. Popowsky. 1949. Anal. Chem. 21:732-737.

Hoberman, H. D. and D. Rittenberg. 1943. J. Biol. Chem. 147:211-227.

LeGall, J., D. V. Dervartanian, E. Spilker, J. P. Lee and H. D. Peck. 1971. Biochim. Biophys. Acta 234:525-530.

Mathews, R., S. Charlton, R. H. Sands and G. Palmer. 1974. J. Biol. Chem. 249:4326-4328.

Mortenson, L. E. and J. S. Chen. 1974. In "Microbial Iron Metabolism" (Neilands, J. B., ed.) p. 231-282. Academic Press, New York.

Mortenson, L. E., and J. S. Chen. 1975. In "Microbial Production and Utilization of Gases ($H_2$, $CH_4$, CO)". Göttingen, West Germany, In Press.

Multani, J. S. and L. E. Mortenson. 1972. Biochim. Biophys. Acta 256:66-70.

Nakos, G. and L. E. Mortenson. 1971. Biochim. Biophys. Acta 227:576-583.

Palmer, G., H. Brintzinger and R. W. Estabrook. 1967. Biochemistry 6:1658-1664.

Que, L., R. H. Holm and L. E. Mortenson. 1975. J. Am. Chem. Soc. 97:463-464.

Thauer, R. K., B. Kaufer, M. Zahringer and K. Jungermann. 1974. Eur. J. Biochem. 42:447-452.

Wu, J. Y. and J. T. Yang. 1970. J. Biol. Chem. 245:212-218.

Yagi, T. 1970. J. Biochem. 68:649-657.

# SEQUENCE INVESTIGATION OF THE *CLOSTRIDIUM PASTEURIANUM* NITROGENASE: THE PARTIAL AMINO ACID SEQUENCE OF AZOFERREDOXIN

Masaru Tanaka, Mitsuru Haniu, Kerry T. Yasunobu and Leonard E. Mortenson*
Dept. of Biochem-Biophys, University of Hawaii Medical School and * Dept. of Biological Sciences Purdue University, West Lafayette, Indiana

Our laboratory has been investigating the primary structures of various iron-sulfur proteins (1) with various long term goals in mind. These are: 1) to provide essential data for structure-function investigations; 2) to provide evolutionary yardsticks for measuring the age of the iron-sulfur containing organisms; and provide necessary structural data for the X-ray crystallographers interested in determining the complete structures of the Fe-S proteins. Thus far, the Fe-S proteins sequenced have had relatively low molecular weights and perhaps it is time to extend the studies to proteins of greater complexity and with higher molecular weights. The protein chosen in the present investigation was the *Clostridium pasteurianum* nitrogenase and more specifically the azoferredoxin component.

The methods for the isolation, the chemical characterization and functional properties of nitrogenase have been documented in review articles (2-7). The current concept of the sequence of the molecular events occuring during $N_2$ reduction have been recently summarized by Burris and Orme-Johnson (6) and is shown in Fig. 1.

Dalton and Mortenson (2) have reported the molecular weight of azoferredoxin component of nitrogenase to be 55,000 and a subunit molecular weight of 27,500. The two identical subunits were found to be identical chemically and to contain 4 iron atoms, 4 acid-labile sulfide atoms, and 12 cysteine residues per molecule of protein. Molybdoferredoxin, the other nitrogenase component, has a molecular weight of 220,000 and is composed of two dissimilar subunits of molecular weight 60,000 and 50,000. Molybdoferredoxin is reported to contain 2 molybdenum atoms, 24 iron atoms, 24 atoms of sulfide and about 40 cysteine residues per molecule (8,9).

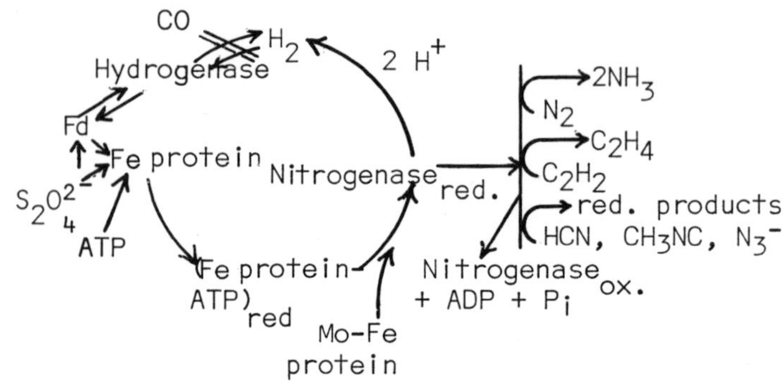


Fig. 1 The current concept of the sequence of molecular events in $N_2$ reduction. (Taken from Reference 6).

In the present comminication, we wish to report the partial sequence of the azoferredoxin component of nitrogenase.

## EXPERIMENTAL SECTION

Azoferredoxin was isolated from C. pasteurianum strain W5 cells by the previously described purification procedure described by Zumft and Mortenson (10).

The procedures used for the preparation of the S-β-carboxymethyl-cysteinyl-azoferredoxin, (Cm-ferredoxin), amino acid composition and sequence determinations have been described in previous reports from our laboratory (for example, see Refs. 11 and 12) and will not be described in this report.

## RESULTS

### Amino Acid Composition of Azoferredoxin

The amino acid composition of the C. pasteurianum azoferredoxin has been determined by Chen et al (8) and their result is summarized in Table I. The subunit has been reported to contain 261 amino acids (Mol. wt. 27,500) and the basic amino acid content was reported to be $lysine_{16}$, $histidine_{2}$, and $arginine_{12}$.

### Strategy of Sequence Determination

The following procedures were adopted to determine the complete amino acid sequence of the protein: 1) Determination of the $NH_2$ and COOH-terminal amino acids by conventional end-group analyses; 2) Determination of the $NH_2$-terminal sequence of the CM-azoferredoxin by use of the Protein Sequencer; 3) Determination of the sequences of the tryptic peptides; and 4) the isolation of the cyanogen bromide fragments which would provide overlapping sequences in order that the tryptic peptides could be ordered to yield the complete sequence. Each of these steps will be considered in the sections to follow in this article.

TABLE I

Amino Acid Composition of the *Clostridium pasteurianum* Azoferredoxin Component of Nitrogenase*

| Amino Acid | Azoferredoxin subunit | |
|---|---|---|
| | Prior analysis | From Peptides (this paper) |
| Lysine | 16 | 16 |
| Histidine | 2 | 2 |
| Argine | 12 | 12 |
| Aspartic acid + Asparagine | 23 | 22 |
| Threonine | 12 | 13 |
| Serine | 12 | 13 |
| Glutamic acid + Glutamine | 35 | 35 |
| Proline | 9 | 9 |
| Glycine | 31 | 32 |
| Alanine | 20 | 20 |
| Cysteine | 6 | 6 |
| Valine | 17 | 19 |
| Methionine | 8 | 11 |
| Isoleucine | 17 | 20 |
| Leucine | 26 | 26 |
| Tyrosine | 9 | 12 |
| Phenylalanine | 6 | 5 |
| Tryptophan | 0 | 0 |
| Total Residues | 261 | 273 |
| Molecular weight | 27,500 | 28,850 |

* Dalton and Mortenson (2).

## End Group Determinations of the Protein

One step of Edman degradation of the Cm-protein clearly showed that methionine was the $NH_2$-terminal amino acid. Both hydrazinolysis and carboxypeptidase A digestion yielded leucine in high yields.

## $NH_2$-Terminal Sequence of Cm-azoferredoxin

When the protein derivative was analyzed in the Beckman Model 890 Protein Sequencer, it was possible to establish the sequence of the first 41 amino acids from the $NH_2$-terminal end of the protein. The residues identified are shown in Fig. 2 and the Pth-amino acids liberated during Edman degradation were identified by gas-liquid chromatography, by thin layer chromatography and by amino acid analysis of the acid hydrolyzed Pth-amino acids.

## Isolation and Sequence of the Tryptic Peptides

The detailed isolation procedure of the tryptic peptides will be described elsewhere. In general the tryptic peptides were purified by Sephadex G-50 chromatography and ion exchange column chromatography, usually in that order. The tryptic peptides isolated and their sizes are summarized in Table II. Our preliminary data indicate that there are 25 major peptides and the total amino acid residues were 273 rather than the value of 261 shown in Table I. However, there was agreement concerning the total number of lysine and arginine residues, 16 and 12 respectively, present in the 25 tryptic peptides.

TABLE II

Summary of Tryptic Peptides Derived from Azoferredoxin*.

| Peptide | No. of residues | Peptide | No. of residues |
|---|---|---|---|
| T-32b | 20 | T-64d | 5 |
| T-34b | 40 | T-64e1 | 6 |
| T-44d | 17 | T-64e2 | 16 |
| T-47 | 23 | T-64f | 9 |
| T-52 | 16 | T-65c | 5 |
| T-53 | 10 | T-65d | 4 |
| T-54c | 18 | T-65e | 7 |
| T-54d | 14 | T-65f | 9 |
| T-56 | 13 | T-66c1 | 2 |
| T-64a1 | 3 | T-66c2 | 3 |
| T-64a2 | 5 | T-66d | 7 |
| T-64b | 5 | T-66e | 12 |
| T-64c | 4 | | |

*A total of 25 peptides were isolated in which their total amino acid content added up to 273 residues. In the experiment, 110 mg of protein was digested at 34° at pH 8.0 for 24 hours in the presence of 2% trypsin by weight.

After purification, the sequence of the tryptic peptides were determined by conventional sequencing techniques. The quantitative data will be presented elsewhere and only the sequence of the peptides are presented in Table III. The isolation of only 25 tryptic peptides, when the protein contains 16 lysine and 12 arginine residues, can be readily explained when the sequence of the tryptic peptides are examined. Peptide T-32b contained an arginyl-glutamyl bond and this type of bond is known to be slowly cleaved by trypsin. In addition, peptide T-54c contained an amino-terminal lysine residue as well as a prolyl-lysyl-prolyl sequence which is known to be resistant to trypsin cleavage. Peptide T-54d also contained an amino-terminal lysine as well as a COOH-terminal lysine residue. Thus, only 25 tryptic peptides are obtainable from the protein. Since peptide T-56 was the only peptide lacking basic amino acids, it was obviously the COOH-terminal peptide and it contained a COOH-terminal leucine residue which was determined to be the COOH-terminal residue of the intact protein.

TABLE III

Amino Acid Sequence of the Tryptic Peptides Derived from Cm-Azoferredoxin

| Peptide Number | Amino Acid Sequence |
|---|---|
| T-32b | Ser-Val-Leu-Asp-Thr-Leu-Arg-Glu-Glu-Gly-Glu-Asp-Val-Glu-Leu-Asp-Ser-Ile-Leu-Lys |
| T-34b | Gly-Ile-Ile-Thr-Ser-Ile-Asn-Met-Leu-Glu-Gln-Leu-Gly-Ala-Tyr-Thr-Asp-Asp-Leu-Asp-Tyr-Val-Phe-Tyr-Asp-Val-Leu-Gly-Asp-Val-Val(Ser/Cys)Gly-Gly-Phe-Ala-Met-Pro-Ile-Arg |
| T-44d | Gln-Thr-Val-Ile-Glu-Tyr-Asp-Pro-Thr-Cys-Glu-Gln-Ala-Glu-Glu-Tyr-Arg |
| T-47 | Ala-Gln-Glu-Ile-Tyr-Ile-Val-Ala-Ser-Gly-Glu-Met-Met-Ala-Leu-Tyr-Ala-Ala-Asn-Asn-Ile-Ser-Lys |
| T-52 | Cys-Val-Glu-Ser-Gly-Gly-Pro-Glu-Pro-Gly-Val-Gly-Cys-Ala-Gly-Arg |
| T-53 | Thr-Ile-Met-Val-Val-Gly-Cys-Asp-Pro-Lys |
| T-54c | Lys-Val-Asp-Ala-Asn-Glu-Leu-Phe-Val-Ile-Pro-Lys-Pro-Met-Thr-Glx-Glu-Arg |
| T-54d | Lys-Val-Ala-Asn-Glu-Tyr-Glu-Leu-Leu-Asp-Ala-Phe-Ala-Lys |
| T-56 | Leu-Glu-Glu-Ile-Leu-Met-Gln-Tyr-Gly-Leu-Met-Asp-Leu |
| T-64a1 | Glu-Gly-Lys |
| T-64a2 | Ala-Asp-Ser-Thr-Arg |
| T-64b | Ala-Glu-Ile-Asn-Lys |
| T-64c | Gly-Ile-Gln-Lys |
| T-64d | Gly-Gly-Ile-Gly-Lys |
| T-64e1 | Ser-Pro-Met-Val-Thr-Lys |
| T-64e2 | Ser-Thr-Thr-Thr-Gln-Asn-Leu-Thr-Ser-Gly-Leu-His-Ala-Met-Gly-Lys |
| T-64f | Leu-Gly-Gly-Ile-Ile-Cys-Asn-Ser-Arg |
| T-65c | Ser-Gly-Gly-Val-Arg |
| T-65d | Glu-Leu-Ala-Arg |
| T-65e | Gln-Val-Ala-Ile-Tyr-Gly-Lys |
| T-65f | Leu-Leu-Leu-Gly-Gly-Leu-Ala-Gln-Lys |
| T-66c1 | Met-Arg |
| T-66c2 | Tyr-Ala-Lys |
| T-66d | Glu-Gly-Tyr-Gly-Gly-Ile-Arg |
| T-66e | Glu-Leu-Gly-Ser-Gln-Leu-Ile-His-Phe-Val-Pro-Arg |

## Cyanogen Bromide Fragments

In order to obtain overlapping sequences, another method of protein cleavage, besides trypsin cleavage, is required and at the present time, we are attempting to separate the cyanogen bromide fragments. Until this phase of the research is completed, the sequence studies cannot be completed. However, a cyanogen bromide fragment has been isolated in a pure form which contains residues 34-108 (fragment, CNBr-1d) and sequence analysis of this peptide in the Protein Sequencer established the amino acid sequence from residues 34-97. This is the extent of the sequence investigations on the CNBr fragments at the present time.

Partial Reconstruction of the Amino Acid Sequence of Azoferredoxin

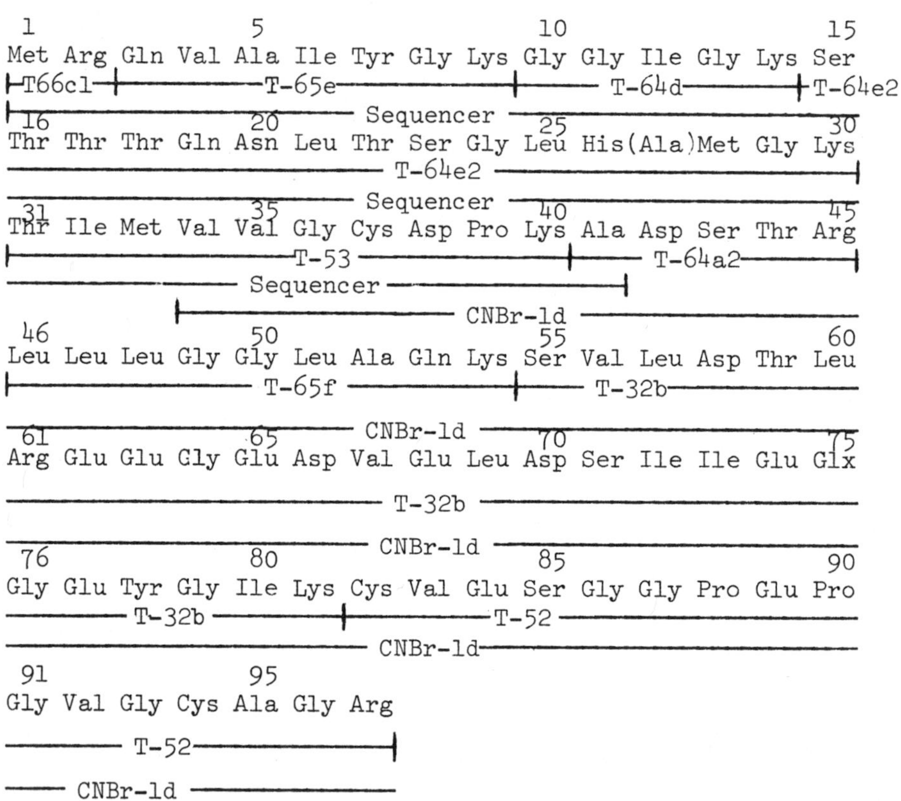


Fig. 2. Reconstruction of $NH_2$-Terminal Amino Acid Sequence of <u>Clostridium pasteurianum</u> Azoferredoxin

The Protein Sequencer analysis of the intact protein established the sequence of residues 1-41 and showed that peptide T-66c1, T-65e, T-64d, T-64e2 and T-53 existed in this order from the $NH_2$-terminal end of the protein (Fig. 2). Sequence determination of CNBr-1d in the Protein Sequencer eludicated the sequence from residues 34-83 and showed that peptides T-53, T-64a2, T-65f, T-32b and T-52 followed peptide T-64e2. Sequence determination of peptide T-52 itself extended the known sequence to residue 97 as shown in Fig. 2.

## DISCUSSION

As soon as the sequences of the cyanogen bromide fragments are completed, it will be possible to place all of the tryptic peptides and the complete amino acid sequence of the azoferredoxin will be determined. At the present stage, it is possible to present the partial sequence from the $NH_2$-terminal position to residue 97. The amino acid composition of the tryptic peptides suggests that there are 12 more amino acids in the protein than was found from the amino acid composition data (2). However, the total number of residues of lysine and arginine appear to be correct since there is agreement between the amino acid composition and the sequence data. There are six cysteine residues in the azoferredoxin moiety. In the part of the protein sequenced thus far, cysteine residues were found in positions 37, 82 and 94. There may be three additional cysteine residues in peptides T-34b, T-44d and T-64f. The cysteine residues are important since in iron-sulfur proteins most of them are ligands of iron and are structural components of active center of the protein. In azoferredoxin, the cysteine residues are distributed randomly throughout the protein molecule and not in the Cys-XXX-YYY-Cys or Cys-XXX-YYY-ZZZ-Cys pattern, where the triple X, Y, and Zs represent amino acids, found in the bacterial ferredoxins. It also appears that there is no sequence homology between azoferredoxin and the other iron-sulfur proteins sequenced to date. However, some of the above mentioned conclusions must be considered tentative until the complete amino acid sequence of azoferredoxin has been determined. The prelimimary $NH_2$- and COOH-terminal sequences of azoferredoxin previously referred to in reference 13 must be modified since they are not in agreement with the presently reported sequence data.

## ACKNOWLEDGEMENTS

The research project was supported in part by research Grants GM 16784, GM 16228 from the National Institutes of Health and Grant GB 43448 from the National Science Foundation and grants to L.E.M. from the National Science Foundation (GB-22629) and National Institutes of Health (AI 04865).

## REFERENCES

1. Yasunobu, K. T., and Tanaka, M. (1973). in Iron-Sulfur Proteins (Lovenberg, W., ed.), Academic Press, New York, N. Y., Vol II, pp. 27-130.
2. Dalton, H., and Mortenson, L.E. (1972). Bacteriol. Reviews 36, 231-260.
3. Hardy, R.W.F., Knight, E., Jr., McDonald, C.C., and D'Eustachio A.J. (1965). in Non-Heme Iron Proteins (San Pietro, A., ed.), The Antioch Press, Yellow Springs, Ohio, pp. 275-282.
4. Hardy, R.W.F., Burns, R.C., and Parshall, G.W. (1971). Advan. Chem. Soc. 100, 219-247.
5. Orme-Johson, W.H. (1973). Ann. Rev. Biochem. 42, 159-204.
6. Burris, R.H., and Orme-Johnson, W.H. (1974), in Microbial Iron Metabolism (Neilands, J.B. ed.), Academic Press, New York, N. Y., pp. 187-209.
7. Zumft, W.G., and Mortenson, L.E. (1975). Biochim. Biophys. Acta 416, 1-52.
8. Chen, J.-S-, Multani, J.S., and Mortenson, L.E. (1973). Biochim. Biophys. Acta. 310, 51-59.
9. Huang, T.C., Zumft, W.G., and Mortenson, L.E. (1973). J. Bacteriol. 113, 884-890.
10. Zumft, W.G., and Mortenson, L.E. (1973). Eur. J. Biochem. 35, 401-409.
11. Tanaka, M., Haniu, M., Yasunobu, K.T., Evans, M.C.W., and Rao, K.K. (1975). Biochem. 14, 1938.
12. Tanaka, M., Haniu, M., Yasunobu, K.T., Rao, K.K., and Hall, D.O. (1975). Biochemistry 14, 5535.
13. Mortenson, L.E. (1972). Methods Enzymol. 24, 455.

# ON THE NATURE OF AN INTERMEDIATE THAT IS FORMED DURING THE ENZYMATIC CONVERSION OF PHENYLALANINE TO TYROSINE

Seymour Kaufman

Laboratory of Neurochemistry, National Institute of Mental Health, Bethesda, Maryland 20014

## INTRODUCTION

The enzymatic conversion of phenylalanine to tyrosine proceeds according to equations 1 and 2 [1].

(1) Phenylalanine + tetrahydropterin + $O_2$ → tyrosine + quinonoid dihydropterin + $H_2O$

(2) DPNH(TPNH) + $H^+$ + quinonoid dihydropterin → $DPN^+(TPN^+)$ + tetrahydropterin

Reaction 1 is catalyzed by phenylalanine hydroxylase and reaction 2 by dihydropteridine reductase.

Under our usual assay conditions, i.e., at pH 6.8, the rate of tyrosine formation is a linear function of hydroxylase concentration. At pH 8.0, and at low concentrations of tetrahydrobiopterin (biopterin-$H_4$), however, the rate of tyrosine formation per mg hydroxylase decreases with increasing concentrations of hydroxylase [2]. Furthermore, this peculiar decrease in the specific activity of the hydroxylase is seen with biopterin-$H_4$, but not with 6,7-dimethyl-5,6,7,8-tetrahydropterin ($DMPH_4$) [2].

Under these special assay conditions (pH 8.0-8.2, low biopterin-$H_4$ concentrations and high hydroxylase concentrations) the hydroxylation reaction is markedly stimulated by a protein that we have isolated in homogeneous form from rat liver, phenylalanine hydroxylase stimulating protein (PHS). In the presence of PHS, the specific activity of the hydroxylase is constant even under the special assay conditions [3]. We have shown that PHS stimulates the hydroxylation reaction by decreasing the apparent $K_m$ of the hydroxylase for biopterin-$H_4$ and not by increasing $V_{max}$ [4].

A detailed analysis of the kinetics of the hydroxylation reaction under conditions where PHS stimulates led to a scheme for the reaction that involves the formation of a free intermediate, designated S' [4]. At pH 6.8, the non-enzymatic breakdown of S' to products (P) is not rate-limiting. At pH 8, on the other hand, the conversion of S' to P is relatively slow and imposes a limit on the overall reaction rate. According to this formulation, the role of PHS is the catalysis of the conversion of S' to P.

To explain why this abnormal kinetic behavior (and hence, the dependence of the rate of the reaction on the presence of PHS) is not observed with $DMPH_4$, or with either $DMPH_4$ or biopterin-$H_4$ at lower pH values, we have assumed that the non-enzymatic transformation of S' to P is more rapid at lower pH values and that the S' that is formed in the presence of $DMPH_4$ is even more unstable than is the intermediate formed in the presence of biopterin-$H_4$. Under these conditions, therefore, S' never accumulates and does not impose a limit on the rate of the overall reaction.

Although these earlier results with PHS suggested the existence of an intermediate in the hydroxylation reaction, they provided little information about the structure of the intermediate. Since the stability of S' appears to vary with the structure of the pterin, however, it seems unlikely that S' is a simple precursor of the amino acid product, tyrosine. For example, some of our kinetic data would be consistent with S' being the epoxide of phenylalanine, but one would then expect that the same S' would be formed in the presence of different pterins. Since this does not seem to be the case, it seems more likely that S' is a pterin derivative. It could be either a precursor of the quinonoid dihydropterin product or a complex involving the pterin, oxygen, and phenylalanine, a complex which then breaks down to give tyrosine and the dihydropterin. In either case, PHS would be the enzyme that catalyzes this transformation of S' to the product or products of the hydroxylation reaction.

In the present paper, direct evidence will be presented for the existence of an intermediate in the hydroxylation reaction. The properties of the intermediate are consistent with those of S' in the above model. Furthermore, the present results indicate that the intermediate is a precursor of the quinonoid dihydropterin.

## RESULTS AND DISCUSSION

Based on our previous kinetic analysis of the hydroxylation reaction, we have tried to obtain more direct evidence for the formation of an intermediate. Since, in the presence of low concentrations of biopterin-$H_4$, S' appears to behave like an inhibitor, one of the approaches that we tried was to carry out the hydroxylation reaction in the presence of Sephadex-G50. Under these conditions, small molecules such as S' should be diluted into a larger volume than are the protein molecules, and, if S' were indeed an inhibitor, Sephadex should stimulate the reaction. The results of a typical experiment are shown in Table I. It can be seen that under conditions where the hydroxylation reaction is stimulated by PHS, Sephadex also stimulates. The extent of stimulation (19%) is in reasonable agreement with the maximum that could have been expected on the basis of the extra dilution of small molecules (28%).

We have also looked for spectral evidence for the formation of an intermediate under conditions where our earlier experiments indicated that it accumulates: i.e., pH values around 8.1-8.2, high hydroxylase concentrations, and relatively low biopterin-$H_4$ concentrations, and, of course, in the absence of PHS.

Figure 1 shows the results of a typical experiment which was carried out by determining the absorption spectrum of a reaction mixture containing all of the compounds of the hydroxylase system, including the enzymatic tetrahydropterin regenerating system (dihydropteridine reductase and TPNH), but in the absence of phenylalanine. The control cuvette contained the identical mixture, but lacked, in addition, biopterin-$H_4$. The initial spectrum (0 min), therefore, is that of biopterin-$H_4$, with a peak at around 300 nm. After repeated scans had shown that there was no change in the spectrum under these conditions, the hydroxylation reaction was started by the addition of exactly the same amount of phenylalanine to both cuvettes. As can be seen, there is a very rapid change (within one minute) in the spectrum, characterized by a large increase in absorbance between 240 and 260 nm. One can also see that there is a considerable fall in absorbance at 300 nm, the region of the peak for biopterin-$H_4$. There is also an increase in absorbance

Table I

Effect of Sephadex G-50 on Rate of Hydroxylation Reaction

| CONDITIONS | STIMULATION percent |
|---|---|
| pH 6.8, no PHS | 0 |
| pH 8.2, with PHS | 1.3 |
| pH 8.2, no PHS | 19 |

Included volume equal to 28% of total volume

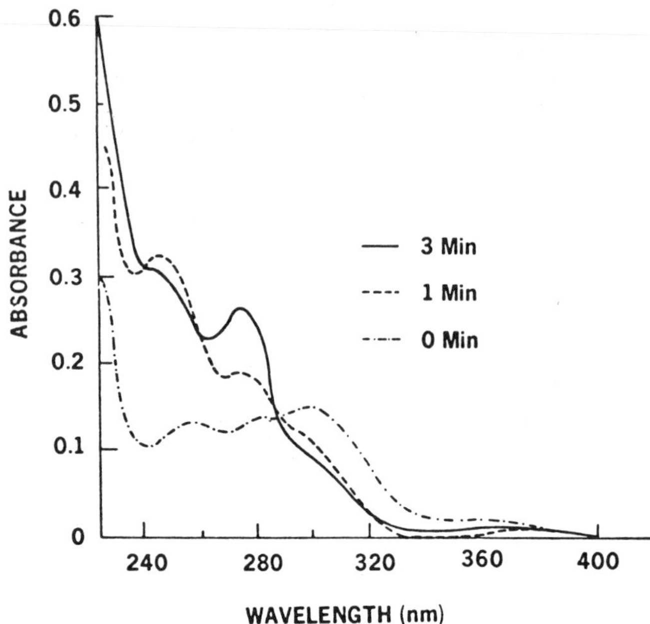


Fig. 1. Spectral changes during the enzymatic conversion of phenylalanine to tyrosine. The complete reaction mixture contained the following components (in µmoles) in a final volume of 1.0 ml: potassium phosphate, pH 8.2, 30; TPNH, 0.08; phenylalanine, 1.2; biopterin-$H_4$, 0.015; glucose-6-phosphate, 6.0; phenylalanine hydroxylase, 90% pure, 80 µg; dihydropteridine reductase, catalase, glucose-6-phosphate dehydrogenase, in excess. Temperature, 22° to 23°. The control cuvette contained the same components without biopterin-$H_4$.

at 270-275 nm, which is due to tyrosine formation. In subsequent scans, the only change that one sees is a continuous increase in absorbance at 275 and 230 nm due to the ongoing formation of tyrosine. These last changes continue until the oxygen in the cuvette is depleted, at which point, the hydroxylation reaction comes to a halt. These spectral changes in the region 240-260 nm are due, we think, to the formation of an intermediate. It should be emphasized that they are not due to the formation of either the 7,8-dihydropterin or the quinonoid dihydropterin; both of these compounds have higher absorption in the 320-330 region than does the tetrahydropterin [5,6], and, as can be seen, there is a decrease in absorbance in this region.

The decrease in absorbance above 285 nm indicates that some of the biopterin-$H_4$ has disappeared and therefore, that the new compound

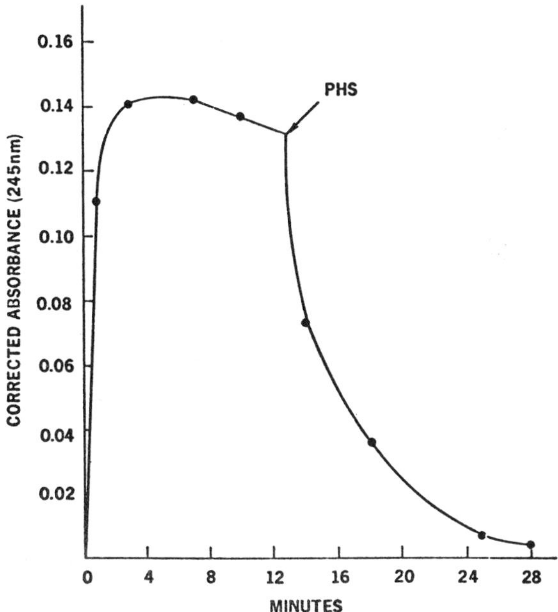


Fig. 2. The effect of addition of PHS on the absorbance changes at 245 nm. The components of the reaction mixture were the same as those in the experiment described in Fig. 1 except that the amount of biopterin-$H_4$ used was 0.010 μmoles in a final volume of 1.0 ml. PHS, where added, was 3 μg of the pure protein [3].

is formed from biopterin-$H_4$. These findings support the conclusion that the intermediate is a pterin derivative. Furthermore, from the magnitude of the spectral changes, (and the amounts of biopterin-$H_4$ present initially), one can conclude that a large fraction of the biopterin-$H_4$ has been converted to this new compound.

The next question of interest was what effect will the addition of PHS have on this intermediate? In Fig. 2, we have plotted the absorbance changes at 245 nm, corrected for the contribution at this wavelength due to the tyrosine being formed. It can be seen that prior to the addition of PHS, the absorbance is relatively constant, suggesting that steady state conditions have been approximated (the slow decrease in absorbance at 245 nm is probably due to instability of the hydroxylase under these conditions, since the rate of tyrosine formation also falls with time - see Table II). The addition of PHS leads to a rapid decrease to its initial value of the absorbance at 245 nm. This effect of PHS indicates that the spectral intermediate is the substrate for PHS (or that it is in rapid equilibrium with the substrate); i.e., the spectral intermediate is S'.

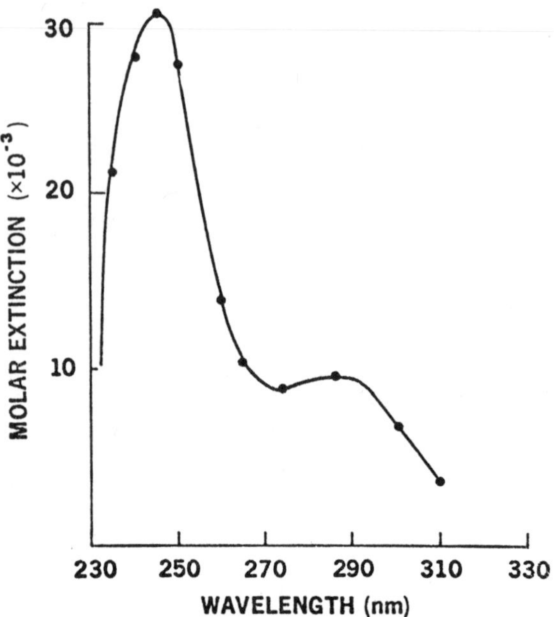


Fig. 3. The absorption spectrum of the intermediate. The reaction mixture was the same as that described in the legend of Fig. 1. At each wavelength, the observed changes in absorption have been corrected (using known extinction coefficients) for the contribution due to the conversion of phenylalanine to tyrosine and for the disappearance of biopterin-$H_4$.

Other results are consistent with this conclusion. Much less (about 1/4 as much) of the intermediate accumulates at pH 6.8. At pH 8.2, if PHS is included in the original reaction mixture, there is very little accumulation of the intermediate.

By correcting at each wavelength for the contribution due to tyrosine formation (and for that due to the loss of biopterin-$H_4$) we have constructed an approximate absorption spectrum of the intermediate and this is shown in Fig. 3.

The spectrum resembles to some extent that of 7,8-dihydropterins at this pH value, the main difference being that the latter compounds have three absorption maxima in the ultraviolet region (maxima at 229, 279, and 324 nm with approximate extinction coefficients of 26,000, 10,000 and 5,600 [7]. It can be seen that compared to 7,8-dihydropterins,both absorption peaks in the intermediate are shifted to longer wavelengths by approximately 15 nm. The amount of the intermediate that is being formed, a value that allowed us

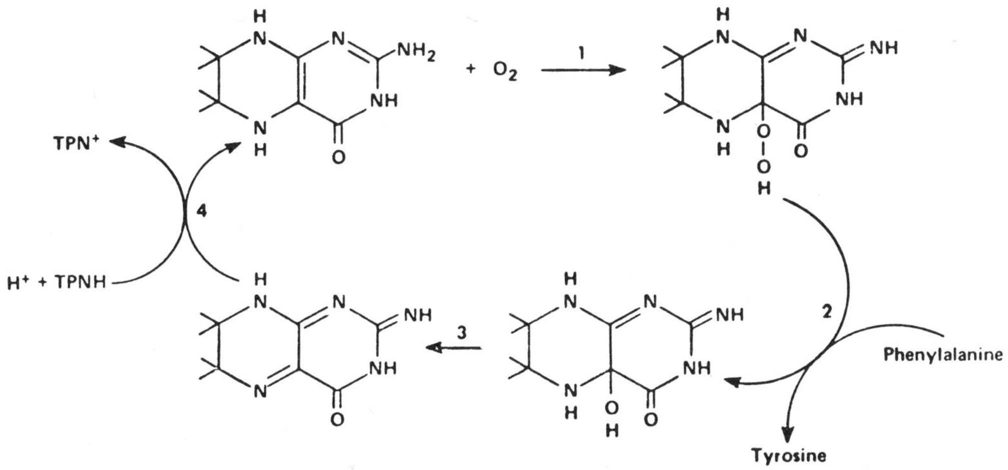

Fig. 4. Hypothetical mechanism for the enzymatic conversion of phenylalanine to tyrosine involving a tetrahydropterin hydroperoxide.

to estimate extinction coefficients, was calculated from the amount of biopterin-$H_4$ that disappeared. This value, in turn, was calculated from the loss in absorbance at 300-320 mμ. Under these conditions, about half of the added biopterin-$H_4$ had been converted to the intermediate at steady state: in this particular experiment, 15 nmoles biopterin-$H_4$ were added initially and about 8 nmoles of S' were formed.

As already mentioned, our previous kinetic analysis indicated that S' was either a precursor of the dihydropterin or a complex of the pterin, oxygen and the amino acid.

The relatively large fraction of biopterin-$H_4$ that is converted to the intermediate under these special conditions suggested that it might be possible to distinguish between the two general structures for the intermediate.

An attractive candidate for S' is shown in the hydroxylation mechanism that is outlined in Fig. 4 [1]. This scheme, which assumes that the active hydroxylating agent is a hydroperoxide of the tetrahydropterin, is a detailed version of one that we proposed in 1962

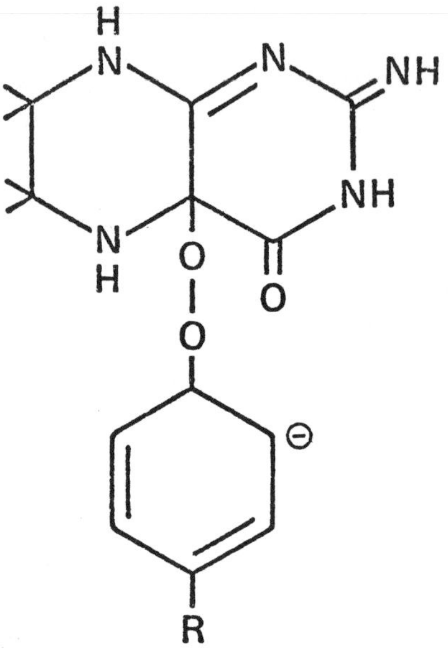

Fig. 5. Structure of a hypothetical intermediate in the hydroxylation reaction.

[8]. It is supported by the recent work of Mager and Berends, who have obtained evidence for the formation of hydroperoxides during the non-enzymatic oxidation of tetrahydropterins [9,10].

As can be seen, there is an intermediate between the putative tetrahydropterin hydroperoxide and the quinonoid dihydropterin. This intermediate, a hydrated precursor of the latter compound, could be S'. According to this model, it is necessary to assume that reaction 3 (Fig. 4) is rapid enough to be non-rate limiting at pH 7.0. At pH 8, however, it would become rate-limiting in the absence of PHS. The role of PHS in the hydroxylation reaction would be the catalysis of reaction 3 (Fig. 4).

The second general structure for S' that we are considering, a derivative between the hydroperoxide and phenylalanine, is shown in Fig. 5. If such a compound were formed, the role of PHS would be to catalyze its cleavage to tyrosine and quinonoid dihydropterin (*via* the hydroxytetrahydropterin). According to this idea, reaction 3 (Fig. 4) would never be rate-limiting.

The first mechanism (i.e., S' is the hydrated precursor of the dihydropterin) predicts that at early times there should be a discrepancy between the amounts of tyrosine and the amounts of dihydropterin that are formed, the discrepancy being caused by the amount of S' that accumulates; i.e., tyrosine should equal the sum of dihydropterin and S'. By contrast, if S' were a complex of tetrahydropterin, oxygen, and phenylalanine, tyrosine and quinonoid dihydropterin should always be formed in equal amounts.

We have examined the ratio of products formed during very early times of the hydroxylation reaction. For most of the experiments, dihydropterin formation was measured indirectly by providing an excess of dihydropterin reductase and DPNH and actually measuring the formation of $DPN^+$ from DPNH. We have shown that 1 mole of $DPN^+$ is formed for each mole of quinonoid dihydropterin formed throughout the course of the reaction under these conditions.

We have found that there is a marked discrepancy between tyrosine formation and DPNH oxidation during early times of incubation (Table II), the ratio being about 3:1. As expected, the discrepancy is made up by the amount of S' that accumulates: for the first minute, about 11 nmoles of tyrosine, 3 nmoles of $DPN^+$ and about 8 nmoles of S' are formed. The S' concentration appears to reach a constant, steady state value, which is maintained for at least 4 minutes. By contrast, in the presence of PHS, the ratio of tyrosine and $DPN^+$ formed is close to 1.0

It should be noted that after the first few minutes, the rates of tyrosine and $DPN^+$ formation are about the same. Since the concentration of S' remains relatively constant, the absolute amount of tyrosine formed continues to exceed the amount of $DPN^+$ formed, but the ratio of these two products approaches 1.0. With longer incubation times, and greater accumulations of the two products, therefore, the deviation of the ratio from 1.0 (and the effect of PHS on the ratio) would be difficult to detect experimentally.

Although not shown in the Table, when $DMPH_4$ is substituted for biopterin-$H_4$, the tyrosine/$DPN^+$ ratio is close to 1.0 under the conditions used in the experiment, indicating that the corresponding intermediate does not accumulate to the same extent when $DMPH_4$ is used.

From the concentration of S' and the measured rate of formation of $DPN^+$ (or tyrosine), one can calculate that the rate constant ($k_3$) for the breakdown of S' (formed from biopterin-$H_4$) is about 0.01-0.012 per second, which gives a half-life for the intermediate of about 1 minute at 22° at pH 8.2.

These results fit in with the first model which assumes that S' is a precursor of the dihydropterin and do not support the idea that S' is a derivative of tyrosine or a precursor of tyrosine. Specifically, we are proposing that S' is the 4a-hydroxytetrahydropterin shown in Fig. 4. We have assigned the hydroxy group to the

Table II

Products Formed During Early Times of Phenylalanine Hydroxylation

| Time min | S' nmoles | Tyrosine nmoles | $DPN^+$ nmoles | Tyrosine/$DPN^+$ |
|---|---|---|---|---|
| 1 | 8.49 | 10.7 | 3.2 | 3.32 |
| 2 | 8.49 | 20.4 | 10.5 | 1.94 |
| 3 | 8.90 | 26.5 | 16.1 | 1.65 |
| 4 | 8.71 | 32.0 | 21.6 | 1.48 |
| 5 | 8.22 | 37.5 | 27.2 | 1.38 |
| 0.5* | - | 42.3 | 41.7 | 1.01 |

*With PHS (3 μg) present from the start of the incubation.

The complete reaction mixture contained the following components (in μmoles) in a final volume of 1.0 ml: potassium phosphate, pH 8.2, 30; phenylalanine, 1.2; DPNH, 0.06; biopterin-$H_4$, 0.015; phenylalanine hydroxylase, 90% pure, 53 μg; dihydropteridine reductase and catalase in excess. Temperature, 22° to 23°. Biopterin-$H_4$ was omitted from the control cuvette. The amount of tyrosine formed was determined by the nitrosonapthol procedure [11]. The amount of DPNH oxidized and the amount of biopterin-$H_4$ that disappeared (equated with the amount of S' formed) were estimated from the observed changes in absorbance at 360 nm and 310 nm, using known extinction coefficients.

4a rather than to the 8a position of the pteridine ring because of the effect of substituents in the 6 position on the stability of the intermediate. As already mentioned, there is less accumulation of the intermediate starting with $DMPH_4$ than there is starting with biopterin-$H_4$. Our interpretation of these results is that the dihydroxypropyl side chain on carbon 6 of biopterin stabilizes the intermediate. From models of biopterin, it seems unlikely that the hydroxypropyl substituent at carbon 6 can interact intramolecularly with a group at the 8a position, whereas it can readily interact with a group at the 4a position and therefore stabilize the intermediate.

Before concluding, there is a variation of the hydroxylation scheme in Fig. 4 that should be mentioned because it suggests another possible structure for S'. Recently, Hamilton [12] has proposed that the tetrahydropterin hydroperoxide, itself, is not the hydroxylating species in pterin-dependent monooxygenases, but rather the open-ring (opened between nitrogen 5 and carbon 4a) vinylogous ozone compound that is formed from the hydroperoxide. According to this proposal, one oxygen atom of the ozone derivative is transferred to the substrate to give the hydroxylated product. The other product is the pterin-derived compound, a 2-amino-4,5-dioxopyrimidine, which would then undergo ring closure to form the para-quinonoid dihydropterin.

Although not stated explicitly in Hamilton's proposal, the conversion of the dioxopyrimidine to the quinonoid dihydropterin would probably involve the formation of the carbinolamine compound (i.e., the 4a-hydroxytetrahydropterin shown in Fig. 4) as an intermediate. Therefore, the role of PHS in this modified scheme could be precisely the same as the one already discussed, namely, catalysis of the conversion of the 4a-hydroxytetrahydropterin to the quinonoid dihydropterin. However, this modified scheme raises the possibility that the intermediate that we have detected during phenylalanine hydroxylation could be the dioxopyrimidine. Indeed, the absorption spectrum that we have derived and depicted in Fig. 3 could be that of the dioxopyrimidine, or the 4a-hydroxytetrahydropterin, or a mixture of both compounds.

In a summary, the present results demonstrate that there is an intermediate in the enzymatic hydroxylation of phenylalanine. The intermediate is a precursor of the quinonoid dihydropterin. PHS catalyzes its conversion to the latter compound. Furthermore, if the structure that we are proposing for the intermediate, i.e., a hydrated form of the dihydropterin, proves to be correct, it would provide strong, if indirect, evidence for the involvement of a hydroperoxide of the tetrahydropterin as the hydroxylating agent in the enzymatic conversion of phenylalanine to tyrosine.

## REFERENCES

1. Kaufman, S.: The phenylalanine hydroxylating system from mammalian liver (Vol. 35 of Advances in Enzymology, ed. by Meister, A.), 245-319, John Wiley & Sons, Inc., 1971.

2. Kaufman, S.: A protein that stimulates rat liver phenylalanine hydroxylase. J. Biol. Chem. 245, 4751-4759 (1970).

3. Huang, C., Max, E.E., and Kaufman, S.: Purification and characterization of phenylalanine hydroxylase stimulating protein from rat liver. J. Biol. Chem. 248, 4235-4241 (1973).

4. Huang, C.Y., and Kaufman, S.: Studies on the mechanism of action of phenylalanine hydroxylase and its protein stimulator. J. Biol. Chem. 248, 4242-4251 (1973).

5. Kaufman, S.: Studies on the mechanism of the enzymatic conversion of phenylalanine to tyrosine. J. Biol. Chem. 234, 2677-2682 (1959).

6. Kaufman, S.: The nature of the primary oxidation product formed from tetrahydropteridines during phenylalanine hydroxylation. J. Biol. Chem. 236, 804-810 (1961).

7. Pfleider, W., and Zondler, H.: Synthese und eigenschaften blockierten 7,8-dihydropterine. Chem. Ber. 99, 3008-3021 (1966).

8. Kaufman, S.: Aromatic hydroxylation (in Oxygenases, ed. by Hayaishi, O.), 129-179, Academic Press, Inc., New York, 1962.

9. Mager, H.I.X., and Berends, W.: Hydroperoxides of partially reduced quinoxalines, pteridines and (iso)alloxazines: Intermediates in oxidation processes. Rev. Trav. Chim. 84, 1329-1343 (1965).

10. Mager, H.I.X., Addink, R., and Berends, W.: Coupled and decoupled processes in the autooxidation of partially reduced pteridines and flavins. Rev. Trav. Chim. 86, 833-851 (1967).

11. Udenfriend, S., and Cooper, J.R.: The chemical estimation of tyrosine and tyramine. J. Biol. Chem. 196, 227-233 (1952).

12. Hamilton, G.A.: Chemical models and mechanisms for oxygenases (in Molecular Mechanisms of Oxygen Activation, ed. by Hayaishi, O.), 405-451, Academic Press, Inc., New York, 1974.

# SOME PROPERTIES OF BOVINE PINEAL TRYPTOPHAN HYDROXYLASE

Arata Ichiyama*, Hiroyuki Hasegawa*, Chiharu Tohyama**, Chieko Dohmoto** and Tohru Kataoka**
Department of Biochemistry, Hamamatsu University School of Medicine* and Faculty of Medicine, University of Tokyo**
Handa-cho, Hamamatsu-shi 431-31, Japan*
and Hongo, Bunkyo-ku, Tokyo 113, Japan**

The conversion of L-tryptophan to L-5HTP is the first step in the biosynthesis of serotonin in the brain and of melatonin in the pineal gland and is catalyzed by tryptophan hydroxylase. Since the first demonstration by Grahame-Smith that a specific tryptophan hydroxylase exists in the brain (1), the enzyme has been studied at various laboratories including our own (2-11) and is now accepted as a pterin-dependent aromatic amino acid monooxygenase (12).

Three pterin-dependent aromatic amino acid monooxygenases so far described, phenylalanine, tyrosine and tryptophan hydroxylases, have been shown to exhibit striking similarities in so many of their properties that they have been regarded as a family of enzymes and a common reaction mechanism has been anticipated (12-16). In addition to the common properties, however, each enzyme shows some characteristic properties, the detailed description of which may help deeper understanding about the physiological function, regulatory mechanisms and also the reaction mechanism of respective enzyme.

Recently, we have found and reported that bovine pineal tryptophan hydroxylase is almost inactive unless the enzyme preparation is preincubated in the presence of dithiothreitol or other sulf-

---

Abbreviations used are : 5HTP, 5-hydroxytryptophan; $BpH_4$, 5,6,7,8-tetrahydrobiopterin; $DMPH_4$, 6,7-dimethyl-5,6,7,8-tetrahydropterin; $6MPH_4$, 6-methyl-5,6,7,8-tetrahydropterin.

This investigation has been supported in part by the Scientific Research Fund of the Ministry of Education of Japan and the Naito Foundation Research Grant.

hydryl reagents under anaerobic conditions (17). Taking advantage of the observation that the inactive form of tryptophan hydroxylase is more stable than the activated enzyme, the enzyme was purified as the inactive form about 1,000-fold from the bovine pineal gland (18). The enzyme activity was determined each time after the enzyme was activated. In this paper, some characteristic properties of the newly purified pterin-dependent aromatic amino acid monooxygenase, bovine pineal tryptophan hydroxylase, will be presented.

## EXPERIMENTAL PROCEDURE

*Determination of Tryptophan and Phenylalanine Hydroxylase Activities (Fig. 1).* The enzyme preparation was first activated by preincubation with 30 mM dithiothreitol and 50 μM $Fe^{2+}$ at pH 8.1 under anaerobic conditions. Enzyme was added to the preincubation mixture in a Thunberg tube and air in the tube was replaced 3 times with $N_2$ while the tube was immersed in an ice-bath. The tube was then transferred to a water-bath of 30° and was incubated for 30 min. The hydroxylase reaction was started by adding 50 μl-aliquot of the activated enzyme to 0.1 ml of the reaction mixture which contained 11.25 nmoles of L-tryptophan-1-$^{14}C$ (usually 170,000 cpm), 120 nmoles of tetrahydropterin cofactor and 500 nmoles of dithiothreitol unless otherwise stated. The reaction was carried out at 37° and pH 7.0 for 1 min and stopped by the addition of excess o-phenanthroline followed by immersion in a boiling water-bath. In experiments presented in Figs. 7 and 8, 150 nmoles of NADPH and excess dihydropteridine reductase were included in the reaction mixture. To the heated reaction mixture was then injected a decarboxylase mixture which contained 0.2 μmole each of EDTA-$Na_2$ and pyridoxal phosphate, 0.1 μmole of D,L-5HTP as carrier and 3 to 7 milliunits of a partially purified

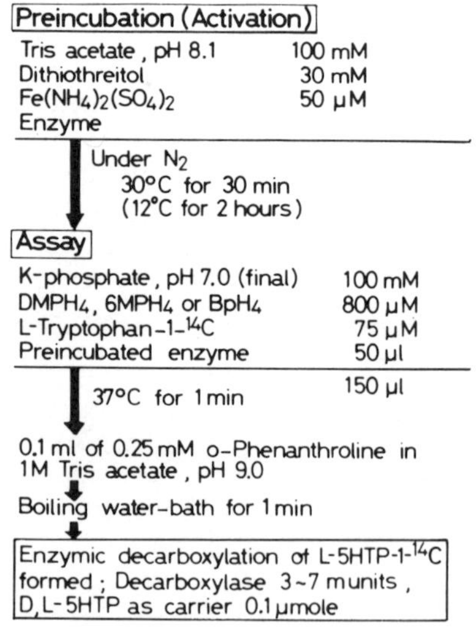


Fig. 1. Outline of the assay of tryptophan hydroxylase.

hog kidney aromatic L-amino acid decarboxylase. The decarboxylation reaction was carried out at 37° and pH 9.0 for 30 min and the reaction was stopped by perchloric acid. $^{14}CO_2$ evolved was trapped in alkali and the radioactivity was determined by a liquid scintillation counter. Under these specified conditions L-5HTP was quantitatively converted to $CO_2$ and serotonin while only negligible amounts of L-tryptophan-1-$^{14}$C were decarboxylated. In all experiments, control incubation was run simultaneously in the absence of tetrahydropterin cofactor or with a boiled enzyme preparation and experimental data were corrected for the control value. Aromatic L-amino acid decarboxylase was partially purified from the hog kidney according to the method of Christenson, Dairman and Udenfriend (19) up to the step of ammonium sulfate fractionation. One unit of the decarboxylase was defined as the amount which catalyzes the conversion of 1 μmole of L-5HTP to $CO_2$ per min under conditions described above. When phenylalanine hydroxylase activity was determined, L-tryptophan-1-$^{14}$C was replaced by 300 μM L-phenylalanine-(ring-4-$^3$H) and the formation of THO was measured by a slight modification of the method of Guroff and Abramovitz (20).

Purification of Tryptophan Hydroxylase. Tryptophan hydroxylase was purified from the bovine pineal gland essentially as described previously (17,18). A representative result of the purification is presented in Table I. Some experiments were carried out with the Sephadex G-200 fraction but most kinetic studies were conducted with a partially purified enzyme, for the preparation of which DEAE cellulose step was omitted and P-cellulose and Sephadex G-200 were replaced by CM-Sephadex (C-50) and Sephadex G-100, respectively. The purification achieved by this procedure was approximately 400-fold. The hydroxylapatite fraction was also used for some purposes.

Other Materials and Methods. Biopterin was isolated from the skin of bullfrog according to the method of Fukushima and Akino (21) and reduced to the tetrahydro-form by a modification of the method

Table I. Purification of tryptophan hydroxylase.

| Enzyme Preparations | Volume | Enzyme Activity | Yield | Protein | Specific Activity |
|---|---|---|---|---|---|
| | (ml) | (nmoles/min) | (%) | (mg) | (units/mg) |
| Homogenate | 1,300 | 11,040 | 100 | 50,700 | 0.22 |
| Digitonin Sup. | 1,820 | 9,397 | 85 | 24,600 | 0.38 |
| Amm. Sulf. (25-57 %) | 270 | 7,171 | 65 | 5,990 | 1.20 |
| Hydroxylapatite | 125 | 3,548 | 32 | 1,100 | 3.23 |
| P-Cellulose | 46.5 | 2,521 | 23 | 393 | 6.42 |
| DEAE-Cellulose | 5.3 | 748.5 | 6.8 | 33.8 | 22.14 |
| Sephadex G200 | 4.6 | 400.2 | 3.6 | 1.7 | 235.40 |

Bovine pineal gland : 400 g

This data are taken from Nukiwa et al (18).

of Bobst and Viscontini (22) with platinium black as catalyst. $DMPH_4$ and $6MPH_4$ were obtained from Calbiochem. The concentration of tetrahydropterin was determined spectrophotometrically. Molar extinction coefficients used were : $\epsilon$(265-267 nm in 0.1 N HCl) = 16,000 and $\epsilon$(298-300 nm in 0.1 M K-phosphate, pH 7.0 immediately after neutralization) = 10,500 (23).

L-Tryptophan-1-$^{14}C$ and D,L-p-chlorophenylalanine-$^{3}H$(G) were obtained from New England Nuclear. Tryptophan-$^{14}C$ was purified by partition chromatography on a Sephadex G-25 column as described previously (11). The purification of p-chlorophenylalanine-$^{3}H$ was achieved by Sephadex G-10 column chromatography; the commercial sample was applied on a column of Sephadex G-10 (1.1 x 35 cm) equilibrated with $H_2O$ and the elution was performed with $H_2O$ at room temperature at a flow rate of about 15 ml per hour. By this procedure, p-chlorophenylalanine was successfully separated from impurities and also from phenylalanine, tyrosine and m-chloro-p-tyrosine (see Fig. 6). L-Phenylalanine-(ring-4-$^{3}H$) was obtained from the Radiochemical Centre, Amersham and purified according to the method of Guroff and Abramovitz (20) except that the purified preparation was dissolved in $H_2O$ and stored frozen at -20° without adding ethanol; no appreciable degradation of phenylalanine-$^{3}H$ occurred for more than months under these conditions.

## RESULTS

Activation of Bovine Pineal Tryptophan Hydroxylase. The time course of the activation of bovine pineal tryptophan hydroxylase is shown in Fig. 2. In this experiment, tryptophan and phenylalanine hydroxylase activities of the enzyme preparation (Hydroxylapatite fraction) before the preincubation was 1.0 to 1.5 % of the total activities. When the enzyme was preincubated anaerobically at 12° with 30 mM dithiothreitol for the indicated periods, gradual increases of both tryptophan and phenylalanine hydroxylase activities occurred resulting in an 100-fold activation.

The velocity of the activation was a function of the concentration of dithiothreitol and was accelerated by low concentrations of $Fe^{2+}$ (Fig. 2) (17,24). The effect of $Fe^{2+}$ was substituted by some reducing agents such as borohydride and hydrosulfite but not by other metal ions. The effect of dithiothreitol was substituted in part only by other sulfhydryl reagents; when dithiothreitol was omitted, no activation occurred even in the presence of both $Fe^{2+}$ and borohydride (17,24). Under aerobic conditions, the activation occurred as rapidly as under anaerobic conditions but the activated enzyme was more unstable and the activation was followed by a relatively rapid inactivation.

This activation of bovine pineal tryptophan hydroxylase was a time-, temperature- and pH-dependent process suggesting that some changes may occur in the enzyme protein associated with the activation and since a common property among the activators so far found

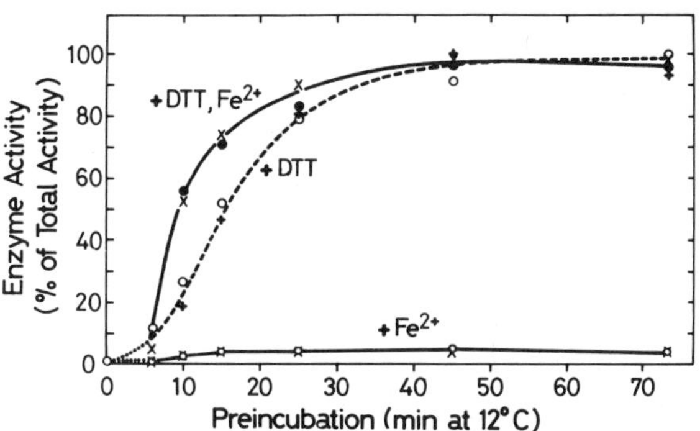


Fig. 2. Activation of bovine pineal tryptophan hydroxylase. The preincubation was carried out under conditions similar to that described under "Experimental Procedure". At 0-time, enzyme was added to the preincubation mixture in a Thunberg tube and air in the tube was replaced with $N_2$ while the tube was immersed in an ice-bath. After 5 min, the tube was transferred to a water-bath of 12°. $Fe^{2+}$ was added to the mixture at 6 min. Where indicated, either dithiothreitol (DTT) or $Fe^{2+}$ was omitted. ● and o represent tryptophan hydroxylase activity and x and + show phenylalanine hydroxylase activity. Data are taken from Nukiwa et al (18).

was their reducing ability, we have speculated that the activation may represent the reduction of enzyme protein associated with conformational changes (17).

In the following experiments, the enzyme was first fully activated regardless of the mechanism of the process and the properties of the activated enzyme were examined.

Time Course of the Tryptophan Hydroxylase Reaction. The time

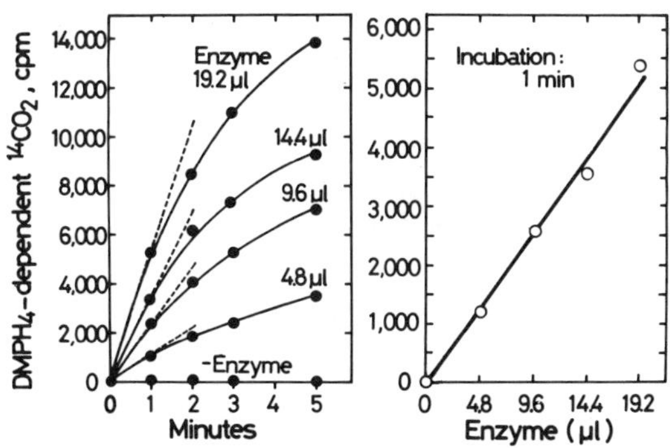


Fig. 3. Time course of tryptophan hydroxylation.

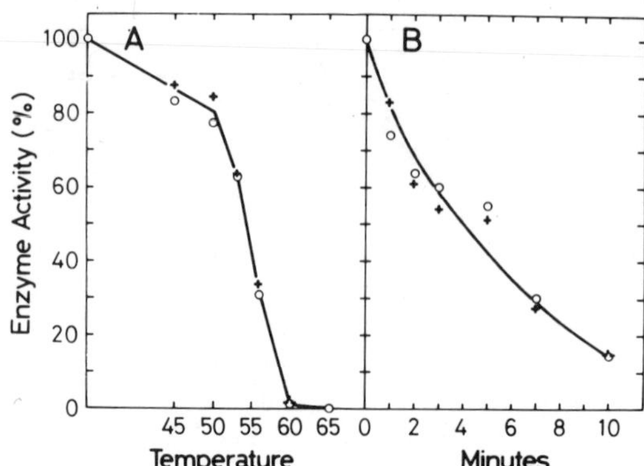


Fig. 4. Heat treatment of tryptophan hydroxylase. A partially purified enzyme (Hydroxylapatite fraction) was subjected to heat treatment at various temperature for 2 min (A) or at 53° for various periods indicated (B). The enzyme was then activated and tryptophan (o) and phenylalanine (+) hydroxylase activities were determined.

course of the tryptophan hydroxylase reaction was quasi-linear only for 1 min and proceeded linearly with the amount of the activated enzyme (Fig. 3). The time course of the phenylalanine hydroxylation was almost similar with that of the tryptophan hydroxylation. Addition of NADH or NADPH and dihydropteridine reductase purified from the sheep liver according to the method of Kaufman and Levenberg (25) up to the step of calcium phosphate gel treatment was without any effect under these conditions.

Substrate Specificity of the Enzyme. The purified enzyme catalyzed the hydroxylation of both L-tryptophan and L-phenylalanine at a comparable rate as described previously (18). The maximum velocity of the phenylalanine hydroxylation was about 2.5 times higher than

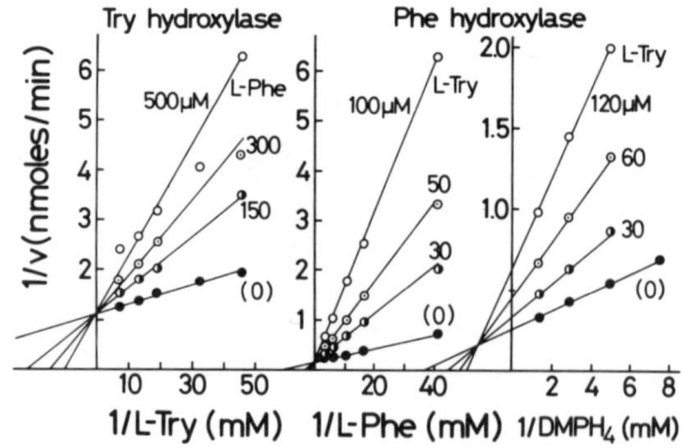


Fig. 5. Inhibition by alternative substrate.

that of the tryptophan hydroxylation when $BpH_4$ was used as cofactor. Tyrosine hydroxylase activity as measured by the method of Nagatsu, Levitt and Udenfriend (26) was less than one-hundredth of that of tryptophan hydroxylase, whether the reaction was carried out at pH 7.0 or at pH 6.0.

Tryptophan and phenylalanine hydroxylase activities showed almost identical behavior upon purification, activation (Fig. 2), heat treatment (Fig. 4), gel filtration chromatographies and sucrose density gradient ultracentrifugation. In good agreement with Lovenberg, Jequier and Sjoerdsma (10), L-phenylalanine inhibited the tryptophan hydroxylation competitively with respect to L-tryptophan and L-tryptophan inhibited the phenylalanine hydroxylation competitively with L-phenylalanine. The inhibition pattern by the alternative substrate was non-competitive with respect to tetrahydropterin cofactor (Fig. 5). All results thus indicated that both enzyme activities are catalyzed by the same enzyme.

In addition to L-tryptophan and L-phenylalanine, p-chlorophenylalanine, an inhibitor of the tryptophan hydroxylation in the brain, was found to serve as a substrate of this enzyme. When $^3H$-labeled p-chlorophenylalanine was incubated with an activated enzyme in the presence of $BpH_4$ and then radioactive compounds were separated by a Sephadex G-10 column chromatography, approximately 15 % of the radioactivity was recovered in the tyrosine fraction and 24 % in the m-chloro-p-tyrosine fraction (Fig. 6). Radioactive tyrosine and m-chloro-p-tyrosine were identified by paper chromatography with 5 different solvent systems; n-butanol : acetic acid : $H_2O$ (12 : 3 : 5), isopropanol : $NH_4OH$ : $H_2O$ (8 : 1 : 2), n-butanol : ethyl methyl ketone : $H_2O$ : diethylamine (40 : 40 : 20 : 4), isoamyl alcohol : pyridine : $H_2O$ (7 : 7 : 6) and phenol : $H_2O$ (4 : 1, w/w). Approximately 20 % of the radioactivity from the m-chloro-p-tyrosine fraction moved to a position not exactly corresponding to m-chloro-p-tyrosine upon paper chromatography especially with phenol : $H_2O$ as solvent. In the control experiment in which tetrahydropterin cofactor was omitted or the boiled enzyme preparation was used, more than 98 % of the radioactivity was recovered as p-chlorophenylalanine (Fig. 6).

The stoichiometry of the p-chlorophenylalanine hydroxylation catalyzed by bovine pineal tryptophan hydroxylase is shown in Table II. Experiments were carried out with $BpH_4$ and $DMPH_4$ as cofactor and in each case, the decrease of p-chlorophenylalanine-$^3H$ was accounted for by the formation of tyrosine-$^3H$ and m-chloro-p-tyrosine-$^3H$ (plus small amounts of unknown compound). It was also interesting to note that the ratio between the formation of tyrosine and m-chloro-p-tyrosine (plus unknown) appeared to vary with the structure of the tetrahydropterin cofactor used; with $BpH_4$ approximately 40 % of the radioactive products was tyrosine whereas only less than 25 % was recovered as tyrosine with $DMPH_4$ (Table II). When the time course of the hydroxylation of p-chlorophenylalanine was followed, the formation of tyrosine and m-chloro-p-tyrosine roughly paralleled

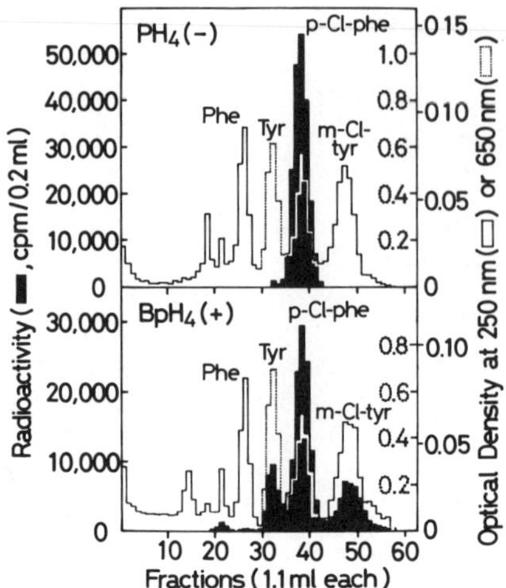


Fig. 6. p-Chlorophenylalanine as substrate of bovine pineal tryptophan hydroxylase. Reactions were carried out under standard assay conditions for 15 min except that tryptophan-$^{14}$C was replaced by 114 μM D,L-p-chlorophenylalanine-$^{3}$H(G) (1,250,000 cpm). The reaction was stopped by immersing the tube in a boiling water-bath and an aliquot of the reaction mixture was subjected to Sephadex G-10 column chromatography together with authentic phenylalanine, tyrosine, p-chlorophenylalanine and m-chloro-p-tyrosine as carriers. The chromatographic separation of authentic samples were monitored by the optical density at 250 nm and that at 650 nm after reaction with Folin-Ciocalteau reagent. $PH_4$(-) represents that the reaction was conducted in the absence of tetrahydropterin cofactor.

Table II. Stoichiometry of p-chlorophenylalanine hydroxylation.

| Assay Conditions | | Radioactivity recovered (cpm) | | | |
|---|---|---|---|---|---|
| System | Cofactor | pCl-phe | mCl-tyr+Tyr | mCl-tyr | Tyr |
| 0-time | $DMPH_4$ | 248,870 | 3,632 | 2,284 | 1,349 |
| Complete | $DMPH_4$ | 212,029 | 40,897 | 30,678 | 10,519 |
| | Difference | -36,841 | +37,265 | +28,394 | +9,172 |
| | (pmoles) | (-508) | (+514) | (+392) | (+129) |
| $-PH_4$ | — | 235,934 | 3,399 | 920 | 2,479 |
| Complete | $BpH_4$ | 142,423 | 93,766 | 55,021 | 38,745 |
| | Difference | -93,511 | +90,367 | +54,101 | +36,266 |
| | (pmoles) | (-1,290) | (+1,247) | (+746) | (+500) |

The formation of volatile compound (possibly THO) was 6335 cpm (2.6 %) in the complete system with $BpH_4$ as cofactor.

with each other indicating that tyrosine and m-chloro-p-tyrosine are simultaneously formed from p-chlorophenylalanine.

Effects of Substrate and Cofactor Concentrations. When tryptophan hydroxylase reaction was carried out with $BpH_4$, $6MPH_4$ and $DMPH_4$ as cofactor at various tryptophan concentrations, $BpH_4$ was shown to be most effective (Fig. 7). The apparent Km for L-tryptophan was between 10 and 20 μM, irrespective of the structure of the cofactor used. When L-phenylalanine was the substrate, marked substrate inhibition was observed only when naturally occurring $BpH_4$ was used (Fig. 7). Km for L-phenylalanine was slightly higher than that for L-tryptophan and was 30 to 40 μM with $BpH_4$ and $6MPH_4$ as cofactor and approximately 70 μM with $DMPH_4$.

Fig. 8 shows the effects of the cofactor concentration on tryptophan and phenylalanine hydroxylase activities. With L-tryptophan as substrate, $BpH_4$ was most effective but the apparent Km for $BpH_4$ and $6MPH_4$ was as high as 110 to 120 μM. When L-phenylalanine was the substrate, the effects of all tetrahydropterin cofactors tested were similar and the apparent Km values for $BpH_4$, $6MPH_4$ and $DMPH_4$ were 75 μM, 93 μM and 175 μM, respectively.

Since the Km values for $BpH_4$ obtained appeared to be significantly higher than the expected tissue concentration of this cofactor, the effect of protein kinase was examined. Protein kinase was partially purified from the bovine brain according to the method of Miyamoto et al (27) and incubated with the tryptophan hydroxylase preparation in the presence of ATP, cyclic AMP and $Mg^{2+}$ for 10 min before, during and after the activation of the hydroxylase. However, none of the conditions tested affected the hydroxylase activity

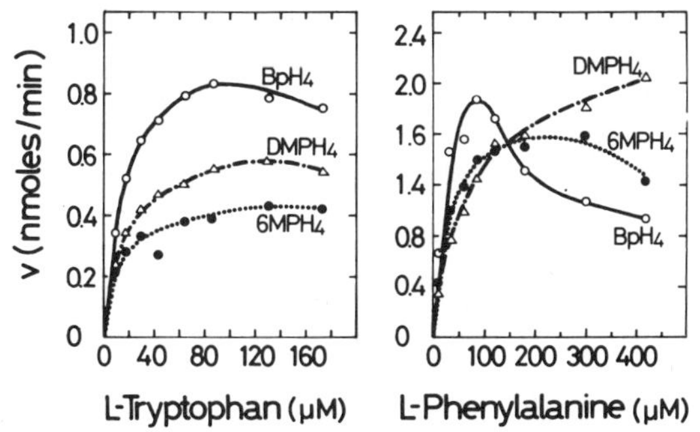


Fig. 7. Tryptophan and phenylalanine hydroxylase activities as a function of substrate concentration. Reactions were carried out under the standard assay conditions except that the concentration of $6MPH_4$ was 0.3 mM when L-tryptophan was substrate. Apparent Km values for L-tryptophan and L-phenylalanine calculated from this experiment were 14 and 70 μM (cofactor: $DMPH_4$), 11 and 37 μM (cofactor :$6MPH_4$) and 16 and 32 μM (cofactor:$BpH_4$), respectively (18).

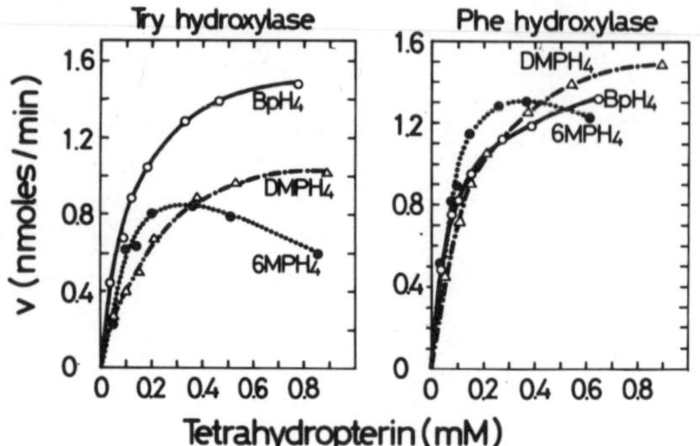


Fig. 8. Tryptophan and phenylalanine hydroxylase activities as a function of tetrahydropterin concentration. Reactions were carried out under the standard assay conditions except that the concentration of L-phenylalanine was 84 μM with $BpH_4$ as cofactor and 240 μM when $6MPH_4$ was used.

or the Km value.

Inhibitory Effects of the Reaction Products. When the effects of the reaction products were examined, only L-5HTP, the product of tryptophan hydroxylation, was significantly inhibitory for both tryptophan and phenylalanine hydroxylations (fig. 9). Although both L-tryptophan and L-phenylalanine are hydroxylated at a comparable rate and apparent Km for these two amino acids were in the same order, the inhibitory effect of L-tyrosine was detectable only at much higher concentrations (Fig. 9). D-5HTP was without any inhibitory effect.

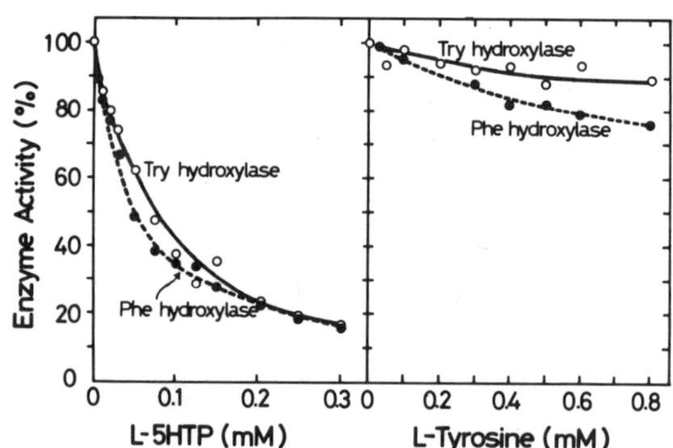


Fig. 9. Effects of L-5HTP and L-tyrosine on tryptophan and phenylalanine hydroxylase reaction. Reactions were performed with $DMPH_4$ as cofactor. When L-5HTP was added to the reaction mixture, the amount of carrier 5HTP added to the decarboxylation mixture was changed so that 5HTP-$^{14}C$ formed was converted quantitatively to $CO_2$.

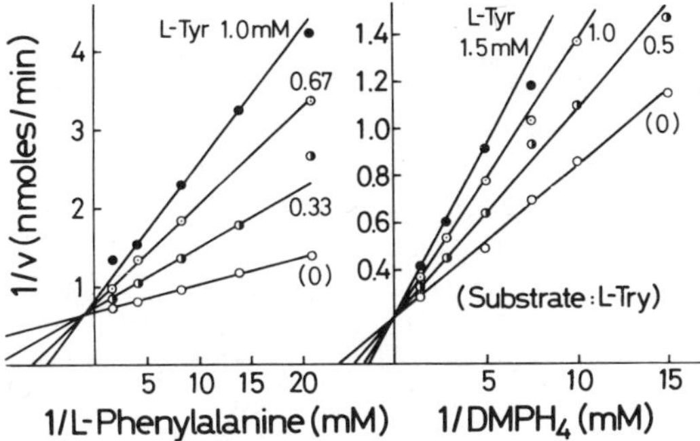


Fig. 10. Inhibition pattern of L-tyrosine.

When high concentrations of L-tyrosine were used and the inhibition pattern was examined, the inhibition was shown to be competitive with respect to tetrahydropterin cofactor and most probably non-competitive with respect to amino acid substrates (Fig. 10). In contrast, L-5HTP inhibited the reaction competitively with respect to L-tryptophan and uncompetitively with respect to $DMPH_4$ (Fig. 11). The inhibition by L-5HTP and L-tyrosine thus appeared to differ not only quantitatively but also qualitatively.

Various tryptophan and phenylalanine derivatives were examined for their inhibitory effect. In case of tryptophan derivatives, both carboxyl and amino groups of the alanine side-chain were found to be required for the inhibition and hence end-products of the biosynthetic pathway or their metabolites, serotonin, melatonin, 5-hydroxyindole acetic acid and so on, were without any inhibitory effect. Catecholamines or their precursor, L-DOPA, were only slightly inhbitory at 1 mM or higher concentrations. L-5HTP was the only naturally occurring inhibitory compound so far detected.

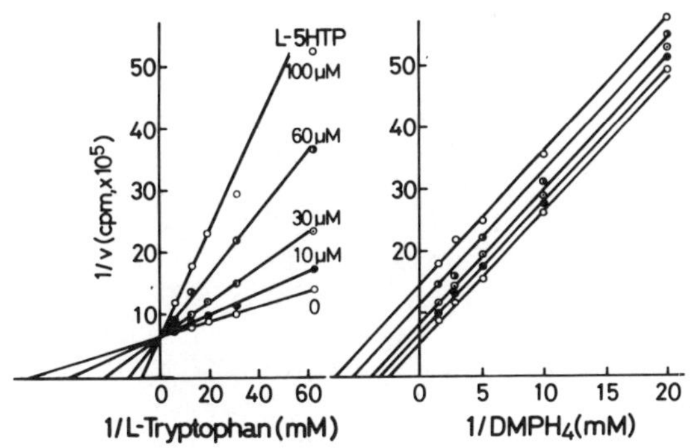


Fig. 11. Inhibition pattern of L-5HTP.

## DISCUSSION

The properties of bovine pineal tryptophan hydroxylase described above and previously (17,18) were clearly distinguishable from those of rabbit brain tryptophan hydroxylase which was recently purified and characterized in the Kaufman's laboratory (12,14,16). The molecular weight and the sedimentation constant ($s_{20,w}$) of the rabbit brain enzyme were reported to be 230,000 and 10.86 (16) whereas the values of the bovine pineal enzyme were approximately 30,000 and 3.07, respectively (18). Although an apparent turnover number (moles of tryptophan hydroxylated per mole of enzyme per min) of as low as less than 1 may be estimated for the brain enzyme from the data presented by Tong and Kaufman (16), the number of the bovine pineal enzyme can not be less than 7 when calculated from the data shown in Table I. Both the brain and pineal tryptophan hydroxylases catalyzed phenylalanine hydroxylation as well but the Km of the pineal enzyme for phenylalanine with $BpH_4$ as cofactor was approximately one-ninth of that reported for the brain enzyme (Fig. 7 and reference 16). The Km of the brain enzyme for tryptophan was shown to vary with the structure of the tetrahydropterin cofactor used and the reported values were significantly higher than the relatively constant Km values of the pineal enzyme; with $DMPH_4$ as cofactor, the Km of the brain enzyme was approximately 20 times higher than that of the pineal enzyme (Fig. 7 and reference 14). Some evidence so far obtained indicated that the properties of bovine pineal tryptophan hydroxylase were rather resembled to those of the rabbit brain enzyme and not to the bovine pineal enzyme (28). It is evident that the hydroxylation of tryptophan in the brain leading to the biosynthesis of serotonin, a putative transmitter of the serotonergic neuron, and the reaction in the pineal gland leading to the formation of melatonin, an antigonadotrophic indoleamine hormone, are catalyzed by different tryptophan hydroxylases and are probably under separate regulatory mechanisms. Bovine pineal tryptophan hydroxylase has also been differentiated from rat liver phenylalanine hydroxylase, another enzyme which catalyzes hydroxylations of both tryptophan and phenylalanine (12,15,29), by a number of criteria (18).

When respective properties of the bovine pineal enzyme were concerned, however, many of them were shared, at least qualitatively, with some other pterin dependent aromatic amino acid monooxygenases. The substrate specificity of bovine pineal tryptophan hydroxylase was not restricted to its possible physiological substrate, L-tryptophan, but extended to catalyze phenylalanine hydroxylation at a comparable rate; this property was shared with adrenal tyrosine hydroxylase (30) and brain tryptophan hydroxylase (16). The ability of the bovine pineal enzyme to catalyze the hydroxylation of p-chlorophenylalanine was shared with phenylalanine hydroxylases from rat liver and pseudomonas species (31,32). This property of the pineal enzyme was compatible with the <u>in vivo</u> observation made by Deguchi and Barchas (33) that the administration of p-chlorophenylalanine significantly decreased serotonin levels in the pineal as well as in the brain but the pineals of p-chlorophenylalanine-treated rats retained tryptophan hydroxylase and only the affinity of this enzyme was reduced. Even the striking activation of bovine pineal tryptophan hydroxylase by the preincubation with dithiothreitol appeared to be not a specific property of this enzyme. Though quantitatively less significant tyrosine hydroxylase activity in a partially purified bovine adrenal preparation was shown to increase approximately 2-fold following preincubation with $Fe^{2+}$ and 2-mercaptoethanol (34). Hosoda and Glick (4) detected tryptophan hydroxylase activity in a cell-free preparation from neoplastic murine mast cells only after dialysis against 0.1 M 2-mercaptoethanol under anaerobic conditions. Phenylalanine hydroxylase from pseudomonas species was shown to be almost totally inactive unless it was preincubated for several min with either $Fe^{2+}$, $Hg^{2+}$, $Cu^{1+}$, $Cu^{2+}$ or $Cd^{2+}$ (35,36). In this case, the activation of the bacterial hydroxylase occurred in the absence of sulfhydryl reagents at a slightly acidic pH. The discrimination at a molecular level of the active and inactive states of such enzymes may be useful for the understanding of the reaction mechanism of pterin-dependent amino acid monooxygenase.

## SUMMARY

Bovine pineal tryptophan hydroxylase is a pterin-dependent aromatic amino acid monooxygenase with a broad substrate specificity and with low Km values for the amino acid substrates, L-tryptophan and L-phenylalanine. p-Chlorophenylalanine, an inhibitor of the tryptophan hydroxylation in the brain, also serves as a good substrate of the bovine pineal enzyme. The full activity of this enzyme is detected *in vitro* only after preincubation with dithiothreitol under reductive conditions. The enzyme is profoundly and more or less specifically inhibited by L-5HTP, the product of tryptophan hydroxylation, suggesting a possible regulatory role of this hydroxylated amino acid. The Km of the enzyme for tetrahydrobiopterin, a presumed natural cofactor, is significantly higher than the expected tissue concentration if the cofactor is assumed to be uniformly distributed in the tissue. These properties of bovine pineal tryptophan hydroxylase are distinguishable from those of the hydroxylase in the brain indicating that the hydroxylation of tryptophan in the brain leading to the synthesis of serotonin and the reaction in the pineal gland leading to the formation of melatonin are catalyzed by different tryptophan hydroxylases and are probably under separate regulatory mechanisms.

## REFERENCES

1. Grahame-Smith, D. G. (1964) Biochem. Biophys. Res. Communs. 16, 586-592.

2. Gal, E. M., Armstrong, J. C., and Ginsberg, B. (1966) J. Neurochem. 13, 643-654.

3. Green, H., and Sawyer, J. L. (1966) Anal. Biochem. 15, 53-64.

4. Hosoda, S., and Glick, D. (1966) J. Biol. Chem. 241, 192-196.

5. Grahame-Smith, D. G. (1967) Biochem. J. 105, 351-360.

6. Lovenberg, W., Jequier, E., and Sjoerdsma, A. (1967) Science 155, 217-219.

7. Grahame-Smith,D. G. (1968) Adv. Pharmacol. 6A, 37-42.

8. Green, H. (1968) Adv. Pharmacol. 6A, 18-19.

9. Ichiyama, A., Nakamura, S., Nishizuka, Y., and Hayaishi, O. (1968) Adv. Pharmacol. 6A, 5-17.

10. Lovenberg, W., Jequier, E., and Sjoerdsma, A. (1968) Adv. Pharmacol. 6A, 21-35.

11. Ichiyama, A., Nakamura, S., Nishizuka, Y., and Hayaishi, O. (1970) J. Biol. Chem. 245, 1699-1709.

12. Kaufman, S., and Fisher, D. B. (1974) in "Molecular Mechanisms of Oxygen Activation" (Hayaishi, O., ed) pp 285-369, Academic Press, New York.

13. Friedman, P. A., Lloyd, T., and Kaufman, S. (1972) Molecular Pharmacol. 8, 501-510.

14. Friedman, P. A., Kappleman, A. H., and Kaufman, S. (1972) J. Biol. Chem. 247, 4165-4173.

15. Fisher, D. B., and Kaufman, S. (1973) J. Biol. Chem. 248, 4345-4353.

16. Tong, J. H., and Kaufman, S. (1975) J. Biol. Chem. 250, 4152-4158.

17. Ichiyama, A., Hori, S., Mashimo, Y., Nukiwa, T., and Makuuchi, H. (1974) FEBS Letters 40, 88-91.

18. Nukiwa, T., Tohyama, C., Okita, C., Kataoka, T., and Ichiyama, A. (1974) Biochem. Biophys. Res. Communs. 60, 1029-1035.

19. Christenson, J. G., Dairman, W., and Udenfriend, S. (1970) Arch. Biochem. Biophys. 141, 356-367.

20. Guroff, G., and Abramowitz, A. (1967) Anal. Biochem. 19, 548-555.

21. Fukushima, T., and Akino, M. (1968) Arch. Biochem. Biophys. 128, 1-5.

22. Bobst, von A., and Viscontini, M. (1966) Helv. Chim. Acta 49, 875-884.

23. Personal communication from Dr. T. Fukushima, Faculty of Science Tokyo Metropolitan University, Setagaya-ku, Tokyo, Japan.

24. Hori, S. (1975) Biochim. Biophys. Acta 384, 58-68.

25. Kaufman, S., and Levenberg, B. (1959) J. Biol. Chem. 234, 2683-2688.

26. Nagatsu, T., Levitt, M., and Udenfriend, S. (1964) J. Biol. Chem. 239, 2910-2917.

27. Miyamoto, E., Petzold, G. L., Kuo, J. F., and Greengard, P. (1973) J. Biol. Chem. 248, 179-189.

28. Unpublished data.

29. Renson, J., Weissbach, H., and Udenfriend, S. (1962) J. Biol. Chem. 237, 2261-2264.

30. Shiman, R., Akino, M., and Kaufman, S. (1971) J. Biol. Chem. 246, 1330-1340.

31. Kaufman, S. (1961) Biochim. Biophys. Acta 51, 619-621.

32. Guroff, G., Kondo, K., and Daly, J. (1966) Biochem. Biophys. Res. Communs. 25, 622-627.

33. Deguchi, T., and Barchas, J. D. (1972). Molecular Pharmacol. 8, 770-779.

34. Petrack, B., Sheppy, F., Fetzer, V., Manning, T., Chertock', H., and Ma, D. (1972) J. Biol. Chem. 247, 4872-4878.

35. Guroff, G., and Ito, T. (1965) J. Biol. Chem. 240, 1175-1184.

36. Guroff, G., and Rhoads, C. A. (1967) J. Biol. Chem. 242, 3641-3645.

EVIDENCE FOR PARTICIPATION OF NADH-DEPENDENT REDUCTASE IN THE REACTION OF BENZOATE 1,2-DIOXYGENASE (BENZOATE HYDROXYLASE)

Hitoshi Fujisawa, Mutsuo Yamaguchi and Takashi Yamauchi

Department of Biochemistry, Asahikawa Medical College

Asahikawa 071-01, Japan

## INTRODUCTION

Benzoate 1,2-dioxygenase which has so far been referred to as benzoate hydroxylase catalyzes the double hydroxylation of benzoate, namely the incorporation of two hydroxyl groups into benzoate, resulting in the formation of 1,2-dihydro-1,2-dihydroxybenzoate (DHB). Both oxygen atoms of the two hydroxyl groups introduced into the substrate are derived from the molecular oxygen (1,2).

COOH NADH + $H^+$ NAD$^+$ COOH OH OH H

$O_2$

Benzoic acid DHB

Double hydroxylation reaction was first demonstrated by Kobayashi *et al.* with anthranilate hydroxylase which catalyzes the formation of catechol from anthranilic acid (3). They demonstrated by experiments with $^{18}O$ that both atoms of oxygen incorporated in two hydroxyl groups of catechol are exclusively derived from molecular oxygen. Benzene hydroxylase was also shown to catalyze the incorporation of both atoms of molecular oxygen into benzene (4). Both of them were shown to be resolved into two protein components (5,6).

The reaction mechanism of single hydroxylation by monooxygenases has been extensively studied in many laboratories (7,8), but the study of the reaction mechanism of double hydroxylation by dioxygenases has not progressed, since the enzyme systems catalyzing double hydroxylations are extremely unstable and pure enzymes have not been available for use in studying the reaction mechanism.

In this symposium we describe some of our recent studies suggesting that benzoate 1,2-dioxygenase consists of two protein components and one component which is a non-heme iron protein may function as a terminal oxygenase and the other component which is a flavin adenine dinucleotide (FAD)-containing non-heme iron protein may function as the reductase of the terminal oxygenase in this system. Preliminary accounts of a portion of this work have already appeared (9).

## MATERIALS AND METHODS

*Pseudomonas arvilla* C-1 (10) grown in a medium containing sodium benzoate as an inducer was used as a source of benzoate 1,2-dioxygenase system. NADH and $NAD^+$ were products of Oriental Yeast Company, Japan. Cytochrome *c* and streptomycin sulfate were obtained from Boehringer Mannheim. DHB was prepared biologically by the use of *Alcaligenes eutrophus* strain B9 (1). This strain was kindly donated by Dr. B. F. Johnson, Department of Bacteriology and immunology, University of California, Berkeley.

Benzoate 1,2-dioxygenase was assayed spectrophotometrically by measuring the decrease in absorbance at 340 nm of NADH or polarographically by determining the oxygen consumption (11). The assay system contained, in a final volume of 2.0 ml, 2.0 μmole of FAD, 170 μmoles of Tris-HCl buffer (pH 8.0), and a suitable amount of enzyme solutions. The reaction was started by adding benzoate and the initial rate was recorded at 25°.

NADH-cytochrome *c* reductase was assayed spectrophotometrically by measuring the reaction of ferricytochrome *c* by NADH at 550 nm. The assay system contained, in a final volume of 2.0 ml, 0.25 μmole of NADH, 0.08 μmole of cytochrome *c*, 170 μmoles of Tris-HCl (pH 8.0), and a suitable amount of enzyme.

Protein was determined by the method of Lowry *et al*. (12). All spectrophotometric measurements were carried out with a Hitachi 356 recording spectrophotometer. Sodium dodecyl sulfate (SDS) polyacrylamide gel electrophoresis was carried out as described by Laemmli (13).

## RESULTS AND DISCUSSION

Benzoate 1,2-dioxygenase system obtained from *Pseudomonas arvilla* grown in a medium containing benzoate as an inducer was separated into two fractions by ammonium sulfate fractionation as

Table I. Requirements of Two Components for Benzoate 1,2-Dioxygenase Activity

| Component added | Benzoate 1,2-dioxygenase (units) |
|---|---|
| A(40-50 % ammonium sulfate), 2.9 mg | 0 |
| B(60-70 % ammonium sulfate), 3.4 mg | 0 |
| A, 2.9 mg + B, 3.4 mg | 0.11 |

Table II. Coinduction of Benzoate 1,2-Dioxygenase and NADH-Cytochrome c Reductase Activities by Benzoate

| Carbon source added | Cytochrome c reductase (units/mg) | Benzoate 1,2-dioxygenase (unit/mg) | |
|---|---|---|---|
| | | 40-50 % amm. sulf. fraction | 60-70 % amm. sulf. fraction |
| Glucose | 0.03 | 0 | 0 |
| Succinate | 0.02 | 0 | 0 |
| Benzoate | 1.43 | 0.025 | 0.021 |

*Pseudomonas arvilla* was grown on media containing the carbon source indicated and cells were harvested at late logarithmic phase. Activity of NADH-cytochrome c reductase was determined with the crude extract. Activity of benzoate 1,2-dioxygenase was determined after separation by ammonium sulfate fractionation. The activity of 40-50 % ammonium sulfate fraction was determined in the presence of 3.4 mg of 60-70 % ammonium sulfate fraction obtained from the cells grown on the benzoate medium. The activity of 60-70 % ammonium sulfate fraction was determined in the presence of 2.9 mg of the 40-50 % ammonium sulfate fraction obtained from the cells grown on the benzoate medium.

shown in Table I. Neither the 40-50% saturated ammonium sulfate fraction nor the 60-70% saturated ammonium sulfate fraction alone had significant benzoate 1,2-dioxygenase activity and both of them were required for the activity. The 40-50% ammonium sulfate fraction had NADH-cytochrome c reductase activity and it was induced by the addition of benzoate into the culture medium as well as benzoate 1,2-dioxygenase activity as shown in Table II. This observation suggested the possibility that NADH-cytochrome c reductase may participate in the reaction of benzoate 1,2-dioxygenase. Therefore, we tried to purify the NADH-cytochrome c reductase.

The purification procedure of NADH-cytochrome c reductase is summarized in Table III. All purification procedures were carried out in Tris-HCl buffer (pH 8.0) containing 0.5 mM dithiothreitol, because this enzyme was extremely unstable in the absence of sulfhydryl reagents. Overall purification was about 640-fold with 20% yield. Benzoate 1,2-dioxygenase activity was measured in the presence of a fixed amount of 60-70% ammonium sulfate fraction. The ratio of activities of NADH-cytochrome c reductase and benzoate 1,2-dioxygenase did not vary throughout the purification procedure, supporting the possibility of participation of NADH-cytochrome c reductase in the reaction of double hydroxylation of benzoate.

The purified NADH-cytochrome c reductase which was homogeneous as judged by SDS-polyacrylamide gel electrophoresis as shown in Figure 1 was an FAD-containing non-heme iron protein and catalyzed

Table III. Purification of NADH-Cytochrome c Reductase

| Fraction | Protein (mg) | Specific activity (units/mg) Cytochrome c redictase (A) | Specific activity (units/mg) Benzoate 1,2-dioxygenase (B) | A/B | Yield (%) |
|---|---|---|---|---|---|
| Crude extract | 22,970 | 1.43 | ---* | | 100 |
| Streptomycin | 20,720 | 1.52 | ---* | | 96 |
| Amm.sulf.(40-50%) | 6,685 | 3.00 | 0.019 | 158 | 61 |
| Acid treat.(pH4.5) | 1,844 | 9.36 | 0.059 | 159 | 52 |
| Sephadex G-200 | 179 | 93.2 | 0.59 | 158 | 43 |
| DEAE-Sephadex | 19 | 496 | 3.16 | 157 | 29 |
| Sephadex G-100 | 7.2 | 911 | 5.77 | 158 | 20 |

Assay of benzoate 1,2-dioxygenase was carried out in the presence of 2.0 mg of 60-70 % ammonium sulfate fraction. It was not possible to achieve levels of the 60-70 % ammonium sulfate fraction which are nonlimiting in the assay of the enzyme. Consequently, the assay of benzoate 1,2-dioxygenase was emperical. Under the assay conditions the activity of benzoate 1,2-dioxygenase was proportional to the amount of the protein added. The yield was calculated from the activity of cytochrome c reductase.

* Activities of benzoate 1,2-dioxygenase could not be determined because these preparations contained the protein of 60-70 % ammonium sulfate fraction.

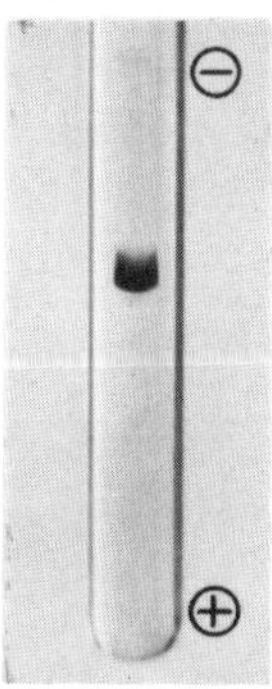


Figure 1. SDS-polyacrylamide gel electrophoresis of the purified NADH-cytochrome c reductase. The enzyme (60 μg of protein) was treated with SDS and 2 mercaptoethanol and submitted to electrophoresis with a 7.5 % separating gel. Migration was from top to bottom. The gels were stained with Amino black 10B.

the reduction of artificial electron acceptors such as ferricyanide, methylene blue and 2,6-dichlorophenolindophenol as well as cytochrome c by NADH.

The absorption spectrum of NADH-cytochrome c reductase is shown in Figure 2. It has a broad absorption in the visible range with peaks at around 460 nm and 335 nm. The absorption spectrum in the visible range may be due to both of flavin and non-heme iron in the enzyme, since it contains one mole of FAD and two gram atoms of iron per mole of enzyme.

The 60-70% ammonium sulfate fraction had the activity which catalyzed the double hydroxylation of benzoate in the presence of NADH-cytochrome c reductase as described before. It was also

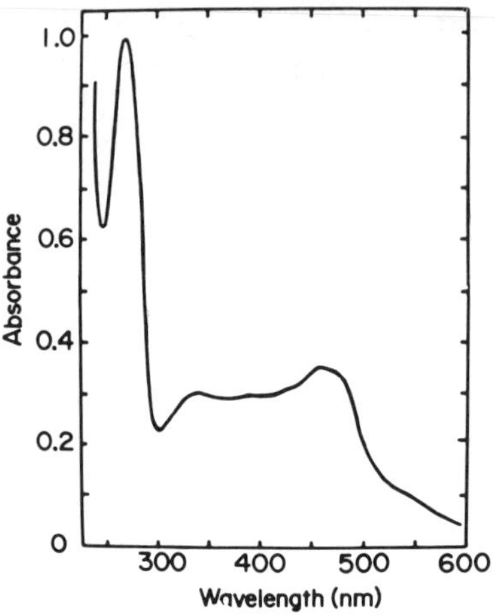


Figure 2. Absorption spectrum of NADH-cytochrome c reductase. The concentration of the enzyme was 1.2 mg per ml in 0.1 M Tris-HCl buffer (pH 8.0) containing 0.5 mM dithiothreitol.

Table IV. Analytical Properties of Benzoate 1,2-Dioxygenase System

| | FAD-containing non-heme iron protein (NADH-cytochrome c reductase) | Non-heme iron protein (80 % pure) |
|---|---|---|
| M.W. | 37,000 | 250,000-300,000 |
| FAD | 0.9 mole/mole of enzyme | - |
| Iron | 2.1 g atoms/mole of enzyme | 3.3 moles/100,000g protein |
| Heme | - | - |

purified about 45-fold with a yield of 20% by streptomycin treatment, ammonium sulfate fractionation, pH 4.5 precipitation, Sephadex G-200 gel filtration, DEAE-Sephadex column chromatography, and Sephadex G-100 gel filtration. All purification procedures were carried out in the presence of 20 mM benzoate and 0.5 mM dithiothreitol, since the activity was very unstable in the absence of substrate. The purified preparation was at least about 80% pure upon polyacrylamide gel electrophoresis.

Some analytical properties of the enzymes are summarized in Table IV. Analyses by Sephadex G-200 gel filtration and SDS-polyacrylamide gel electrophoresis indicated that NADH-cytochrome c reductase was a single polypeptide of molecular weight of 37,000 which contained one mole of FAD and two atoms of iron. The iron content was determined by colorimetric analysis and atomic absorption spectroscopy. The molecular weight of the enzyme catalyzing double hydroxylation of benzoate in the presence of NADH-cytochrome c reductase was roughly determined to be approximately 300,000 by Sephadex G-200 gel filtration. The iron content of the 80% pure preparation was determined to be 33 nmoles per mg of protein by

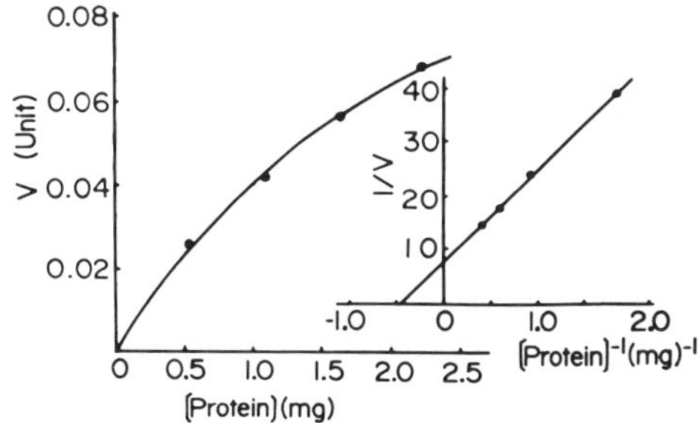


Figure 3. Effect of the concentration of non-heme iron protein on the rate of double hydroxylation of benzoate. Benzoate 1,2-dioxygenase activity was determined at different concentrations of the non-heme iron protein in the presence of 1.9 ug of the FAD-containing non-heme iron protein under the standard assay conditions. Reciprocal velocities are plotted as functions of reciprocal concentrations of the non-heme iron protein in the right figure.

atomic absorption spectroscopy. There was no sign of FAD and heme present in this preparation as judged by the absorption spectrum.

Two different proteins, a non-heme iron protein and an FAD-containing non-heme iron protein which had NADH-cytochrome c reductase activity were required for the reaction of double hydroxylation of benzoate as described above. Then, to examine the relation between these two proteins during the reaction, the effects of the protein concentrations on the rate of double hydroxylation of benzoate were studied. When the velocity of double hydroxylation of benzoate with a fixed amount of FAD-containing non-heme iron protein in the presence of excess amounts of substrates was plotted against the concentration of non-heme iron protein, a hyperbolic relationship was obtained and the Lineweaver-Burk plots yielded a completely straight line as shown in Figure 3. A similar result was observed when the experiments were carried out under different concentrations of the FAD-containing non-heme iron protein with a fixed concentration of the non-heme iron protein. These results suggest that direct interaction between NADH-cytochrome c reductase and the non-heme iron protein is required for double hydroxylation of benzoate.

In order to investigate the mode of the interaction of these two proteins required for the double hydroxylation of benzoate, we examined whether the NADH-dependent reductase activity itself of the FAD-containing non-heme iron protein was necessary for double hydroxylation or not. When 2,6-dichlorophenolindophenol (DCIP) which can serve as an electron acceptor of NADH-cytochrome c reductase was added to the reaction mixture during the reaction of double hydroxylation of benzoate, benzoate-dependent oxygen consumption was significantly inhibited (Figure 4). When the reduction of DCIP, which was monitored spectrophotometrically by measuring the absorbance at

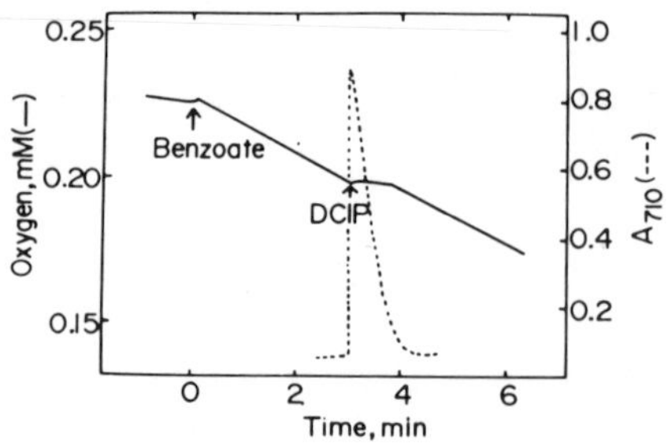


Figure 4. Inhibition of double hydroxylation of benzoate by DCIP. The reaction mixture contained 1 μmole of NADH, 0.2 μmole of $FeSO_4$, 0.05 μmole of FAD, 2 μmoles of benzoate, 6.7 μg of NADH-cytochrome c reductase, and 960 μg of crude non-heme iron protein in a total volume of 2.0 ml. Benzoate 1,2-dioxygenase activity was measured polarographically. The reaction was started by the addition of benzoate. DCIP was added at the time as indicated by the arrow at the final concentration of 0.4 mM. Reduction of DCIP by the enzyme was monitored spectrophotometrically by measuring the absorbance at 710 nm in parallel experiments under identical experimental conditions.

710 nm, was completed, benzoate-dependent oxygen consumption was completely recovered. As the velocity of the reduction of cytochrome c by the enzyme was too fast to be analyzed, DCIP was used in this experiment. This result indicates that the NADH-dependent reductase activity itself may participate in the reaction of double hydroxylation of benzoate and it furthermore suggests the possibility that NADH-dependent reductase may catalyze the reduction of non-heme iron protein during double hydroxylation reaction.

The native non-heme iron protein also has a broad absorption in the visible range with a peak at around 460 nm and a shoulder at around 560 nm (Figure 5). When NADH and a catalytic amount of FAD-containing non-heme iron protein, namely NADH-cytochrome c reductase was added to the substrate amount of the non-heme iron protein under anaerobic conditions, the absorption in the visible range decreased and a new absorption spectrum with a peak at around 520 nm appeared as shown in Figure 5. When the air was introduced, the spectrum reverted again to the original spectrum. The NADH-dependent reduction of non-heme iron protein catalyzed by NADH-cytochrome c reductase supports the possibility that the non-heme iron protein may function as a terminal oxygenase, namely benzoate 1,2-dioxygenase and NADH-cytochrome c reductase may function as the NADH-benzoate 1,2-dioxygenase reductase.

When sodium dithionite was used as a terminal oxygenase reducing agent in place of NADH-cytochrome c reductase, a similar spectral change was observed (Figure 5). When a limiting amount of dithionite was added to the substrate amount of the non-heme iron protein in the presence of benzoate under anaerobic conditions and then the air was introduced, a significant amount of product, 1,2-

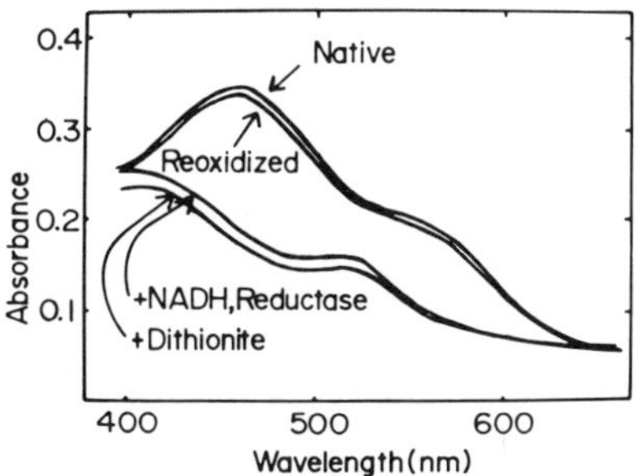


Figure 5. Absorption spectra of non-heme iron protein. The incubation mixture contained 10 μmoles of benzoate, 3.5 mg of highly purified non-heme iron protein, 40 nmoles $FeSO_4$ and 100 μmoles of Tris-HCl buffer (pH 8.0) in a final volume of 1.0 ml(Native). The enzyme was reduced by the addition of 0.125 μmole of NADH and 10 μg of NADH-cytochrome c reductase (+NADH, Reductase) or by the addition of 0.1 μmole of sodium dithionite (+Dithionite) under anaerobic conditions. The reduced enzyme was reoxidized by shaking with air (Reoxidized).

dihydro-1,2-dihydroxybenzoate, was produced. This preliminary finding provides evidence to support the contention that the non-heme iron protein is a terminal oxygenase and its reduction is essential for double hydroxylation of benzoate.

In summary, benzoate 1,2-dioxygenase system which catalyzes double hydroxylation of benzoate consists of two protein components, a non-heme iron protein and a FAD-containing non-heme iron protein. The results of above-described experiments and similarities to mono-oxygenase systems present in microsome, mitochondria, and Pseudomonads when taken together, have lead to the tentative reaction scheme shown below:

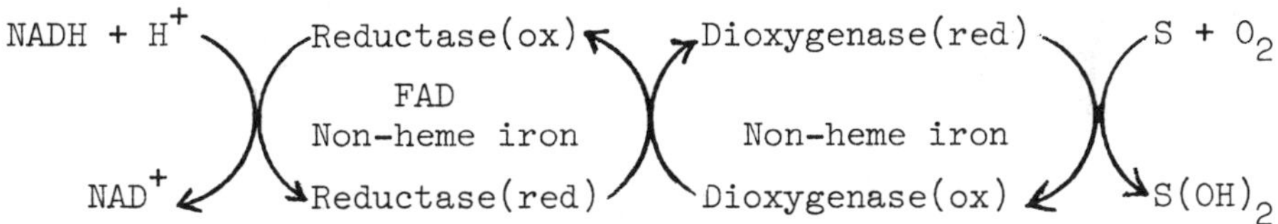


NADH reduces the FAD-containing non-heme iron protein and then, the reduced form of FAD-containing non-heme iron protein reduces the non-heme iron dioxygenase and then its reduced form catalyzes double hydroxylation of benzoate. In other words, non-heme iron protein functions as benzoate 1,2-dioxygenase and FAD-containing non-heme iron protein functions as NADH-benzoate 1,2-dioxygenase reductase.

Recently, Axcell and Geary reported that benzene-oxidizing system (benzene hydroxylase) obtained from the species of Pseudomonas consisted of three protein components and they speculated the involvement of NADH-dependent reductase in benzene-oxidizing reaction (14).

ACKNOWLEDGEMENTS

This work has been supported in part by the Scientific Research Fund of the Ministry of Education of Japan, and by the Naito Research Grant for 1975.

REFERENCES

1. Reiner, A.M., and Hegeman, G.D. (1971) Biochemistry 10, 2530-2536.
2. Reiner, A.M. (1971) J. Bacteriol. 108, 89-94.
3. Kobayashi, S., Kuno, S., Itada, N., and Hayaishi, O. (1964) Biochem. Biophys. Res. Commun. 16, 556-561.
4. Gibson, D.T., Cardini, G.E., Maseles, G.C., and Kallio, R.E. (1970) Biochemistry 9, 1631-1635.
5. Kobayashi, S., and Hayaishi, O. (1970) in Methods in Enzymology (Kolowick, S.P., and Kaplan, N.O., Eds.) Vol. XVII, pp. 505-510, Academic Press, New York.
6. Gibson, D.T., Koch, J.R., and Kallio, R.E. (1968) Biochemistry, 7, 2653-2662.
7. Hayaishi, O. (1974) in Molecular Mechanisms of Oxygen Activation (Hayaishi, O., Ed.) pp. 1-28, Academic Press, New York and London.
8. Gunsalus, I.C., Pederson, T.C., and Sligar, S.G. (1975) Ann. Rev. Biochem. 44, 377-407.
9. Yamaguchi, M., Yamauchi, T., and Fujisawa, H. (1975) Biochem. Biophys. Res. Commun. 67, 264-271.
10. Kojima, Y., Fujisawa, H., Nakazawa, T., Kanetsuna, F., Taniuchi, H., Nozaki, M., and Hayaishi, O. (1967) J. Biol. Chem. 242, 3270-3278.
11. Hagihara, B. (1961) Biochim. Biophys. Acta 46, 134-142.
12. Lowry, O.H., Rosebrough, N.J., Farr, A.L., and Randall, R.J. (1951) J. Biol. Chem. 193, 265-275.
13. Laemmli, U.K. (1970) Nature 227, 680-685.
14. Axcell, B.C., and Geary, P.J. (1975) Biochem. J. 146, 173-183.

# SUBUNIT STRUCTURE OF NONHEME IRON-CONTAINING DIOXYGENASES*

Mitsuhiro Nozaki+, Ryotaro Yoshida, Chieko Nakai, Masayoshi Iwaki+, Yukikazu Saeki and Hiroyuki Kagamiyama+

Department of Medical Chemistry, Kyoto University Faculty of Medicine, Kyoto, Japan; Department of Biochemistry, Osaka University School of Medicine, Osaka, Japan; and Department of Biochemistry, Shiga University of Medical Science, Shiga, Japan

Dioxygenases are enzymes that catalyze the incorporation of two atoms of molecular oxygen into various substrates. These enzymes have been discovered in all types of living organisms and shown to perform a variety of functions. Among these, cleavage of the aromatic ring is one function that appears to depend largely, perhaps entirely, upon this type of enzyme. Cofactors involved in these enzymes are nonheme iron, heme or copper (1). The indole ring-cleaving enzyme, tryptophan 2,3-dioxygenase, is known to contain heme as a cofactor (2). A flavonol-cleaving enzyme, quercetinase, has been reported to be a copper protein (3). With the exception of these two enzymes, most of the other dioxygenases, if not all, contain nonheme iron as the cofactor. Among these, some enzymes contain the ferrous form of iron and some, the ferric form.

For example, when the catechol ring is cleaved by individual enzymes, two different modes of ring fission have so far been demonstrated. One is between the two hydroxylated carbon atoms to form muconic acid derivatives, which is referred to as intradiol cleavage, and the other between one hydroxylated carbon atom and an adjacent carbon carrying hydrogen, and is referred to as extradiol cleavage (4). It is of interest that the intradiol enzymes that have so far been purified and characterized contain the ferric form of iron, whereas the extradiol enzymes contain ferrous iron (1).

---

* This work has been supported by a Grant-in-aid for Scientific Research from the Ministry of Education, Science and Culture, Japan.

+ Present address: Department of Biochemistry, Shiga University of Medical Science, Shiga, Japan.

Although extensive studies on the reaction mechanisms of these enzymes have been carried out in our laboratory, detailed protein chemical analyses of this type of enzyme have never been done. In this Symposium, we would like to focus our attention on the subunit structure of the ferric iron-congaining dioxygenases.

## Protocatechuate 3,4-dioxygenase

Protocatechuate 3,4-dioxygenase is a typical example of an intradiol enzyme and catalyzes the cleavage of the benzene ring of protocatechuic acid to form β-carboxymuconic acid (Equation 1).

$$\text{HOOC-}C_6H_3\text{(OH)(OH)} + O_2 \longrightarrow \text{HOOC-}C_4H_3\text{(COOH)(COOH)} \qquad (1)$$

This enzyme was obtained in a crystalline form by Fujisawa and Hayaishi and shown to contain eight atoms of ferric iron per mole of enzyme based on a molecular weight of 700,000 (5). The fact that the enzyme is dissociated into homogeneous subunits of approximately 90,000 daltons by alkaline treatment, together with an electron microscopic observation has led us to conclude that the enzyme consists of eight identical subunits, each of which contains one atom of ferric iron and a single substrate binding site (6).

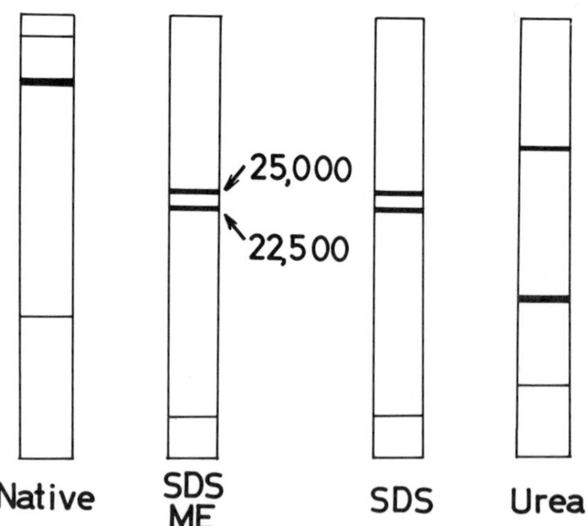


Fig. 1. Gel electrophoreses of protocatechuate 3,4-dioxygenase.

However, the 90,000 dalton subunit has now been found to dissociate further into smaller non-identical subunits upon acrylamide disc gel electrophoresis in the presence of either SDS or urea, as shown in Figure 1. In the absence of SDS or urea, the native enzyme gave one distinct main band and a faint, slow moving band. The bands near the bottom of the gels in this and later figures are not protein bands but those of the marker dye. The faint band was not observed when the sample was incubated with iodoacetamide and mercaptoethanol or when electrophoresis was carried out in the presence of sodium dodecyl sulfate (SDS), indicating that the faint band is a polymerized form of the enzyme. When the native enzyme was treated with 1% each of SDS and mercaptoethanol and electrophoresed in the presence of 0.1% each of SDS and mercaptoethanol, two distinct bands with molecular weights of 22,500 and 25,000, respectively, were observed. When the incubation and electrophoresis were carried out in the absence of mercaptoethanol, essentially identical results were obtained, indicating that disulfide linkages are not involved in the interaction of the subunits. When the enzyme treated with 8 M urea was subjected to disc gel electrophoresis in the absence of SDS, two clearly separated bands were also observed.

In an attempt to separate these subunits, isoelectric focusing and chromatography of the enzyme on SP-Sephadex C-25 have been carried out in the presence of 6 M urea (Fig. 2). Upon isoelectric focusing in the presence of urea, the enzyme was separated into two major fractions, as shown in Figure 2 A. The isoelectric points of these fractions were approximately 5.2 and 9.5, and hereafter they will be referred to as the α and β subunits, respectively. Likewise, as shown in Figure 2 B, these subunits were also separated by SP-Sephadex chromatography in the presence of urea at pH 5.8.

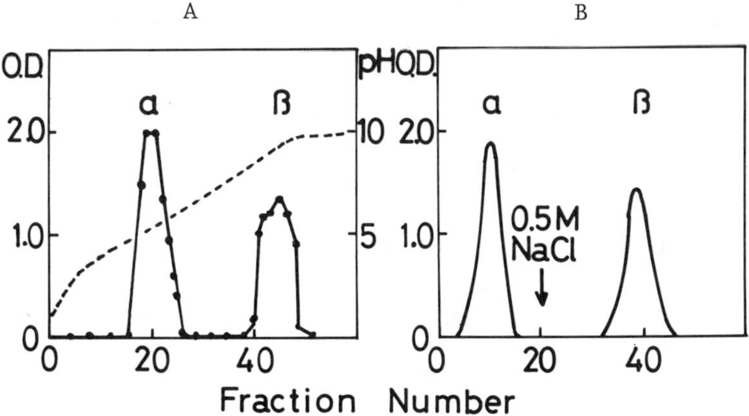


Fig. 2. Isolation of subunits by isoelectric focusing (A) and SP-Sephadex (B). Solid lines, O.D. at 280 nm; Dotted line, pH.

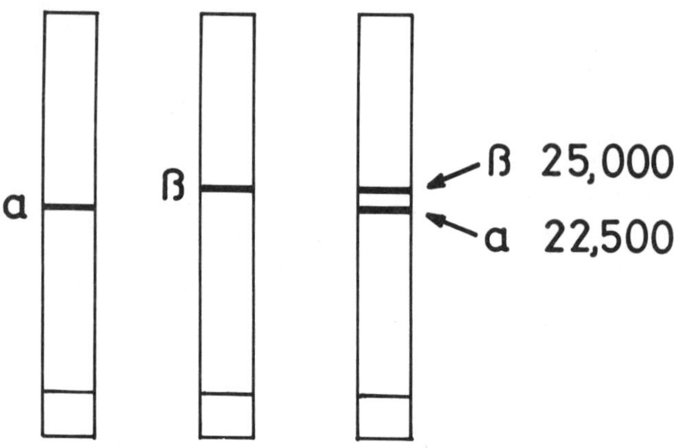


Fig. 3. SDS gel electrophoreses of the separated subunits.

Each fraction thus separated gave a single band on SDS gel electrophoresis, as shown in Figure 3. The acidic α subunit corresponded to the 22,500 dalton band, and the basic β subunit to the 25,000 dalton band. A mixture of the two fractions gave two distinct bands.

Amino acid compositions of each subunit and the native enzyme are shown in Table I. There are distinct differences between the two subunits. As would be expected from the isoelectric points, the α subunit contained more acidic amino acids than the β subunit did, and the β subunit contained more basic amino acids than the α subunit did. It should be noted that the α subunit contained one methionine and two cysteine residues, while the β subunit contained two methionine and four cysteine residues. The amino acid composition of the native enzyme obtained by actual analyses agreed quite well with that calculated from the compositions of the α and β subunits, based on the assumed subunit structure of $(\alpha_2\beta_2)_8$. These results confirmed the assumption that the native enzyme consists of 8 protomers, each of which is composed of a pair of two non-identical subunits $(\alpha_2\beta_2)$.

Results of analyses of the $NH_2$-terminal sequences and COOH-terminal amino acid residues are summarized in Table II. The $NH_2$-terminal sequences were determined by an automatic sequencer using the Edman degradation procedure (7). The $NH_2$-terminal residue of the native enzyme, as well as those of the α and β subunits, were found to be proline, with recoveries of more than 60%. This was confirmed with the manual Edman degradation (8) and the cyanate methods (9).

Table I

Amino acid compositions of native protocatechuate 3,4-dioxygenase and its subunits[a/]

| | α | β | $(\alpha_2+\beta_2)$X8 | Native[b/] |
|---|---|---|---|---|
| Lys | 5 | 8 | 208 | 182 |
| His | 5 | 7 | 192 | 194 |
| Arg | 11 | 14 | 400 | 351 |
| Asp | 24 | 23 | 752 | 718 |
| Thr | 10 | 9 | 304 | 286 |
| Ser | 5 | 9 | 224 | 205 |
| Glu | 24 | 16 | 640 | 513 |
| Pro | 13 | 21 | 544 | 477 |
| Gly | 14 | 18 | 512 | 460 |
| Ala | 19 | 15 | 544 | 477 |
| Val | 11 | 9 | 320 | 265 |
| Met | 1 | 2 | 48 | 46 |
| Ile | 12 | 15 | 432 | 364 |
| Leu | 19 | 17 | 576 | 459 |
| Tyr | 7 | 8 | 240 | 188 |
| Phe | 9 | 9 | 288 | 247 |
| 1/2 Cys | 2 | 4 | 96 | 95 |
| Trp | 6 | 4 | 160 | 160 |

a/ The values are expressed as number of residues on the basis of the assumption that the native enzyme has a molecular weight of 700,000 daltons; the α subunit, 22,500 daltons; and the β subunit, 25,000 daltons.

b/ Data were taken from reference (5).

Table II

Summary of the sequence analysis of protocatechuate 3,4-dioxygenase and its subunits

| Steps | 1 | 2 | 3 | 4 | 5 | 6 | 7 | 8 | 9 | 10 | 11 | 12 | 13 | 14 | | COOH-terminal |
|---|---|---|---|---|---|---|---|---|---|---|---|---|---|---|---|---|
| | | Ile | | Leu | Leu | Pro | Glx | Thr | | | | | | | | |
| Native enzyme | Pro | | Glx | | | | | | | | | | | | ----- | Phe |
| | | Ala | | Asx | Asx | Ala | Arg | Phe | | | | | | | | |
| α Subunit | Pro | Ile | Glu | Leu | Leu | Pro | Glu | Thr | Pro | Ser | Glx | Thr | Ala | Gly | ----- | Phe |
| β Subunit | Pro | Ala | Gln | Asp | Asn | Ala | Arg | Phe | Val | Ile | Arg | Asx | Arg | Asx | ----- | ? |

The $NH_2$-terminal sequence analyses of the native enzyme revealed that in each of the first eight degradation steps except for the first and third, a pair of amino acids was obtained. $NH_2$-terminal sequence analyses of the separated subunits were also performed. The amino acids released from the $\alpha$ and $\beta$ subunits at each step coincided with those released from the native enzyme.

Determinations of the COOH-terminal residue were performed by three different methods (10), including digestion by carboxypeptidases (11) hydrazinolysis (12) and tritium labeling (13). All these determinations resulted in the same answer. Phenylalanine was found to be the COOH-terminal amino acid for the native enzyme and the $\alpha$ subunit. However, the COOH-terminal residue of the $\beta$ subunit was not detected by any of these methods.

It is of interest that the $NH_2$-terminal residues of the two non-identical subunits were both determined to be proline, which is rarely an $NH_2$-terminal residue, especially in bacterial enzymes.

As was mentioned above, the $\alpha$ and $\beta$ subunits contained one and two methionine residues, respectively, and these methionines were not released by the Edman degradation up to 14 steps. This result may facilitate the determination of the total amino acid sequence of these subunits, because the peptides are cleaved specifically at methionine residues by cyanogen bromide treatment.

## Pyrocatechase

The focus of this discussion will now be shifted to results obtained with another intradiol dioxygenase, pyrocatechase, which was the first enzyme proved to be a dioxygenase by Hayaishi and his colleagues (14). This enzyme catalyzes the conversion of catechol to *cis,cis*-muconic acid with the insertion of both atoms of molecular oxygen (Equation 2). This enzyme has been reported to have a molecular weight of 90,000 and to contain 2 atoms of ferric iron per mole of enzyme (15). However, more careful analyses have recently revealed that the enzyme actually has a molecular weight of approximately 60,000 and contains one atom of ferric iron.

$$\text{catechol (1,2-dihydroxybenzene)} + O_2 \longrightarrow \textit{cis,cis}\text{-muconic acid (HOOC-CH=CH-CH=CH-COOH)} \qquad (2)$$

The purified enzyme gave a single band on acrylamide gel electrophoresis. However, when the enzyme was treated with SDS, it was also found to dissociate into two non-identical subunits with molecular weights of 30,000 and 32,000, respectively, as determined by SDS gel electrophoresis (Fig. 4). Likewise, two bands were observed when the enzyme was treated with 8 *M* urea.

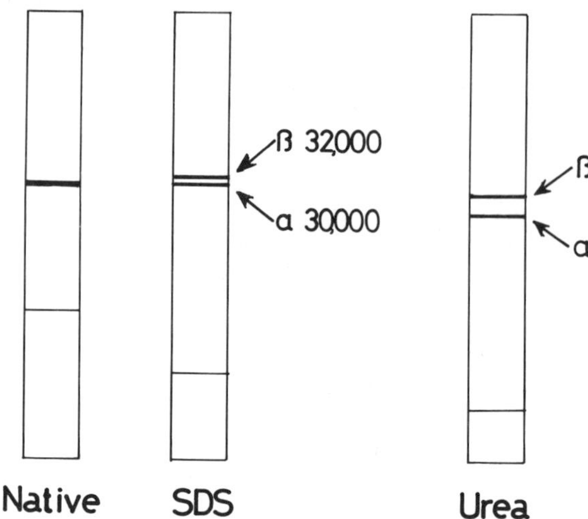


Fig. 4. Gel electrophoreses of pyrocatechase.

As shown in Fig. 5, these subunits could be separated by chromatography on CM-cellulose. Each separated fraction gave a single band on SDS acrylamide gel electrophoresis, whereas a mixture of the two fractions gave two distinct bands, suggesting that, like protocatechuate 3,4-dioxygenase, pyrocatechase also consists of two types of subunits.

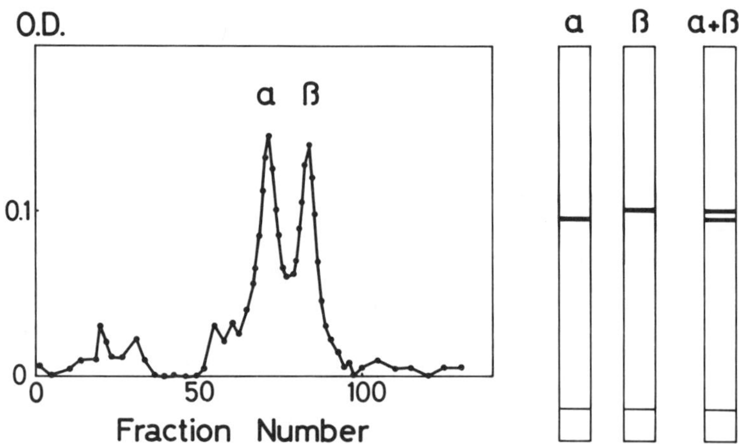


Fig. 5. Isolation of pyrocatechase subunits by CM-cellulose (left) and SDS disc gel electrophoreses of the separated subunits (right).

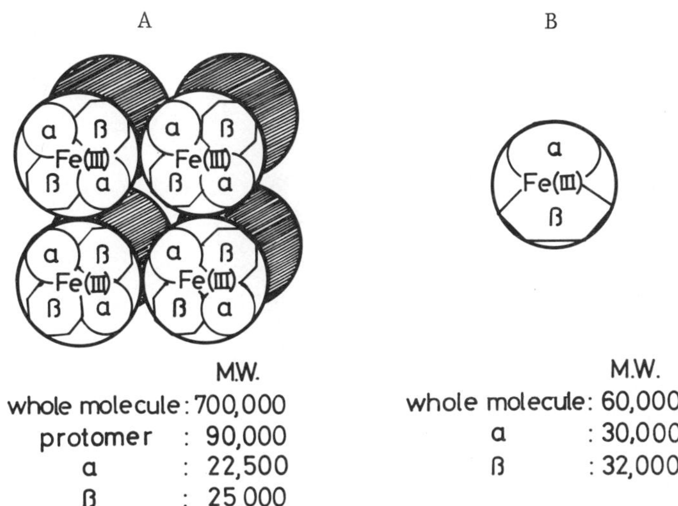


Fig. 6. Proposed subunit structures of protocatechuate 3,4-dioxygenase (A) and pyrocatechase (B).

## Conclusion

In summary, protocatechuate 3,4-dioxygenase consists of eight identical protomers, each containing two pairs of two non-identical subunits, $(\alpha_2\beta_2)_8$. This protomer $(\alpha_2\beta_2)$ appears to contain one atom of ferric iron, forming one active site of the enzyme. Likewise, pyrocatechase consists of two non-identical subunits $(\alpha\beta)$. One atom of ferric iron is present per pair of the subunits to form one molecule of the enzyme, which has a single active site (Fig. 6).

Total sequence analyses of these two enzymes and comparison of their sequences are now in progress in our laboratory.

## Acknowledgements

The authors would like to express their appreciation to Prof. O. Hayaishi for his guidance and continuous encouragement during the course of this investigation. Thanks are also due to Dr. H. Wada of Osaka University for making available to us the sequencer provided by the Toyo Rayon Scientific Foundation, to Dr. C. Y. Lai for his valuable discussion, and to Drs. C. M. Edson and S. C. Hubbard for their critical readings of the manuscript.

## References


(1) Nozaki, M. in Molecular Mechanisms of Oxygen Activation. Ed. by O. Hayaishi, Academic Press, Inc., New York, 1974, p. 135

(2) Nozaki, M., and Ishimura, Y. in Microbial Iron Metabolism. Ed. by J. B. Neilands, Academic Press, Inc., New York, 1974, p. 417

(3) Oka, T., and Simpson, F. J. (1971) Biochem. Biophys. Res. Commun. 43, 1

(4) Nozaki, M., Kotani, S., Ono, K., and Senoh, S. (1970) Biochim. Biophys. Acta 220, 213

(5) Fujisawa, H., and Hayaishi, O. (1967) J. Biol. Chem. 243, 2673

(6) Fujisawa, H., Uyeda, M., Kojima, Y., Nozaki, M., and Hayaishi, O. (1972) J. Biol. Chem. 247, 4414

(7) Edman, P., and Begg, G. (1967) Eur. J. Biochem. 1, 80

(8) Edman, P. (1956) Ann. N. Y. Acad. Sci. 88, 761

(9) Stark, G. R., and Smyth, D. G. (1963) J. Biol. Chem. 238, 214

(10) Narita, K. in Protein Sequence Determination. Ed. by S. B. Needleman, Springer-Verlag, 1970, p. 25

(11) Hayashi, R., Moore, S., and Stein, W. H. (1973) J. Biol. Chem. 248, 2296

(12) Braun, V., and Schroeder, W. A. (1967) Arch. Biochem. Biophys. 118, 241

(13) Matsuo, H., Fujimoto, Y., and Tatsuno, T. (1966) Biochem. Biophys. Res. Commun. 22, 69

(14) Hayaishi, O., Katagiri, M., and Rothberg, S. (1955) J. Amer. Chem. Soc. 77, 5450

(15) Kojima, Y., Fujisawa, H., Nakazawa, A., Nakazawa, T., Kanetsuna, F., Taniuchi, H., Nozaki, M., and Hayaishi, O. (1967) J. Biol. Chem. 242, 3270

# IRON-SULFUR PROTEINS, THE MOST NUMEROUS AND MOST DIVERSIFIED COMPONENTS OF THE MITOCHONDRIAL ELECTRON TRANSFER SYSTEM

Helmut Beinert

Institute for Enzyme Research, University of Wisconsin

Madison, Wisconsin 53706

## I. INTRODUCTION

The field of iron-sulfur (Fe-S) proteins has developed rapidly in the past ten years, in depth and in breadth, as is best witnessed by the appearance within the past few years of three volumes of a treatise exclusively devoted to these compounds (1). The great number and variety of recently discovered plant, bacterial and mammalian Fe-S proteins and of the reactions catalyzed by them, exemplify this progress. This presentation will focus on the last mentioned area, namely that of Fe-S proteins in mammalian systems, more precisely on those occurring in the electron transfer system of mitochondria.

The information available on these Fe-S proteins has been emerging rapidly from a few laboratories and is scattered in the literature (2-13). We will therefore try to assess here what is reasonably well established at this time and where the future may take us.

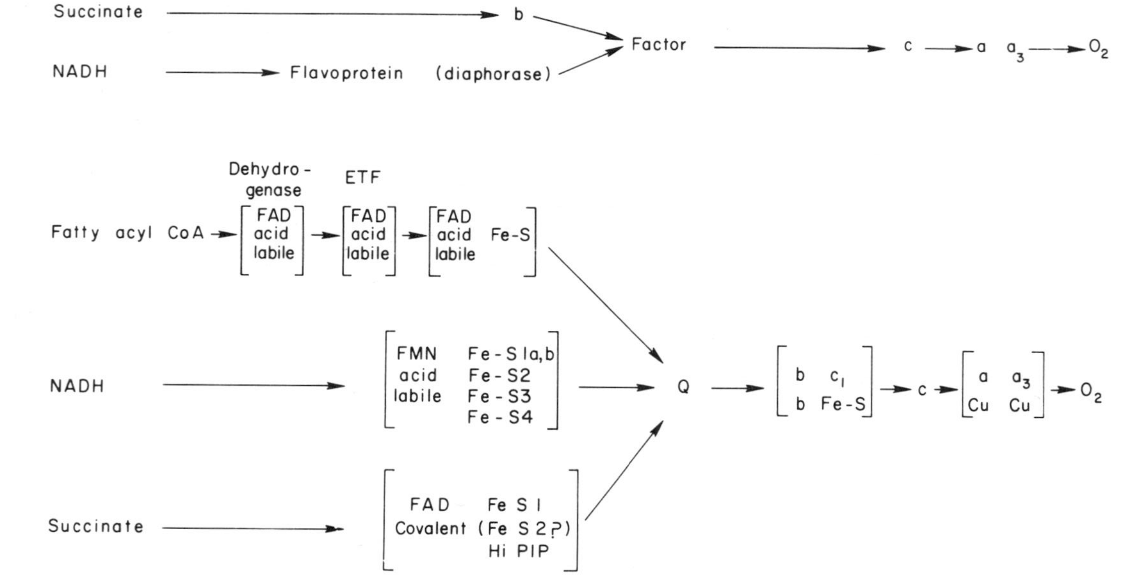


Scheme 1. Components of the electron transfer system of mammalian mitochondria as depicted in 1950 (top) and in 1975 (bottom).

## II. IRON-SULFUR CENTERS OF THE MITOCHONDRIAL ELECTRON TRANSFER SYSTEM.

### 1) "The Respiratory Chain"

Scheme 1 shows the so-called "respiratory chain" of 25 years ago compared to what we might picture today. Needless to say, things have not become simpler. There are two points which such a comparison brings out: First, the components known mainly from the classical work of Keilin, namely the cytochromes, are concentrated at the oxygen side of the system, whereas the more recently discovered components, such as Q and the Fe-S proteins or Fe-S centers* are more numerous toward the substrate side.

Secondly, counting species only, not total quantities, we recognize that there are more Fe-S centers than cytochromes, and if we count all the more recently discovered components, including ubiquinone (Q) and copper, the cytochromes are considerably outnumbered in terms of species and total capacity for electron uptake.

Before we discuss in more detail implications of these developments and of the points just raised, we will summarize the present state of knowledge as to where and how many Fe-S centers are present in the mitochondrial electron transfer system.

### 2) NADH Dehydrogenase

There are 4 to 5 Fe-S centers associated with NADH dehydrogenase. They can be differentiated by their EPR signals and their behavior on oxidation reduction. The midpoint potentials range from about -400 to +20 mVolt (cf. 6). The uncertainty as to whether there are 4 or 5 centers stems from the fact that the center of lowest midpoint potential and the only center detectable at liquid nitrogen temperature, namely, so-called center 1, appears to have two components, 1a and 1b, with different properties (2, 4,6) but very similar EPR spectra. However, the concentration of each of the subspecies present appears to be less than one half that of the acid labile FMN, so that there seems to be heterogeneity (4). Therefore, although there are 5 different species of Fe-S centers present, there are at best 4 equivalents of Fe-S centers associated with every flavin of NADH dehydrogenase. Center 2, the Fe-S center of highest midpoint potential, is the only one about which no serious ambiguities exist at this time. Al-

*The term centers is preferred here because this does not imply that the Fe-S groups involved are located on proteins or subunits different from those bearing other known (or unknown) components.

though there is general agreement that there are two additional Fe-S centers present, there is still uncertainty about their EPR characteristics and midpoint potentials (4,6,10). Center 3 is defined as the center that, on reduction, exhibits a resonance at g=2.1; it has a relatively high midpoint potential. The position of the other resonances is uncertain (4,6). Simulations of EPR spectra have suggested that the additional resonances of center 3 are at g=1.94 and 1.88 and those of center 4 at g=2.04, 1.93 and 1.86 (4). This assignment, however, is hard to reconcile with the midpoint potential values reported for center 3 (-240 mVolt) and 4 in Complex I (-400 mVolt), since the line at g=1.86 does not behave as expected of a resonance belonging to a Fe-S center of low midpoint potential (10). It has been reported that there are two additional Fe-S centers 5 and 6 (6). However, the existence or quantitative significance of these centers has become very doubtful from several lines of evidence (4,5). At very low temperatures (1.5 to 2.6°K) and low microwave powers the signals attributed to these centers cannot be detected.

## 3) Succinate Dehydrogenase

There are 2 or 3 Fe-S centers associated with the succinate dehydrogenase system (7,9,14,15). Two different Fe-S centers of the ferredoxin type with similar EPR spectra but widely differing midpoint potentials can be observed. Center 1 ($E_{m,7}$ = +30 mVolt) is reducible by succinate, whereas center 2 ($E_{m,7}$ = ≤260 mVolt) is only reduced by dithionite. In addition, a high potential type of Fe-S center is associated with succinate dehydrogenase. This center is not observed by EPR in most solubilized succinate dehydrogenase preparations, even if no iron and labile sulfur is lost on solubilization. In succinate ubiquinone reductase (Complex II) Center 1 and the Hipip type center occur at an approximate 1:1 stoichiometry with bound flavin, whereas there is always less of center 2. Since center 2 is not reduced by succinate and in most preparations appears to occur in amounts considerably less than that of the bound flavin, its significance is questionable.

## 4) ETF Dehydrogenase

This Fe-S flavoprotein was not known until very recently, although the low field line of its EPR signal had been recognized for some time (11,16). The EPR signal is only seen below ∿40°K (13). This protein contains acid labile FAD and four iron and labile sulfur atoms per flavin in a single ferredoxin type Fe-S center. It is found in mitochondria as well as in submitochondrial particles and has to be solubilized by detergents. This indicates that the protein is bound to the membrane. It is reduced with a half-time of ∿ 5 msec by the reduced form of the electron transferring flavoprotein (ETF) (17,18) of the fatty acyl CoA dehydrogenase system and presumably represents the link of this pathway to the electron transfer system.

### 5) Fe-S Center of the Cytochrome $bc_1$ Segment

This protein has been known for over 10 years (19) since the broad EPR signal of its reduced form is readily detected at liquid nitrogen temperature. It has one of the highest midpoint potentials observed for any compound of the ferredoxin type, viz. +280 mVolt. This value is independent of pH (20). The protein is located on the oxygen side of the antimycin block. It is of interest that similar ferredoxin type proteins of high midpoint potential have been found in photosynthetic bacteria and chloroplasts (21, 22).

### 6) Fe-S Proteins Solubilized on Sonication of Mitochondria.

In addition to the centers or proteins mentioned above or shown in Scheme 1 there are additional Fe-S centers present in mitochondria. Two of those that seem to be present in quantities of the order of the components discussed above are a high potential Fe-S protein (12), apparently different from that associated with the succinate dehydrogenase system (12,14,23), and a ferredoxin type Fe-S protein occurring in the outer mitochondrial membrane (2,3,24). The latter protein is found as a contaminant in submitochondrial particles, indicating that these particles, as routinely prepared, contain appreciable amounts of outer membrane fragments (3). When in the membrane, the protein is reduced by NADH and NADPH, a property which is lost on purification of the protein.* There is no information on the function of either one of these proteins. The Hipip type protein can only be detected by EPR at $\leq 30°K$ whereas the outer membrane protein is readily observed at liquid nitrogen temperature.

## III. SOME ASPECTS OF THE PROPERTIES AND FUNCTION OF IRON-SULFUR CENTERS OF THE ELECTRON TRANSFER SYSTEM

Now we will return to a discussion of the scheme given above and some of the implications of the information contained in it and in the descriptive section just given.

---

*Personal communication from Dr. Bäckström.

### 1) Clustering of Iron-sulfur Centers, or Iron-sulfur Centers and Flavin, and Electron Transfer

To the first point made above (Section II, 1), the following considerations are pertinent: Single Fe-S centers are in general one electron acceptors and donors--if we neglect here their ability to undergo two electron changes under specific conditions (see III,3)--as are the cytochromes. However, all Fe-S centers located at the substrate side of the system occur in groups of several centers and/or are associated with a flavin prosthetic group. While in the region of the system, where the b and c cytochromes occur, only transfer of single electrons is required, at the substrate side electrons are fed in in pairs from the substrates. As a two electron acceptor of the quinoid type, flavin by itself has the capability of "splitting" electron pairs by shuttling appropriately between its three possible oxidation states, but the provision by nature of Fe-S centers at every now known entrance point of electrons from substrate into the electron transfer system would suggest that the splitting of electron pairs and introduction of single electrons into the system can be more effectively accomplished by Fe-S centers in conjunction with flavin. This seems to be particularly well exemplified by the mode of connection of the fatty acyl dehydrogenation system to the cytochrome chain. There is a relay system of three flavoproteins involved: One dealing directly with substrate, one transferring electrons between this substrate linked dehydrogenase and the next carrier in the row, namely the third flavoprotein, which contains a Fe-S center and which seems to connect into the cytochrome b-Q region (cf. Scheme 1).

Since in NADH and succinate dehydrogenases at least, there are Fe-S centers of varied midpoint potentials available, one can visualize that these may be useful in tuning in exactly at the potential levels required between the respective donor substrate and the flavin acceptor system with which they communicate, namely either the flavoquinone/flavosemiquinone or the flavosemiquinone/flavohydroquinone couple, which may have very different midpoint potentials. Some of the multiple centers may, of course, also be expected to be tunable to the ultimate acceptor of higher potential such as, e.g., Q. The same considerations apply to this quinoid system. By rapid kinetic studies it has thus far not been possible to observe differences in the rates of reduction of the Fe-S centers and the flavin. In the case of milk xanthine oxidase, for which the most extensive data are available, it was concluded that the electron carriers in a single active site are in very rapid equilibrium and that the Fe-S centers, by exchanging electrons forth and back, presumably serve the purpose of restoring in these groups (molybdenum and flavin in xanthine oxidase) which interact directly with the primary reductant (substrate) or oxidant ($O_2$), the oxidation state required for most efficient interaction with substrate or oxygen (25).

### 2) Iron-sulfur Centers Versus Cytochromes, Their Quantitative Distribution

To the second point above, (Section II, 1), it is useful to add some quantitative information. Although by species the Fe-S centers outnumber the cytochromes, their total quantity and hence total capacity for electron uptake is less than that of the cytochromes. From analytical data on acid labile FMN and FAD and on covalently bound FAD, which give us an estimate of NADH-, ETF- and succinate dehydrogenases, respectively, and from the information on the number of Fe-S centers present, as discussed above, we can approximate the total capacity (26) for electron uptake of all Fe-S centers together. Within the uncertainties mentioned above, we have 4 to 5 (NADH dehydrogenase) + 2 to 3 (succinate dehydrogenase) + 1 (ETF dehydrogenase) + 1 (Rieske-Fe-S protein), i.e., maximally 10 and minimally 8 Fe centers. If we take for sonicated particles (ETP) labile FMN as 0.15, labile FAD as 0.1, and covalent FAD as 0.2 nmole per mg of protein and the Rieske protein as stoichiometric to $\underline{c_1}$ and multiply with the numbers of Fe-S centers associated with these species we arrive at:

$$0.15x4 + 0.1x1 + 0.2x2 + 0.4x1 \ \mu\text{moles per mg protein} = 1.5 \text{ nmoles per mg protein}$$

Not counting cytochrome $\underline{c}$ since these data are for submitochondrial particles, from which variable quantities of cytochrome $\underline{c}$ have been lost, we arrive at this count for cytochrome $\underline{a}$, $\underline{a_3}$, $\underline{b}$ and $\underline{c_1}$:

$$0.8 + 0.8 + 0.8 + 0.4 \text{ nmoles per mg protein} = 2.8 \text{ nmoles per mg protein}$$

### 3) Oxidation States and Oxidation-reduction Midpoint Potentials of Iron-sulfur Centers

Other aspects which invite a comparison between the Fe-S proteins and the cytochromes are the span of midpoint potentials covered by both classes of compounds and their capability of one-vs. two-electron transfer. Not even considering the Fe-S proteins of plant origin, which are known to have very low midpoint potentials ($\lesssim$-500 mVolt), we find that the Fe-S centers of the mitochondrial electron transfer system span a range of -400 to + 300 mVolt, at least 200 mVolt more than what we know for the cytochromes of the same system. In addition, Fe-S proteins have the intrinsic capability of occurring in three oxidation states (27), namely that of the high potential Fe-S proteins, such as Hipip of Chromatium (28), that of oxidized ferredoxins and that of reduced ferredoxins. Although this has been verified for model systems (29) and by subjecting natural Fe-S proteins to special conditions (30), there is no evidence to date that the Fe-S centers of the electron transfer system make use of this capability. Reversible interconversions between the three possible oxidations states have been observed with bacterial Fe-S proteins (31-33), although usually only a partial, not a stoichiometric, conversion can be achieved. However, the possibility of a sequential two electron transfer exists and has to be kept in mind in future research on Fe-S centers, particularly, since high potential type Fe-S centers do occur in mitochondria in addition to the ferredoxin types. It has been shown that it is constraint by the protein that determines which oxidations states are accessible (30). Thus, it seems possible that through changes in conformation via external influences, higher or lower oxidation states may come into play.

## 4) Rates of Oxidation-reduction of Iron-sulfur Centers

Although initially there was some doubt whether the Fe-S centers were indeed reacting at rates which would justify their inclusion in the main path of electron transfer, all kinetic studies on rates of reduction or reoxidation of Fe-S centers which have been done to date have shown that these reaction rates lie in the same range as those of the flavin and cytochrome components. An interesting observation is that the reaction rates of the individual Fe-S centers of any single system, such as NADH or succinate dehydrogenase, are of very similar magnitude. On rapid mixing with NADH, the order of reduction of centers 1, 2 and 3 of NADH dehydrogenase is that seen during titration (10). On reoxidation with Q-1 the order is the reverse (5). Thus, it appears that what we observe is merely a rapid titration. This would mean that the electron distribution within any single active center of the dehydrogenase is very rapid so that the rates we see are merely governed by the oxidation-reduction potentials of the components and the rates of entry or exit of electrons from the active center. As mentioned above, this conclusion was arrived at with the enzyme xanthine oxidase on the basis of much more extensive data (25) and may well apply more generally to active centers incorporating multiple Fe-S and flavin groups.

### 5) Involvement of Iron-sulfur Centers in Energy Conservation

NADH dehydrogenase with its multiple Fe-S centers of different midpoint potentials exemplifies yet another significant feature of electron transfer, particularly with a view toward coupling of energy production to electron transfer. The midpoint potential of center 1 of NADH dehydrogenase is very close to that of the reductant, viz. the NADH/NAD couple, whereas the midpoint potential of center 2 is very close to that of the natural oxidant, ubiquinone, so that electrons may enter and exit the dehydrogenase at an equipotential level. Thus the oxidative step proper takes place within the structure of the enzyme over a span of 300 to 350 mVolt. There have been suggestions that certain Fe-S centers of the dehydrogenase may be specifically involved in oxidative phosphorylation at Site 1 (34). Work with *Candida utilis* has shown that a complete NADH dehydrogenase with all Fe-S centers is needed for phosphorylation at Site 1 to occur, but no particular component could be singled out as being the site or key factor in energy conservation (35). Also, a number of observations that shifts of midpoint potentials occur on addition of ATP have remained contradictory and are not readily explained to date (26).

## IV. IMPLICATIONS FOR THE CONSTITUTION OF THE MITOCHONDRIAL ELECTRON TRANSFER SYSTEM.

Despite much uncertainty which remains concerning the number, exact location and function of Fe-S proteins and centers of the electron transfer system--which incidentally is equally true for the cytochromes--the discovery of the multiple Fe-S centers of the electron transfer system and what we know to date about their reaction capabilities, calls for a careful reappraisal of traditionally held concepts of the structure and function of the electron transfer system and of structural and functional relationships between its components.

In view of the number and variety of components the idea of a linear "chain" of all components becomes unrealistic. One is rather inclined to visualize a network composed of domains, within which electron transfer may be very rapid ($<1$ ms), but which among themselves communicate at rates in the range of those generally measured for electron transfer ($\sim 1$ to 100 ms) in mitochondria. It is well possible that in some instances these domains coincide with what we now call the "complexes" of the respiratory chain. There may then still be something akin to a linear sequence namely a sequence of domains, rather than individual carriers. This picture, of course, comes very close to the often invoked one of components "on the main chain" and components on more or less rapidly "equilibrating sidepaths". However, the picture of components "on the main chain" and "on sidepaths" in my opinion does not properly express the intimacy of connection between these two types of components, namely that, as in NADH or succinate dehydrogenase, they are all part of one and the same structure and that electron transfer between them is for most purposes instantaneous. It is quite reasonable to assume that only one specific component of each domain--or if it were, "complex"--communicates directly with other domains or complexes, and these "gate keepers" would then be the ones considered "on the main chain" according to previous terminology.

## V. OUTLOOK

The most obvious task ahead in the area under discussion is a more thorough definition of a number of the less well explored Fe-S centers. This must include information on the concentration, at which they occur, their association with other components and their function. In addition, of course, the search for new, additional Fe-S components will continue and we must keep in mind that EPR spectroscopy may detect only a fraction of those present. If Fe-S centers should occur in tightly packed clusters, their resonances may become undetectable through mutual interactions. Bacterial nitrogenase, where under most conditions the majority of centers is undetectable by EPR, constitutes an example of this phenomenon. Nevertheless, reaction conditions may be found, under which one or the other center may become detectable, e.g., when one of the interacting partners undergoes a reaction and thus changes its state.

While strong interactions of paramagnets will render them undetectable by EPR, weak interactions will not do so, but will rather lead to additional resonances which may be exploited for pinpointing such weak interactions. It will be a foreseeable development that, as the primary resonances from the electron carriers of the electron transfer system become established, minor, thus far neglected features in the spectra will be studied which may, in a number of instances, signal such weak interactions. Examples of this are already at hand in the interaction of ubisemiquinone with an unknown paramagnetic species (36) and of the Fe-S centers of succinate dehydrogenase (37). Analogous examples from soluble Fe-S proteins,viz., clostridial ferredoxin (38) and xanthine oxidase are also known (39).

Finally, the exact location in or on the membrane of the various Fe-S centers of the mitochondrion is or is likely to become a topic of great concern, particularly in view of contemporary concepts of the relationship of electron and ion transport and energy conservation.

## REFERENCES


1. "Iron-Sulfur Proteins", Vol. 1, 1973; Vol. II, 1973; Vol. III, in press. (W. Lovenberg, ed.), Academic Press, New York.

2. S.P.J. Albracht, Biochim. Biophys. Acta 347 (1974), 183-192.

3. S.P.J. Albracht and H.-G. Heidrich, Biochim. Biophys. Acta 376 (1975), 231-236.

4. S.P.J. Albracht and G. Dooijewaard (1976). In "Electron Transfer Chains and Oxidative Phosphorylation", (E. Quagliarello, et al., ed.) North Holland Publishing Co., Amsterdam, in press.

5. H. Beinert and F. J. Ruzicka (1976). In "Electron Transfer Chains and Oxidative Phosphorylation", (E. Quagliarello, et al. ed.) North Holland Publishing Co., Amsterdam, in press.

6. T. Ohnishi, Biochem. Biophys. Acta 387 (1975), 475-490.

7. T. Ohnishi, D. B. Winter, J. Lim, and T. E. King, Biochem. Biophys. Res. Commun. 53 (1973), 231-237.

8. T. Ohnishi, J. S. Leigh, C. I. Ragan, and E. Racker, Biochem. Biophys. Res. Commun. 56 (1974), 775-782.

9. T. Ohnishi, D. Winter, J. Lim, and T. E. King, Biochem. Biophys. Res. Commun. 61 (1974), 1017-1025.

10. N. R. Orme-Johnson, R. E. Hansen, and H. Beinert, J. Biol. Chem. 249 (1974), 1922-1927.

11. N. R. Orme-Johnson, R. E. Hansen, and H. Beinert, J. Biol. Chem. 249 (1974), 1928-1939.

12. F. J. Ruzicka and H. Beinert, Biochem. Biophys. Res. Commun. 58 (1974), 556-563.

13. F. J. Ruzicka and H. Beinert, Biochem. Biophys. Res. Commun. 66 (1975), 622-631.

14. H. Beinert, B.A.C. Ackrell, E. B. Kearney, and T. P. Singer, Biochem. Biophys. Res. Commun. 58 (1974), 564-572.

15. H. Beinert, B.A.C. Ackrell, E. B. Kearney, and T. P. Singer, Eur. J. Biochem. 54 (1975), 185-194.

16. T. Ohnishi, D. F. Wilson, T. Asakura, and B. Chance, Biochem. Biophys. Res. Commun. 46 (1972), 1631-1638.

17. H. Beinert (1963). In "The Enzymes", Vol. VII, pgs. 467-476, Academic Press, Inc.

18. F. L. Crane, and H. Beinert, J. Biol. Chem. 218 (1956), 717-731.

19. J. S. Rieske, R. E. Hansen, and W. S. Zaugg, J. Biol. Chem. 239 (1964), 3017-3022.

20. J. S. Leigh, Jr., and M. Erecinska, Biochim. Biophys. Acta 387 (1975), 95-106.

21. R. Malkin and P. J. Aparicio, Biochem. Biophys. Res. Commun. 63 (1975), 1157-1160.

22. R. C. Prince, J. G. Lindsay, and P. L. Dutton, FEBS Letters 51 (1975), 108-111.

23. T. Ohnishi, W. J. Ingledew, and S. Shiraishi, Biochem. Biophys. Res. Commun. 63 (1975), 894-899.

24. D. Bäckström, I. Hoffström, I. Gustafsson, and A. Ehrenberg, Biochem. Biophys. Res. Commun. 53 (1973), 596-602.

25. J. S. Olson, D. P. Ballou, G. Palmer, and V. Massey, J. Biol. Chem. 249 (1974), 4363-4382.

26. H. Beinert, Iron-sulfur proteins, Vol. III, (W. Lovenberg, ed.), Academic Press, New York, in press.

27. C. W. Carter, Jr., J. Kraut, S. T. Freer, R. A. Alden, L. C. Sieker, E. Adman, and L. H. Jensen, Proc. Nat. Acad. Sci. U.S. 69 (1972), 3526-3529.

28. K. Dus, H. De Klerk, K. Sletten, and R. G. Bartsch, Biochem. Biophys. Acta 140 (1967), 291-311.

29. T. Herskovitz, B. A. Averill, R. H. Holm, J. A. Ibers, W. D. Phillips, and J. F. Weiher, Proc. Nat. Acad. Sci. U.S. 69 (1972), 2437-2441.

30. R. Cammack, Biochem. Biophys. Res. Commun. 54 (1973), 538-554.

31. D. F. Erbes, R. H. Burris, and W. H. Orme-Johnson, Proc. Nat. Acad. Sci. U.S. (1975), in press.

32. W. V. Sweeney and J. C. Rabinowitz, and D. C. Yoch, J. Biol. Chem. 250 (1975), 7842-7847.

33. D. C. Yoch, W. V. Sweeney, and D. I. Arnon, J. Biol. Chem. 250 (1975), 8330-8336.

34. T. Ohnishi, Biochem. Biophys. Acta 301 (1973), 105-128.

35. J. G. Cobley, S. Grossman, T. P. Singer, and H. Beinert, J. Biol. Chem. 250 (1975), 211-217.

36. F. J. Ruzicka, H. Beinert, K. L. Schepler, W. R. Dunham, and R. H. Sands, Proc. Nat. Acad. Sci. U.S., 72, (1975), 2886-2890.

37. T. Ohnishi, J. S. Leigh, C. I. Ragan, and E. Racker, Biochem. Biophys. Res. Commun. 56 (1974), 775-782.

38. R. Mathews, S. Charlton, R. H. Sands, and G. Palmer, J. Biol. Chem. 249 (1974), 4326-4328.

39. D. J. Lowe, R. M. Lynden-Bell, and R. C. Bray, Biochem. J. 130 (1972), 239-249.

# COMPOSITION AND ENZYMATIC PROPERTIES OF THE MITOCHONDRIAL NADH- AND NADPH-UBIQUINONE REDUCTASE (COMPLEX I)

Youssef Hatefi

Department of Biochemistry
Scripps Clinic and Research Foundation
La Jolla, California 92037

## COMPOSITION AND ENZYMATIC PROPERTIES OF COMPLEX I

Complex I is one of the five component enzyme complexes of the mitochondrial electron transport-oxidative phosphorylation system (Fig. 1) (1-3). The enzymatic reactions catalyzed by complex I are summarized in Table I and the composition of the preparation in terms of flavin, nonheme iron (Fe), acid-labile sulfide (S*), ubiquinone, and phospholipids is given in Table II. The resolution of complex I by chaotropic agents (4-6) results in solubilization of ∿ 20% protein containing flavin and approximately 50% of the Fe and S* content of complex I. The remainder of Fe and S* plus lipids are associated with the water-insoluble fraction (Fig. 2). Upon fractionation with ammonium sulfate, the soluble fraction yields an iron-sulfur protein with epr characteristics of FeS* center 2 of complex I (see below), plus a flavoprotein containing approximately 4 g atoms of Fe and 4 moles of S* per mole of FMN. The latter catalyzes the oxidation of NADH by various quinones and ferric compounds, including cytochrome *c* (Table III).

Polyacrylamide gel electrophoresis of complex I and its chaotrope-solubilized fractions in the presence of sodium dodecylsulfate and mercaptoethanol has revealed the presence of multiple polypeptides in complex I, ranging in molecular weight from 8,000 to 75,000 (Table IV) (9). A majority of the well-defined polypeptides appear

---

Abbreviations: Epr, electron paramagnetic resonance; Q, ubiquinone; $K_3$, 2-methylnaphthoquinone (menadione); DCIP, 2,6-dichloroindophenol; AcPyADH, reduced 3-acetylpyridine adenine dinucleotide; FeS1,2,3,4, iron-sulfur centers 1,2,3 and 4, respectively; DOCA, deoxycholate.

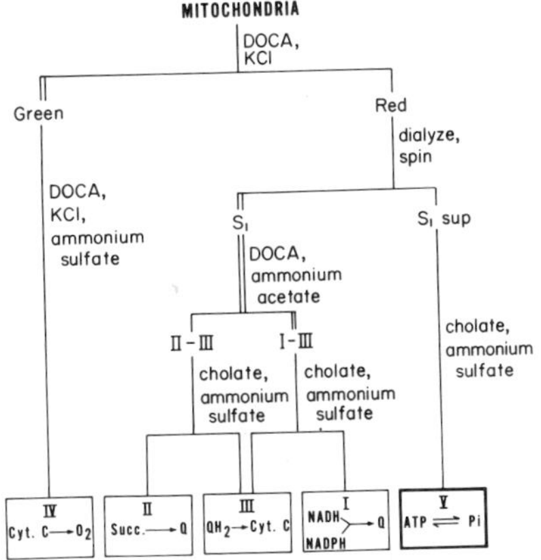


Fig. 1. Scheme showing the fractionation of beef-heart mitochondria into enzyme complexes I, II, III, IV and V (from ref. 2).

TABLE I

Enzymatic Properties of Complex I

1. NADH → Q, $K_3Fe(CN)_6$, NAD

2. NADPH → Q, $K_3Fe(CN)_6$, NAD

3. Q reduction is inhibited by rotenone, piericidin A, barbiturates, Demerol, mercurials, NADH > 0.2 mM, rhein (NADH competitive inhibitor)

4. $K_m^{NADH} = 7\ \mu M$, $K_m^{NADPH} = 0.5\text{-}1\ mM$

5. pH optima: NADH oxidation, pH 7
NADPH oxidation, pH < 6

6. Oxidative phosphorylation and energy-linked transhydrogenation (NADH → NADP) in presence of appropriate coupling factors

For details, see ref. 1, 7 and 14.

TABLE II

Composition of Complex I

| Component | Concentration (per mg protein) |
|---|---|
| FMN (acid extractable) | 1.4-1.5 nmoles |
| Nonheme iron | 23-26 ng atoms |
| Acid-labile sulfide | 23-26 nmoles |
| Ubiquinone-10 | 4.2-4.5 nmoles |
| Cytochromes | < 0.1 nmole |
| Lipids | 0.22 mg |

From Hatefi *et al.* (1).

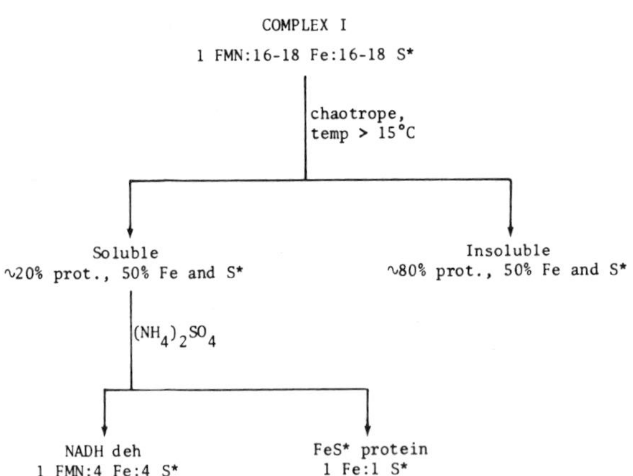


Fig. 2. Scheme showing the resolution of complex I in the presence of chaotropic agents.

TABLE III

Enzymatic Properties of NADH Dehydrogenase

1. NADH → Quinones, ferric compounds, NAD

2. Inhibited by mercurials, NADH > 0.2 mM in presence of 2-electron acceptors

3. $K_m^{NADH}$ → ferricyanide, cyt.*c* : 65 μM
   → Q, $K_3$, DCIP : 133 μM

For details, see ref. 7 and 8.

TABLE IV

Polypeptide Composition of Complex I

| Mol. wt. x 1000 | Chaotrope solubilized | FeS* Protein | NADH deh | Molar ratio |
|---|---|---|---|---|
| 75 | + | + | | 1 |
| 53 | + | + | | }2 |
| 53 | + | | + | |
| 42 | | | | }2 |
| 39 | | | | |
| 33 | | | | 1 |
| 29 | + | + | | 1 |
| 26 | + | | + | 1 |
| 25 | | | | |
| 23.5 | | | | |
| 22 | | | | |
| 20.5 | | | | |
| 18 | | | | |
| 15.5 | + | | | |
| 8 | | | | |

Courtesy: Dr. C.I. Ragan (9).

to occur in complex I in equimolar ratios. The soluble FeS* protein preparation contains 3 polypeptides, and the soluble NADH dehydrogenase appears to be composed of two subunits of molecular weights 53,000 and 26,000. The larger subunit of NADH dehydrogenase appears to be distinct from the 53,000 dalton component found in the soluble FeS* protein fraction (9).

Electron paramagnetic resonance studies on NADH-treated complex I at near liquid helium temperatures have revealed that complex I contains at least four major types of FeS* center, designated FeS* centers 1, 2, 3 and 4 (10-11). The epr signals for the latter two centers overlap. According to Orme-Johnson *et al.* (11), the low-field ($g_z$) signal of center 4 is about 0.003 g greater than that of center 3. The center 1 signal appears to be composed of two components, which have been designated FeS* centers 1a and 1b (12). At 4°, all centers are reduced with NADH within 6 msec, and on the basis of electron consumption the molarity of each of the centers 2, 3 and 4 appears to be comparable to that of complex I flavin (11). Titration of complex I with graded amounts of NADH suggested to Orme-Johnson *et al.* (10,11) that the order of the oxidation-reduction potentials ($E_m$) of these centers is $3 \geq 2 >> 4 > 1$. According to Ohnishi (12), the $E_m$ values at pH 7.2 are: center 1a, -380 ± 20 mV; center 1b, -240 ± 20 mV; center 2, -20 ± 20 mV; center 3, -240 ± 20 mV; center 4, -410 ± 20 mV. Other results are more in agreement with the conclusions of Orme-Johnson *et al.* regarding the relative $E_m$ values of centers 2 and 3.

## DIRECT NADPH OXIDATION BY COMPLEX I

Until recently, it was generally believed that the mitochondrial electron transport system is not capable of oxidizing NADPH directly, but can do so by way of transhydrogenation from NADPH to NAD as catalyzed by the membrane-bound transhydrogenase enzyme. Our studies showed that submitochondrial particles can oxidize NADPH at appreciable rates ($\geq$ 250 nmoles$\cdot$min$^{-1}\cdot$mg$^{-1}$ protein at pH 6.0 and 30°) in the absence of effective amounts of NAD (13,14). Both the NADPH dehydrogenase and the NADPH → NAD transhydrogenase activities fractionated mainly into complex I and exhibited similar responses to pH (Fig. 3) and to inhibition by palmitoyl coenzyme A (14,15). Epr studies on NADPH-treated complex I showed that centers 2 and 3 (Fig. 4 and 5) were reduced by NADPH, but center 1 was not detectably reduced (14). Light absorption of NADPH-treated complex I *minus* NADH-treated complex I showed a flavin-type spectrum, indicating that under the conditions used the flavin of complex I was reduced mainly by NADH (14). Thus, it appeared that at neutral pH electrons from NADPH were distributed mainly among the components of complex I which had potentials near that of ubiquinone (i.e., FeS* centers 2 and 3), while electrons from NADH reduced the flavin and the low-potential FeS* centers as well.

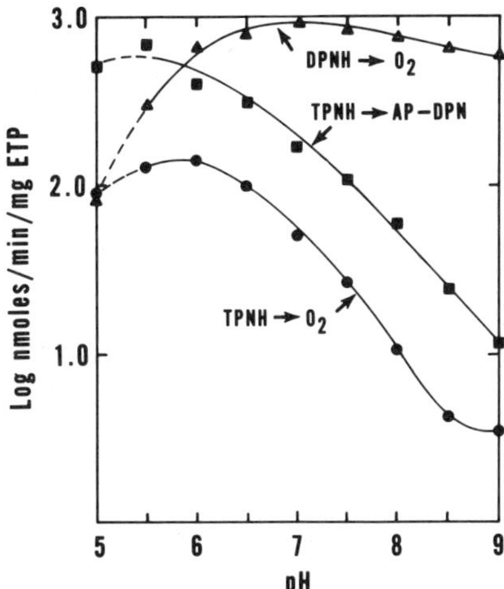


Fig. 3. pH dependence of NADH oxidase, NADPH oxidase and NADPH → NAD transhydrogenase activities of submitochondrial particles from beef heart. For details see ref. 14.

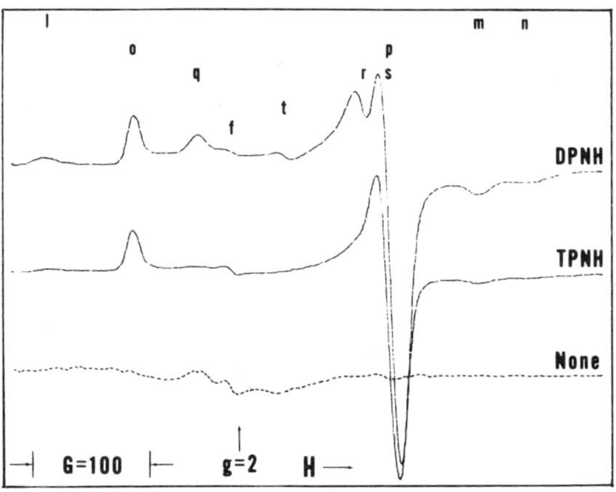


Fig. 4. First derivative epr spectra of complex I reduced with NADH and NADPH. FeS* centers 1 (q, r, s), 2 (o, p), and 3 + 4 (l, m, n). For details see ref. 14.

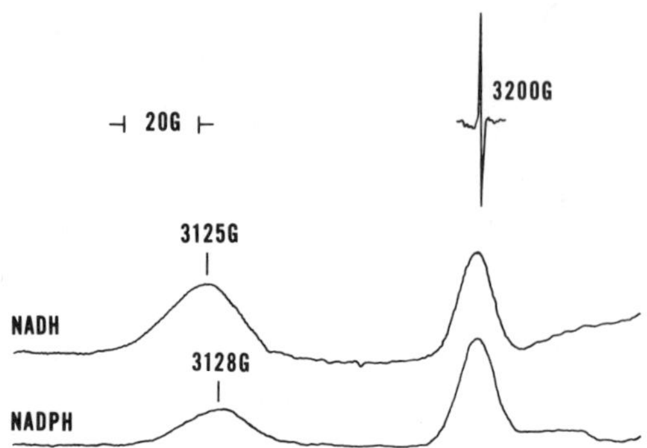


Fig. 5. First derivative epr spectra of NADH- and NADPH-treated complex I aligned against a marker at 3200 G to show the field positions of the g ≃ 2.1 signals of iron-sulfur centers 3 + 4. Conditions: 0.4 ml of complex I at 45 mg protein/ml of 0.25 M sucrose containing 50 mM Tris-HCl, pH 7.5. Where indicated 10 μl of 50 mM NADH or NADPH were added. Microwave frequency, 9.198 GHz; power, 2 mW; modulation, 4 G; temperature, 13°K. From ref. 16.

More recently, it was shown that treatment of submitochondrial particles or complex I at 0° with trypsin completely destroyed transhydrogenation from NADPH to NAD with little (∿ 10%) or no effect on the rates of NADH or NADPH oxidation (15). The trypsin effect was shown to be due probably to one or more arginyl residues at or near the active site of the transhydrogenase enzyme. Thus, treatment of submitochondrial particles with the specific arginine-binding reagent butanedione in the presence of borate buffer at pH 9.0 (Fig. 6) destroyed the NADPH → NAD transhydrogenase activity with little or no effect on NADH and NADPH oxidase activities (Fig. 7). This inhibition could be prevented when the particles were treated with appropriate amounts of NAD plus NADP prior to the addition of butanedione (15).

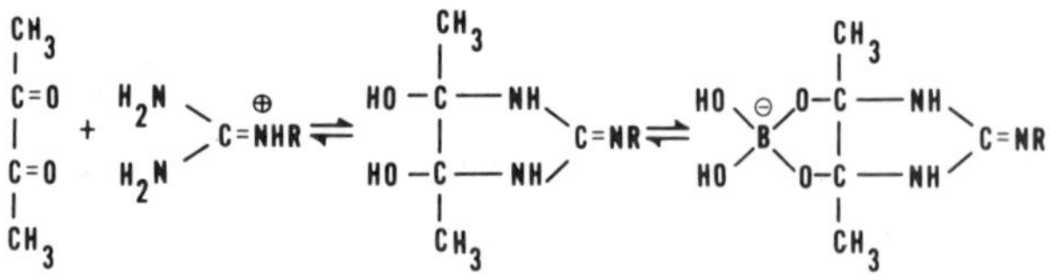

$$CH_3-C(=O)-C(=O)-CH_3 + (H_2N)_2C{=}NHR^{\oplus} \rightleftharpoons [HO-C(CH_3)-NH]_2{>}C{=}NR \rightleftharpoons (HO)_2B^{\ominus}[O-C(CH_3)-NH]_2{>}C{=}NR$$

Fig. 6. Interaction of butanedione with protein arginyl residues in the presence of borate buffer.

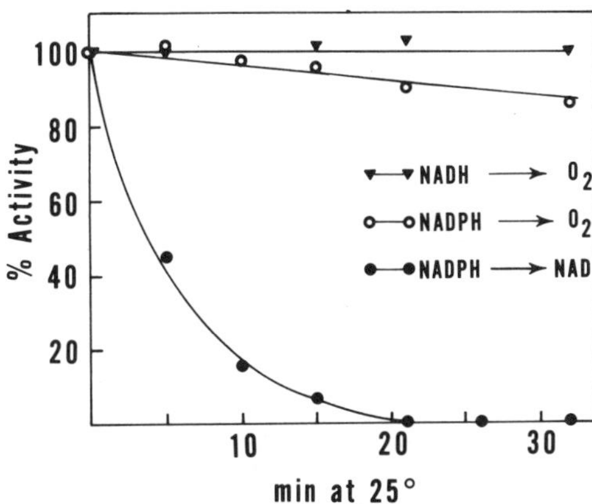


Fig. 7. Effect of preincubation of submitochondrial particles with 10 mM butanedione in the presence of pH 9.0 borate buffer on the oxidation rates of NADH, NADPH, and NADPH by NAD. For details see ref. 15.

Earlier rapid-freeze epr experiments on AcPyADH-treated complex I had shown that, contrary to expectation, the reduction sequence of the FeS* centers was not in accordance with their increasing reduction potentials. The reduction sequence was 2, 3 + 4, 1, i.e., the component with a potential near zero mV appeared to have been most rapidly reduced (10,11). Thus, it was deemed desirable to see whether under appropriate conditions FeS* center 1 could be reduced by NADPH now that trypsin treatment of complex I removed the complication of any transhydrogenase involvement due to a possible presence of traces of NAD either bound to complex I or added as a contaminant of NADPH. The results showed that, indeed, in the presence of rotenone and at pH 6.5 (which is more favorable than neutral pH for NADPH oxidation) FeS* center 1 could be partially reduced with NADPH (Fig. 8) (16). Control epr experiments also showed that treatment of complex I for 75-90 min at 0° with trypsin had no detectable effect on the reduction of FeS centers by either NADH or NADPH. The sequence of reduction of the iron-sulfur centers of complex I by NADPH is not known. The only conclusion that is admissible at this time is that regardless of whether the reductant is NADH, AcPyADH, or NADPH, the electrons appear to reach all the centers of complex I, possibly even FMN when NADPH is the reductant. The difference in the extent of the reduction of the centers might be a reflection of the rate of electron entry into complex I (NADH > AcPyADH >> NADPH at pH 7-8; NADH > AcPyADH ≥ NADPH at pH 6-6.5) and the slow leakage of electrons to oxygen under the conditions of epr experiments.

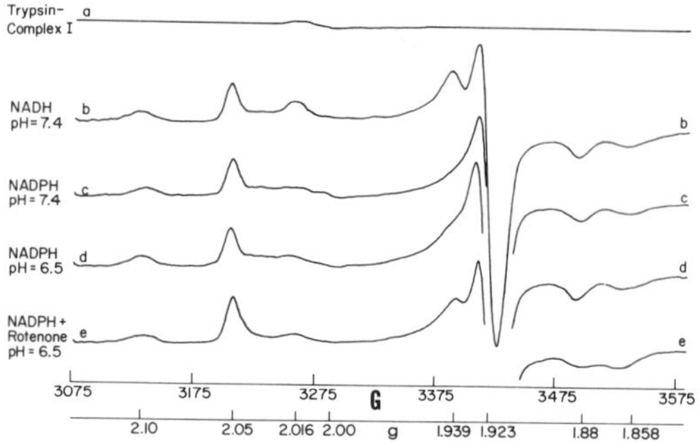


Fig. 8. First derivative epr spectra of trypsin-treated complex I reduced with NADH and NADPH at pH 7.4 and 6.5. Conditions: 0.35 ml of trypsin-treated complex I at 35 mg protein/ml of 0.25 M sucrose containing 100 mM potassium phosphate at the pH values shown; microwave frequency, 9.206 GHz; modulation, 5 G; power, 2 mW; gain, 10; and temperature, 13°K. Where indicated 20 μl of 30 mM NADH or NADPH and 10 μl of 5 mM rotenone in ethanol were added. The approximate g values for the signals of FeS* centers 1, 2, and 3 + 4 are respectively as follows: Center 1 at 2.02, 1.939, 1.923; center 2 at 2.05, 1.926, 1.922; and centers 3 + 4 at 2.1, 1.88, 1.86.

## ENERGY COUPLING PROPERTIES OF COMPLEX I

It was shown by Schatz and Racker (17) that submitochondrial particles could synthesize ATP at the expense of NADH oxidation by added ubiquinone-1. Subsequently Ragan and Racker (18) demonstrated that under appropriate conditions a preparation from mitochondria, designated hydrophobic proteins, could catalyze oxidative phosphorylation (P/O < 0.5) linked to electron transfer from NADH to ubiquinone. As compared to submitochondrial particles, this preparation was essentially free of cytochrome oxidase and had a low NADH-ubiquinone reductase activity. Addition of complex I to the hydrophobic proteins preparation increased the rates of oxidation (NADH → ubiquinone-1) and phosphorylation (but not the P/O ratio). A similar tactic was used by Ragan and Widger (19) to demonstrate energy-linked transhydrogenation by complex I in the presence of hydrophobic proteins, phospholipids and ATP. Our studies showed that phosphorylating submitochondrial particles yield comparable P/O values of 2.4 to 2.9 with either NADH or NADPH as substrate. Thus it appears that complex I is capable of (a) energy conservation at the expense of either NADH or NADPH oxidation; and (b) energy transfer to the mitochondrial ATP synthesizing system (e.g., complex V). Also, in the presence of appropriate factors complex I can transfer energy from ATP to the transhydrogenase enzyme system which resides in complex I itself.

Gutman *et al.* (20) have shown that in piericidin A-blocked submitochondrial particles, addition of ATP results in uncoupler-sensitive reverse electron transfer from FeS* center 2 to NAD. These results have suggested that coupling site 1 in complex I or the comparable region of the respiratory chain is located on the substrate side of FeS* center 2. As stated above, complex I appears to contain two FeS* centers with $E_m$ values near that of FMN and the nicotinamide nucleotides (-310 mV), and two FeS* centers with $E_m$ values near that of ubiquinone. The increase in $E_m$ between these two regions is roughly about 300 mV, which for a two-electron transfer system corresponds to $\Delta G^{\circ\prime} \approx 14$ kcal, a quantity sufficient for the synthesis of 1 mole of ATP. Thermodynamically, therefore, coupling site 1 in complex I appears to correspond to the step concerned with electron transfer from FMN, $FeS_1^*$ and $FeS_4^*$ to $FeS_2^*$, $FeS_3^*$ and ubiquinone (Fig. 9). An obvious problem however, is that the FeS* centers are single-electron transfer entities, and single electron transfer over a span of 300 mV would not yield enough energy for ATP synthesis. Since the existence of a device for storage and summation of energy quanta is not very likely, the above considerations suggest that electron transfer between the low- and high-potential components of complex I is paired. This would be possible if the FeS* proteins across the potential gap contained more than one FeS* center and acted as two-electron oxidation-reduction systems.

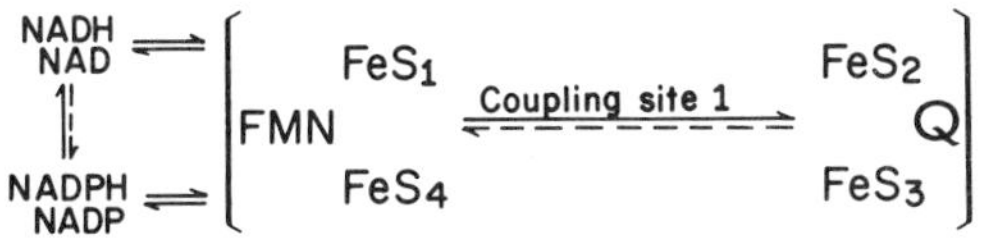

Fig. 9. Scheme showing the possible arrangement of complex I components relative to (a) the sites of NADH and NADPH oxidation, and (b) coupling site 1. Dashed arrows indicate energy-requiring reactions.

## Acknowledgements

The author wishes to thank Dr. I.C. Ragan for providing a preprint of the manuscript from which the data of Table IV were extracted. The epr studies were performed in collaboration with Dr. A.J. Bearden, Donner Laboratory, University of California, Berkeley. The work of the author's laboratory was supported by grants USPHS AM08126 and CA13609 and NSF GB-43470.

## REFERENCES


1. Hatefi, Y., Haavik, A.G., and Griffiths, D.E. (1962) *J. Biol. Chem. 237,* 1676-1680.
2. Hatefi, Y., Hanstein, W.G., Galante, Y., and Stiggall, D.L. (1975) *Fed. Proc. 34,* 1699-1706.
3. Hatefi, Y., and Stiggall, D.L. (1976) in *The Enzymes* (Boyer, P.D., ed.) Vol. 13, pp. 175-297.
4. Hatefi, Y., and Stempel, K.E. (1967) *Biochem. Biophys. Res. Commun. 3,* 301-308.
5. Davis, K.A., and Hatefi, Y. (1969) *Biochemistry 8,* 3355-3361.
6. Hatefi, Y., and Hanstein, W.G. (1974) in *Methods in Enzymol.* (Fleischer, S., and Packer, L., eds.) Vol. 13, pp. 770-790.
7. Hatefi, Y., and Stempel, K.E. (1969) *J. Biol. Chem. 244,* 2350-2357.
8. Hatefi, Y., Stempel, K.E., and Hanstein, W.G. (1969) *J. Biol. Chem. 244,* 2358-2365.
9. Ragan, C.I., *Biochem. J.,* in press.
10. Orme-Johnson, N.R., Orme-Johnson, W.H., Hansen, R.E., Beinert, H., and Hatefi, Y. (1971) *Biochem. Biophys. Res. Commun. 44,* 446-452.
11. Orme-Johnson, N.R., Hansen, R.E., and Beinert, H. (1974) *J. Biol. Chem. 249,* 1922-1927.
12. Ohnishi, T. (1975) *Biochim. Biophys. Acta 387,* 475-490.
13. Hatefi, Y. (1973) *Biochem. Biophys. Res. Commun. 50,* 978-983.
14. Hatefi, Y., and Hanstein, W.G. (1973) *Biochemistry 12,* 3515-3522.
15. Djavadi-Ohaniance, L., and Hatefi, Y. (1975) *J. Biol. Chem. 250,* 9397-9403.
16. Hatefi, Y., and Bearden, A.J., submitted.
17. Schatz, G., and Racker, E. (1966) *J. Biol. Chem. 241,* 1429-1438.
18. Ragan, C.I., and Racker, E. (1973) *J. Biol. Chem. 248,* 2563-2569.
19. Ragan, C.I., and Widger (1975) *Biochem. Biophys. Res. Commun. 62,* 744-749.
20. Gutman, M., Singer, T.P., and Beinert, H. (1972) *Biochemistry 11,* 556-562.

# FACTORS CONTROLLING THE TURNOVER NUMBER OF SUCCINATE DEHYDROGENASE: A NEW LOOK AT AN OLD PROBLEM

B.A.C. Ackrell, E.B. Kearney, P. Mowery,
and T.P. Singer

*Molecular Biology Division, Veterans Administration Hospital, San Francisco, CA 94121 and Department of Biochemistry and Biophysics, University of California, San Francisco, CA 94143*

H. Beinert

*Institute of Enzyme Research, University of Wisconsin, Madison, WI 53706*

A.D. Vinogradov

*Department of Animal Biochemistry, Moscow State University, Moscow, USSR*

and

G.A. White

*Agriculture Canada, London, N6A 5B7, Ontario, Canada*

Identification of the rate-limiting step in the catalytic cycle of succinate dehydrogenase and the related problem of defining the factors which determine its turnover number have for a long time occupied the interest of many investigators. During the past two years a collaboration among several laboratories has been initiated in order to use a variety of experimental approaches to these problems. The present report summarizes the progress which has been achieved.

## TURNOVER NUMBERS OF SUCCINATE DEHYDROGENASE IN DIFFERENT PREPARATIONS FROM HEART

It has been known for many years that the turnover number of the enzyme (usually expressed per mole of covalently bound flavin) is much higher in mitochondria and submitochondrial particles like ETP than in simpler membrane fragments, e.g., Complex II, or in soluble enzyme preparations. We have in the past adopted as standard of reference for full activity the turnover number obtained in the PMS-DCIP assay at 38° at infinite PMS concentration (1). In this assay the turnover number of ETP was found to be 18,000 ± 1,000 for several years (2). More recent values obtained for ETP are somewhat higher (20,000 ± 1,000), perhaps because the "batch activation" with nitrate (3) we now use is superior to activation with succinate in the assay cuvettes previously used (1). The reduction of DCIP can also be mediated in membrane preparations by $Q_1$ or the Q analog, DPB, instead of PMS, with equal activity. In fresh soluble enzyme preparations, extracted anaerobically in the presence of succinate, virtually the same turnover number is obtained in the reduction of either PMS or ferricyanide at $V_{max}$ with respect to the oxidant, if the rate of reduction of ferricyanide is extrapolated from measurements at very low dye concentrations (4,5). This type of ferricyanide reduction is characterized by a low $K_m$ (< 200 μM) and is distinct from the reaction observed at higher ferricyanide concentrations. It is highly labile, and disappears rapidly on exposure of the enzyme to air. This reaction is not observed in membranes, possibly because the reaction site is buried in the membrane where ferricyanide cannot penetrate.

The turnover numbers of the different succinate dehydrogenase preparations used in this study are compared in Table I. In addition to ETP, the reference material, two other membrane preparations are included: the Keilin-Hartree preparation and Complex II, for which turnover numbers of 10,000 to 12,000 have been reported (6,7); certain samples of Complex II isolated by the authors have given values as high as 14,000. The values given for soluble preparations are those obtained immediately after rapid isolation by procedures designed to avoid inactivation. It may be seen that the turnover number of such soluble preparations is the same (or nearly the same) as that of the particle from which they are extracted. Thus, extraction of a Complex II sample of 10,000 turnover number yields an SDB with the same turnover number (7); Keilin-Hartree preparations (turnover number ∿ 10,000 to 13,000) extracted with butanol yield an enzyme of 8-10,000 turnover number.

[1] Abbreviations: Q, coenzyme Q; DCIP, 2,6-dichlorophenolindophenol; DPB, 2,3-dimethyoxy-5-methyl-6-pentyl-1,4-benzoquinone; PMS, phenazine methosulfate; TTF, thenoyltrifluoroacetone.

TABLE I

Turnover Numbers of Succinate Dehydrogenase from Beef Heart

| Preparation | Turnover Number* |
|---|---|
| ETP particles | 20,000 ± 1,000 |
| Keilin-Hartree particles (K-H) | 10,000 - 13,000 |
| Complex II particles | 12,000 ± 2,000 |
| Davis-Hatefi SDB | ∿ 10,000 |
| Butanol-extracted, reconstitutively active, soluble enzyme from K-H | 8,000 - 10,000 |
| Butanol-extracted, reconstitutively active, soluble enzyme from ETP | 13,500 ± 500 |

*Moles succinate ox./min./mole of covalently bound flavin at 38° in the PMS-DCIP assay at $V_{max}$.

It is important to note that the same butanol extraction procedure applied to ETP yields a higher turnover number (13,000 to 14,000), which is, however, lower than in ETP itself (Table I).

It should be emphasized that the fall in turnover number which accompanies solubilization of the enzyme from the membrane is clearly distinct from inactivation of the enzyme, which can occur on exposure of soluble preparations to air. The main question we wish to consider in this paper is the reason for these differences in turnover number.

Two alternative explanations will be considered. The first is that these differences are due to inactivation of the enzyme. Thus, a particulate or soluble sample of succinate dehydrogenase with a turnover number of 10,000 could be a mixture of approximately equal amounts of active and dead enzyme, with both forms contributing to the flavin content of the preparation on which the turnover number is based. A variant of this idea is that a whole spectrum of enzyme molecules may be present, with varying degrees of catalytic activity, and 10,000 represents their average turnover number.

The second alternative is that the environment of the intact membrane is essential for the high turnover number (∿ 20,000), and that the decline in turnover number to ∿ 10,000 on solubilization or on processing the membrane to simpler forms such as Complex II could be the result of removing the catalytically competent enzyme from its lipid milieu. The notion that some membrane component, perhaps $Q_{10}$, is essential for full activity is consistent with the facts that soluble preparations of the enzyme are lipid-free, and that Complex II, which also has a low turnover number, is severely depleted with respect to $Q_{10}$. As discussed below, it has in fact been suggested that $Q_{10}$ modulates succinate dehydrogenase activity (8) or that it serves as a second (alternate or additional) reaction site for PMS, increasing the rate of reoxidation of the enzyme (9). The low turnover number of Keilin-Hartree preparations (Table I) is not readily explainable in terms of this hypothesis, since the isolation of these particles does not involve procedures which are apt to deplete their lipid content. In this type of preparation it is possible that some inactivated enzyme is also present; this is in agreement with the finding that soluble enzyme prepared by butanol extraction from Keilin-Hartree preparations has a lower turnover number than that extracted from ETP.

## ACTION OF CYANIDE, TTF, CARBOXIN, AND OF Q DEPLETION

Many years ago Giuditta and Singer (10) reported that the incubation of Keilin-Hartree preparations with cyanide results not only in complete inactivation of electron transport from succinate dehydrogenase to the respiratory chain but also leads to about 50% loss of succinate-PMS activity. The catalytic activity of soluble preparations was unaffected by cyanide. Since the turnover number of the soluble enzyme, isolated by procedures available at that time, was close to that of cyanide-treated Keilin-Hartree particles, it was suggested that there may be two reaction sites for PMS in membranes, one of which is inactivated or lost on extraction and on incubation with cyanide. The target of cyanide was proposed to be a nonheme iron component of succinate dehydrogenase (10).

The action of cyanide on ETP is illustrated in the left side of Fig. 1. The cyanide-treated sample shows 50% loss of succinate dehydrogenase activity at $V_{max}$ with respect to PMS. TTF, an iron chelator, has long been known to inhibit electron transport between succinate dehydrogenase and Q (11). The effect of TTF on succinate-PMS activity is indistinguishable from the action of cyanide (8): in ETP 50% of the activity is inhibited, but soluble preparations are not affected. The effect of TTF can be produced also by a class of compounds called carboxins, first shown to be inhibitors of succinate oxidation in the corn smut *Ustilago* (12).

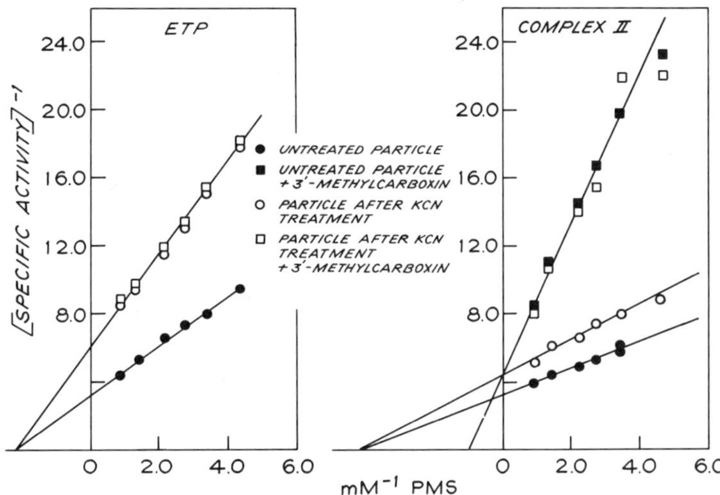


Fig. 1 Effect of 3-methylcarboxin and of incubation with cyanide on PMS-DCIP activity. Left, ETP, turnover number = 19,000 at 38°; right, Complex II, turnover number = 13,200. Incubation with cyanide (0.02 M) was for 1 hr at 30°, pH 8.0.

As expected from the similarity of their effects, the action of cyanide and TTF or carboxin on the mammalian system are not additive; i.e. cyanide-treated particles are not inhibited further by TTF or carboxin. These facts are illustrated on the left side of Fig. 1 where 3-methylcarboxin is the inhibitor used. We have found $I^{50}$ values between $10^{-6}$ and $10^{-7}$ M for several carboxins in ETP preparations from beef heart (Table II).

A few years ago Rossi et al. (8) reported that on extraction of $Q_{10}$ from inner membranes 50% of the succinate-PMS activity is lost, along with all succinoxidase activity, and that on reincorporation of $Q_{10}$ in the extracted membrane these activities are restored. The Q-depleted preparations, showing decreased succinate dehydrogenase activity, are not inactivated by cyanide or TTF. These effects are illustrated in Fig. 2 for an ETP preparation from our own studies.

The fact that the reincorporation of $Q_{10}$, without adding other lipids, restores the full succinate dehydrogenase activity to pentane-extracted preparations implicates $Q_{10}$ as the membrane component required for the high turnover number characteristic of inner membrane preparations. One interpretation of the effect of the reversible removal of $Q_{10}$ on succinate-PMS activity has been that endogenous $Q_{10}$ is the reaction site of PMS in the membrane and that the higher activity results from more rapid reoxidation of

TABLE II

Inhibition of Succinate-PMS Reaction by 1,4 Oxathiin Derivatives in ETP

| Inhibitor | $I^{50}$ (μM)* |
|---|---|
| 3'-Octoxycarboxin | 0.02 |
| 4'-Benzoylcarboxin | 0.65 |
| 4'-Carboethoxycarboxin | 1.50 |
| 4'-(1-Butoxy)carboxin | 0.70 |
| 4'-(1-Hexoxy)carboxin | 0.04 |
| 3'-Phenoxycarboxin | 0.05 |
| 5,6-Dihydro-2-methyl-1,4-oxathiin-3-(1-decylcarboxamide) | 0.09 |
| 5,6-Dihydro-2-methyl-1,4-oxathiin-3-carboxanilide (carboxin) | 0.7 |
| 3'-Methylcarboxin | 0.4 |
| 4'-(1-Butyl)carboxin | 0.16 |
| 4'-Phenoxycarboxin | 0.06 |

*Concentration giving 50% inhibition of the inhibitable activity in the PMS-DCIP assay at 38°.

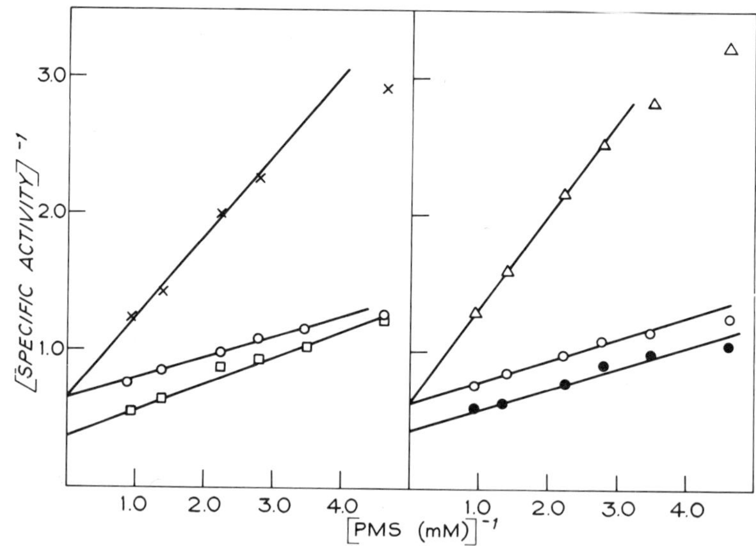


Fig.2 Effect of removal and restoration of $Q_{10}$ on PMS-DCIP activity. Left side: □, lyophilized ETP; o, same after exhaustive extraction with pentane; x, ETP inhibited by 0.5 m<u>M</u> TTF. Right side: o, pentane-extracted ETP; ●, same after adding back $Q_{10}$; Δ, pentane-extracted ETP + 0.5 m<u>M</u> TTF.

the reduced enzyme when this route is operative. The reason why cyanide, TTF, and carboxin reduce the succinate-PMS activity of membranes is that they interrupt electron transport between the enzyme and the Q pool (9). Another interpretation, suggested by Rossi et al. (8), is that $Q_{10}$ may play a regulatory role in succinate dehydrogenase activity (8).

Whatever the mechanism whereby added $CoQ_{10}$ restores full succinate dehydrogenase activity in extracted particles, these observations and the effect of cyanide, TTF, and of carboxin are more readily reconciled with the interpretation that the low turnover number in Complex II and in soluble preparations is due to removal of the enzyme from the membrane environment than that it is due to inactivation of the enzyme. An extrapolation of this conclusion is that the sensitivity of succinate dehydrogenase activity to cyanide, TTF, or carboxin in a given preparation reflects the extent to which the native membrane environment ($Q_{10}$ content?) is preserved. In samples of Complex II which have a turnover number of ∿ 10,000, succinate-PMS activity is unaffected by these agents, provided that activity is calculated at $V_{max}$ with respect to the oxidant, although the apparent $K_m$ for PMS is increased in their presence (Fig. 3).

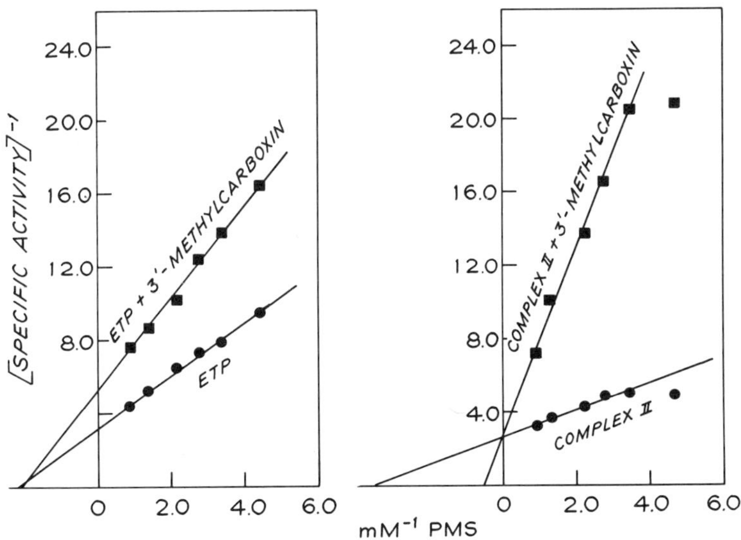


Fig. 3 Comparison of effect of 3-methylcarboxin on ETP and Complex II. Left side: ●, untreated ETP (turnover number = 18,700); ■, ETP + 0.84 μM 3-methylcarboxin. Right side: ●, untreated Complex II (turnover number = 10,800); ■, Complex II + 0.84 μM 3-methylcarboxin.

In samples of Complex II having a higher turnover number cyanide, TTF, and carboxin do cause loss of activity, even at $V_{max}$ with respect to PMS, lowering the turnover number to ∿10,000 (Fig. 1, right side). In contrast to Compex II, in Keilin-Hartree preparations, which also have a turnover number close to 10,000 the succinate-PMS activity is ∿50% inactivated by cyanide as in ETP. This supports the conclusion reached above that such preparations contain inactivated succinate dehydrogenase.

## EPR STUDIES OF THE CATALYTIC CYCLE

A study of the kinetics of the catalytic cycle by EPR in combination with rapid freezing appeared to be a logical approach to defining the reasons for the wide range of turnover numbers of the enzyme in different preparations. Thus, if the main reason for these differences were in the mechanism of the reoxidation of the reduced enzyme (as is implied in the hypothesis of 2 reaction sites for PMS), this would become apparent on comparing the rates of reoxidation of the reduced Fe-S components of the enzyme by PMS in ETP and in soluble preparations or Complex II. Likewise, if Q were the second PMS site, EPR should reveal differences in the rates of reoxidation in normal, Q-depleted, and Q-replenished ETP. Further, kinetic studies with EPR might detect the presence of inactive forms of the enzyme in low turnover number preparations and perhaps localize the site of action of TTF and carboxins.

As it turned out, some of these expectations have been fulfilled, others have not. Unfortunately, kinetic studies with EPR on submitochondrial particles like ETP proved unfeasible, because of the large electron sink provided by the Q pool and the cytochromes. Although the rates of reduction of the EPR-detectable components by succinate could be followed if the reoxidation of the enzyme by Q was blocked with TTF or carboxin (3), we found subsequently that these inhibitors destroy some of the Fe-S components. This precludes, of course, meaningful studies on TTF- or carboxin-inhibited membranes. Even serial pentane extraction (8) until all succinoxidase activity disappeared apparently left enough Q in the particles to interfere with kinetic studies with EPR. As a result, our studies have been restricted to Complex II and various types of soluble preparations of the enzyme, all of which have a lower turnover number than intact membranes.

We have reported (3,13) that Complex II contains one mole each of Hipip and of the ferredoxin type (g=1.94) Fe-S center per mole of covalently bound flavin, in addition to a lower amount of a second ferredoxin type component, which is reduced by dithionite but not by succinate, and which has been referred to as "center 2" in the literature. Soluble preparations had the same composition, except

TABLE III

Stoichiometry of EPR - Detectable Iron-Sulfur Components in Succinate Dehydrogenase Preparations

| Preparation | Activation | Temperature of observation | Electron equiv. per mole of bound flavin recovered in | | | |
|---|---|---|---|---|---|---|
| | | | High-potential iron protein g=2.01 | Succinate g=1.94 | $Na_2S_2O_4$ g=1.94 | Difference: $Na_2S_2O_4$-succ. ("center 2") |
| | % | K | equiv./mol. | | | |
| Complex II | 100 | 6 | 1.18 | 0.95 | 1.19 | 0.24 |
| | | 13 | 1.14 | 1.03 | 1.26 | 0.23 |
| Soluble enzymes* | | | | | | |
| Type 3 | 88 | 13 | 0.167 | 1.09 | 1.35 | 0.26 |
| Type 3 | 100 | 13 | 0 | 0.79 | 1.37 | 0.58 |
| Type 1 | 100 | 13 | 0 | 0.87 | 1.46 | 0.59 |
| Reconstitutively active enzyme from Keilin-Hartree preparation | 100 | 13 | 0.30 | 1.1-1.2 | 1.75 | 0.65-0.55 |
| " | 100 | 13 | 0.35 | 0.88 | 1.29 | 0.41 |

* Type 3 = perchlorate extracted from Complex II (7); Type 1, extracted from acetone powder of ETP (16).

that the Hipip center was present in much lower amounts or not at all.

In these experiments the soluble enzyme was always prepared in the absence of succinate, although we were aware that this resulted in loss of reconstitution activity, because the presence of even traces of substrate precluded studies on the kinetics of the reduction by succinate. Ohnishi et al. subsequently reported that reconstitutively active soluble preparations, extracted with succinate, retained the Hipip component and that such preparations also contained a 1:1 ratio of "center 2" and flavin (15), in contrast to ETP, Complex II, or other soluble samples. The kinetics of the Hipip in soluble preparations were not examined by these workers, however; thus it was not known whether the Hipip was catalytically competent (14).

We have now studied some reconstitutively active, soluble, preparations, isolated from Keilin-Hartree particles, with regard to the stoichiometry of the three types of Fe-S components. As shown in Table III, the Hipip component was indeed present in such samples immediately after isolation but its concentration was only 30 to 40% of that of the covalently bound flavin, in contrast to the 1:1 ratio in Complex II. Confirming our earlier results (3), the ferredoxin component reduced by dithionite but not by succinate ("center 2") was also present in amounts considerably less than stoichiometric with the flavin.

TABLE IV

Rates of Reduction of Ferredoxin and Hipip Fe-S Centers in Reconstitutively Active Succinate Dehydrogenase* by Succinate

| Expt. | Temp. | Ferredoxin center: Electron equiv. per mole of flavin in reduced form at | | | | | Hipip center: Electron equiv. per mole of flavin detectable in oxidized form at | | | | |
|---|---|---|---|---|---|---|---|---|---|---|---|
| | °C | (ms) 0 | 6 | 25 | 100 | 230 | (ms) 0 | 6 | 25 | 100 | 230 |
| 1 | 18 | 0.20 | | | 0.65 | 0.65 | 0.18 | | | 0.16 | 0.13 |
| 2 | 17 | 0.23 | 0.51 | | 0.78 | | 0.20 | 0.15 | | 0.10 | |
| 3 | 15 | 0.17 | | 0.61 | | 0.61 | 0.24 | | 0.24 | | 0.16 |

* Prepared from ETP by butanol extraction in the presence of succinate (17). Turnover time in the catalytic assay was 25-35 ms at 16°.

As to kinetics, in both Complex II and perchlorate-extracted soluble preparations only 60 to 80% of the ferredoxin type (g=1.94) Fe-S center was reduced by succinate within the turnover time calculated from the PMS-DCIP assay at the same temperature. Even less - on the average no more than 50% - of the Hipip component in Complex II was reduced by succinate within the turnover time of the enzyme; the rest reacted very slowly or not at all (3).

As shown in Table IV, the kinetics of the reduction of the ferredoxin center by succinate in fresh, reconstitutively active samples was found to be much the same as in soluble preparations isolated without succinate: some 60% was reduced within the turnover time of the enzyme. The preparations contained only a small amount of the Hipip component, of which only a small fraction (in one preparation) or none reacted rapidly. Therefore, by no means all of the Hipip observable in reconstitutively active samples seems to be catalytically competent.

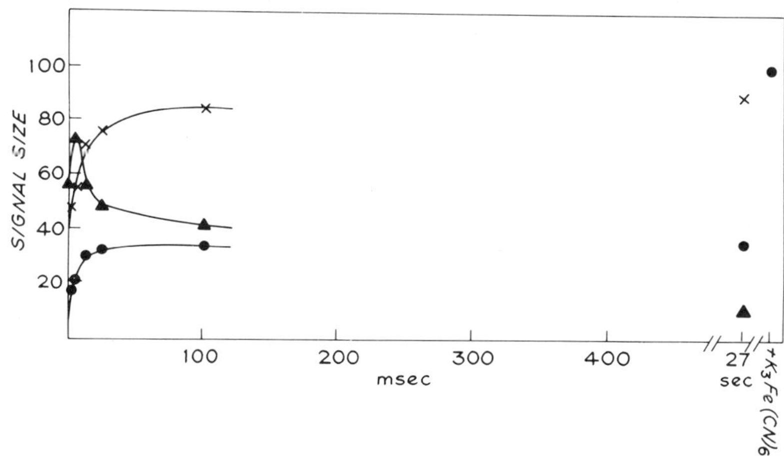


Fig. 4 Kinetics of reoxidation of EPR detectable components of Complex II by DPB. Signal size is expressed on ordinate as % reoxidized by DPB for the Hipip and g=1.94 centers and as % maximal signal for the radical. The 100% value for the Hipip is that observed in the fully oxidized control enzyme; for the g=1.94 it is the signal in the enzyme reduced by dithionite to the level where "center 1" is completely and "center 2" not significantly reduced, and for the radical it is the highest value reached during reoxidation of the enzyme by DPB in the presence of TTF. Symbols: X, g=1.94; •, Hipip; ▲, radical. Temperature was 0°.

Examination of the kinetics of the reoxidation of the reduced enzyme has confirmed the conclusion that a part of the ferredoxin center and a substantial fraction of the Hipip center in Complex II react too slowly to be of catalytic significance. As seen in Fig. 4, following reduction of Complex II by succinate, the Q analog, DPB, reoxidized rapidly only ∿80% of the g=1.94 center and less than 40% of the Hipip center, but subsequent addition of 5mM ferricyanide resulted in essentially complete reoxidation. Since, under the experimental conditions, reoxidation by ferricyanide may be regarded as a chemical rather than enzymatic event, this experiment again indicates that a significant part of the ferredoxin type Fe-S center and at least half of the Hipip center are not fuctional in the preparation under these conditions.

These experiments seem to suggest that a significant fraction of the succinate dehydrogenase molecules in the Complex II and in the soluble preparation examined are not catalytically competent. This finding assumes considerable interest if one considers that all the preparations studied were of the "low turnover number" type.

We have also studied the rate at which PMS reoxidizes the EPR-detectable components of the enzyme in Complex II preparations reduced by succinate or dithionite. Reoxidation of all components was complete in 6 msec at 0°, well below the turnover time of the enzyme. These data strongly suggest that in the usual PMS assay of the dehydrogenase reoxidation by the dye cannot be rate-limiting. It follows that the decreased turnover number of the enzyme in cyanide inactivated, carboxin or TTF inhibited, or in pentane-extracted membranes cannot be due to the loss of a second reaction site for PMS.

## ACTION OF CARBOXINS

Since carboxin derivatives and TTF convert succinate dehydrogenase in inner membrane preparations from the "high turnover number" type of enzyme to one of low turnover number and since they are generally useful in selectively blocking electron flux between the dehydrogenase and the Q pool, it was of interest to try to pinpoint their site of action by rapid kinetic methods using EPR. Fig. 5 shows typical results for the action of 3-methylcarboxin on Complex II. In addition to the kinetic effects to be described, this inhibitor causes a slow autoreduction of the preparation, with a half-time of about 1 hr at 4°, and some destruction of the Fe-S centers, particularly of the Hipip type. The latter effect is seen on comparing the signal levels in the control and 3-methylcarboxin-treated samples at zero time in the reduction experiment

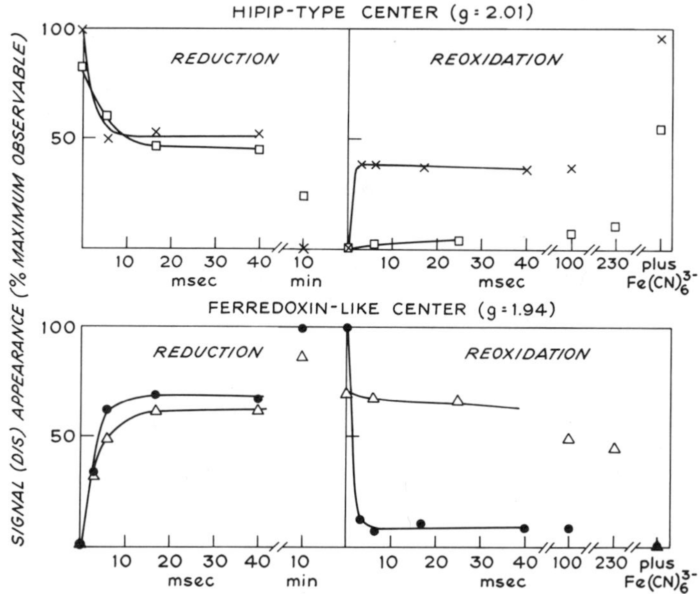


Fig. 5 Effect of 3-methylcarboxin on kinetics of the reduction and reoxidation of the g-1.94 and Hipip centers. In measuring reduction the oxidized enzyme was rapidly mixed with succinate (0.1 M final conc.) at 15°. In reoxidation experiments the enzyme, reduced with 1.8 mM succinate, was shot at 15° against DPB to give 0.2 mM final concentration. Symbols: X, Hipip control; □, Hipip + 25 μM 3-methylcarboxin; •, g=1.94 control; Δ, g=1.94 + 25 μM 3-methylcarboxin.

and after ferricyanide treatment in the reoxidation experiment. It is clear that the Hipip center is more extensively destroyed than the ferredoxin center and that it is particularly evident in the reoxidation experiment.

The kinetic data in Fig. 5 seem to bear out this interpretation. Reduction of the oxidized ferredoxin center by succinate is not significantly inhibited, while reduction of the Hipip center is slightly inhibited. Reoxidation of both centers in the reduced enzyme by DPB is, however, severely inhibited. It seems possible that the inhibition of the reoxidation by DPB of the g=1.94 component may reflect an obligatory oxidation route via the Hipip center, rather than a direct effect on the ferredoxin component. (It should be noted that since an excess of succinate had to be used to achieve complete reduction, the reoxidation must compete with reduction by the residual succinate. This does not interfere with the control, since reoxidation by DPB is complete within 3 msec, being much faster than the reduction, but it does lead to extremely slow rates in the inhibited sample).

These experiments have important implications on the reason why TTF and carboxins lower the turnover number of the enzyme in intact membranes. It could be argued that since TTF and carboxins destroy the Hipip component of the enzyme and severely inhibit its reoxidation by DPB, the effect of these inhibitors on the turnover number of succinate dehydrogenase in the PMS assay is not necessarily due to blocking electron transport to the Q pool but might be the direct result of their effect on the Hipip center. There is no evidence, however, that the Hipip center is essential for full PMS activity (i.e., TN ∿21,000) or that it even participates in this activity. It is possible that in the PMS assay, even in ETP, electrons are accepted from the flavin and/or ferredoxin centers, since both of these components are reoxidized by PMS in 1.5 msec at 0° in reduced Complex II (13).

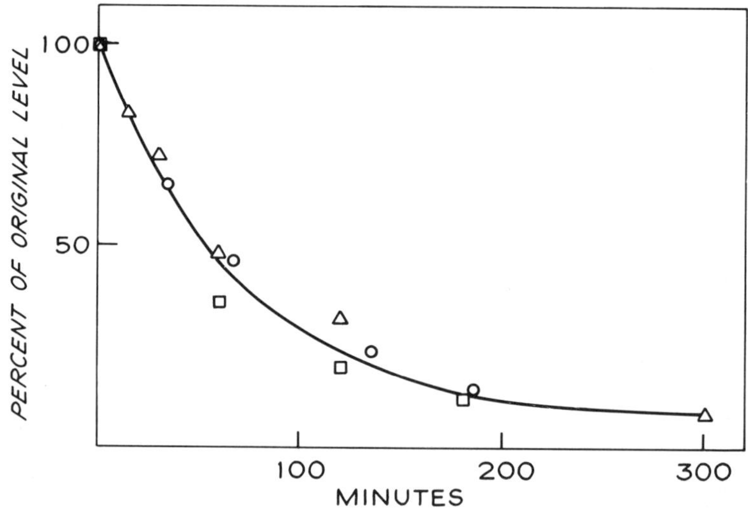


Fig. 6 Decay of reconstitutively active enzyme at 0° on contact with air. Symbols: o, reconstitution activity; □, "low $K_m$" ferricyanide reductase activity; Δ, Hipip EPR signal.

## "LOW Km" FERRICYANIDE REDUCTASE ACTIVITY, HIPIP, AND RECONSTITUTION ACTIVITY

Before presenting direct evidence that the membrance enviroment increases succinate dehydrogenase activity we must consider some properties of the "low $K_m$" ferricyanide reductase activity and its relation to the Hipip center. As mentioned earlier, this activity is seen only in soluble preparations extracted anaerobically with succinate and it decays rapidly on exposure to air (4,5). Because of this marked instability it has been suggested that both the site involved in this reaction and reconstitution activity may require the integrity of the same component in the enzyme. This is supported by the fact that at a series of pH values decay of both activities occurs at the same rate (5). It has also been suggested (14) that the Hipip center and reconstitution activity decay at similar rates. It was clearly of interest to compare all three parameters. As shown in Fig. 6, "low $K_m$" ferricyanide reductase, reconstitution activity, and the Hipip signal decay at indistinguishable rates. It would be premature to conclude, however, that both of the activities measured here are functions of the Hipip center, since preparations of the type used in these experiments are practically completely active in reconstitution tests (i.e., are nearly fully reincorporated into alkali-treated preparations) and have maximal "low $K_m$" ferricyanide activity (5) but have nevertheless lost most of the Hipip signal in isolation, and what is left appears to be enzymatically inert (Tables III and IV). Concurrently with the loss of ferricyanide reductase and reconstitution activities, succinate-PMS activity also decays, but this loss levels off at 50-60% of the original, while the former activities disappear completely (5).

The loss of these various activities is reversed on incubation of the air-inactivated enzyme with inorganic $Fe^{+2}$, $S^{-2}$, and mercaptoethanol (Table V). Previous studies in other laboratories have shown that reconstitution activity is reversed by these treatments (18,19). This experiment supports the hypothesis that the activity of the enzyme measured at low ferricyanide concentrations and reconstitution activity are intimately interrelated and that both are expressions of how "healthy" a soluble, purified preparation is, i.e., to what extent it has escaped preparative damage. This conclusion is further supported by studies on the recombination of the soluble enzyme with alkali-treated membranes.

## EFFECT OF INCORPORATION OF THE DEHYDROGENASE INTO MEMBRANES ON ITS CATALYTIC ACTIVITY

The last experimental approach we would like to present offers direct evidence for the fact that the membrane environment

TABLE V

Restoration of Catalytic and Reconstitution Activities of Air-Inactivated Enzyme

| | Catalytic activity (μmoles succ./min/mg) | | Reconstitution activity |
|---|---|---|---|
| | 100 μM $Fe(CN)_6^{-3}$ | PMS ($V_{max}$) | |
| Freshly extracted enzyme | 5.2 | 13.0 | Active |
| Same, after 4 hr in air at 0° | 0.17 | 6.5 | Inactive |
| Same, after incubation with $Fe^{+2}$, $S^{-2}$, and mercaptoethanol | 5.1 | 16.0 | Active |

Reconstitutively active enzyme (15 mg/ml) from Keilin-Hartree preparations was aged at 0° in 50 m$\underline{M}$ $P_i$, pH 7.8, with 0.1 mM EDTA present. To 1 ml aged enzyme succinate, ferrous ammonium sulfate, sodium sulfide, and mercaptoethanol were added at 20, 3, 4, and 700 m$\underline{M}$ final concentration. After incubation under argon for 10 min at 30°, the enzyme was cooled, and anaerobically passed through a Sephadex G-25 column equilibrated with 50 m$\underline{M}$ Tris-sulfate, 50 μ$\underline{M}$ EDTA, 10 m$\underline{M}$ succinate, and 100 μ$\underline{M}$ DTT, pH 7.8. Assays were conducted at 23°.

increases the turnover number of the enzyme. Despite extensive studies in several laboratories on the effect of incorporating soluble succinate dehydrogenase into alkali-treated inner membrane preparations, it has apparently not been noted that along with restoration of succinoxidase activity and increased stability, this treatment increases the turnover number of the enzyme in the PMS-DCIP assay. Perhaps the reason is that in most studies Keilin-Hartree preparations were used, the turnover number of which is only slightly higher than that of the soluble preparations extracted from it (Table I). For this reason we have used ETP both as source of the soluble enzyme and of the alkali-treated membrane accentuating the differences in turnover number. The flow sheet below illustrates a typical experiment.

It is seen that on mixing an alkali-treated ETP with about 1/3rd of the amount of soluble succinate dehydrogenase (flavin

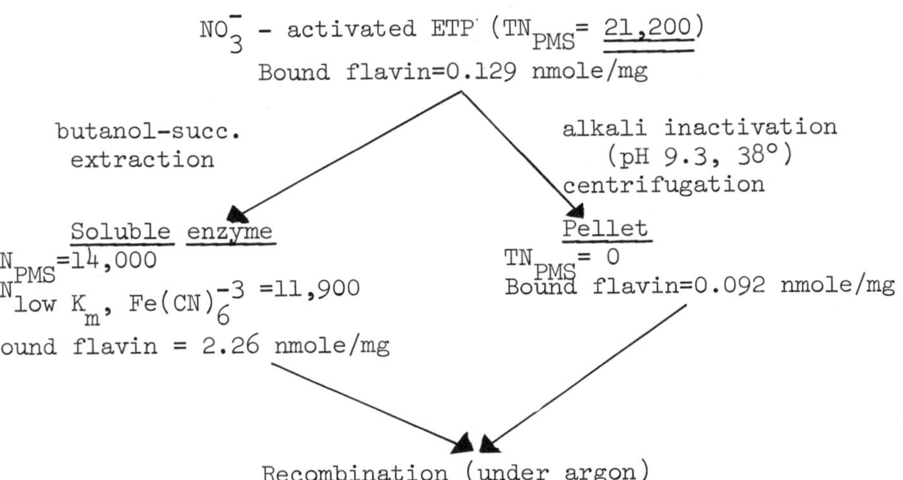


Recombination (under argon)

137 total activity units of soluble SD (= 8.94 nmoles of bound flavin)

\+

alkali-treated pellet (11.6 nmoles bound flavin)

Incubate 10 min under argon

↓

Total activity = 214 units (56% increase)

↓ centrifuge

**Supernatant**

Total activity = 17.6 units
Total bound flavin=1.82 nmoles
$TN_{PMS}$=9,670
$TN_{low\ K_m,\ Fe(CN)_6^{-3}}$ = 0

**Reconstituted pellet**

(after washing in centrifuge)
Total bound flavin=15.7 nmoles
Bound flavin content=0.154 nmoles/mg, including 0.092 nmoles/mg "dead" enzyme
$TN_{PMS}$= 21,450 (based on incorporated flavin)

basis) present prior to alkali inactivation, thus ensuring that the particle is in excess, the total PMS-DCIP activity increases by 56%. This is clear proof that the membrane, or some component of it, increases the turnover number of the dehydrogenase. Further evidence for this is seen in comparing the turnover number in the PMS-DCIP assay of the original particle (21,200), the soluble enzyme (14,000), and the reconstituted particle (21,450, based on newly acquired flavin).

Another important feature of this experiment is that, although the alkali-treated ETP was in large excess over the added soluble enzyme, some 15% of the latter failed to recombine and this part had a lower turnover number in the PMS assay and lacked "low $K_m$" ferricyanide activity. In some other experiments a substantially larger fraction (~40%) of the added enzyme failed to recombine and this part again lacked "low $K_m$" ferricyanide activity. This is further evidence that the ability to recombine (i.e., reconstitution activity) and "low $K_m$" ferricyanide reductase are intimately interrelated. The data also show that even in the most carefully prepared, strictly fresh soluble preparations a fraction of the enzyme is preparatively modified and that fraction has a low turnover number in the PMS assay. Correcting for the PMS activity of the uncombined fraction, the turnover number at 38° of the reconstitutively active soluble enzyme is at least 15,000. It follows that soluble preparations from heart with a lower turnover number are likely to contain some inactivated or modified enzyme.

## CONCLUSIONS

The data presented show that the turnover number of succinate dehydrogenase, as measured in the PMS-DCIP assay, is a sensitive indicator of both preparative modification and of the environment of the enzyme. The best evidence that the environment of the relatively intact inner membrane increases the turnover number of the enzyme is provided by the reconstitution experiment just presented. The effect of reversible removal of the $Q_{10}$ from membranes on the turnover number is further evidence for this and points to $Q_{10}$ as the component involved. Of the two possibilities considered to account for the increased activity of the enzyme in membranes, i.e., modulation of the activity by $Q_{10}$ and providing a second reaction site for the electron acceptor, the former appears more likely. The reason for this tentative conclusion is that in rapid kinetic EPR studies reoxidation of the reduced enzyme by PMS is far too fast to be rate-limiting even in "low turnover number" preparations depleted in $Q_{10}$. The former explanation would also account more readily for the fact that the extraction of $Q_{10}$ from membranes reversibly alters the $K_m$ for succinate (7).

Evidence that the low turnover number of at least some preparations indicates the presence of inactive or modified (i.e., less active) forms has come from many lines: from the detection of slowly reacting Fe-S components in kinetic studies with EPR in Complex II and in all soluble preparations, from the partition of the reconstitutively active soluble enzyme with alkali-treated particles yielding a fraction of low turnover number which is unable to combine, and from the different turnover numbers that the butanol-succinate extraction yields when applied to different particles. It would appear, in fact, that even the membrane preparations used in this study, except, perhaps, for ETP, contain varying amounts of dead or modified enzyme. It appears, therefore, that both removal from the membrane environment and inactivation of the protein may contribute to the low turnover number of purified samples of the enzyme.

The situation is complicated by the fact that it is difficult to calculate from available data what fraction of the enzyme in a given preparation is catalytically competent but is less active because of removal from the lipid (Q?) environment, and what fraction is inactivated. Since the reincorporable part of the soluble enzyme, devoid of $Q_{10}$, has a turnover number of 15,000, this might be viewed as a minimum value for the active form of the enzyme when separated from the membrane environment. If $Q_{10}$ is the membrane component which increases succinate dehydrogenase activity, extraction of the quinone should not lower the turnover number to a value less than 15,000. Yet, as seen in Fig. 2 (also in ref. (8)), pentane extraction causes 45 to 50% decrease in the turnover number of ETP (i.e., to 10,000-11,000), which is 85 to 90% reversed on reincorporating $Q_{10}$. The same low value for turnover numbers is obtained on treatment with TTF, carboxin, or cyanide. While the action of these reagents may be due, at least in part, to effects on Fe-S centers, the result of pentane extraction cannot be readily explained on this basis. Conceivably, the catalytic mechanisms of the enzyme in soluble preparations and in pentane extracted or cyanide treated membranes are different, so that the turnover numbers cannot be extrapolated from the soluble enzyme to its membrane bound form.

## ACKNOWLEDGEMENTS

The investigation was supported by the National Institutes of Health (1 PO 1 HL 16251 and GM 12394) and by the National Science Foundation (GB 36570X). H.B. is the recipient of a Research Career Award (5-K06-GM-18442) from the Institute of General Medical Sciences. A.D.V. was supported by the US-USSR collaborative project on Myocardial Metabolism.

## REFERENCES

1. Singer, T. P. (1974) *Methods of Biochemical Analysis* *22*, 123-175

2. Singer, T.P., and Cremona, T. (1964) in *Oxygen in the Animal Organism* (Dickens, F. and Neil, E., eds), pp 179-214, Pergamon Press, Oxford.

3. Beinert, H., Ackrell, B.A.C., Kearney, E.B., and Singer, T.P. (1975) *Eur. J. Biochem.* 54, 185-194.

4. Vinogradov, A.D., Gavrikova, E.V., and Goloveshkina, V.G. (1975) *Biochem. Biophys. Res. Communs.* *65*, 1264-1269.

5. Vinogradov, A.D., Ackrell, B.A.C., and Singer, T.P. (1975) *Biochem. Biophys. Res. Commun.* 67, 803-809.

6. Singer, T.P. (1966) in *Comprehensive Biochemistry*, Vol. 14 (Florkin, M., and Stotz, E.H.,eds.) pp 127-198, Elsevier Publishing Co. Amsterdam.

7. Davis, K.A. and Hatefi, Y. (1971) *Biochemistry* *10*, 2509-2516.

8. Rossi, E., Norling, B., Persson, B. and Ernster, L. (1970) *Eur. J. Biochem.* 16, 508-513.

9. Singer, T.P., Kearney, E.B., and Kenney, W.C. (1973) *Advan. Enzymol.* *37*, 189-272.

10. Giuditta, A., and Singer, T.P. (1959) *J. Biol. Chem.* *234*, 661-671.

11. Ziegler, D.M. (1961) in *Biological Structure and Function*, Vol. II. (Goodwin, T.W. and Lindberg, O.) pp 253-264, Academic Press, New York.

12. White, G.A. (1971) *Biochem. Biophys. Res. Commun.* 44, 1212-1219.

13. Singer, T.P., Beinert, H., Ackrell, B.A.C., and Kearney, E.B. (1975) in *Proceedings of the Tenth FEBS Meeting, Paris* (Raoul, Y., ed) pp 173-185, ASP, Amsterdam.

14. Ohnishi, T., Winter, D.B., Lim, J., and King, T.E. (1974) *Biochem. Biophys. Res. Commun.* *61*, 1017-1025.

15. Ohnishi, T., Leigh, J.S., Winter, D.B., Lim, J., and King, T.E. (1974) *Biochem. Biophys. Res. Commun.* 61, 1026-1035.

16. Coles, C.J., Tisdale, H.D., Kenney, W.C., and Singer, T.P. (1972) Physiol. Chem. Phys. 4, 301-316.

17. King, T.E. (1967) Methods Enzymol. 10, 322-331.

18. Baginsky, M.L. and Hatefi, Y. (1968) Biochem. Biophys. Res. Commun. 32, 945-950.

19. King, T.E., Winter, D., and Steele, W. (1972) in Structure and Function of Oxidation-Reduction Enzymes (Åkeson, Å., and Ehrenberg, A., eds.) pp 519-532, Pergamon Press, Oxford.

# BIOCHEMICAL AND EPR PROBES FOR STRUCTURE-FUNCTION STUDIES OF IRON SULFUR CENTERS OF SUCCINATE DEHYDROGENASE

Tsoo E. King, Tomoko Ohnishi, Daryl B. Winter*, and J. T. Wu†
Dept. of Chemistry and Laboratory of Bioenergetics, State University of New York, Albany, N.Y., and
Dept. of Biochemistry and Biophysics, University of Pennsylvania, Philadelphia, Pa.

Since the discovery of succinate dehydrogenase, this enzyme has remained as one of the, if not the, most colorful enzymes essential for mitochondrial electron transport. Perhaps because it is tenaciously attached to the membrane and "uniquely" reacts with electron acceptors, reports on the serious attempts of its solubilization did not appear until 1955. It is true that Tsou in 1950 definitively established the non-identity of succinate dehydrogenase and cytochrome *b* in the respiratory chain (1); this establishment dispersed and clarified a major confusion which otherwise might have hindered progress further. Nonetheless, even today the structure-function relationship of succinate dehydrogenase is still largely unknown. This paper will briefly review our work on a specific facet of the dehydrogenase, *i.e.* the iron-sulfur centers by biochemical and EPR studies.

## 1. FORMULATION OF SOLUBLE SUCCINATE DEHYDROGENASE AS AN IRON-SULFUR-FLAVOPROTEIN

In the Third International Congress of Biochemistry in Brussels interestingly two important papers, from two laboratories more than 10,000 miles apart representing then the two most polarized countries, on the solubilization and purification of this enzyme were read. The investigators are Singer, Kearney and Zastrov from

*Present address, Dept. of Biochemistry, University of Oregon Medical School, Portland.
†Present address, Dept. of Biochemistry, University of Utah Medical School, Salt Lake City.

then the E. B. Ford Institute in Detroit, and Wang, Tsou, and Wang from Academia Sinica, Shanghai**. They used different methods of solubilization; the Detroit group employed Tris buffer extraction of acetone powder of heart mitochondria, whereas the workers from the People's Republic of China used a butanol reaction of submitochondrial particles preincubated with cyanide and succinate. (Butanol and acetone powder had both been used prior to these two groups for purification of succinate dehydrogenase.) The acceptors used were also different--Singer _et al._ (2) introduced phenazine methosulfate, a rather exotic electron acceptor, although first applied in studying oxidation reactions by Dickens and McIlwain (3-5) more than 15 years earlier. The Chinese school, on the other hand, had taken advantage of a rather mundane chemical, ferricyanide. However, several important similarities of the purified soluble enzymes emerge. Among them, both preparations (6,7) do not react directly with cytochrome _c_, or other commonly used artificial electron acceptors such as methylene blue and 2,6-dichlorophenol-indophenol, contain a covalently bound FAD (acid non-extractable flavin)(8), and 2 or 4 iron (the preparations by Singer _et al._) or 4 iron (the Chinese preparation) atoms per mole of flavin.

The first detection of sulfide was attributed to the smell described by Massey (9) as: "When succinic dehydrogenase [Singer preparation] was precipitated with TCA [trichloroacetic acid] or denatured by boiling, it was noticed that a strong odor resembling $H_2S$ was given off." Although he did not interpret the source of the odor directly, the paper went further "...the liberation of -SH groups as a result of protein denaturation is a fairly general phenomenon of protein chemistry [quoting F. W. Putnam's review in _The Proteins_ edited by Neurath and Bailey, 1953]." If Massey implies SH groups of protein as the cause, it is quite natural, for aside from the Putnam review, the deterioration of protein (egg for example) frequently releases the $H_2S$ odor. Perhaps, "the major credit for the discovery of the first iron-sulfur protein" as one (10) of the organizers and his co-worker of the present symposium aptly put it, "goes to Mortensen _et al._ [Valentine and Carnahan], Valentine _et al._ [Jackson, Wolfe, and Brill], and Wolfe _et al._ [Wolin and Wolin]...." in ferredoxins in 1962-63.

The existence of nonheme iron in mitochondria was first discovered by the Green School (11) deduced from simple straightforward reasoning. They found that the total iron in submitochondrial particles is much more than the summation of iron

**Because of the prevailing situation the abstract was distributed separately by the Organization Committee of the Congress in the meeting.

derived from cytochromes***. At about the same time Green and co-workers were working on a project of metalloflavoproteins, such as xanthine oxidase (e.g. Ref. 13a). Mahler and Elowe in the Madison Laboratory as early as in 1953 reported the existence of 4 atoms of iron per mole of flavin in DPNH-cytochrome c reductase derived from mitochondria while the reductase itself does not contain heme (14, 14a). Perhaps due to the influence of these findings, the aforementioned investigators of the solubilization of succinate dehydrogenase examined the nonheme iron content.

The serious application of EPR spectroscopy in biological systems started perhaps in the early Fifties; Melvin Calvin and Barry Commoner in this country were among the earlier workers with custom-built instruments. Indeed, Commoner contributed a paper dealing with the EPR investigation of flavin semiquinone of succinate dehydrogenase. In 1959, Beinert and Sands applied the EPR technique for studying nonheme iron (15) and copper (16) of mitochondrial enzymes. As a matter of fact, soluble DPNH-cytochrome c reductase (15) and cytochrome oxidase (16) were the first enzymes studied for iron and copper-respectively. After the availability of commercial instruments, application of this technique has been suddenly blooming in spite of the cost of the machinery, fortunately at a time when funding agencies were in the golden era after the stimulati[on] from sputnik. At any rate, as far as iron is concerned, flavin was suspected at one time to co-exist with the "$g$ = 1.94" signal. Even when the ferredoxin studies were relatively well-established, this signal from the mitochondrial source was still considered to be associated more with flavin than with sulfide. We think that not until discussions at ISOX-I in 1964 (17) came the final divorce of flavin from the "g = 1.94" signal and at the same time the associatio[n] with the labile sulfide was firmly established in animal systems by inductive reasoning from the results obtained with a number of enzymes. Since then Beinert and his co-workers have richly contributed to this field. We are certain Dr. Beinert at this meeting will enlighten us more formally or informally.

***Around that time (~1953), the term ETP (electron transport particle), was not coined and the term, submitochondrial particles was rather infrequently used by the Madison group. Analysis of "hemin" (i.e. the total cytochromes)(2.2 nmoles per mg protein) and "non-hemin Fe" (5.2 nmoles per mg) was thus made in "DPNH oxidase" (11). DPNH oxidase in a way (but not "completely") may be considered lexicographically a precursor of EPT. In fact, it retains a considerable amount of succinate oxidase. Later on the term, ETP, was coined and it was prepared by only a slightly different way from the way "DPNH oxidase" was. ETP was, perhaps, first used in 1956 by Mackler and Green (12), who reported two preparation methods while the work of Crane, Glenn, and Green (13) dealt with the properties of this preparation.

This is not the place to chronicle the glorious development of the iron-sulfur protein field, even for mitochondrial enzymes, so we will restrict ourselves to presenting some of our humble work on iron-sulfur centers of succinate dehydrogenase. Since this is likewise not a conventional review, we'll quote results from other laboratories only when pertinent. Since the quotes inevitably may be incomplete we ask for your charitable acquittance.

## 2. EARLY WORK ON THE RECONSTITUTIVE ACTIVITY OF SUCCINATE DEHYDROGENASE

The complete papers (6,7) for the solubilization and the purification of succinate dehydrogenase mentioned in the previous section were not available until the end of 1956 and early 1957. When one (TEK) of the authors was in Professor Keilin's laboratory in 1957, the methods of solubilization used by these two schools were compared, and it was not much of a decision to select the Chinese method since the starting material was the Keilin-Hartree preparation, rather than the heart mitochondria acetone powder method used by the reported Americans. We immediately confirmed the properties of the Wang *et al.* (7) preparation in the first trial and found that the enzyme was not only active toward ferricyanide but also phenazine methosulfate, as expected. Actually, the specific activity even at fixed levels of acceptors was determined to be greater with phenazine methosulfate than ferricyanide. However, the aim of working on succinate dehydrogenase was not to obtain a purer enzyme or enzyme of higher activity.

Now let us go back to the early Fall of 1949 in Madison. While working at the Biochemistry Department, I (TEK) had the privilege to have a rather extensive conversation with Professor David Green of the Enzyme Institute. The Enzyme Institute then was only about 2 or 3 years old but was an *avant-garde* research laboratory for studying respiratory enzymes; many biochemists then looked up to it as a sort of mecca in the field. Professor Green, with the first Paul Lewis Award (1946) in hand, was in the most vigorous mood both intellectually and experimentally.

In the conversation, I put forward an "idea" to him. Briefly, I reasoned that organization was a natural phenomenon characteristic to the universe from a submolecular or even subatomic level to the entire galaxy. The patterns of subatomic particles in the atom are in good order, not random or chaotic. Likewise, when one crystallizes ammonium sulfate, for example, the crystals formed show only certain patterns regardless of the methods or solvents used in crystallization. It is obvious that this phenomenon occurs also in biological systems as they have definite forms. Even then, it was quite certain that intracellular respiration was a manifestation of a well-organized structure. If so, then individual components could be purified and then put them back together to resume the original organization and, of course, the manifestation of the specific organization is the function. But my idea did not apparently impress Professor Green, at least that was what I sensed. At that time he was busy working on the isolation of coenzyme A as well as on the "Cyclophorase". The latter study was one of the important prefatory investigations along with those of Leloir, Lehninger, and others on the systematic studies of the biochemistry and morphology of the mitochondria.

It was 8 years later that I presented literally the same reasoning to the late Professor David Keilin, although less ambitiously than was said in Madison. Professor Keilin most enthusiastically received it. Actually, a few years before, I had already briefly discussed it with him on at least two occasions*. According to him, the earlier work of Tissieres in his laboratory on cytochrome _c_ was quite in line with my idea. The project was settled by starting with succinate dehydrogenase, perhaps partly because the Brussels reports were still fresh in my mind and one of the methods is actually very simple, that is, the method developed by the Chinese--two of the authors having worked directly under the Professor. Other reasons for selecting succinate dehydrogenase rather than the more characterized DPNH-cytochrome _c_ reductase were manifold. Intuitively, by then at least, the insensitivity of the reductase toward antimycin A bothered me after staying in Madison's Biochemistry Department because I was so familiar with the pioneer work of van Potter and F. M. Strong about this antibiotic. Furthermore, reading a 1956 paper entitled "Is Mahler's Soluble $CoIH_2$-Cytochrome _c_ Reductase an Artifact?" (18) convinced me succinate dehydrogenase might be a better choice. It may be emphasized that, in spite of these reactions then, the Mahler enzyme was the first well-defined enzyme purified of mitochondrial origin which met at least the Dixon and Webb idea about an enzyme [_cf_. Ref. 19. Even the famous cytochrome oxidase preparations were not very well defined since the 6-fold purified preparations of Lucille Smith and Elmer Stotz was isolated later (19a)].

A great deal of time was then spent to obtain a system which would retain all of the other respiratory components but was free from, or depleted of, succinate dehydrogenase. Although a paper by Keilin and Hartree (20) reported a particle which is devoid of succinate dehydrogenase activity, the adoption by another person 17 years later is quite another matter. Dr. Ted Hartree was in

*In most American laboratories the subject of projected research supported by extramural funds can be grossly altered only through undue difficulty.

America that Fall of 1957. Eventually a particle was obtained without succinate to phenazine methosulfate or ferricyanide activity by using a principle similar to that of Keilin and Hartree (20). The inactivation and the removal of succinate dehydrogenase (later on the actual removal was found) from the Keilin-Hartree preparation is dependent on the temperature, pH, and other parameters (cf. Ref. 21, p. 173, and cross references cited).

However, when the Wang's succinate dehydrogenase was mixed with the dehydrogenase-depleted particle, no coupling occurred despite the variation of the two components in all feasible proportion. No oxygen consumption took place as we hoped. Consequently, experiments on the function of the particle were exhaustively checked again. The activity of cytochrome oxidase in the particle was only slightly impaired. Cytochromes b and $c_1$ showed the same absorption spectra as those found in the untreated sample but there was no convenient method then, as a matter of fact now, to quantitatively determine their electron transport activities*.

Naturally, we used also the Detroit method that was not easy to do in Cambridge, even with help from the Biochemistry Department. During the preparation, all of the centrifuges in the Molteno Institute were occupied. Despite such efforts, the dehydrogenase would not function with the particles. In the course of these few weeks, despite the number of frustrations and many pages of negative results, my belief of possible success was not changed one iota. Indeed, I learned a great deal about the solubilization and preparation of succinate dehydrogenase. From the very beginning I was bothered by the presence of cyanide in the preincubation mixture although I knew that Tsou had a "penchant" for cyanide. Several years before he had studied the cyanide inactivation of succinate-methylene blue reductase and the cyanide reaction with cytochrome c. But, I could not figure out just what the purpose was of putting cyanide in the mixture for the solubilization of succinate dehydrogenase. It was practically impossible to communicate with him even from England. In a way, the British post then seemed suspicious and, indeed, reluctant to transmit messages to China--actually, it was perhaps almost as bad as it was in America. I was influenced by the report of Dr. Heinz Fraenkel-Conrat (see Ref. 23) that cyanide could react with disulfide to form -SH and -SCN as Mauthner had previously reported (24). In the meanwhile,

*The reduced ubiquinone ($QH_2$) to cytochrome c activity can only reveal limited information as we later found that succinate dehydrogenase does not reconstitute with Complex III (i.e. Madison's $QH_2$-c reductase) but reconstitutes with the "cytochrome b-$c_1$ complex" or our $QH_2$-c reductase to re-form antimycin A sensitive succinate-cytochrome c reductase (22).

TABLE I. A Chronical of Some Soluble, Lipid-Free Succinate Dehydrogenase Preparations

| Enzyme or method No. and abbreviation | Solubilization agent and starting material | Year of report | Flavin: Fe:S | Reconstitutive activity[3] | Iron sulfur Center S-3 signal | Main[4] references for preparation |
|---|---|---|---|---|---|---|
| 1, AA-SDH | Tris extraction of acetone powder of mitochondria | 1955 | 1:4[2] (or 1:2) | - | - | 1 |
| 2 | Butanol extraction of HMP[1] preincubated with cyanide and succinate | 1955 | 1:4[2] | - | ? | 2 |
| 3, BS-SDH | Same as (2) (or reductase[1]), but without cyanide in preincubation mixture | 1958 | 1:8:8 | + | + | 3 |
| 4, B-SDH | Same as (3) but without preincubation or without cyanide and succinate in preincubation | 1962 | 1:8:8 | - | - | 4 |
| 5, AS-SDH | Alkaline (pH ∿10-11.5) treatment of HMP (or reductase) preincubated with succinate | 1962 | 1:8:8 | + | + | 4 |
| 6, A-SDH | Same as (5) but without preincubation | 1962 | 1:8:8 | - | - | 4 |

| | | | | | | |
|---|---|---|---|---|---|---|
| 7, CN-SDH | Cyanide (∿20-40 mM) extraction of HMP | 1967 | 1:6:4 | - | - | 5, 6 |
| 8, PS-SDH | Extraction of Complex III preincubated with succinate and dithiothreitol with a high (0.4-0.8 M) concentration of perchlorate | 1970 | 1:8:8 | + | + | 7 |

[1] HMP is the abbreviation for the Keilin-Hartree heart muscle preparation and the reductase is for succinate-cytochrome c reductase, antimycin A sensitive.

[2] Sulfide was not determined.

[3] Refers to the activity after physical recombination with the cytochrome b-$c_1$ complex or the "alkali-treated" submitochondrial particles to form succinate-cytochrome c reductase or the succinate oxidase respiratory chain, respectively (see for examples Refs. 21, 22). All enzymes listed are active toward phenazine methosulfate and ferricyanide.

[4] References are:


1. Bernath, P., and Singer, T. P. (1962) Meth. Enzymol. 5, 597-614.
2. Wang, T. Y., Tsou, C. L., and Wang, Y. L. (1956) Scientia Sinica 5, 73-85.
3. King, T. E. (1967) Meth. Enzymol. 10, 634-641.
4. King, T. E. (1963) J. Biol. Chem. 238, 4037-4051.
5. King, T. E., Winter, D., and Steele, W. (1972) in Oxidation Reduction Enzymes (Åkeson, Å., and Ehrenberg, A., eds.) pp. 519-532, Pergamon Press, Oxford and New York; and also Winter, D. B., Ohnishi, T., Wu, J. T., and King, T. E., in preparation.
6. Cerletti, P., Zanetti, G., Testolin, G., Rossi, C., and Osenga, G. (1971) in Flavins and Flavoproteins (Kamin, H., ed.) pp. 629-640, University Park Press, Baltimore.
7. Davis, K. A., and Hatefi, Y. (1971) Biochemistry 13, 2509-2516.

Professor Keilin already had given me two Hopkins papers (25,26) to read. I reasoned then, which was most probably of course incorrect, that if the Hopkins sulfhydryl groups had formed a disulfide bond, the latter could react with cyanide according to the Mauthner reaction. I discussed this aspect with an extremely well-versed protein chemist in the Department of Biochemistry _i.e._ about the disulfide reaction with cyanide. From a purely chemical consideration _per se_, he stated that the reaction was very reasonable and also that it is unlikely that succinate would prevent the reaction.

My conclusion was then very simple, _i.e._ to try the Wang procedure again, but without cyanide in the preincubation mixture since I could do it, anyway, up to the calcium gel elution stage, in less than two hours and there was a plentiful supply of the Keilin-Hartree preparation. The very first preparation of succinate dehydrogenase prepared in this way by a "minor" modification reacted beautifully with the dehydrogenase-depleted particle to give the first reconstituted succinate oxidase chain (27), in the autumn of 1957. The "minor" modification was, of course, the deletion of cyanide in the incubation mixture. Since then, several other modifications of the method of solubilization have appeared, but the principle remains the same. The preincubation with succinate was essential to have reconstitutively active preparations; cyanide, even in the presence of succinate, prevented the isolation of active samples. Although certain other suitable reducing agents (27) may replace succinate, succinate was still the best. Without preincubation of the starting material (that is, submitochondrial particles alone) with a suitable reducing agent reconstitutively inactive preparations result. Table I recounts the properties of some of the preparations of both active and inactive succinate dehydrogenases.

Incidentally, close scrutiny of the function and to a lesser extent the structure as partly summarized in the table will give the I.U.B. Enzyme Commission for the Nomenclature something to seriously ponder and it may cast a doubt about the wisdom to simply generalize succinate dehydrogenase as "succinate:(acceptor) oxidoreductase, a flavoprotein (FAC) containing iron" only under a number of 1·3·99·1. First, the true "acceptor" in intracellular respiration for succinate dehydrogenase is unknown. The differentiation of reconstitutively inactive and active enzymes may critically question the verisimilitude or validity of the generic name, succinate dehydrogenase. As a matter of fact, the same token of reasoning may be applied to DPNH dehydrogenase as No. 1·6·99·3.

## 3. THE "g = 1.94" SIGNAL FOR NONHEME IRON

Using DPNH or succinate as electron donors, Beinert and Lee (29) in their studies of the submitochondrial particles, the mitochondria, and whole heart, have attributed the signals at g values of 1.94 and 2.01 at temperature of -178°C to be due to a ferrous iron ligand configuration of nonheme iron. They concluded: "...the observed EPR signal is very specific for the particular structure of the new component, as we have been unable to reproduce the signal under the same conditions using model $Fe^{++}$ systems." Further studies in subsequent years, of course, have proved Beinert and Lee (29) were correct in spite of the transient criticisms by some theoreticians exhorting that g values of the iron were impossibly low according to "theoretical calculation."

However, at that time the small problem was rather uncertain. The evidence presented did not actually permit Beinert and Lee (29) to decide that these signals (g = 1.94, g = 2.01) were due to one or two entities. Since reconstitutively active succinate dehydrogenase, at least by *a priori* consideration, should be most close to the enzyme in particles, we were then privileged to collaborate with Dr. H. S. Mason and systematically examine the EPR behavior of this soluble, active enzyme in 1961 (29a). Two of the first EPR spectra of succinate dehydrogenase are reproduced in Fig. 1. At the same time, we analyzed more than 80 spectra.

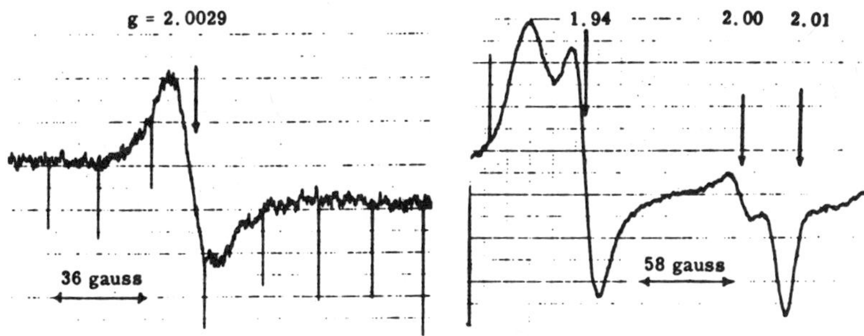


Fig. 1. EPR spectra of Center S-1 in succinate dehydrogenase (reconstitutively active, BS-SDH, No. 3 of Table I)(King, Howard, and Mason 1961 (29a)). Left figure is at room temperature. Modulation amplitude was 8 gauss; scan speed, 46 gauss/min.; sample cell volume, 0.075 ml. Right figure is at liquid nitrogen temperature. Modulation amplitude was 8 gauss; scan speed, 74 gauss/min.; sample cell volume, 0.2 ml containing 2.6 mg protein in 0.1 M phosphate buffer, pH 7.8; $5.8 \times 10^{-2}$ M succinate. Enzyme activity (zero time) was 21 µmoles succinate oxidized/min/mg enzyme protein.

From this investigation, one or two salient points emerged. First, results from the statistical analysis show that the ratio of signal magnitudes of $\underline{g}$ = 1.94 to $\underline{g}$ = 2.01 is constant for the reconstitutively active enzyme in the presence of various concentrations of substrate, inhibitors and reducing agents. In contrast, the ratio of signals at 1.94 to 2.00 vary greatly with conditions. The $\underline{g}$ = 2.00 signal is due to the semiquinone of the dehydrogenase flavin as first reported by Commoner and Hollacher (30) in heart homogenates. These facts suggest that $\underline{g}$ = 1.94 and $\underline{g}$ = 2.01 are indeed from one entity.

Secondly, the signal $\underline{g}$ = 2.00 in the purified enzyme at either room temperature or $-196^{o}$ was present in the absence of succinate and fumarate. Although at $25^{o}$ this signal is maximal at a ratio of succinate:fumarate = 1.7, according to Commoner and Hollacher (30) and confirmed by us, at $-196^{o}$ all signals (semiquinone and iron) increase with increasing concentrations of succinate and decrease with increasing concentrations of fumarate. Does temperature play a role in this disparity? At any rate, the significance of this observation is not very clear even now. Nonetheless, the fact that the semiquinone signal is not decreased by increasing the concentration of succinate up to more than 1,000 times the concentration of the flavin present but is completely abolished by dithionite does seem to suggest the absence of the completely reduced form or flavin$\cdot H_2$ in the reaction. On the other hand, our recent preliminary observations (unpublished) show $\underline{n}$ = 2 in the potentiometric titration of the semiquinone signal. We are in the progress of resolving the contradiction of these two disparate observations.

Shortly after we had established that soluble succinate dehydrogenase contains a ratio of flavin:Fe:S = 1:8:8 (see the next section), we isolated from succinate dehydrogenase a flavin-free nonheme iron protein with about 38 natoms Fe per mg of protein (31). This nonheme iron protein is quite resistant to proteolytic enzyme and exhibits practically the same EPR behavior as the intact enzyme with respect to the nonheme moiety. By consideration of this observation, together with those points summarized in a discussion (31a) following a paper on succinate dehydrogenase at a Flavin and Flavoproteins Symposium in 1965, we suggest the existence of at least 2 species of nonheme iron in the dehydrogenase (cf. also pp. 226-228 of Ref. 21, and also 31b). Practically all those present, who are working in this field, at the symposium objected to our view (as epitomized by what DerVartanian (32) said: "I do not believe in two species of Fe-S."). However, two physical and theoretical chemists, Drs. J. P. Colpa and J. D. W. Van Voorst, concurred with our view but their views were somewhat different from ours as Colpa summarized (33): "One kind of iron atom may be linked with sulfur and gives the ESR signal at $\underline{g}$ = 1.94. Another kind of iron atom may be chelated to the flavin part of the molecules which gives the *g* = 2.00 signal. It seems to me that this model is completely consistent with the experimental data and gives a better understanding of them than the 'single iron atom model' proposed in the lecture [paper by DerVartanian]."

## 4. THE OCCURRENCE OF LABILE SULFIDE IN SOLUBLE SUCCINATE DEHYDROGENASE

Through the influence of studies on ferredoxins, we examined the labile sulfide and found the latter exists in the same ratio as iron, *i.e.* flavin:Fe:S = 1:8:8 for reconstitutively active succinate dehydrogenase from a number of different batches of the enzyme (34). However, the Amsterdam group in E. C. Slater's laboratory (35) reported only about one-half the amount of sulfide in their succinate dehydrogenase from porcine heart (*i.e.* flavin:Fe:S = 1:8:4) as we found in the bovine enzyme (34). Their enzyme was active, "*viz.* 3900 moles succinate oxidized per mole of flavin per minute at 25$^{o}$ at infinite succinate and the acceptor concentrations." (35) The disparity is due most probably to the difficulty in sulfide determination (*cf.* Ref. 36) or inadvertent loss of some sulfide, rather than a species difference although only ferricyanide activity and no reconstitution experiment was reported in that communication. Their procedure of preparation actually follows our method with succinate in the preincubation mixture. In the butanol method, even in the absence of succinate in the preincubation mixture, the reconstitutively inactive enzyme, as we later found, also contained 8 Fe and 8 sulfide per flavin (37). Our unpublished results of succinate dehydrogenase from pig heart likewise showed the ratio of flavin:Fe:S = 1:8:8.

Although all these preparations tested were not pure, the practically pure succinate dehydrogenase obtained by Davis and Hatefi (38) from particulate succinate-ubiquinone reductase (Complex II) also showed a ratio of 1:8:8. Moreover, they have cleaved the dehydrogenase into two nonidentical subunits; the one with flavin shows a flavin:Fe:S = 1:4:4 for subunit of molecular weight of 70,000 and the other contained no flavin but a Fe:S = 4:4 per mole of the subunit of molecular weight of 27,000. Similar results have been obtained in less pure preparations by Righetti and Cerletti (39). Other workers have also reported a ratio of 1:8:8 (*e.g.* see Ref. 40).

In summary, all the soluble preparations may be divided into two classes with respect to the ratio of flavin:Fe:S. One contains the ratio of 1:8:8, while the other shows less iron and sulfide than the former class (*cf.* Table I). All reconstitutively active enzyme preparations possess the ratio of 1:8:8 but not all preparations with the ratio of 1:8:8 are reconstitutively active.

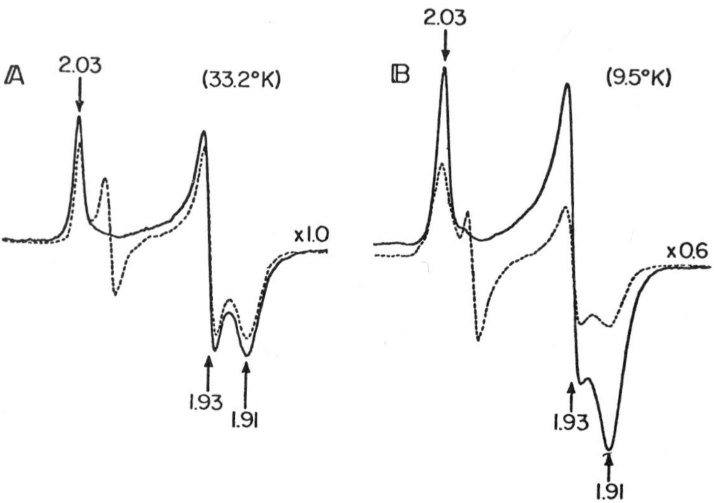


Fig. 2. EPR demonstration for Centers S-1 and S-2 in reconstitutively active succinate dehydrogenase at 33.2°K and 9.5°K (Ohnishi *et al.*, in preparation). The concentration of succinate dehydrogenase (reconstitutively active, BS-SDH, No. 3) was 18.7 mg protein equivalent to 75 nmoles flavin per ml in 50 mM phosphate buffer, pH 7.8. The enzyme was reduced by 10 mM succinate (dashed line) and a slight excess of solid dithionite (solid line), respectively. Field modulation frequency used was 100 kHz; modulation amplitude, 5 gauss; microwave frequency, 9.13 GHz; time constant, 0.3 sec; scanning rate, 500 gauss per minute; and microwave power, 1 mW.

## 5. Fe-S CENTERS S-1, S-2, AND S-3 IN SOLUBLE SUCCINATE DEHYDROGENASE

More than ten years after the experimental finding of "*g* = 1.94" signal, the existence of another species of iron-sulfur center has been demonstrated in succinate dehydrogenase, *i.e.* in addition to the classical "*g* = 1.94" signal of Beinert and co-workers (15, 28, see also Ref. 41). In order to avoid confusion, we have called Beinert's "*g* = 1.94" moiety as Center S-1 and the newly found center as Center S-2 (42, 43). The demonstration of S-2 is not straightforward. Spectra A of Fig. 2 shows the recent version of Center S-1 in reconstitutively active succinate dehydrogenase taken at 33.2°K. Indeed, above 20°K the only EPR active species of succinate-reduced enzyme is S-1 and the semiquinone of the flavin. A temperature profile at microwave power of 1 mW for S-1 is depicted in Fig. 3. The EPR spectra of Center S-1 is readily saturated at temperatures below 20°K. Upon the further lowering of the temperature, the

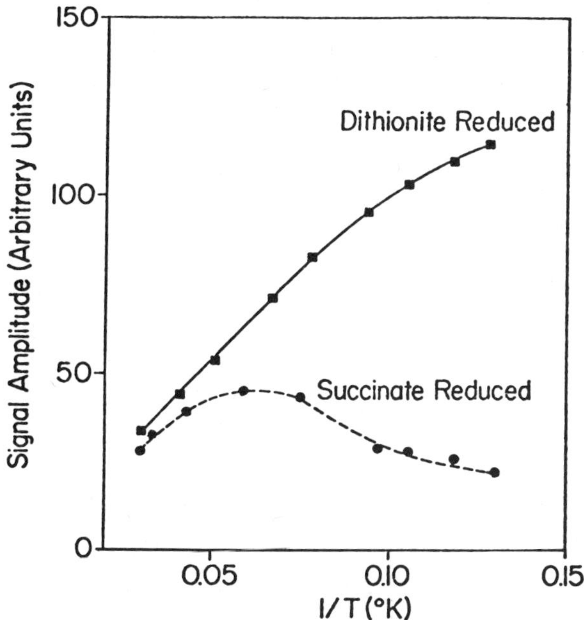


Fig. 3. A temperature profile of the EPR signals of Centers S-1 and S-2 in succinate dehydrogenase (Ohnishi et al., in preparation). The conditions were the same as those of Fig. 2. Signal amplitude was obtained from the height between the central peak and $g_x$ peak of the spectra.

intensity of the signals from dithionite (see below) reduced enzyme is enhanced and additional EPR signals emerge at the same field position but with a different line shape. These characteristics become very distinct at 9.5°K as shown in Spectra B of Fig. 2. In short, Center S-1 can be reduced by either succinate or dithionite, whereas the new signal, S-2, only by dithionite.

Comparison of the line shapes observed in the spectra of the enzymes reduced either with succinate or dithionite reveals that S-1 and S-2 show similar g values ($g_z$ = 2.03, $g_y$ = 1.93, and $g_x$ = 1.91) but differ in their detailed line shapes. This difference may arise from broadening of the central portion of the dithionite reduced spectrum--an approximately 10 gauss change in peak to peak width. This behavior relative to the power saturation of the principal resonance signal ($g_y$) of succinate or dithionite reduced enzymes examined at various temperatures suggests that the relaxation time ($T_1$) of S-2 is shorter than that of S-1; thus, S-2 resonance absorbance can be detected at lower temperature than that for S-1 (at high temperatures S-2 would give too broad signals to be discerned). At a relatively higher temperature (for example, 22°K) signals from succinate-reduced and dithionite-reduced dehydrogenase saturate at about the same level of microwave power, viz. about 1 mW, due to a small spin contribution from Center S-2 under these conditions. However, at 10°K for example, the signals of succinate-reduced enzyme (only Center S-1 is paramagnetic) are already saturated at the lowest of microwave power presently attainable of 0.05 mW, while signals of dithionite-reduced enzyme (i.e. both Centers S-1 and S-2 are paramagnetic) are not saturated up to 1.0 mW. This fact suggests that power saturation of Center S-1 spins is relieved by the nearby Center S-2 spins. Similar behavior can be observed in a wide range of temperatures, namely between 20 and 6°K. Desaturation of Center S-1 together with concomitant broadening of Center S-2 spectra indicates the occurrence of spin-spin interaction between Centers S-1 and S-2 and that these centers are located within 10 Å of each other. When the temperature is further decreased to, say, about 5°K, Center S-2 is partially saturated in the whole power range examined. On the other hand, the signal amplitude of Center S-1 does not show any change with increasing microwave power due evidently to inhomogeneous line broadening (44). The difference between Centers S-1 and S-2 with regard to power saturation is also depicted in Fig. 4 to show the spin-spin interaction between these two centers.

These characteristics as well as perhaps the low $E_m$ value of S-2 (about -400 mv for the soluble enzyme) have caused concern to some investigators (e.g. Ref. 45) who question the validity or the existence of Center S-2. We realize that it is not impossible that only S-1 really occurs in succinate dehydrogenase and the reaction of the enzyme with dithionite may alter the relaxation behavior of S-1 because the spin relaxation mechanisms could be different at different temperatures so that S-2 would show as a separate iron-sulfur center. This doubt, however, does not seem to be reconciled with the quantitative consideration of spin concentration by double integration of the EPR spectra under non-saturating conditions for both centers. Integration results (43-45) give ∿0.9 equivalents for S-1 per flavin at >20°K and ∿1.8 equivalents for S-1 plus S-2 in a wide range of temperatures tested. It seems reasonably safe to conclude that S-1 and S-2 are present in approximately one to one ratio to flavin. The existence of Center S-2 in situ is further discussed in Section 7.

Beinert and co-workers (46) have demonstrated that another iron-sulfur center of the so-called HiPIP type ("High Potential Iron-sulfur Protein," cf. Ref. 47) exists in a particulate succinate-ubiquinone reductase ("Complex II" is the name originally coined by Green in 1956 and was published in a number of papers with his co-workers, cf. Ref. 48). However, Beinert et al. were not able to detect this HiPIP signal in soluble succinate dehydrogenase preparations with either flavin:Fe:S = 1:8:8 or 1:4:4 (46).

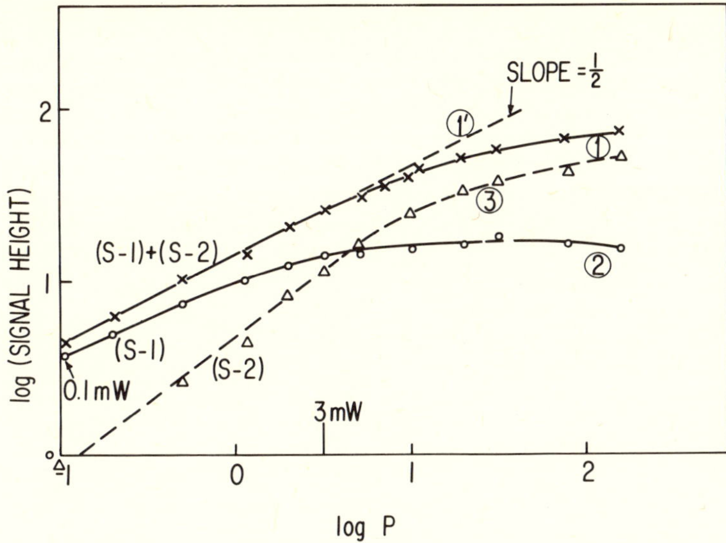


Fig. 4. Spin-spin interaction between Centers S-1 and S-2 of succinate dehydrogenase (BS-SDH) demonstrated from power saturation behavior of their EPR signals. Reconstitutively active enzyme (18.7 mg protein/ml equivalent to 72 µmoles flavin/ml) was dissolved in 50 mM phosphate buffer (pH 7.8). The enzyme was reduced with 10 mM succinate or with a slight excess of solid dithionite. EPR operating conditions are the same as in the legend of Fig. 2. The X-axis is the log of the input microwave power in mW. The sample temperature was 15.6°K.

We have examined reconstitutively active succinate dehydrogenase for the HiPIP signal under the condition used by Beinert *et al.* (46). As we expected, this signal is seen not only in our antimycin-A sensitive *particulate* succinate-cytochrome *c* reductase preparation (both intact and reconstituted from succinate dehydrogenase (BS-SDH, No. 1 of Table I, for example), with our purified ubiquinone-cytochrome *c* reductase (the *b*-$c_1$ complex) preparations and also in reconstitutively active, *soluble* succinate dehydrogenase (49,50). Spectra A of Fig. 5 illustrates the typical spectra of Center S-3 for Complex II. It can be immediately seen that these two spectra are not exactly identical. The particulate preparation exhibits a spectrum which is centered at *g* = 2.01 with peak to peak width of about 25 gauss. The soluble preparation reveals some additional signals with peaks of about 10 gauss away from those of Center S-3 in the particulate reductase which appear to arise from a slightly modified Center S-3. Indeed, less reconstitutively active soluble preparations show more intensified peaks at *g* = 2.03 and 2.00

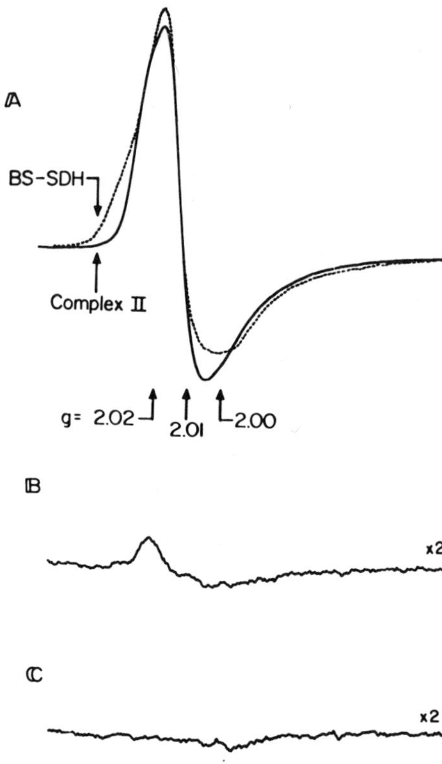


Fig. 5. EPR spectra of Center S-3 of succinate dehydrogenase and Complex II (Ohnishi, Lim, Winter and King, in preparation). EPR operating conditions were the same as described in the legend of Fig. 3, except that the microwave power used was 0.5 mW. (A) Solid line stands for Complex II (succinate-ubiquinone reductase) at about 19 mg protein per ml and was oxidized by 105 μM ferricyanide plus 50 μM phenazine methosulfate. Dashed line stands for soluble reconstitutively active succinate dehydrogenase (BS-SDH, No. 3)(18 mg protein per ml) and was oxidized with 100 μM ferricyanide plus 10 μM phenazine methosulfate. Sample temperature was 8°K. (B) Preparation A after cyanide reaction. (C) The succinate dehydrogenase solubilized with cyanide (CN-SDH, No. 6, 9.7 mg equivalent to 50 nmoles of flavin per ml).

relative to central $g$ = 2.01 of the unmodified preparation. Further modification of S-3 causes complete disappearance of S-3 signal as expected. It has been found that the decay of the reconstitutive activity parallels the decay of the intensity of unmodified S-3 signal which shows approximately a half time in the neighborhood of 30 to 50 minutes (50). The half time of reconstitutibility is also about 30 to 50 minutes (51). Figure 6 depicts the parallelism of the decay of S-3 and the reconstitutive activity of succinate dehydrogenase at the levels of the oxidizing agents which the enzyme can tolerate before EPR experiment. On the other hand, the enzymic activity of succinate-cytochrome c reductase is stable for days. There are no Center S-3 signals in all reconstitutively inactive preparations, without exception, thus far. This applies to the B-SDH, aged BS-SDH, the cyanide solubilized dehydrogenase (CN-SDH, No. 7), the reconstitutively active enzyme after cyanide reaction, and the dehydrogenase extracted with Tris buffer (AA-SDH, No. 1).

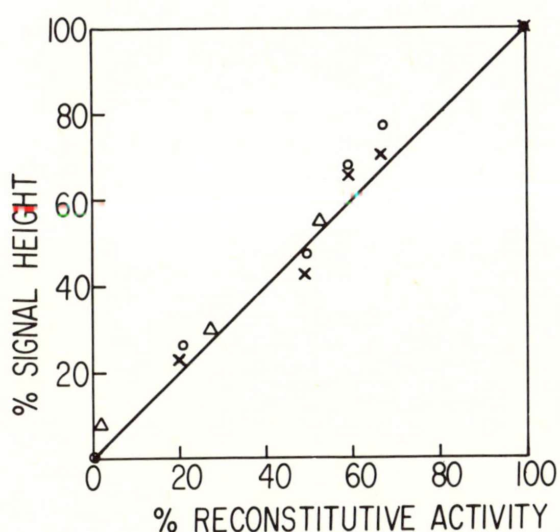


Fig. 6. Correlation between Center S-3 and reconstitutive activity of succinate dehydrogenase. The enzyme (BS-SDH, No. 3) was aged at $0^o$ in air. The decay of S-3 and reconstitutive activity was obtained from zero time to 120 minutes for various points shown. Two completely separate batches of the enzyme were used. One batch of different age was reacted with 100 μM ferricyanide plus 10 μM phenazine methosulfate (O) and 200 μM ferricyanide plus 15 μM phenazine methosulfate (X). Another batch of different age was oxidized by 100 μM ferricyanide plus 10 μM phenazine methosulfate (Δ). The EPR condition was the same as that used in Fig. 5.

Center S-3 in particulate as well as soluble preparations has a very short relaxation time, thus the signal is detectable only at temperatures below 15$^{o}$ at microwave power of 1 mW. The signal intensity increases as a function of temperature (T), which is linear against 1/T to about 7$^{o}$K. In the soluble preparation, the Center S-3 signal possesses a similar temperature profile but signals from the modified S-3 are detectable at higher temperatures and saturate readily within the temperature range of 20 to 7$^{o}$K.

All these results indicate that the main signal centered at $\underline{g}$ = 2.01, under the conditions tested, belongs to the unmodified Center S-3. The presence of modified signals is obviously due to the time required for isolation and purification. Once the dehydrogenase is dissociated from the membrane, it becomes very fragile as revealed by the EPR signal of S-3. Moreover, Center S-3 can be demonstrated by the EPR technique only in the presence of the oxidizing agents, ferricyanide plus phenazine methosulfate. Both are quite inimical to the reconstitutive activity of the dehydrogenase during prolonged standing. Reconstitutively active succinate dehydrogenase possesses a low ferricyanide-saturation site (52-54) in addition to the site conventionally measured (cf. Refs. 7, 55) which shows a $K_m$ for ferricyanide of about 10 mM with the fresh enzyme. The low saturation site with $K_m$ for ferricyanide of about 50 μM (unpublished data) is quite sensitive to oxidizing agents. When the dehydrogenase is in the membrane linked conformation, the labile site may be most probably "hidden" or well protected.

It should be mentioned that Center S-3 does not exist in our cytochrome $\underline{b}$-$\underline{c}_1$ complex (i.e. our ubiquinone-cytochrome $\underline{c}$ reductase) preparation. A HiPIP type signal previously reprqted in Complex III (46) is likely due to contamination with Complex II and its preparation requires extreme caution. So far Center S-3 has been found only in succinate-cytochrome $\underline{c}$ reductase, Complex II, and reconstitutively active succinate dehydrogenase.

## 6. EPR PROPERTIES OF RECONSTITUTIVELY INACTIVE SUCCINATE DEHYDROGENASE

In addition to the complete absence of EPR detectable Center S-3, the EPR characteristics of S-1 and S-2 in the reconstitutively inactive succinate dehydrogenase preparations also differ from that of the active enzyme. The spin content of Center S-2 per flavin is somewhat lower in the inactive than in the active enzyme. Resonance absorbance of Center S-2 can be demonstrated at higher temperature in the inactive enzyme, for example even at 25$^{o}$K, but the $\underline{g}$ values in both dehydrogenases are similar. Hence, modification of the active enzyme to the inactive forms appears to accompany some

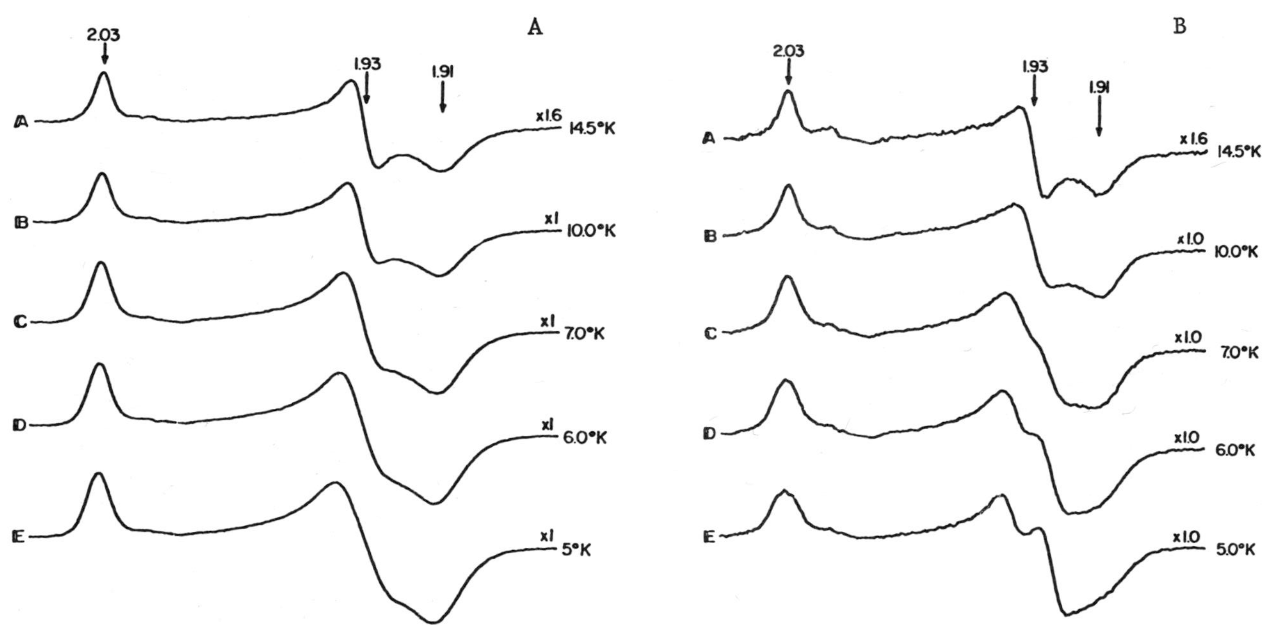


Fig. 7. EPR spectra of dithionite reduced succinate dehydrogenase at different temperatures (Ohnishi *et al.*, in press). The concentration of reconstitutively active succinate dehydrogenase (BS-SDH, No. 3) in Fig. 7A was 18.7 mg protein equivalent and 75 nmoles of flavin per ml and that of the inactive enzyme (AA-SDH, No. 1) in Fig. 7B was 30.2 mg protein equivalent and 60 nmole of flavin per ml. The enzyme in 0.1 M phosphate buffer, pH 7.4 was reduced with a slight excess of solid dithionite. EPR conditions used were microwave power, 0.2 mW; modulation amplitude, 5 gauss; time constant, 0.3 sec; scanning rate, 100 gauss/min; microwave frequency, 9.13 $GH_z$ and the temperatures as indicated. The numbers x1.0 and x1.6 are magnification factors of the spectra.

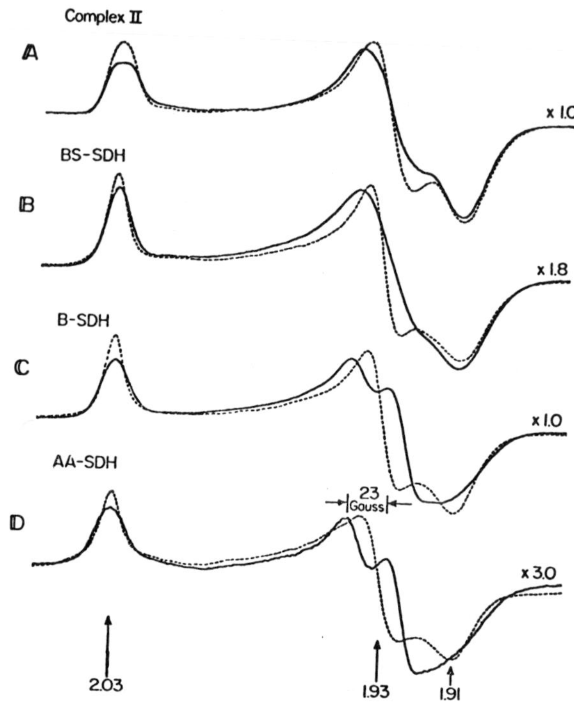


Fig. 8. EPR spectra of dithionite reduced succinate dehydrogenase showing the splitting of $g_y$ peak at a low temperature in the inactive enzyme (Ohnishi *et al.*, in press). EPR operating and other conditions were the same as those in Fig. 7 except the microwave power used was 0.5 mW and temperature was at 10°K (dashed line) and at 5°K (solid line). (A) Complex (II), a particulate succinate-ubiquinone reductase (19 mg protein equivalent and 70 nmoles of flavin per ml). (B) Reconstitutively active dehydrogenase (BS-SDH, No. 3, 18.7 mg equivalent and 75 nmoles of flavin per ml). (C) Reconstitutively inactive dehydrogenase (B-SDH, No. 4, 18.7 mg protein equivalent and 88 nmoles of flavin per ml); and reconstitutively inactive dehydrogenase prepared by extraction of acetone powder of "submitochondrial particles," (instead of "mitochondria" as AA-SDH, No. 1, 30.2 mg protein equivalent and 60 nmoles of flavin per ml).

alteration of environment of Center S-2 and shift the temperature dependence of relaxation behavior of S-2 towards higher temperature (44).

Figure 7 summarizes the differences in the line shapes of S-1 and S-2 as affected by temperature. Spectra C to E of Fig. 7A shows the broadening of the central signal at $7^{o}K$ even under non-saturating power condition, 0.2 mW. Below $7^{o}K$, a major line shape modification occurs (Spectra D and E of Fig. 7B) in the inactive, but not in the active, enzyme. The central resonance undergoes a splitting of approximately 23 gausses and signals on both sides simultaneously became broadened. The fact that this line shape modification also occurs in the inactive preparation which contains only 4 iron per flavin (AA-SDH, No. 1 of Table I) would indicate the modification may be a result of spin-spin interaction between Centers S-1 and S-2. Due to anisotropic change of the spectra, it is likely to arise mainly from dipole-dipole interaction of S-1 and S-2 which may be at most 9 to 10 Å apart; closer distances would be possible if one considers the angular factor (cf. Refs. 59, 60).

Likewise at a power setting of 0.5 mW at temperatures above $8^{o}K$, no significant differences in the line shape has been detected for Center S-2 in the particulate succinate-ubiquinone reductase (Complex II), the reconstitutively active and inactive soluble succinate dehydrogenases. However, upon lowering the temperature below $7^{o}K$, the line shape of the inactive preparations also becomes altered (cf. Fig. 8, spectra C and D). The $g_y$ resonance undergoes a splitting of approximately 23 gausses while $g_x$ and $g_z$ signals are broadened simultaneously. This kind of splitting has been observed in all inactive soluble preparations tested, including the 8 Fe and 4 Fe enzymes. On the other hand, in the particulate preparation, such as intact or reconstituted succinate-cytochrome c reductase or succinate-ubiquinone reductase, no splitting has been observed, even at a temperature as low as $4.2^{o}K$.

## 7. $E_m$ VALUES OF IRON-SULFUR CENTERS OF SUCCINATE DEHYDROGENASE

Potentiometric titration using EPR signals as a monitoring criterion gives $E_m$ values (see Refs. 42-44, 49, 50) of Centers S-1, S-2, and S-3 as summarized in Table II. $E_m$ of S-1 and S-2 have been found to be approximately -5 and -400 mV in the soluble enzymes.

The fragility of Center S-3 prevents the direct determination of the standard potential of this iron-sulfur center in the soluble enzyme. The $E_m$ value of +60 mv listed in Table II is obtained by potentiometric titration of Complex II or succinate-cytochrome c reductase. A similar value, approximately 60 mv for $E_{m,7.2}$, has also been obtained by potentiometric titration with the succinate-

TABLE II. Mean $E_{m,7.4}$ Values in mv of Centers S-1, S-2, and S-3 of Succinate Dehydrogenase in Soluble and Particulate Forms

| | Soluble | Particulate |
|---|---|---|
| S-1 | -5 | -5 |
| S-2 | -400 | -260 |
| S-3 | * | +60 |

*The signal is fragile during the titration.

fumarate pair for Center S-3 in submitochondrial particles of pigeon heart (Ingledew, Ohnishi, and Chance, to be published). Our results for the redox potentials of the flavin in the dehydrogenase by EPR technique is in progress. Recently, Kearney and co-workers (56) using a different method have reported $E_{m,7.4}$ to be -60 to -90 mv. However, the preliminary results from our laboratories by the redox titration of the flavin free radical signal, using succinate-fumarate couple, suggest that $E_{m,7.4}$ value of flavin is closer to that of Center S-1 rather than the values reported by Kearney *et al.* (56).

It is likely that electrons from succinate may be first transferred to the flavin, then to S-1 and finally to S-3, at least from a purely thermodynamic consideration if one assumes that the standard potential of succinate-fumarate is the same in the solution as it is on the protein surface. The $E_m$ value for Center S-2 in the soluble preparations is remarkably low, -400 mv, lower than that of DPNH/DPN pair, although the failure of the latter can be due to the activation barrier. Center S-2 in intact succinate-cytochrome *c* reductase shows a higher $E_m$ value *viz.* -260 mv. When the soluble dehydrogenase is reincorporated into the cytochrome *b-c*$_1$ complex, its $E_m$ value reverts back to -260 mv. Even with the latter value, it is difficult to visualize the electron transport pathway involving S-2. One cannot explain the possible decrease of $E_m$ for succinate/fumarate pair as a cause when the substrate is attached to the protein surface, because Center S-2 can be demonstrated only by dithionite reduction and not by succinate or DPNH.

However, that S-2 is not an artifact can be reasoned, *inter alia* (see Ref. 50), by its existence in heart mitochondria (57) and submitochondrial particles of yeast (58). In these two cases, S-2 can be demonstrated also by dithionite. It is very possible that Center S-2 may not be directly involved in electron transfer but in some structural way.

## 8. REACTIONS OF CYANIDE WITH SUCCINATE DEHYDROGENASE

Cyanide reactions with succinate dehydrogenase are complicated. The reaction with particle bound succinate dehydrogenase has been shown by Tsou (1) to cause a slow, temperature dependent irreversible inactivation of succinate oxidation by methylene blue. This inactivation, usually called the Tsou effect, can be prevented by the presence of succinate or dithionite, but not competitive inhibitors such as fumarate or malonate, in the course of the cyanide reaction. However, this cyanide reaction does not affect succinate oxidation by ferricyanide or phenazine methosulfate (61,63). As mentioned earlier, one of the first two methods of solubilization of succinate dehydrogenase is the butanol extraction of submitochondrial particles which are preincubated with succinate and cyanide (7). The dehydrogenase thus obtained is reconstitutively inactive. It is active only toward certain artificial electron acceptors, such as ferricyanide and phenazine dyes. Remarkably, the dehydrogenase solubilized by the same method except for the omission of cyanide in the preincubation mixture is reconstitutively active (28,51).

When $^{14}C$ labeled cyanide is used for the reaction with the soluble enzyme, radioactivity is incorporated into the protein. Likewise, when the submitochondrial particles are first reacted with $^{14}C$ labeled cyanide, extensively washed, and then subjected to the usual preparation method (Method No. 3 of Table I), the enzyme thus solubilized and purified also contains radioactivity (64,65) and is reconstitutively inactive. Moreover, in contrast to the beneficial action of succinate in the Tsou effect, the $^{14}C$ incorporation is not affected by the presence of the substrate during the cyanide reaction (64). The kinetics of $^{14}C$ incorporation into succinate dehydrogenase protein differs from that of the submitochondrial particles. It has been found that considerable incorporation occurs even at the apparent "zero" time either in the dehydrogenase or in the submitochondrial particles (64). Indeed, the reconstitutively active enzyme rapidly, much more rapidly than the Tsou effect, becomes inactive when it reacts with cyanide (63,66).

All of these observations reflect the complexity of the reactions of cyanide with succinate dehydrogenase and have not been satisfactorily explained. However, it has been found that cyanide can cause the dissociation of succinate dehydrogenase from submitochondrial particles at neutral pH. Indeed, a simple method has been worked out for the preparation of succinate dehydrogenase up to an acid-nonextractable flavin content of 7.0 nmoles per mg and a specific activity of 16 µmoles succinate oxidized per mg per minute at 23° (65). On a flavin basis, there is a 45-fold increase in purification. The purified enzyme shows 6-7 iron and 4 sulfide per flavin. The dehydrogenase thus prepared also contains two non-identical subunits of about 70,000 and 27,000 daltons, respectively.

TABLE III. Cyanide Solubilization of Succinate Dehydrogenase from Heart Submitochondrial Particles

| No. | Conditions of solubilization | Supernatant: Color | Supernatant: % of acid nonextractable flavin solubilized |
|---|---|---|---|
| 1. | 30 mM cyanide, 2 hrs. | Amber | 45 |
| 2. | 30 mM cyanide, 2 hrs; then 5 mM TTA, 2 min. | Amber | 45 |
| 3. | 1 mM TTA, 2 min; then 30 mM cyanide | Colorless | 4 |
| 4. | 5 mM TTA, 2 min; then 30 mM cyanide | Colorless | 3 |
| 5. | 1 mM TTA, 2 min, centrifuged; then 30 mM cyanide, 2 hrs. | Colorless | 0 |
| 6. | 10 mM succinate, 1 hr, or 5 mg dithionite, 5 min; then 30 mM cyanide, 2 hrs. | Colorless | 4 |

The submitochondrial particles (the Keilin-Hartree heart muscle preparation) contained about 10 mg per ml at pH 7.6. 2-Theonyltrifluoroacetone (TTA) in ethanoic solution was used, the final concentrations of ethanol in the systems were less than 1%. The ethanol "blanks" were zero. In systems 1 to 4, the mixtures were centrifuged and the acid-nonextractable flavin in the supernatant fractions was determined. In system 5, the mixture after TTA treatment was centrifuged and the residue was washed once with phosphate-borate buffer and finally resuspended in the same buffer. The cyanide reaction was conducted at 26° for 2 hrs. In systems 6, the heart muscle preparation was previously incubated with succinate or dithionite as indicated; after incubation, but without centrifugation, the neutralized cyanide was added.

The solubilization is a function of the temperature, time of the reaction, and cyanide concentration but is not affected by pH. Succinate or dithionite prevents solubilization* (Table III). But the most interesting aspect of cyanide solubilization is the complete inhibition by prior, but not subsequent, reaction of the submitochondrial particles with 2-theonyltrifluoroacetone (TTA) at as low a final concentration as 1 mM. Moreover, the TTA reacted particles, even after the apparent removal of the inhibitor by repeated dilution and centrifugation, yields no acid-nonextractable flavin when further incubated with cyanide (65, 67).

*In the original method of Wang *et al*. (7), the solubilizing agent is butanol; the Wang method involves preincubation of the submitochondrial particles with cyanide and succinate.

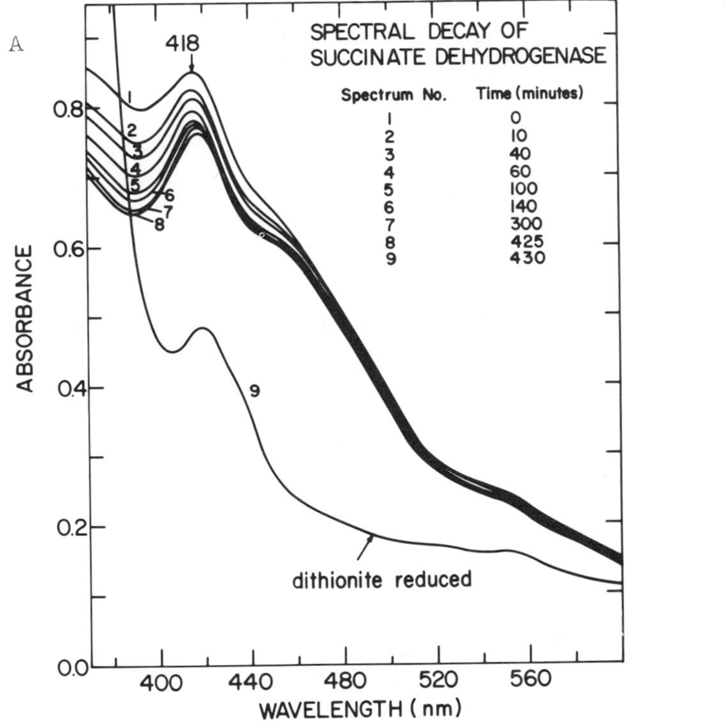


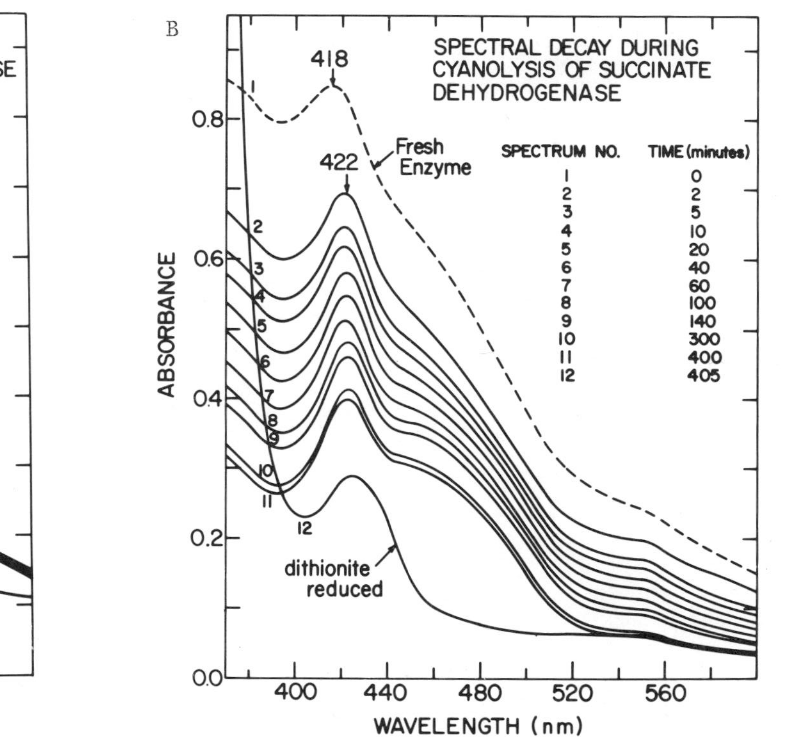


Fig. 9. Spectral decay during cyanide reaction of reconstitutively active succinate dehydrogenase (BS-SDH, No. 3)(Winter, Ohnishi, Wu, and King, in preparation). (A) In the absence of cyanide, the system contained 5 mg enzyme per ml (0.1 M carbonate-bicarbonate buffer, pH 10 used). (B) Same as A except in the presence of 50 mM KCN; the spectral decay in the absence of cyanide was already subtracted from the observed spectra, i.e. the difference spectra of the system with cyanide minus that without cyanide. Temperature was about 23$^{o}$ and the light path of the cuvette was 1 cm.

Reaction of cyanide with reconstitutively active succinate dehydrogenase rapidly quenches the spectrum of the enzyme over the entire visible region examined. In confirming the previous results (68,69), the spectrum of the dehydrogenase decays with time even in the absence of cyanide as shown in Fig. 9A. However, the spectral decay from the cyanide reaction is much more pronounced. The net quenching is depicted in Fig. 9B. The rate and the extent of the spectral decrease from the cyanolysis is a function of pH, while pH practically does not affect the spectral decay of the enzyme tested between 7 and 10 for as long as 300 minutes in the absence of cyanide. In the presence of 50 mM cyanide for example, the pseudo-first order constant for the initial spectral change is about $3 \times 10^{-3}$ and $23 \times 10^{-3}$ $Min^{-1}$ at pH 8 and 9, respectively (65).

In the reaction of cyanide with the reconstitutively active, soluble enzyme, free thiocyanate is also liberated. The amount of free thiocyanate liberated is about the same under aerobic and anaerobic conditions. Although the presence of succinate slightly decreases thiocyanate production, competitive inhibitors (such as fumarate and malonate) and TTA do not show any effect. As shown in Table IV, the rate of the thiocyanate production is directly proportional to the disappearance of labile sulfide over many hours. On the other hand, there is no simple relationship between thiocyanate liberation and the disappearance of nonheme iron.

More than 95% of the labile sulfide can be removed from the dehydrogenase at pH 3.2. The labile sulfide-depleted dehydrogenase does not yield thiocyanate upon reaction with cyanide. These results, together with observed similarities in the properties of succinate dehydrogenase solubilized by cyanide and the dehydrogenase treated with cyanide (65), indicate that the free thiocyanate is, indeed, derived from labile sulfide or the iron-sulfur clusters rather than other moieties of the enzyme. The attack of cyanide upon the iron-sulfur centers can account for the diminished sulfide as well as iron content (i.e. 1:6:4 for flavin-Fe:S) of the dehydrogenase solubilized by cyanide. This explanation is further borne out by the observation of the existence of free thiocyanate in the crude extract of submitochondrial particles after cyanide solubilization.

The cyanide reaction with reconstitutively inactive succinate dehydrogenase (B-SDH, No. 4 of Table I) liberates free thiocyanate too and causes the decrease of labile sulfide and nonheme iron. The behavior is quite different, especially in the initial phase of the reaction, from the reaction with the reconstitutively active enzyme. The disappearance of labile sulfide occurs approximately twice as fast as the formation of free thiocyanate and the nonheme iron is also released more rapidly (65).

TABLE IV. Parallelism of Thiocyanate Liberation and Decrease of Labile Sulfide and Nonparallelism with the Disappearance of Nonheme Iron During Cyanide Reaction with Reconstitutively Active Dehydrogenase

| Time (min) | Thiocyanate nmoles/mg (A) | Labile sulfur lost, nmoles/mg (B) | Nonheme iron lost, nmoles/mg (C) | $\frac{A}{B}$ | $\frac{A}{C}$ |
|---|---|---|---|---|---|
| 10 | 3.0 | 2.6 | 0.8 | 1.15 | 3.75 |
| 20 | 4.6 | 4.5 | 1.5 | 1.02 | 3.06 |
| 30 | 5.5 | 6.0 | 2.1 | 0.92 | 2.62 |
| 40 | 6.2 | 7.0 | 2.8 | 0.89 | 2.21 |
| 50 | 7.0 | 7.8 | 3.0 | 0.90 | 2.33 |
| 60 | 7.9 | 8.2 | 4.6 | 0.97 | 1.72 |
| | | | Average | 0.98 | |

Reconstitutively active succinate dehydrogenase (BS-SDH, No. 3, approximately 10 mg/ml) was incubated with 50 mM cyanide at 23° in 70 mM Tris-chloride and 30 mM phosphate buffer, pH 9.0, under argon. At the times indicated, aliquots were plunged into an ice bath and applied immediately to G-50 Sephadex columns (9.9 x 25 cm). The enzyme was eluted by oxygen free, 50 mM phosphate buffer, pH 7.8 and immediately frozen. Other aliquots were added directly to ferric nitrate solution to stop the reaction, and determination of free thiocyanate was made without delay.

Similar EPR spectra have been observed (Spectra B and C, Fig. 5) for the succinate dehydrogenase prepared by cyanide solubilization (CN-SDH, No. 7 of Table I) and the reconstitutively active enzyme (BS-SDH, No. 3) after undergoing the cyanide reaction, as we more or less expected. When only the Center S-1 resonance absorption is present as in the dithionite reduced preparations, both cyanide enzymes display the classical "$g$ = 1.94" signal (indicating rhombic symmetry) as does the reconstitutively active enzyme. This signal can also be observed after succinate reduction. However, the cyanide treated preparations exhibit a pronounced trough around the $g$ = 2 region in contrast to the absence of such trough in other reconstitutively inactive preparations. The midpoint potential of Center S-1 is found to be also about 0 mv at pH 7.4. Likewise, double integration of the S-1 signal of the cyanide preparations gives approximately the same spin equivalent per flavin. At 5 mW microwave power and 9.2°K, power saturation of Center S-1 signals are mostly relieved by the presence of S-2 in the paramagnetic state, and conversely, the central signal of the Center S-2 spectrum is broadened due presumably to the spin-spin interaction between Centers S-1 and S-2. The Center S-2 contents of both the cyanide preparations approximates 0.5 to 0.7 spin equivalent per flavin; this content resembles other reconstitutively inactive preparations.

TABLE V. A Summary of EPR Characteristics of Iron-Sulfur Centers of Reconstitutively Active and Inactive Succinate Dehydrogenase

| Center | Reconstitutively active dehydrogenase | Reconstitutively inactive dehydrogenase |
|---|---|---|
| S-1 | 1. Paramagnetic in reduced (succinate or dithionite) state.<br>2. Rhombic symmetry, $g_z$ = 2.03, $g_y$ = 1.93, $g_x$ = 1.91.<br>3. Detectable at relatively high temperature, even 100°K.<br>4. Readily saturated, especially <20°K.<br>5. S-1 spin:flavin ∿ 1:1. | All the same as the active enzyme except cyanide enzymes show a pronounced trough at $g$ = 2 region |
| S-2 | 1. Paramagnetic in reduced (dithionite only) state.<br>2. Rhombic symmetry, $g_z$ = 2.03, $g_y$ = 1.93, $g_x$ = 1.91.<br>3. Detectable only below 20°K at 1 mW.<br>4. $T_1$ (S-2) < $T_1$ (S-1).<br>5. No splitting of $g_y$ at T = 4.2 to 6°K.<br>6. S-2 spin:flavin ∿ 1:1. | 1. Same<br>2. Same<br>3. Detectable at higher temperatures too, e.g. 22°K.<br>4. $T_1$ (S-2) is longer than in active enzyme.<br>5. Splitting of $g_y$ at T = 4.2 to 6°K.<br>6. S-2 spin < flavin. |
| S-3 | 1. Paramagnetic in oxidized state.<br>2. Relatively symmetrical signal centered at $g$ = 2.01 with peak-peak width of ∿25 gauss.<br>3. Detectable only below 15° at 1 mW.<br>4. Not readily saturated.<br>5. Very labile.<br>6. No splitting at T < 6°. | Not detectable |

When the sample temperature is lowered to near liquid helium temperature, a major modification of the spectral line shape of the dithionite-reduced enzymes is obtained in both cyanide preparations. The $g_y$ peak splits approximately 23 gausses with concomitant broadening of $g_z$ and $g_x$ components. No such splitting has been observed in reconstitutively active enzyme or in intact or reconstituted succinate-cytochrome c reductase under the same conditions. Chemical activation (37) of these cyanide enzymes reverts the EPR characteristics to those of the active enzyme.

It is not clear why the midpoint potential for S-2 of both cyanide preparations shows about 25% of $E_{m,7.4}$ = -260 mv and 75% of $E_{m,7.4}$ = -400 mv from our preliminary experiments, although n values for both components are one. Since Center S-2 for cyanide preparations contains 2 species of different midpoint potentials, an advantage is thus taken by reducing either one or both components. At a temperature below 5°K, a major line shape modification is found in these preparations equilibrated at $E_m$ of either -350 or -455 mv (65).

Table V summarizes some EPR characteristics of iron-sulfur centers of reconstitutively active and inactive succinate dehydrogenases.

## 9. AN INQUIRY INTO SOME ELECTRON TRANSPORT PROBLEMS INVOLVING SUCCINATE DEHYDROGENASE

The salient questions to be answered for structure-function relationship of succinate dehydrogenase would obviously involve *inter alia* the following: What are the structures of the iron-sulfur centers? Are the electrons sequentially transferred from succinate through succinate dehydrogenase to the next member of the respiratory chain, *i.e.* what is the natural acceptor? Why is preincubation of succinate (or to a lesser extent certain other reducing agents) essential for yielding active succinate dehydrogenase? Needless to say, evidence available cannot satisfactorily explain all these and other questions, but some clues have been found which may eventually lead to a solution of the aforementioned and other problems.

*The cyanide reactions*--The results presented above clearly indicate that events from the admixture of cyanide with submitochondrial particles consist of a composite of simultaneous and consecutive reactions. The Tsou effect (1) is only part of this complicated phenomena. It is beyond any reasonable doubt that the Tsou effect involves the dissociation of succinate dehydrogenase from submitochondrial particles as a result of one of the possible reactions that occur with cyanide. This explains the original observation (1) of the inactivation of succinate oxidation by methylene blue (1) but not by phenazine methosulfate or ferricyanide (61). The dissociation of succinate dehydrogenase is prevented by succinate or dithionite (Table I) so that the Tsou effect is prevented by these reducing agents (1).

The fact that 2-theonyltrifluoroacetone (TTA) prevents dissociation of succinate dehydrogenase from submitochondrial particles by cyanide reaction is important. It is known (70, 71, and references cited therein) that TTA at very low concentrations is a rather specific inhibitor of the succinate oxidase chain evidently due to its chelation with the nonheme iron of the succinate dehydrogenase moiety, whereas TTA, on the other hand, is rather ineffective in inhibiting the artificial activity of soluble succinate dehydrogenase (*cf.* Fig. 4 of Ref. 21). These observations suggest that a *particular* non-heme iron moiety (*i.e.* enzymatically sensitive to TTA and the TTA thus bound cannot be easily removed by washing) participates in the linkage between the dehydrogenase and the rest of the respiratory chain. The chelating agent, TTA, reinforces the linkage by coordination between these two segments and thus renders the dehydrogenase undissociable by cyanide (see also the discussion of $E_m$ of S-2 in Section 7). The linkage of succinate dehydrogenase to the rest of the respiratory chain through the coordination with a non-heme iron center was first indicated from the studies of reversible and irreversible alkaline dissociation of succinate dehydrogenase from submitochondrial particles (31b). Actually in many ways reactions of alkali and some of the cyanide reactions with submitochondrial particles relative to succinate dehydrogenase are very similar except that the coordination of cyanide with non-heme iron of the dehydrogenase is much stronger than that with $OH^-$ (*cf.* Ref. 73). Moreover, our unpublished results suggest that after cyanide binds with iron, the conformation of the dehydrogenase changes prior to other chemical reactions (*vide infra*). In addition to the Tsou effect, cyanide reacts with the dissociated dehydrogenase leading to the formation of free thiocyanate. This reaction is only slightly retarded by succinate. These observations explain the formation of reconstitutively *inactive* succinate dehydrogenase from butanol solubilized preparation of heart muscle which had been preincubated with cyanide and simultaneously with succinate as originally reported by Wang, Tsou and Wang (7). In other words, as soon as succinate dehydrogenase is solubilized by butanol (7), cyanide inactivates the reconstitutive activity of the enzyme.

The reaction of cyanide with soluble succinate dehydrogenase obviously occurs through several intermediate steps eventually leading to the release of free thiocyanate. The first and most rapid one is the spectral change. Cyanide anion, rather than HCN, is apparently responsible for spectral decay as witnessed from the fact that increasing the pH during cyanide reaction causes more rapid decay. Since cyanide is responsible for the spectral modification which occurs more rapidly than thiocyanate production, the initial attack may simply be coordination of the cyanide to iron (*cf.* Ref. 73) which results in the rapid change in spectral and reconstitutive properties. The iron, once coordinated with cyanide, may rearrange or cause the sulfide to become more labile to subsequent nucleophilic attack by cyanide.

Both reconstitutively inactive (B-SDH, No. 4) and active (BS-SDH, No. 3) enzymes prepared by the butanol method liberate free thiocyanate during the reaction with cyanide. The mechanism of this type of reaction has recently been under intensive study using simpler iron sulfur proteins such as ferredoxin and adrenodoxin (cf. Refs. 74, 75). Wallace and Rabinowitz (75) have shown that thiocyanate can be formed by the attack of cyanide on mercaptoethanoldisulfide in the presence of inorganic sulfur, but not by the action of cyanide on the free sulfide alone. Hylin and Wood (76) earlier have demonstrated that this is one of the properties of an organic polysulfide, R-S-S-S-R, and that the thiol acts as a "coenzyme" for the transfer of free sulfur to cyanide. Wallace and Rabinowitz (75) have reported, using $^{35}S$-labeled ferredoxin, that the labile sulfur is released before the formation of thiocyanate, indicating the occurrence of a rearrangement of the basic iron-sulfur center prior to the abstraction of labile sulfide by cyanide.

Our studies on the simultaneous measurement of the disappearance of labile sulfide and liberation of thiocyanate from reconstitutively inactive dehydrogenase (B-SDH, No. 4) show that the sulfur is initially liberated approximately twice as fast as thiocyanate is formed. This would tend to confirm the model, proposed by Wallace and Rabinowitz (75), of cyanide causing the release of sulfur, which then reacts with the enzyme forming persulfides and finally thiocyanate. However, the reconstitutively active succinate dehydrogenase (BS-SDH, No. 3) reacts with cyanide differently. During almost the entire course of the reaction, there is a 1:1 correspondence between sulfur lost and thiocyanate formed.

What these multiple reaction pathways indicate about structural differences between these two enzymes are blurred by the complexity of the enzyme; nevertheless, there are a few implied points. First of all, the initial iron-sulfur center attacked by cyanide is most probably Center S-3 which occurs more rapidly than the Tsou effect (1). EPR signals of this center, which are observable in the enzyme containing its full iron-sulfur complement, are missing in preparations exposed to cyanide. Secondly, reconstitutively active and inactive preparations are known to have different conformations from CD experiments (unpublished results). These steric considerations may partially account for the different rates of the disappearance of sulfide. However, it is also known that incubation with succinate can reduce Center S-3 (S-3 shows up only in the oxidized form) as well as render the enzyme active towards reconstitution. Therefore, the redox states of this center may be different in reconstitutively inactive and active enzymes when they are solubilized and, in turn, the enzymes could have different reactivities toward cyanide. It is not likely that simply a different redox state causes such drastic changes in the basic covalent structure of the Fe-S center, since both the HiPIP signals from *Chromatium* and the bacterial ferredoxin have very different midpoint potentials but have quite similar EPR properties (47,89). It is possible, however, that subtle differences in the electronic environment and tertiary structures of B-SDH and BS-SDH result in the trapping of the labile sulfide of the former enzyme by the formation of a trisulfide (R-S-S-S-R), while the sulfide BS-SDH apparently is directly abstracted by cyanide.

*Absolute Requirement of Succinate During Preincubation*--In our early work (*cf*. Ref. 21), it was demonstrated that while succinate preincubation is necessary for the isolation of a reconstitutively active enzyme from submitochondrial particles, other reducing agents, such as NADH, $Na_2S_2O_4$, etc. (28), are capable of replacing succinate during the preincubation step. However, succinate has been found so far to produce the greatest activity, even though other agents possess greater reducing power. Therefore, a corollary to the above question is: What are the special properties of succinate that make it the most effective reconstitution conferring agent, so to speak? To answer these questions it is necessary to examine the sequence of events that may transpire during the preincubation and subsequent solubilization of the active enzyme during the isolation procedure.

When succinate is added to submitochondrial particles it binds to the active site of the enzyme, a property that practically all other reducing agents do not possess. The substrate is capable of displacing endogenous inhibitors (*e.g.* Ref. 79). Even oxaloacetate, which forms a powerful linkage with the active site, is mostly displaced by succinate (80). There may also be a small conformational change associated with this initial binding as witnessed by changes in the circular dichroic spectra (unpublished results). Furthermore, many competitive inhibitors, such as fumarate, malonate, etc., which are powerful "activators,"* do not cause the enzyme to yield reconstitutively active enzyme (28); hence these two phenomena--activation and solubilization of reconstitutively active dehydrogenase--are completely different phenomena.

After binding with succinate, the substrate is oxidized and the electrons thus derived reduce the prosthetic groups of the enzyme. Two of these groups are located in the flavoiron protein subunit

*The literature on "activation" is as voluminous as the explanations, see also the cross references cited in the latest communication by Ackrell *et al*. (79) for earlier papers. Activation is a phenomenon which means that the enzymic activity of succinate dehydrogenase can be increased by incubation of the enzyme with a number of divergent compounds.

(Fp), which now undergoes a significant conformational change. When electrons flow to the smaller iron protein (Ip) subunit, the latter most probably also undergoes a conformational change. However, the complete absence of any Center S-3 EPR signal in the enzyme solubilized in the oxidized state or by cyanide suggests that at least some structural modification occurs around this center, S-3.

With the substrate bound to the active site, the iron-sulfur centers and flavin reduced, and with the subunits in the proper conformation, reconstitutively active succinate dehydrogenase can be solubilized from the membrane. Upon solubilization, several noticeable changes occur. Inter alia, they are: (a) the half reduction potential of Center S-2 decreases from about -260 to -400 mv; (b) the enzyme becomes unstable, and loses rapidly its ability to be reconstituted and to react with lower concentrations of ferricyanide (52) (half life is found to be 30 to 50 minutes at $0^{o}$); (c) the visible spectrum begins to decay (31a,68,69); (d) the EPR spectrum of Center S-3 rapidly decays (50,50a) with a half life of about 30 to 50 minutes at $0^{o}$; and (e) splitting of central resonance of dithionite reduced enzyme begins to develop which is as detectable only at very low temperature.

All reconstitutively inactive dehydrogenase preparations tested, including the aged reconstitutively active enzyme, B-SDH, CN-SDH, and active enzyme modified by reaction with cyanide, can be chemically reactivated to yield a reconstitutively active enzyme (37). We believe that the main site of "repair" of the enzyme by this procedure occurs at Center S-3. This is based on the fact that the dehydrogenase preparations, in which Centers S-1 and S-2 are completely destroyed (i.e. no EPR signal and no artificial activity) by heating or extended aging, cannot be chemically reactivated.

Structure and Localization of Three Iron-Sulfur Centers--It is known that succinate dehydrogenase consists of two subunits (38) of Fp (70,000 daltons) and Ip (27,000 daltons) (see also Ref. 81). The larger subunit contains the covalently bound flavin (acid-nonextractable) and 4 nonheme iron and 4 labile sulfide. The binding studies of oxaloacetate with succinate dehydrogenase have revealed (82,83) that this inhibitor exerts its effect by reacting with the active site sulfhydryls (perhaps Hopkin's -SH*) in the large subunit.

*Hopkin's effect refers to the inactivation of succinate oxidase or dehydrogenase activity by incubating the enzyme of submitochondrial particles with mild oxidizing agents such as oxidized glutathione (G-S-S-G) and the activity is reverted back by the reaction of the inactivated enzyme with reduced glutathione (25,26 and cf. 63). The inactivation effect can be prevented not only by the substrate and succinate, but also by competitive inhibitors such as malonate, fumarate, etc. It is generally agreed that these sulfhydryls are active center residues which are required for substrate binding.

From the known results, it is not unreasonable to consider that all the machinery necessary for the primary oxidation of succinate is located in this subunit (Fp). This argument is strongly supported by the fact that the enzyme solubilized from acetone powder of mitochondrial particles (AA-SDH, No. 1 of Table I) contains only 4 Fe and 4 S per flavin and is active toward artificial electron acceptors. The acetone powder enzyme (AA-SDH) most probably does not contain the smaller subunit of Ip. All these observations considered collectively indicate the larger subunit of 70,000 daltons is responsible for the artificial activity. The other subunit (Ip) with the remaining 4 Fe and 4 S seems to be necessary for transfer of electrons from it to the cytochrome system as briefly mentioned earlier. The microenvironment surrounding Center S-2 affects its midpoint potential greatly and a reversible change in this potential occurs upon solubilization and reconstitution.

The structure of the iron-sulfur clusters in the dehydrogenase is inferred mainly from spectrophotometric similarities with iron-sulfur proteins whose crystallographic structures are more or less known. Hence inorganic sulfur is presumed to be a ligand of the nonheme iron based on the following observations: (a) absorption bands are in the 350 to 600 nm region, (b) the capability of transferring electrons, (c) the paramagnetism observed in one or more of the oxidation states, and (d) the release of sulfide upon acidification. In addition to these properties, the more recent comparisons of nucleophilic abstractions of sulfur from the iron sulfur proteins (75,84) on the one hand with model iron-sulfur compounds whose X-ray crystallographic structures are known (75,85) further suggest that the iron and sulfur are "covalently" bonded to one another in these iron sulfur proteins. Our observations on the nucleophilic abstraction of labile sulfide and concomitant quenching of absorbance spectrum and other properties (*i.e.* (a) through (d) above) all suggest that the dehydrogenase is a protein in which the labile sulfide is directly liganded to the nonheme iron.

Besides the labile sulfide groups, cysteine sulfur is known or assumed to be bonded to the nonheme iron in all iron-sulfur proteins. In the case of rubredoxin from *Clostridium pasteurianum*, the structure determined from 1.5 Å resolution X-ray crystallographic studies show a tetrahedral coordination of the 4 cysteine to the nonheme iron (86). Plant ferredoxins have 2 cysteine residues coordinated per iron, a finding which was deduced from physico-chemical measurements (87), while the iron in bacterial ferredoxin (88) and *Chromatium* HiPIP (89) also have been shown by X-ray crystallography to be bonded to cysteine residues. Our silver titration studies (unpublished observation) have revealed that the nonheme iron of the dehydrogenase is bound to cysteine sulfur since there is an increase of the total sulfhydryl content when nonheme iron and labile sulfide are removed. An increase of 8 to 12 sulfhydryls per flavin would indicate that the nonheme iron of the dehydrogenase is bound to 1 or 2, or perhaps a combination of 1 and 2 cysteine residues per iron. It should be added, however, that the change of sulfhydryl content computed from measuring the SH content before and after nonheme iron and labile sulfide removal may not be precise enough to accurately establish the number of cysteine residues per iron.

Orme-Johnson (90) has described the 3 iron sulfur structures so far recognized. These include the single iron no labile sulfide rubredoxin structure, the binuclear iron clusters with 2 bridging labile sulfur ligands, and the tetranuclear cluster consisting of 4 nonheme iron and 4 labile sulfides arranged as interpenetrating cubes of different size. In succinate dehydrogenase, it would appear that the single iron rubredoxin structure may be eliminated since this structure exhibits a strong $g = 9$ and $g = 4.3$ EPR signals (91). These signals are absent from reconstitutively active succinate dehydrogenase. Therefore, we are left with the binuclear and tetranuclear clusters as probable structures for iron sulfur centers in the dehydrogenase. Since the fully functioning enzyme contains 8 nonheme iron and 8 labile sulfide per molecule, three possible structures for the iron-sulfur clusters may exist, *viz.* (a) four binuclear clusters, (b) two tetranuclear clusters, or (c) one tetranuclear and two binuclear clusters.

From the work summarized in this paper, together with results reported elsewhere (42-44,49,50), we are inclined to think that Fig. 10 may represent the iron sulfur structures in the dehydrogenase. In addition to Center S-3, two distinct EPR detectable iron-sulfur centers are present in the isolated succinate dehydrogenase and are nearly stoichiometrically equivalent to flavin content (*i.e.* S-1:S-2:flavin = 1:1:1). The same stoichiometry is also obtained even in the enzyme (AA-SDH) which contains only 4 Fe-4 S, indicating that both Centers S-1 and S-2 have 2 Fe-2 S structures similar to that of adrenodoxin or spinach ferredoxin (92). Unfortunately, fragility of Center S-3 in the soluble dehydrogenase precludes the determination, at present, of its stoichiometry with respect to the flavin. But integration of the Center S-3 EPR spectra from particulate preparations, such as Complex II and our succinate-cytochrome *c* reductase, shows that Center S-3 also present in a 1 to 1 ratio with the acid nonextractable flavin. Furthermore, Sweeney *et al.* (93) have recently demonstrated that "superoxidation" of 4 Fe-4 S ferredoxins gives rise to EPR spectra centered at $g = 2.01$, similar to that of Center S-3. To our knowledge, this type of EPR absorbance has never been observed in any well characterized iron-sulfur proteins containing less than 4 Fe and 4 S per molecule. These observations suggest that Center S-3 is probably a 4 Fe-4 S cluster not dissimilar to bacterial HiPIP (47)

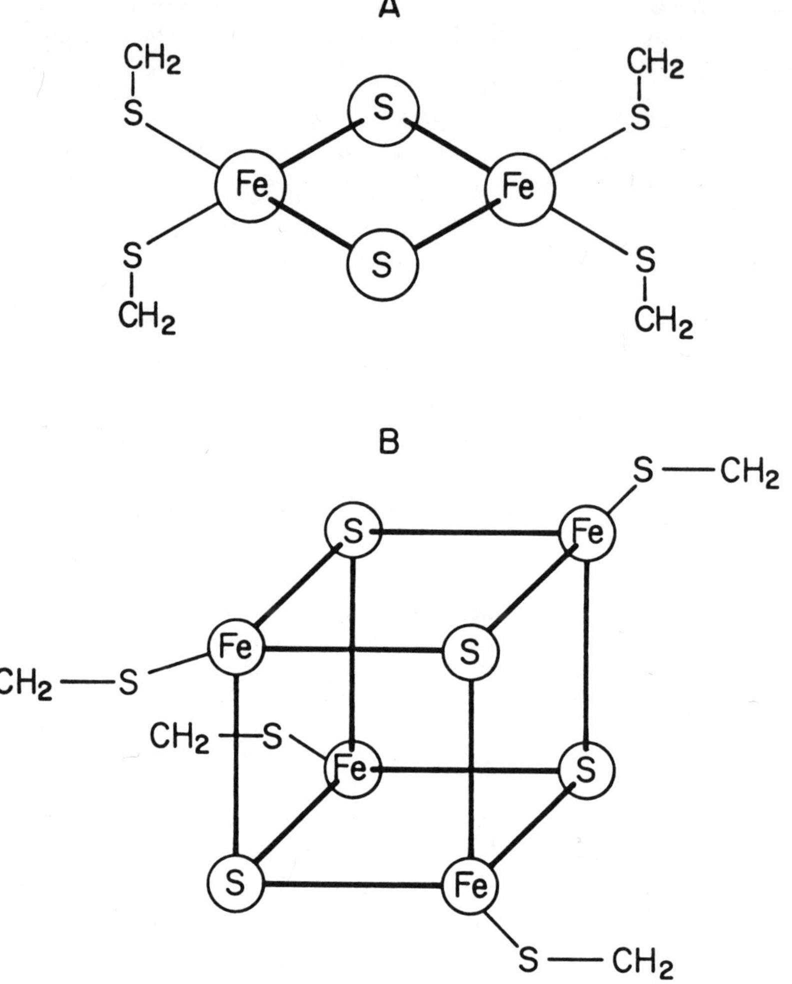


Fig. 10. A proposal for the configuration of iron sulfur centers in succinate dehydrogenase. A. Centers S-1 and S-2 in a configuration of dimeric iron sulfur cluster of the ferredoxin type. B. Center S-3 in a tetrameric iron-sulfur cluster as shown in the high potential iron-sulfur proteins. $-S-CH_2$ symbolizes the bonding with cysteine and S for the labile sulfide.

or to *Bacillus polymyxa* ferredoxins (92). It is possible that the protein environments of these proteins are different from the environment of succinate dehydrogenase so that their signals are less fragile.

The above argument establishes the distribution of two binuclear (Centers S-1 and S-2) clusters in one subunit and the single tetranuclear (Center S-3) cluster in the other. But in which subunit are Centers S-1 and S-2 located? Evidence points to their being located in the Fp subunit. The reconstitutively inactive enzyme extracted from acetone powder of mitochondria (i.e., AA-SDH, No. 1 of Table I), which contains only 4 Fe and 4 S, possesses Centers S-1 and S-2 in nearly stoichiometric amounts relative to the flavin and shows relatively high artificial electron acceptor activity but no reconstitutive capability. Thus, it is reasonably certain that Center S-1 is associated with the transfer of electrons from succinate to ferricyanide or phenazine methosulfate.

In addition, electron transfer within the dehydrogenase molecule appears to occur via a linear or serial pathway. That is, artificial electron acceptor activity is a necessary condition for reconstitutive activity. There are enzyme preparations (e.g. AA-SDH, CN-SDH, and B-SDH) which have artificial activity but no reconstitutive activity; the reverse has never been observed. In other words, Center S-1 is necessary for artificial activity but Center S-1 and Center S-3 are necessary for reconstitution. Since the active site is located in the Fp subunit, and since S-3 and S-1 must be in different subunits, the observation of the linear order of enzymic activity compels us to conclude that Center S-1 and, therefore, also Center S-2 are located in the larger subunit, Fp, and hence Center S-3 is located in the smaller subunit, Ip.

The model, as shown in Fig. 10, is proposed to summarize these conclusions concerning the structure of the succinate dehydrogenase. It can be seen from the figure that the flavin and the 2 ferredoxin-type iron sulfur centers (Centers S-1 and S-2) which are composed of 2 Fe and 2 S each, are located in the 70,000 Fp subunit, and that a single tetranuclear (Center S-3) cluster is located in the small Ip subunit. The intramolecular flow of electrons begins at the active site in the large subunit, continues through Center S-1 and the flavin (the precise sequence of these carriers is yet to be determined) to the small subunit containing Center S-3, and finally through the membrane to the next member of the respiratory chain. Enzyme preparations that contain a "damaged" Center S-3 cannot transfer electrons to the rest of the respiratory chain but may still shunt electrons through Center S-1 or the flavin to artificial electron acceptors.

The structure of Center S-3 depicted in Fig. 11 is tentative and may only exist as such while the enzyme is in its active form. The fact that 2 butanol enzymes exist (i.e. B-SDH and BS-SDH, Nos. 4 and 3, respectively) each containing 8 Fe and 8 S per flavin, but with only the reconstitutively active (BS-SDH, No. 3) preparation

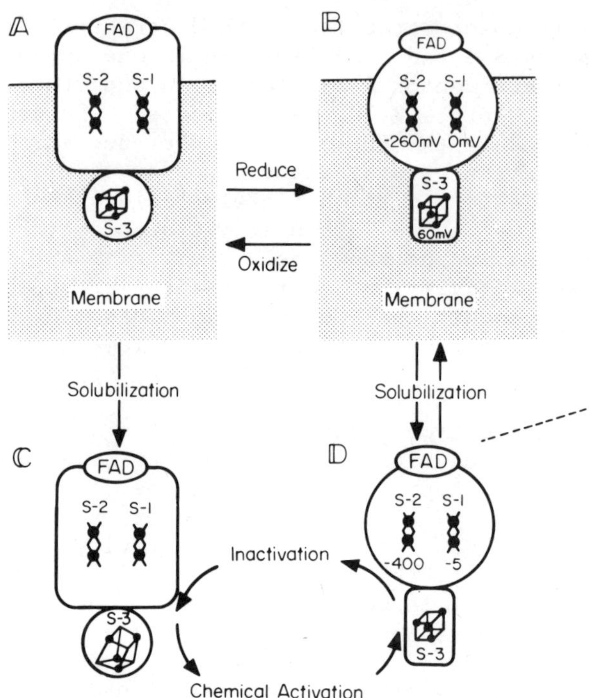


Fig. 11. A diagrammatic illustration of possible changes of conformations in succinate dehydrogenase during oxidation, reduction, solubilization, reconstitution, inactivation and chemical activation. (A) Illustrates the enzyme attached to the mitochondrial membrane in its oxidized state. The only EPR detectable signal observed in this material is from Center S-3. Upon reduction (B), the S-3 signal disappears and two ferredoxin-type centers, S-1 and S-2, with $E_{m,7.4}$ of about 5 mv and -260 mv, respectively, appear. In this reduced state, succinate dehydrogenase may be solubilized to yield (D) causing the $E_{m,7.4}$ of S-2 to decrease to -400 mv. This material may be reconstituted (the reverse of solubilization) to revert back to (B) or inactivated by exposure to air to yield (C). (C) can also be obtained by solubilization of (A) with high pH, butanol, or other agents and in each case splitting of central peak is observed. In (C) the $E_{m,7.4}$ of S-2 is -400 mv and Center S-3 is not EPR detectable. (C), formed by any of the above processes, can be chemically activated to yield (D), in which splitting of central peak is abolished and it becomes reconstitutively active. Form (C') (not shown), which is very similar to Form (C), except containing cyanide as shown by $^{14}C$ experiments, can be formed by cyanide reaction of (D) or by the cyanide solubilization of the dehydrogenase from (A). Likewise, form (C') can be chemically reactivated to Form (D) to displace the $^{14}C$ containing group. The difference shapes of the components symbolizes the different conformational states.

displaying the S-3 signal, demonstrates that some structural perturbation, as shown in the figure, must exist in the inactive enzyme. The differential reactivity of cyanide toward these two preparations also suggests a different structure for S-3 in these functionally different enzymes.

*The Natural Electron Acceptor of Succinate Dehydrogenase*--Thermodynamic data and reconstitution results indicate, beyond a reasonable doubt, the egress of electrons from succinate through succinate dehydrogenase is at Center S-3. But which member of the respiratory chain accepts the emerging electrons? The answer cannot be provided unequivocally. Actually, this point is one of the few uncertainties of the electron pathway in intracellular respiration which has caused considerable controversies. Based on the evidence, mainly from Complex II (that can serve as succinate-ubiquinone reductase) most textbooks or reviews list the natural acceptor as ubiquinone which may be correct in a sense. Soluble succinate dehydrogenase, both reconstitutively active and inactive preparations, cannot use ubiquinone or even $O_2$, as an acceptor under various conditions tested, even in the presence of micellular phospholipids. On the other hand, Complex II is nothing more than a particulate succinate dehydrogenase with some 20 to 30% of lipid. Indeed, from the "purified" Complex II preincubated with succinate, Davis and Hatefi (38) have obtained practically pure soluble succinate dehydrogenase. It is true that this complex also contains tenuously attached cytochromes *b* and some $c_1$ as an impurity. These cytochromes do not undergo oxidation-reduction in the course of the reaction.

The dilemma actually lies in the disparate functions of soluble and particulate succinate dehydrogenases toward ubiquinone*. This Spring, Peter Mitchell has advanced an electron pathway now known as the Q cycle (96). The novel point of this theory lies in the proposal of the Q radical as an intermediate. The sites, as well as reactions, of Q are plural. This hypothesis has not only rejuvenated the importance of electron transport but also reminisced oxidative phosphorylation; after all the latter is a result of mitochondrial electron transport. Without understanding electron transport, it is doubly difficult to clarify mechanism(s) of energy conservation. Less than two months after his announcement, many hours were spent on discussing Mitchell's Q pathway at the Bari International Symposium on Electron Transport and Oxidative Phosphorylation. A more specific scheme recently advanced by

*Although some reports (*e.g.* 94) claim that cytochrome *b* in Complex II serves as a kind of "structural protein" for the function of the complex, the validity of the claim as a structural protein *per se* has been seriously questioned (*e.g.* Ref. 95).

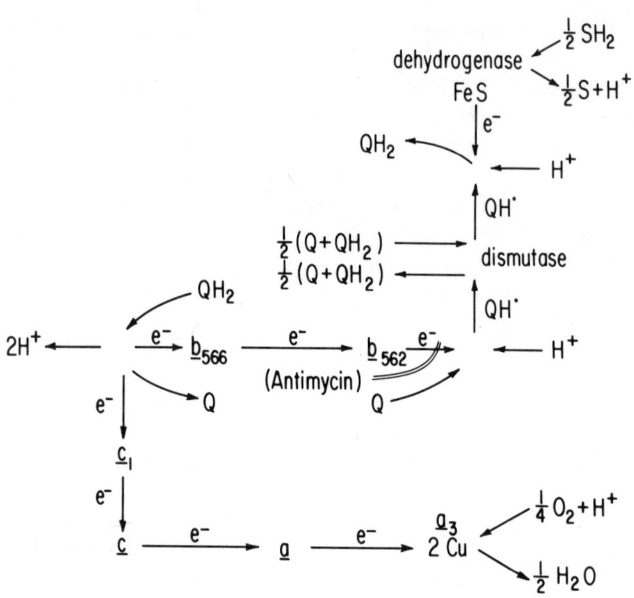


Fig. 12. More specific formulation of Q cycle indicating independent oxidation of Q by the dehydrogenase and reduction of Q by the cytochrome $\underline{b}$-$\underline{c}_1$ complex (P. Mitchell, unpublished).

Mitchell is reproduced in Fig. 12. It may be noticed, this theory proposes the independent oxidation of Q by an Fe-S moiety (S-3 in his mind?) and the reduction of Q by the cytochrome $\underline{b}$-$\underline{c}_1$ complex. The main objection at the Bari Symposium is that the dismutation of Q radicals to fully oxidized or reduced forms is many orders in disfavor of the radical. But this consideration is completely based on the "pure chemistry" in solution. Electron transport in mitochondria and even in Complex II does not occur in solution. The failure of soluble succinate dehydrogenase to react with Q may be precisely because of the differences that exist in solution and the lipid milieu. No one has yet found a suitable phospholipid micelle as well as an "integral" lipid (the term coined by J. S. Singer (cf. 97)) so that the repletion of the soluble succinate dehydrogenase by lipid has been unsuccessful. Finally, it may be said that the components, and sequence, not mentioning mechanisms, from succinate dehydrogenase to cytochrome $\underline{c}_1$ (cf. Ref. 98) remain to be clarified. Is it possible that what Isaac Newton once said "Natura enim simplex est" may not be correct? More likely the state of succinate dehydrogenase research is a point where the key to the problem must be unearthed.

## ACKNOWLEDGMENTS

We are indebted to Dr. Peter Mitchell for the stimulating discussions and for allowing us to use his unpublished material. We also are indebted to Drs. J. L. Glenn, D. E. Green, J. Leigh, B. Mackler, H. R. Mahler, T. Y. Wang, and Y. L. Wang for discussions on various aspects of the subject. The experimental work was supported by grants from the Public Health Service and the National Science Foundation. Finally, the encouragement by and discussion with Dr. B. Chance is deeply appreciated.

## REFERENCES*

1. Tsou, C.L. (1951) Biochem. J. 49, 512-520.
2. Kearney, E.B., and Singer, T. P. (1956) J. Biol. Chem. 219, 963.
3. Dickens, F., and McIlwain, H. (1938) Biochem. J. 32, 1615.
4. Dickens, F. (1936) Biochem. J. 30, 1064 and 1233.
5. McIlwain, H. (1937) J. Chem. Soc. 1704.
6. Singer, T. P., Kearney, E. B., and Bernath, P. (1956) J. Biol. Chem. 223, 599-613.
7. Wang, T. Y., Tsou, C. L., and Wang, Y. L. (1956) Sci. Sinica 5, 73-85.
8. Chi, T. F., Wang, Y. L., Tsou, C. L., Fang, Y. C., and Yu, C. H. (1965) Sci. Sinica 14, 1193-1204 and papers cited.
9. Massey, V. (1957) J. Biol. Chem. 229, 763-770.
10. Yasunobu, K. T., and Tanaka, M. (1973) in Iron-Sulfur Proteins (W. Lovenberg, ed.) Vol. 2, pp. 29-130, Academic Press, New York.
11. Personal communications with Green, D.E.; see also, for example, Mackler, B., Repaske, R., Kohout, P.M., and Green, D. E. (1954) Biochim. Biophys. Acta 15, 437-458.

*Also cf. cross references cited.

12. Mackler, B., and Green D. E. (1956) Biochim. Biophys. Acta 21, 1-6.
13. Crane, F. L., Glenn, J. L., and Green, D. E. (1956) Biochim. Biophys. Acta 22, 475-487.
13a. Mackler, B., Mahler, H. R., and Green, D. E. (1954) J. Biol. Chem. 210, 149-164.
14. Mahler, H. R., and Elowe, D. G. (1953) J. Am. Chem. Soc. 75, 5769-5770.
14a. Mahler, H. R., and Elowe, D. G. (1954) J. Biol. Chem. 210, 165-179.
15. Beinert, H., and Sands, R. H. (1959) Biochem. Biophys. Res. Commun. 1, 171-174.
16. Sands, R. H., and Beinert, H. (1959) Biochem. Biophys. Res. Commun. 1, 175-178.
17. Discussions, Oxidases and Related Redox Systems (T. E. King, H. S. Mason, and M. Morrison, eds.) (1965), especially Vol. 1, pp. 374-380 and 399-417, John Wiley and Sons, New York, New York.
18. Tsou, C. L., and Wu, C. Y. (1956) Sci. Sinica 5, 263-270.
19. Dixon, M., and Webb, E. C. (1958) Enzymes, Longmans, Green, London.
19a. Smith, L., and Stotz, E. (1954) J. Biol. Chem. 209, 819-828.
20. Keilin, D., and Hartree, E. F. (1940) Proc. Roy. Soc. B129, 277-306
21. King, T. E. (1966) in Advances in Enzymology (F.F. Nord, ed.) Vol. 28, pp. 155-236, John Wiley and Sons, New York, New York.
22. Yu, C. A., Yu, L., and King, T. E. (1974) J. Biol. Chem. 249, 4905-4910.
23. Olcott, H. S., and Franckel-Conrat, H. (1947) Chem. Rev. 41, 151-197.
24. Mauthner, J. (1912) Z. Physiol. Chem. Hoppe-Seyl. 78, 28-36.
25. Hopkins, F. G., and Morgan, E. J. (1938) Biochem. J. 32, 611-620.
26. Hopkins, F. G. Morgan, E. J., and Lutwok-Mann, C. (1938) Biochem. J. 32, 1829-1848.
27. Keilin, D., and King, T. E. (1958) Nature, 181, 1520-1522.
28. King, T. E. (1962) Biochim. Biophys. Acta 59, 492-494.
29. Beinert, H., and Lee, W. (1961) Biochem. Biophys. Res. Commun. 5, 40-45.
29a. King, T. E., Howard, R. L., and Mason, H. S. (1961) Biochem. Biophys. Res. Commun. 5, 329-333.
30. Commoner, B., and Hollacher, T. C. (1960) Proc. Nat. Acad. Sci. U.S.A. 46, 405-411.
31. King, T. E. (1965) in Non-Heme Iron Proteins: Role in Energy Conversion (A. San Pietro, ed.) pp. 413-419, Antioch Press, Yellow Springs, Ohio.
31a. King, T. E. (1965) in Flavins and Flavoproteins (E. C. Slater, ed.), Discussion pp. 200-203, Elsevier, Amsterdam.
31b. Wilson, D. F., and King, T. E. (1967) Biochim. Biophys. Acta 131, 265-279.

32. DerVartanian, D. V. (1966) in Flavins and Flavoproteins (E. C. Slater, ed.) Discussion, p. 203, Elsevier, Amsterdam.
33. Colpa, J. P. (1966) in Flavins and Flavoproteins (E. C. Slater, ed.) Discussion, p. 203, Elsevier, Amsterdam.
34. King, T. E. (1964) Biochem. Biophys. Res. Commun. 16, 511-515.
35. Zeylemaker, W. P., DerVartanian, D., and Veeger, C. (1965) Biochim. Biophys. Acta 99, 183-184.
36. King, T. E., and Morris, R. O. (1967) Methods Enzymol. 10, 634-641.
37. King, T. E., Winter, D., and Steele, W. (1972) in Oxidation Reduction Enzymes (A. Akeson and A. Ehrenberg, eds.) pp. 519-532, Pergamon Press, Oxford and New York.
38. Davis, K. A., and Hatefi, Y. (1971) Biochemistry 10, 2509-2516.
39. Righetti, O. O., and Cerletti, O. O. (1971) FEBS Lett. 13, 181-185.
40. Coles, C. J., Tisdale, H. D., Kenny, W. C., and Singer, T. P. (1972) Physiol. Chem. Phys. 4, 301-316.
41. Beinert, H. (1973) in Iron-Sulfur Proteins (W. Lovenberg, ed.) Vol. 1, Chapter 1, Academic Press, New York.
42. Ohnishi, T., Winter, D. B., Lim, J., and King, T. E. (1973) Biochem. Biophys. Res. Commun. 53, 231-237.
43. Ohnishi, T., Leigh, J. S., Winter, D. B., Lim, J., and King, T. E. (1974) Biochem. Biophys. Res. Commun. 61, 1026-1035.
44. Ohnishi, T., Salerno, J. C., Winter, D. B., Lim, J., Yu, C. A., Yu, L., and King, T. E. (in preparation).
45. Beinert, H., and Ruzicka, J. (1975) in Abstracts of International Symposium on Electron Transfer Chains and Oxidative Phosphorylation, Fasano, September 15-18, p. 7.
46. Beinert, H., Ackrell, B. A. C., Kearney, E. B., and Singer, T. P. (1975) Eur. J. Biochem. 54, 185-194.
47. Bartsch, R. G. (1963) in Bacterial Photosynthesis (H. Gest, A. San Pietro, and L. P. Vernon, eds.) pp. 315-326, Antioch Press, Yello Spring, Ohio.
48. Green, D. E., Wharton, D. C., Tzagoloff, A., Rieske, J. S. and Brierley, G. P. (1965) in Oxidases and Related Redox Systems (T. E. King, H. S. Mason, and M. Morrison, eds.) Vol. 2, pp. 1032-1076, John Wiley and Sons, New York, New York.
49. Ohnishi, T., Winter, D. B., Lim, J., and King, T. E. (1974) Biochem. Biophys. Res. Commun. 61, 1017-1025.
50. Ohnishi, T., Lim, J., Winter, D. B., and King, T. E. (in preparation).
50a. Ohnishi, T., Winter, D. B., and King, T. E. (1976) Proc. of International Symposium on Electron Transfer Chains and Oxidative Phosphorylation, Fasano, September 15-18, p. 7
51. King, T. E. (1963) J. Biol. Chem. 238, 4037-4051.
52. Ketterman, Jr., J. (1964) Dissertation, Oregon State University, Corvallis, Oregon.
53. Vinogradov, A. D., Gavrikova, E. V., and Goloveshkiva, V. G. (1975) FEBS Abstract No. 1265.

54. Lim, J., Yu, C. A., Vinogradov, A. D., Ketterman, Jr., J., and King, T. E. (in preparation).
55. King, T. E. (1963) J. Biol. Chem. 238, 4032-4036.
55a. Orme-Johnson, N. R., Hansen, R. E., and Beinert, H. (1974) J. Biol. Chem. 249, 1928-1939.
56. Ackrell, B. A. C., Kearney, E. B., and Edmonason, D. (1975) J. Biol. Chem. 250, 7114-7119.
57. Ohnishi, T. (1975) Biochim. Biophys. Acta 387, 475-490.
58. Ohnishi, T., Cartledge, T. G., and Lloyd, D. (1975) FEBS Lett. 52, 90-94.
59. Leigh, J. S. (1970) J. Chem. Phys. 52, 2608-2612.
60. Mathews, R., Charlton, S., Sands, R. H., and Palmer, G. (1974) J. Biol. Chem. 249, 4326-4328.
61. Keilin, D., and King, T. E. (1958) Biochem. J. 69, 32p.
62. Giuditta, A., and Singer, T. P. (1959) J. Biol. Chem. 234, 666-671.
63. Keilin, D., and King, T. E. (1960) Proc. Roy. Soc., London B152, 163-187.
64. Lee, C. P., and King, T. E. (1962) Biochim. Biophys. Acta 59, 716-718.
65. Winter, D. B., Ohnishi, T., and Wu, J. T. Y. (in preparation)
66. Wang, T. Y., and Wang, Y. L. (1964) Sci. Sinica 13, 1779-1809.
67. Wu, J. T. Y. (1967) Dissertation, Oregon State University, Corvallis, Oregon.
68. DerVartanian, D. V. (1965) Dissertation, University of Amsterdam, The Netherlands.
69. DerVartanian, D. V., and Veeger, C. (1964) Biochim. Biophys. Acta 92, 233-247.
70. Takemori, S., and King, T. E. (1964) Science 144, 852-853.
71. Takemori, S., and King, T. E. (1964) J. Biol. Chem. 239, 3546-3558.
72. King, T. E., and Takemori, S. (1964) J. Biol. Chem. 239, 3559-3569.
73. Ford-Smith, M. H. (1964) The Chemistry of Complex Cyanide, Her Majesty's Stationery Office, London.
74. Petering, D., Fee, J. A., and Palmer, G. (1971) J. Biol. Chem. 246, 643-653.
75. Wallace, E. F., and Rabinowitz, J. C. (1971) Arch. Biochem. Biophys. 146, 400-409.
76. Hylin, J. W., and Wood, J. L. (1959) J. Biol. Chem. 234, 2141-2144.
77. Sieker, L. C., Adams, E., and Jensen, L. H. (1972) Nature 235, 40-45.
78. Carter, C. W., Kraut, J., Freer, S. T., Alden, R. A., Sieker, L. C., Adman, E., and Jensen, L. H. (1972) Proc. Natl Acad. Sci. 69, 3526-3532.
79. Kearney, E. B., Ackrell, B. A. C., and Mayr, M. (1972) Biochem. Biophys. Res. Commun. 49, 1115-1121.
80. Winter, D. B., and King, T. E. (1974) Biochem. Biophys. Res. Commun. 56, 290-295.

81. Hanstein, W. G., Davis, K. A., Ghalambor, M. A., Hatefi, Y. (1971) Biochemistry 10, 2517.
82. Vinogradov, A. D., Winter, D. B., and King, T. E. (1972) Biochem. Biophys. Res. Commun. 49, 441-444.
83. Winter, D. B., and King, T. E. (1974) Biochem. Biophys. Res. Commun. 56, 290-295.
84. Kimura, T. (1971) J. Biol. Chem. 246, 5140.
85. Coucouvanis, D., and Lippard, S. J. (1968) J. Amer. Chem. Soc. 90, 3281.
86. Watenpaugh, K. D., Sieker, L. C., Herriott, J. K., and Jensen, L. H. (1971) Cold Spring Harbor Symp. Quant. Biol. 36, 359-367.
87. Brintzinger, H., Palmer, G., and Sands, R. H. (1966) Proc. Natl Acad. Sci. 55, 397.
88. Jensen, L. H. (1972) Abstr. Metalloenzyme Conf., Oxford, p. 5
89. Carter, Jr., C. W., Freer, S. T., Xuong, N. H., Alden, R. A., and Kraut, J. (1971) Cold Spring Harbor Symp. Quant. Biol. 36, 381-385.
90. Orme-Johnson, W. H. (1973) Ann. Rev. Biochem. 42, 159.
91. Piesach, J., Blumberg, W. E., Lode, E. T., and Coon, M. J. (1971) J. Biol. Chem. 246, 5877.
92. Orme-Johnson, W. H., and Sands, R. H. (1973) in The Iron-Sulfur Proteins (Lovenberg, W., ed.) p. 195, Academic Press, New York, New York.
93. Sweeney, W. V., Bearden, A. J., and Rabinowitz, J. C. (1974) Biochem. Biophys. Res. Commun. 59, 188-194.
94. Bruni, A., and Racker, E. (1968) J. Biol. Chem., 243, 962-971.
95. Schatz, G., and Saltzaber, J. (1971) in Probes of Structure and Function of Macromolecules and Membranes (Chance, B., Yonetani, T., and Mildvan, A. S., eds.) Vol. 1, 437-444, Academic Press, New York, New York.
96. Mitchell, P., FEBS Lett. (1975) 56, 1-6.
97. Singer, J. S., and Nicolson, G. L. (1972) Science 175, 720-732.
98. King, T. E., Yu, C. A., and Yu, L. (1976) in Electron-Transfer Chains and Oxidative Phosphorylation (Quagliariello, E., Palmieri, F., Papa, S., Siliprandi, N., and Slater, E. C., eds) pp. 105-118, North-Holland Publishing Co., Amsterdam.

Postscript--This paper was written starting in the middle of September for the American-Japanese Seminar on Iron- and Copper-Proteins which has been painstakingly organized for nearly two years. It is adopted from several original papers and a longer review, all in preparation--November, 1975.

# HEME $\underline{a}$ AND COPPER ENVIRONMENTS IN CYTOCHROME OXIDASE

Yutaka Orii, Satoshi Yoshida and Tetsutaro Iizuka*

Department of Biology, Faculty of Science, and Department of Biophysics, Faculty of Engineering Science
Osaka University, Toyonaka, Osaka 560 JAPAN

## INTRODUCTION

In 1928, Warburg and Negelein obtained the action spectrum of a respiratory enzyme by utilizing the recovery of a carbon monoxide (CO)-inhibited respiration of yeast suspensions by illumination, and predicted the heme-like nature of the active site of this enzyme (Warburg and Negelein, 1928). Ten years later, from spectral examinations of the effect of several respiratory inhibitors on the heart muscle preparation, Keilin and Hartree postulated the existence of cytochrome $\underline{a}_3$, and considered the identity of this cytochrome with cytochrome oxidase or the respiratory enzyme. However, they reserved the final conclusion, since they failed to observe spectral changes inducible by illumination (Keilin and Hartree, 1939). This discrepancy, however, was reconciled by the finding of Chance that under the atmosphere of CO and $O_2$ in a 1:1 ratio, instead of 100% CO as employed by Keilin and Hartree, the CO complex was photodissociated easily at room temperatures (Chance, 1953). Chance *et al.* also found that at 77°K CO was photodissociated irreversibly from its complex (Chance *et al.*, 1965), and later the recombination of CO was shown to occur with the midpoint temperature of around 180°K as the sample temperature was raised (Yonetani *et al.*, 1973). Contrary to cytochrome oxidase-CO, the CO complexes of other hemoproteins like Hb, Mb, horse raddish peroxidase (HRP) and yeast cytochrome $\underline{c}$ peroxidase (CCP) released the CO almost irreversibly by irradiation at 4.2°K, and one half recombination occurred at 25-30°K (Yonetani *et al.*, 1973).

As a cause for this unusual behavior of cytochrome oxidase-CO, Chance *et al.* speculated that the lipids layers surrounding the oxidase provided a considerably less restricted structure which allowed the displacement of the photodissociated CO sufficiently far from the heme, and Yonetani et al. proposed that part of the protein molecule was inserted between the dissociated CO and the heme to prevent the access of the two reacting species. Although these proposals have not been supported by experimental evidence, the survey of this cause would be helpful in revealing the structure of the active center. Thus, we undertook the following investigations.

## NEW PHOTO- AND THERMOCHROMISM OF CYTOCHROME OXIDASE

Purified bovine heart cytochrome oxidase in solution usually exists as dimer (Orii and Okunuki, 1967; Matsumura et al., 1970; Orii and Yoshikawa, 1973). Therefore, the photochemical behavior of cytochrome oxidase-CO might be altered if the oxidase is converted to the monomer. However, when the monomer was prepared by a treatment with sodium dodecyl sulfate (SDS) (Orii and Okunuki, 1967) no alteration was observed. Monomeric cytochrome oxidase also can be obtained by incubating the oxidase at pH 9 to 10.5 at room temperatures (Matsumura et al., 1970; Orii and Yoshikawa, 1973). The CO complex of thus prepared monomeric oxidase exhibited the same absorption spectrum as that of the intact oxidase both at room and cryogenic temperatures. However, a difference spectrum between an alkali-treated cytochrome oxidase-CO and its reduced form showed unusual spectral changes after the illumination, and eventually the cause was traced to spectral changes occurring in the reference, the reduced form of the alkali-treated oxidase.

Figure 1 illustrates the low temperature absorption spectrum of the above reference with the blank medium as a reference. In addition to the main peaks at 598 and 438 nm, there appeared a sharp and symmetrical peak at 575 nm and a spike-like peak at 428-9 nm. The pH difference spectrum of the reduced oxidase between pH 9.40 and 7.33 showed two major peaks at 575 and 428 nm of nearly the same intensity. The intensity of these two peaks increased as the sample was kept longer between 200° and 250°K, although the 598- and 438-nm peaks in the absolute spectrum never disappeared. The spectrum in Fig. 1 clearly was different from that of a Schiff base of cytochrome oxidase formed between the formyl side group of the heme *a* and an ε-amino group of lysine residues, with a single, assymmetric broad peak at 573 nm (Orii and Okunuki, 1964).

The 575-nm peak appeared whether cytochrome oxidase had been reduced with either dithionite or ascorbate plus cytochrome *c*, and the alkali treatment of the oxidized oxidase failed to produce any spectral change relevant to this new peak.

When reduced cytochrome oxidase exhibiting the 575-nm peak was illuminated at 90°K, the peak diminished instantly, although it reverted to the original level with time. The process of recovery almost obeyed the first order kinetics and its apparent rate

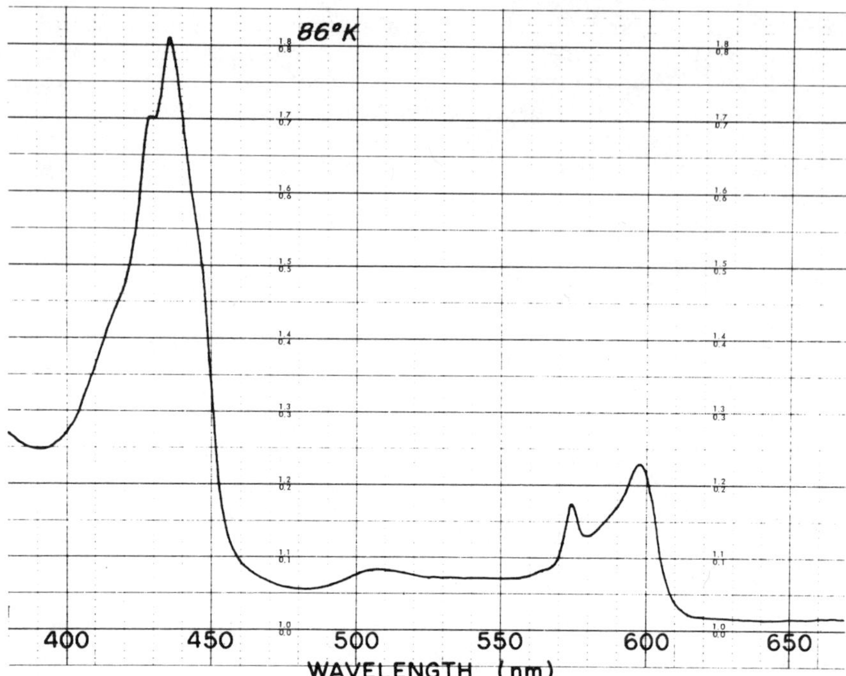


Fig. 1. A low temperature absorption spectrum of reduced cytochrome oxidase incubated at pH 9.43. Cytochrome oxidase (11.2 μM) in 0.05 M sodium phosphate buffer (pH 7.4)-0.25% Emasol 1130-0.1% cholate was reduced with sodium dithionite, and then made to pH 9.43 with 6 N NaOH at room temperatures. After 5 min standing this solution was transferred to the sample compartment of a twin cuvette assembly (Hagihara and Iizuka, 1971), and a blank medium to the reference compartment. The cuvette assembly was dipped into liquid nitrogen, and both the sample and reference solutions were frozen in 2 min. The sample temperature was 86°K.

constant was 8.3 x $10^{-3}$ $sec^{-1}$ (Fig. 2). Between 80° and 95°K such rate constants were determined, and from the Arrhenius plot $\Delta E^{\neq}$ and $\Delta S^{\neq}$ were obtained as 2.6 Kcal/mole and 10.9 eu., respectively. At 26°K, however, the 575-nm peak was photoquenched almost irreversibly as shown in Fig. 3. As the sample temperature was elevated slowly, this peak reappeared and all of the spectra recorded between 52° and 92°K shared clear isosbestic points at 582 and 568 nm in the visible region, suggesting the occurrence of two spectrophotometrically distinguishable species. As shown in Fig. 4, the temperature for 50% recovery was at around 80°K, and this was considerably lower than the midpoint temperature for the recombination of the once photodissociated CO at 170-180°K, initially found by Yonetani _et al._ and

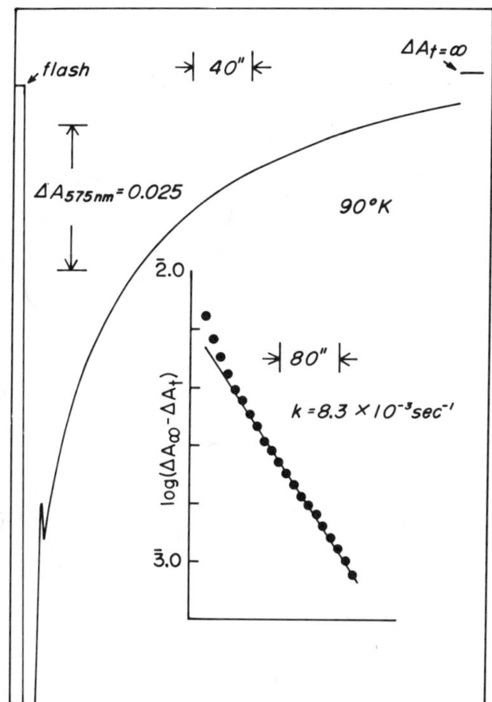


Fig. 2. Restoration of the 575-nm peak after its quenching by illumination at 90°K. Cytochrome oxidase (45.6 μM) was reduced with dithionite for 25 min and the solution was made to pH 9.44 with 6 N NaOH. Five minutes later this sample was frozen in liquid nitrogen. One shot of a photographic flash considerably quenched the 575-nm peak.

confirmed in the present study. Although the reappearance was complete at 100°K, it again disappeared as the sample temperature was raised above that. Based on the assumption that the intensity decrease due to broadening of the absorption peak was practically negligible, fractions of the 575-nm species as well as those of the ordinary species were estimated from the absorbance differences between the 575-nm peak and the isosbestic point at 585 nm. With another assumption that at each temperature the equilibrium was established between these two species, the equilibrium constants were estimated at each temperature. Using these parameters thermodynamic constants such as enthalpy and entropy changes were determined from the plot of log K against the inverse of temperature (Fig. 5).

The 575-nm peak of cytochrome oxidase also appeared even when it was incorporated into membrane constructions. Using heart muscle preparations of the Keilin-Hartree type, we were able to observe

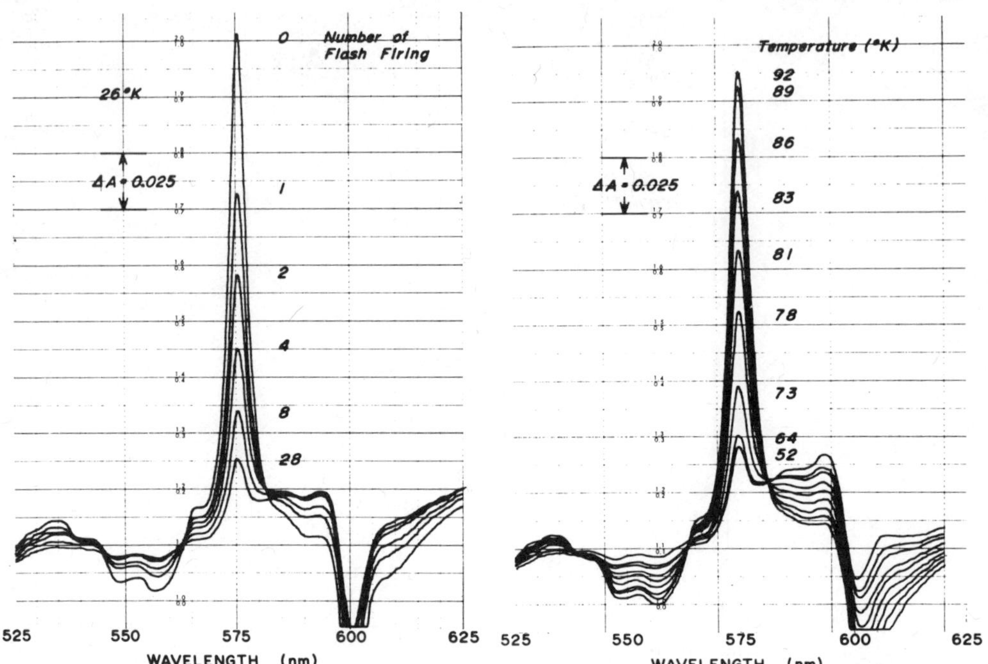


Fig. 3A

Fig. 3B

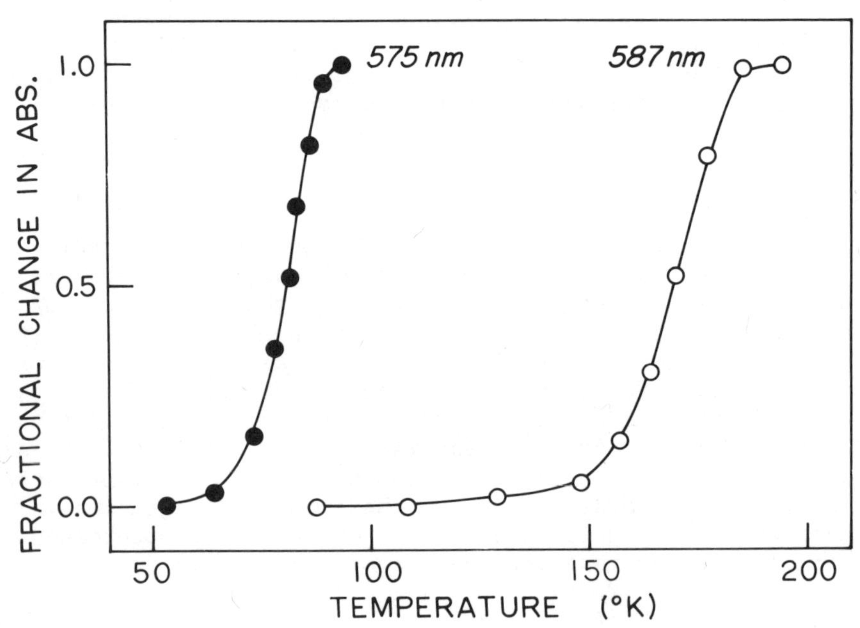


Fig. 4

Fig. 3. Quenching of the 575-nm peak by irradiation at 26°K and its restoration with a rise in temperature. Cytochrome oxidase (29.8 μM) was reduced with dithionite and brought to pH 9.55. This solution was put into the sample compartment of the twin-cuvette assembly and the reduced oxidase at pH 7.33 was in the reference. Both solutions were first frozen in liquid nitrogen and then cooled in liquid hydrogen. The numbers of flash firing are indicated in the figure (A). After almost the complete quenching of this peak the sample temperature was raised with a heating system. During the recording of the spectrum the sample temperature was kept as indecated (B).

Fig. 4. Temperature-dependent restoration of the photoquenched bands at 575 and 587 nm. In separate experiments the 575-nm and 587-nm peaks were quenched by irradiation of more than 50 flashes at 26° and 86°K, respectively. The latter was for cytochrome oxidase-CO. Then the sample temperature was raised as described in the legend for Fig. 3.

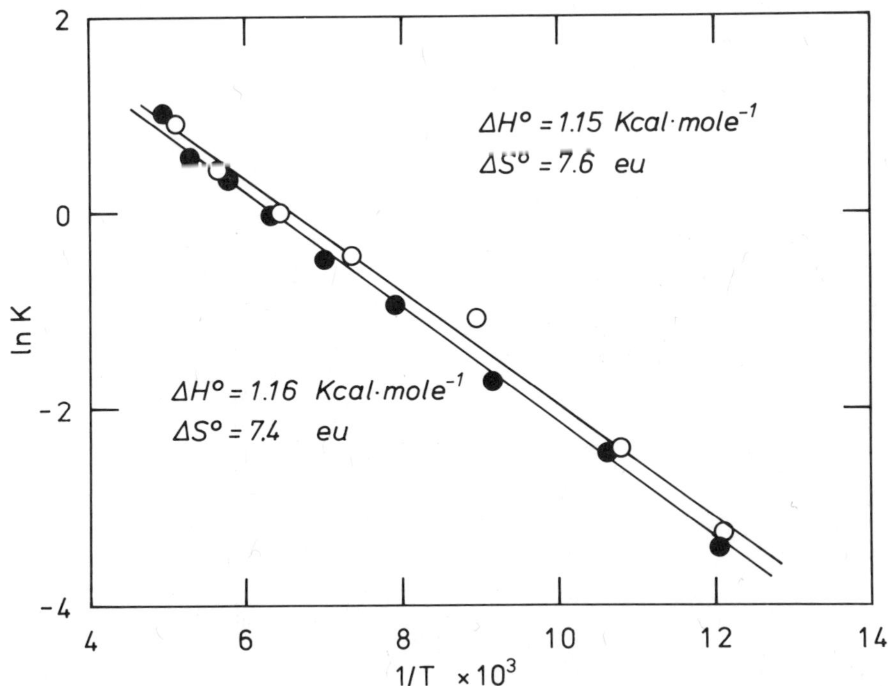


Fig. 5. Equilibrium constants for the formation of the 575-nm species of cytochrome oxidase purified (●) and in submitochondrial particles (○). Reduced cytochrome oxidase was adjusted to pH 9.30, whereas a suspension of dithionite-reduced submitochondrial particles was made to pH 10.0.

the same phenomena as observed with the purified preparation of cytochrome oxidase, and in Fig. 5 the thermodynamic parameters are summarized, too. It is noteworthy that these parameters were almost identical in both cases. Thus, the microenvironments of the heme a in cytochrome oxidase were concluded to be unaltered, as far as they were concerned with the 575-nm peak appearance, even when the enzyme was exposed to serverl treatments which were indispensable for the purification of the enzyme.

The magnitude of the enthalpy changes is smaller than the bond energy for covalent linkages by a factor of around 10, and larger than that for Van der Waals interactions, and falls in an order for the hydrogen bond. Thus, the appearance of the 575-nm peak was concluded to be associated with the hydrogen bond formation.

Among several candidates as a proton donor in the hydrogen bond formation, we preferred one of the seven sulfhydryl groups in cytochrome oxidase (Matsubara et al., 1965), since the pretreatment of the reduced enzyme with p-chloromercuribenzoate (pCMB) prevented appreciably the appearance of the 575-nm peak. The previous

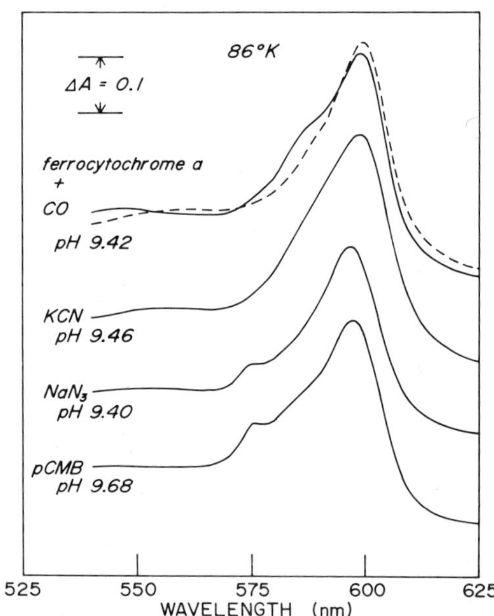


Fig. 6. Effect of several treatments of cytochrome oxidase on the appearance of the 575-nm peak. Before the solution of reduced cytochrome oxidase (21-23 μM) was made alkaline, one of the following ligands had been added. CO gas was bubbled through the solution for 1 min. The concentrations of potassium cyanide, sodium azide and pCMB were 45.5 mM, 50 mM and 238 μM, respectively.

liganding of CO, cyanide and azide to the heme iron was also inhibitory (Fig. 6). Therefore, the access of the two reacting moieties was suggested to be hindered by an incorporation of a foreign substance between the heme plane and the immediate protein environments.

A proton acceptor would be the formyl side group in heme *a*, since the 575-nm peak appearance, at shorter wavelengths than 590 to 600 nm for the ordinary heme *a* peaks, suggested a considerable decrease in the electron withdrawing capacity of the formyl group. In fact, when this was lost by the formation of a Schiff base, cyanhydrin and the addition product of sodium bisulfite (Yanagi *et al.*, 1972; Orri and Iizuka, 1975) there invariably appeared the absorption spectra of the protoheme-type instead of the heme *a*-type. Therefore, the localization of the π-electron(s) in the carbonyl bond towards the carbonyl oxygen, stabilized by the formation of the hydrogen bond, would have resulted in the appearance of the 575-nm peak.

At present it is not certain if the coexistence of the 575-nm peak with the major one at 598 nm would indicate the separate existence of the two species with each peak, or if this unusual spectrum would be due to an only one entity, and this problem is open to further investigations.

## INVOLVEMENT OF COPPER IN THE CARBON MONOXIDE COMPOUND FORMATION

In addition to the CO complexes of Hb, Mb, CCP, HRP, deutero-Mb and meso-Mb, in which the recombination of the photodissociated species was complete at 50°K, those of *Potamilla* chlorocruorin at the prosthetic chlorocruoroheme (Orii, *unpublished*) and of *Pseudomonas* nitrite reductase with the *heme d* moiety (Orii & Shimada, unpublished) were found to belong to these members. Thus, still only cytochrome oxidase-CO remains unusual in that the photodissociation occurs almost irreversibly at the temperature of liquid nitrogen (See Fig. 4). Since this character was still retained with monomeric cytochrome oxidase, prepared either by the SDS or alkali-treatment, we speculated that the copper involved only in this enzyme among all of the hemoproteins examined so far would confer this character on the oxidase-CO complex: If this copper is situated close to the heme *a* and the photodissociated CO is trapped by this, the sample temperature as low as 50°K would not be sufficiently high enough to bring the copper-bound CO to the vicinity of the heme iron.

When this idea was mature in our minds, an interesting paper was reported by Lindsay and Wilson (Lindsay and Wilson, 1974). These authors carried out the redox titration of cytochrome oxidase, soluble and *in situ*., and found that especially in the presence of 30-300 μM CO, the CO complex of the enzyme was titrated as a single species with an n value of 2, and speculated that the invisible copper was involved in the CO complex formation. Thus, we were prompted

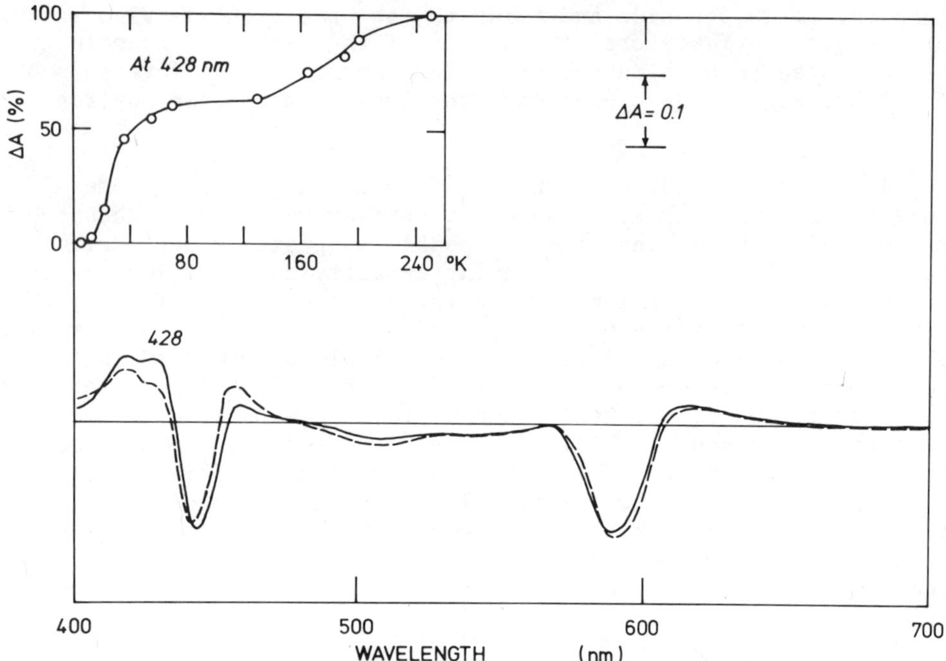


Fig. 7. Spectral change after irradiation of the carbon monoxide complex of monomeric cytochrome oxidase treated with pCMB. Cytochrome oxidase (31.6 μM) was treated with 0.31% SDS and 0.29 mM pCMB at pH 7.4. After the complete reduction with dithionite CO gas was bubbled through the solution for 1 min, and this solution was used as the sample. The reduced enzyme before the CO bubbling was used as the reference. The sample was irradiated by more than 50 flashes at 5°K.

to examine the above-mentioned possibility.

The intrinsic copper in cytochrome oxidase has been reported to be bound to at least one of the seven sulfhydryl groups in the enzyme (Tsudzuki et al., 1967) or to two of them (Kirschbaum and Wainio, 1966). Especially the former investigators revealed that this copper was removed effectively from the protein when pCMB was acted on the depolymerized enzyme with SDS.

By cross-examining the effect of SDS and pCMB on the spectral properties of cytochrome oxidase-CO, we found that when pCMB was acted on the monomeric oxidase its spectrum changed drastically, whereas in the absence of pCMB the monomer was almost the same as the dimer in respect of the spectral characteristics (See Orii and Okunuki, 1967); that is, the reduced form and its CO complex of the pCMB-treated monomer exhibited the absorption maxima at 601 and 441 nm, and at 602 and 434 nm, respectively. Noticeably the height of the reduced Soret peak was lower than that of the oxidized Soret.

Figure 7 illustrates difference spectra between the CO complex and the reduced form at 5°K before and after the illumination. Even after firing of more than 50 flashes only a small absorbance change occurred throughout the wavelength range examined, and the Soret positive peak at 428 nm showed a temperature-dependent change, although the magnitude of this change was too small to acount for the complete photodissociation of CO complex at 5°K. A model compound, heme *a*-imidazole-CO, also showed no positive sign for the photodissociability at 5°K.

Thus, the displacement of the intrinsic copper from its mercaptide with one or two of the sulfhydryl groups apparently shifted the midpoint temperature for the recombination towards lower temperatures, and the significance of the intrinsic copper in the CO complex was indicated. Although it is natural to assume that because of conformational changes what was destroyed upon displacement of the intrinsic copper is the bond between the CO and one of other ligands than the copper and heme iron, this possibility is less likely, since Yonetani *et al*. showed even if the oxygen molecule was bound to the cobalt atom in CoMb or CoHb and to the distal histidine forming an intramolecular bridge, especially the latter through the hydrogen bond formation, the oxygenated compound was photodissociated and restored in the temperature range of 4.2° to 60°K (Yonetani *et al*., 1974).

The negligible photodissociability of the CO complex of free heme *a* was in a marked contrast to the results of Yonetani *et al*. that the photodissociation and recombination of the CO complexes of protoheme and heme *a* occurred at the temperature of liquid helium (Yonetani *et al*., 1973), but in harmony with the suggestion by Kitagawa *et al*.. They pointed out that the bond type between a ligand and heme iron, which is determined by the electronic structure of ligand as well as its geometric arrangement to the heme, possibly determined the easiness of photodissociability (Kitagawa *et al*., 1975). Therefore, in the CO complex of free heme *a* the bond axis between C and O atoms would be perpendicular to the heme plane in favor of the $\sigma$-type bond and less photodissociability is conceivable, whereas in the CO complex of intact cytochrome oxidase, the bond axis would be tilted towards the heme plane due to another coordination of the CO to the intrinsic copper and to a tight space between the heme *a* and its immediate surroundings of the protein origin, and consequently the $\pi$-bond character would prevail in favor of the photodissociability.

The photochemical behavior of a so-called mixed valence cytochrome oxidase-CO complex (Horie and Morrison, 1963) was quite the same as that of the ordinary complex. Since the visible copper was shown to be oxidized (King and Yong, 1973), the copper trap must be ascribed to the invisible copper in the cuprous state. This picture does not contradict the results of the titration studies of King and Yong, and is in accord with a proposal by Wever *et al*. (Wever *et al*., 1974).

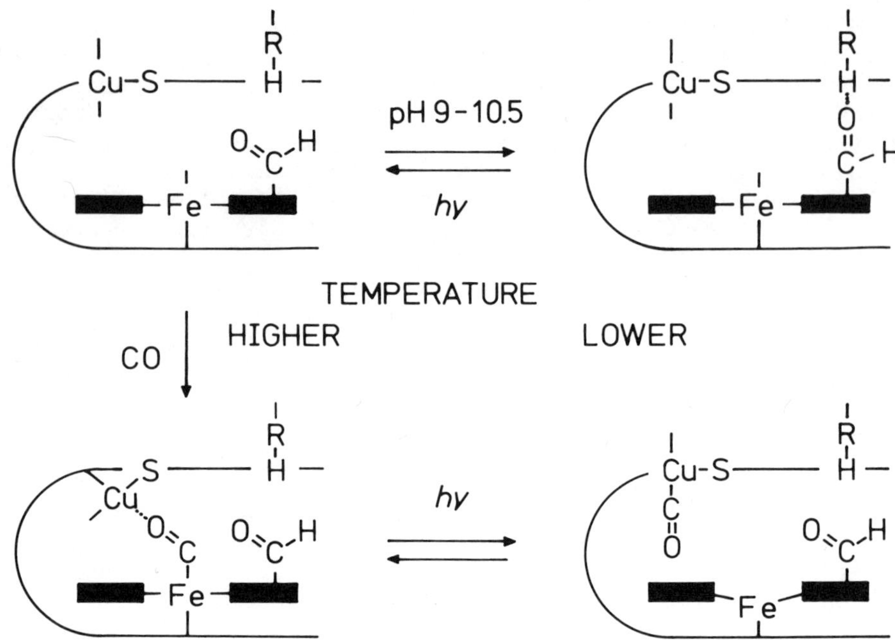


Fig. 8. A schematic model for microenvironments of heme a and copper in cytochrome oxidase at cryogenic temperatures.

Based on the findings in the present studies, a schematic model for microenvironments of heme a and copper in cytochrome oxidase is presented above. The formyl side group was placed near the opening of a heme-copper pocket, since that location was suggested by the solvent perturbation study of cytochrome oxidase (Yamamoto and Orii, 1973).

## REFERENCES

Chance, B., 1953, J. Biol. Chem., 202, 397.

Chance, B., Schoener, B. and Yonetani, T., 1965, ISOX I, 609.

Hagihara, B. and Iizuka, T., 1971, J. Biochem., 69, 355.

Horie, S. and Morrison, M., 1963, J. Biol. Chem., 238, 1855.

Keilin, D. and Hartree, E. F., 1939, Proc. Roy. Soc. London, B127, 167.

Kirschbaum, J. and Wainio, W. W., 1966, Biochim. Biophys. Acta,

118, 643.
Kitagawa, T., Iizuka, T., Saito, M. I. and Kyogoku, Y., 1975, Chem. Lett., 849
Lindsay, J. G. and Wilson, D. F., 1974, FEBS Lett., 48, 45.
Matsubara, H., Orii, Y. and Okunuki, K., 1965, Biochim. Biophys. Acta, 97, 61.
Matsumura, Y., Orii, Y. and Okunuki, K., 1970, Seikagaku (in Japanese), 42, 646.
Orii, Y. and Okunuki, K., 1967, J. Biochem., 61, 388
Orii, Y. and Okunuki, K., 1964, J. Biochem., 55, 37.
Orii, Y. and Yoshikawa, S., 1973, ISOX II, 649.
Orii, Y. and Iizuka, T., 1972, Biochem. Biophys. Res. Comm., 48, 885.
Orii, Y. and Iizuka, T., 1975, J. Biochem., 77, 1123.
Tsudzuki, T., Orii, Y. and Okunuki, K., 1967, J. Biochem., 62, 37.
Warburg, O. and Negelein, E., 1928, Biochem. Z., 202, 202.
Wever, R., Van Drooge, J. H., Van Ark, G. and Van Gelder, B. F., 1974, Biochim. Biophys. Acta, 347, 215.
Yamamoto, T. and Orii, Y., 1973, J. Biochem., 74, 473.
Yanagi, Y., Sekuzu, I., Orii, Y. and Okunuki, K., 1972, J. Biochem., 71, 47.
Yonetani, T., Iizuka, T., Yamamoto, H. and Chance, B., 1973, ISOX II, 401.
Yonetani, T., Yamamoto, H. and Iizuka, T., 1974, J. Biol. Chem., 249, 2168.
King, T. E. and Yong, F. C., 1973, ISOX II, 677.

# HEME INTERACTIONS IN *PSEUDOMONAS* CYTOCHROME OXIDASE

David C. Wharton, Kristina Hill, and Quentin H. Gibson

Department of Biochemistry, The University of Texas Health Science Center, San Antonio, Texas and Section of Biochemistry, Cell and Molecular Biology, Cornell University, Ithaca, New York

## INTRODUCTION

*Pseudomonas* cytochrome oxidase (ferrocytochrome $c_2$:$O_2$ oxidoreductase, EC 1.9.3.2) functions as a terminal electron carrier in the bacterium *Pseudomonas aeruginosa*. The oxidase can donate electrons either to molecular oxygen or to nitrite although it appears that *in situ* it is intended to react with nitrite rather than $O_2$ since the enzyme is only synthesized when the organism is grown anaerobically in the presence of nitrate or nitrite. However, since *Pseudomonas* cytochrome oxidase can catalyze the reduction of $O_2$ to $H_2O$ (a four-electron transfer) it bears some similarity functionally to the cytochrome *c* oxidase of the mitochondrial respiratory chain.

*Pseudomonas* cytochrome oxidase was first purified and studied by Okunuki and his associates (Horio, 1958a; Horio *et al*., 1958) at the University of Osaka. Their purified preparation contained both heme *c* and heme *d* in a crystallizable protein (Yamanaka *et al*., 1962) that, unlike mitochondrial cytochrome *c* oxidase, was water-soluble without the addition of detergents. The elegant studies of Okunuki and co-workers (Horio, 1958b; Horio *et al*., 1961; Yamanaka and Okunuki, 1963a) established that the *Pseudomonas* enzyme was able to accept electrons from two other constituents of the organism, namely a *c*-type cytochrome called cytochrome *c*-551 and a blue-colored, copper-containing protein often referred to as azurin. The oxidase reduced by either of these reductants could then be oxidized either by O2 or by nitrite. These investigators also observed that the classical inhibitors of mitochondrial cytochrome c oxidase, cyanide and carbon monoxide, also inhibited the Pseudomonas enzyme (Horio et al., 1958).

Since the pioneering investigations of Okunuki et al. Pseudomonas cytochrome oxidase has been purified to electrophoretic homogeneity by Kuronen and Ellfolk (1972) and by Gudat et al. (1973) who have utilized newer purification techniques that were not available at the time the early work was performed. Present evidence indicates that Pseudomonas cytochrome oxidase has a molecular weight of about 120,000 and consists of two equivalent subunits (Kuronen and Ellfolk, 1972; Gudat et al., 1973). The intact enzyme contains two heme c groups and two heme d groups (Kuronen, Soraste, and Ellfolk, 1975) both of which contribute to the absorption spectrum as seen in Figure 1. The finding by Nagata et al. (1970) that only one N-terminal amino acid, a lysine residue, is present in the protein and that there are two cysteine residues per 60,000 molecular weight suggests that the two subunits may be identical in their amino acid sequence as well as in their molecular weight. If this were to be the case then each subunit could contain one heme c group and one heme d group. At present, however, there is no other experimental evidence to support this hypothesis.

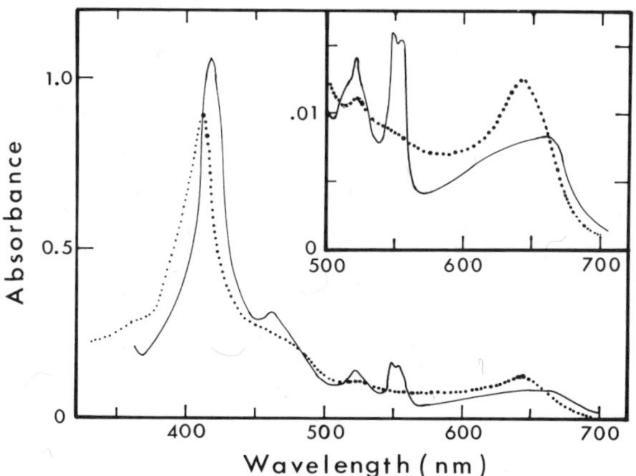


Fig. 1. Absorption spectra of Pseudomonas cytochrome oxidase in 0.05 M Tris-HCl buffer, pH 8.0. ·····, the oxidized preparation; ———, the preparation reduced with $Na_2S_2O_4$.

Since a molecule of Pseudomonas cytochrome oxidase contains four electron-transferring groups and is able to catalyze the reduction of $O_2$ to $H_2O$ it has the potential to serve as a model for the detergent-stabilized mitochondrial cytochrome *c* oxidase. Moreover, its availability in the crystalline state, as well as the finding that the heme *d* moiety can be removed and reconstituted (Yamanaka and Okunuki, 1963b), provides opportunities for solving fundamental questions about the reaction with $O_2$ and the transfer of electrons between heme groups in proteins.

## REACTION WITH COPPER PROTEIN

When reduced copper protein is mixed anaerobically with oxidized Pseudomonas cytochrome oxidase and the reaction is monitored using rapid-reaction spectrophotometric techniques, biphasic absorbance changes are observed at 595, 549, and 424 nm and a monophasic change at 460 nm (Wharton *et al*., 1973). The wavelengths at 549 and 424 nm represent absorption maxima for the ferroheme *c* moiety of the enzyme while those at 595 and 460 nm represent maxima for the cupric form of azurin and the ferroheme *d* of the oxidase, respectively. An example of the biphasic reaction of heme *c* as monitored at 549 nm is presented in Figure 2 which illustrates the spectral change that occurs after mixing 50 μM copper protein with 15 μM oxidase (in terms of the concentration of heme *c*). The rapid phase of the reaction has a half-time of 32 ms and is virtually complete within 200 ms while the slower phase has a half-time of at least 200 ms and requires several seconds to reach completion. The absorbance changes at 424 nm and at 595 nm have similar profiles and half-times. When the absorbance changes at 549 nm are plotted against the changes at 595 nm, straight lines

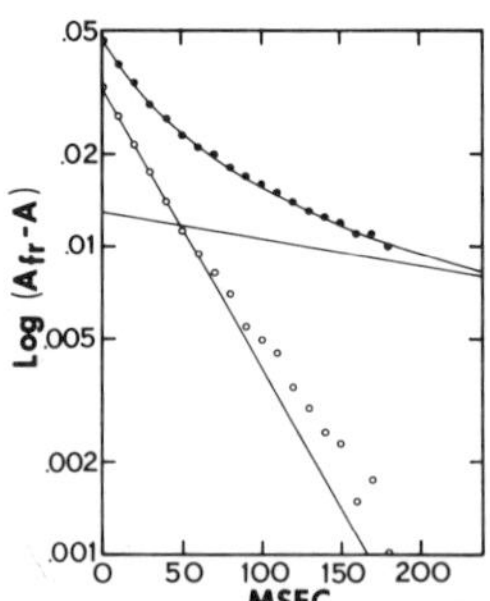


Fig. 2. Reduction of Pseudomonas cytochrome oxidase (15 μM heme *c*) by 50 μM azurin followed at 549 nm. Temperature was 20°C; buffer was 0.1 M phosphate, pH 6.6. ●, data points from the reaction trace; o, points obtained after subtracting the slow phase of the reaction (solid line) from the points of the reaction trace.

are obtained with a slope of about 3. This value is in close agreement with the ratio of the $\Delta\varepsilon_{549\ nm}$ ($\underline{c}^{2+}$ minus $\underline{c}^{3+}$):$\Delta\varepsilon_{595\ nm}$ ($Cu^{1+}$ minus $Cu^{2+}$). The rate and magnitude of these absorbance changes are proportional to the ratio of the concentration of azurin to oxidase.

The total absorbance changes at 460 nm are rather small because the heme d group has a more negative oxidation-reduction potential than either the heme c group or the copper protein and because the heme c is contributing to the absorbance change in the opposite direction. As a result of these problems the data for the absorbance changes at 460 nm during the reduction of the oxidized enzyme by reduced azurin are not particularly good. It seems, however, that the absorbance changes are monophasic and that their rates are comparable to the slower phase of the biphasic reactions observed for heme c and azurin.

In addition to serving as a donor of electrons to Pseudomonas cytochrome oxidase azurin can also accept electrons from the oxidase. When oxidized azurin is mixed anaerobically with reduced Pseudomonas cytochrome oxidase in the stopped-flow spectrophotometer biphasic absorbance changes are again observed at 595, 549 and 424 nm and a monophasic change at 460 nm. As in the case of the transfer of electrons from the reduced azurin to the oxidized enzyme the rate and magnitude of the absorbance changes are proportional to the ratios of the concentrations of azurin to oxidase. Figure 3 illustrates the absorbance changes determined at 549 nm after mixing 60 μM oxidized azurin with 15 μM ascorbate-reduced Pseudomonas cytochrome oxidase (in terms of heme c). The more rapid phase of this reaction has a half-time of 8 ms and is virtually complete within 50 ms while the slower phase has a half-time of 75 ms and requires several seconds to reach completion. Analysis of the data indicates that the rapid portion represents the oxidation of the ferroheme c component of the oxidase by the azurin.

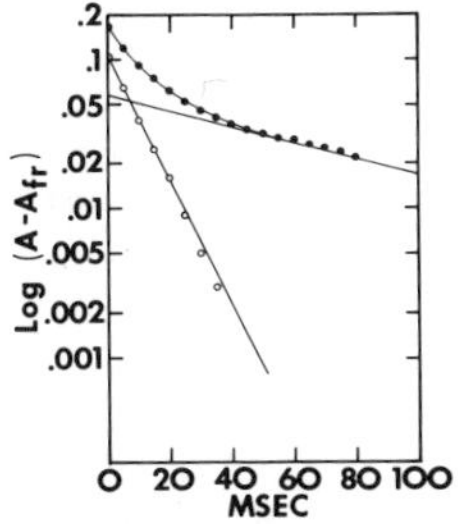


Fig. 3. Oxidation of Pseudomonas cytochrome oxidase (15 μM heme c) by 60 μm azurin followed at 549 nm. Temperature was 20°C; buffer was 0.1 M phosphate, pH 6.6. ●, data points from the reaction trace; o, points obtained after subtracting the slow phase of the reaction (solid line) from the points of the reaction trace.

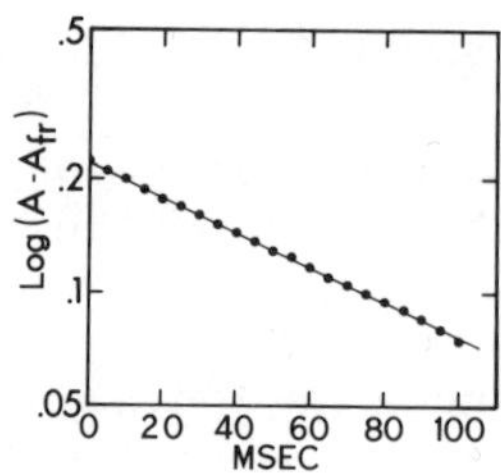


Fig. 4. Oxidation of Pseudomonas cytochrome oxidase (15 μM heme c) by 60 μM azurin followed at 460 nm. Temperature was 20° C; buffer was 0.1 M phosphate, pH 6.6.

The absorbance changes at 460 nm are much greater in the experiments where oxidized azurin is allowed to react with reduced oxidase than when azurin was used as a reductant of the enzyme. This result occurs because the oxidation-reduction potentials of the reactants favor oxidation of the heme d which is represented by the absorbance changes at 460 nm. As seen in Figure 4 these absorbance changes are monophasic and have a half-time of 70 ms, a value that is comparable to the slower changes observed for the heme c moiety and the azurin.

When CO is added to reduced Pseudomonas cytochrome oxidase it combines with the heme d moiety but not the heme c. If the CO complex of the oxidase is mixed with oxidized azurin the azurin becomes reduced by electrons donated from the ferroheme c group while the heme d remains reduced and bound to CO. When the reaction between oxidized azurin and the CO complex of the enzyme is monitored in the stopped-flow spectrophotometer a monophasic absorbance change is now observed at 595, 549 and 424 nm while no change is seen at 460 nm. Figure 5 shows the absorbance change for the heme c group determined at 549 nm after mixing 60 μM azurin with 15 μM Pseudomonas cytochrome oxidase (in terms of heme c). This reaction has a half-time of 7 ms, a value that compares quite favorably with the rapid phase of the uninhibited reaction. The rates observed at 595 and 424 nm are identical.

These results can be interpreted to mean that azurin interacts with Pseudomonas cytochrome oxidase and that a rapid electron transfer occurs between the copper and heme c. The rapid transfer is followed by a slower transfer of an electron between the heme d and

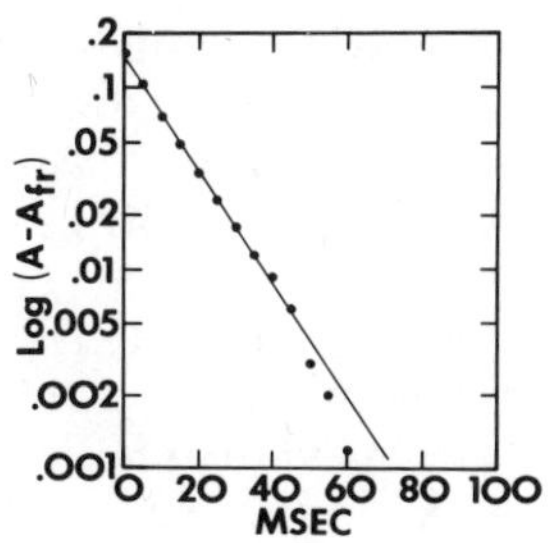


Fig. 5. Oxidation of Pseudomonas cytochrome oxidase (15 μM heme c) by 60 μM azurin followed at 460 nm in the presence of 1 atmosphere of carbon monoxide. Temperature was 20° C; buffer was 0.1 M phosphate, pH 6.6.

the heme c components of the oxidase. The rates of the reaction involving heme d are at least an order of magnitude slower than those involving heme c and azurin. Furthermore, a significantly higher reaction rate is observed between the heme c group and azurin as the concentration of the copper protein is increased. If the reciprocals of the observed pseudo first order rate constants of both the forward and reverse electron transfers are plotted as a function of the reciprocal of the concentration of azurin, a limit of the first-order rate constants for each transfer can be estimated as shown in Figure 6. These values are 29 $s^{-1}$ for the transfer between reduced azurin and oxidized heme c and 120 $s^{-1}$ for the transfer between reduced heme c and the oxidized copper protein.

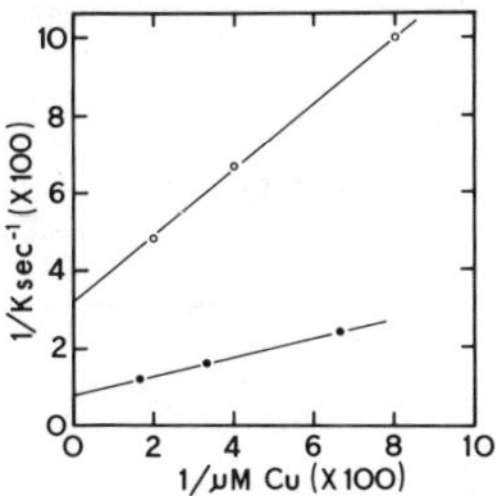


Fig. 6. Double reciprocal plot of pseudo first order rate constants as a function of azurin concentration for the reaction with Pseudomonas cytochrome oxidase (15 μM heme c). o, reaction between reduced azurin and oxidized oxidase; •, reaction between oxidized azurin and reduced oxidase.

The hyperbolic character of the observed rate constants when plotted as a function of azurin concentration suggests that a complex is formed between the oxidase and azurin. Furthermore, the data indicate that the rate of complex formation is considerably faster than the rate of electron transfer since little if any lag in the absorbance changes is observed after mixing the reactants. Similar conclusions were stated recently by Brunori _et al_. (1975) on the basis of temperature-jump studies of the reaction between azurin and _Pseudomonas_ cytochrome oxidase.

## REACTION WITH OXYGEN

The reaction of ascorbate-reduced _Pseudomonas_ cytochrome oxidase with $O_2$ has been examined systematically over a range of wavelengths and at several concentrations of $O_2$. Figure 7 shows the absorbance changes at 549 nm and 460 nm after mixing 12 μM oxidase (in terms of heme _c_) with 142 μM $O_2$ in the stopped-flow spectrophotometer. The faster change at 460 nm proceeds with a half-time of about 85 ms and is nearly complete 400 ms after mixing. The change at 549 nm, on the other hand, proceeds with a half-time of about 150 ms and at 400 ms after mixing the solution still contains a significant amount of unreacted material.

The reaction rates at 460 nm show a linear relationship with the concentration of $O_2$ and calculations indicate that the reaction involving the oxidation of the heme _d_ group is accurately second order with respect to the concentration of $O_2$. The second order rate constant calculated from these data is about $5.7 \times 10^4$ $M^{-1}$ $s^{-1}$ at 20°. Similar rates can be calculated from spectral changes at 660 nm and 645 nm, the α-region of the heme _d_ component.

The oxidation of the ferroheme _c_ moiety of the enzyme, as indicated by the absorbance change at 549 nm, is slower than that

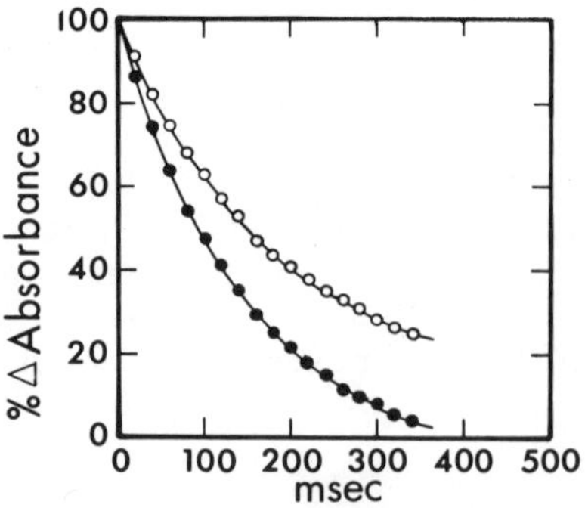


Fig. 7. Oxidation of reduced _Pseudomonas_ cytochrome oxidase (12 μM heme _c_) by 142 μM $O_2$ followed at 460 nm (•) and at 549 nm (o). Temperature was 20° C; buffer was 0.05 M phosphate, pH 6.6.

involving heme *d* and appears to be first order. From the data shown in Figure 7, a first order rate constant of 4.5 $s^{-1}$ can be calculated.

These results indicate that the reaction between *Pseudomonas* cytochrome oxidase and $O_2$ occurs in at least two steps which can be interpreted as follows: one is the oxidation of ferroheme *d* by $O_2$ and the second is the reduction of ferriheme *d* by ferroheme *c*. The first step is accurately second order and proceeds with a rate constant of $5.7 \times 10^4$ $M^{-1}$ $s^{-1}$ at 20° while the second step is slower than the first and is first order. When an excess of an electron donor to the enzyme is present a larger proportion of heme *c* than heme *d* is maintained in the reduced form indicating that the rate-limiting step is the reduction of heme *d* by heme *c*. Identification of the rate-limiting step as the reduction between heme *c* and heme *d* is also supported by the data obtained for the reaction between the oxidase and azurin.

The rates with which *Pseudomonas* cytochrome oxidase reacts with $O_2$ are considerably slower than the rates observed by Gibson *et al*. (1965) for the mitochondrial cytochrome *c* oxidase from beef heart. Thus, the second order rate constant of $5.7 \times 10^4$ $M^{-1}$ $s^{-1}$ for the *Pseudomonas* enzyme is some three orders of magnitude less than that calculated for the mitochondrial oxidase.

The absorbance changes referred to above occur within the first 500 ms after mixing *Pseudomonas* cytochrome oxidase with relatively high concentrations of $O_2$. In addition, there are further and more complicated changes that appear to involve the heme *d* group and which take place between 500 ms after mixing and completion of the reaction several seconds later. The significance of these slower secondary absorbance changes, which account for no more than about 10% of the total change, has not been determined.

Present evidence indicates that *Pseudomonas* cytochrome oxidase contains four potential reducing equivalents, (two heme *c* and two heme *d* groups) (Kuronen and Ellfolk, 1975), when the enzyme is fully reduced. These four equivalents could provide a source of electrons for the reduction of $O_2$ to water. The present results do not enable us to predict a mechanism for the reduction of $O_2$ nor do they differentiate between the pairs of heme *d* and heme *c* in the dimeric unit. However, they do suggest a need for further experiments to search for intermediates in the reduction of $O_2$ as well as to determine how heme *d* on the protein acts in this reduction. It may be pertinent to point out that in some recent studies involving the reaction of *Pseudomonas* cytochrome oxidase with CO, Parr *et al*. (1975) observed biphasic absorbance changes involving the heme *d* group and suggested the presence of cooperativity to explain these changes. Although with our preparation we have not observed a similar biphasic reaction (unpublished results) the possible presence of a cooperative reaction is important and certainly bears closer inspection.

## REMOVAL AND RECONSTITUTION OF HEME D

Yamanaka and Okunuki (1963b) reported the first successful reconstitution of heme d into Pseudomonas cytochrome oxidase after they had earlier removed the heme d using acidified acetone. Reconstitution was carried out by adding an aqueous solution of heme d to the protein at pH 9 and then adjusting the pH to 7 by adding dilute HCl. The reconstituted enzyme had an absorption spectrum that was similar although not identical to the native enzyme and showed 37% of the original oxidase activity. Yamanaka and Okunuki (1963b) also succeeded in reintroducing other heme groups, including protoheme, hematoheme, and heme a from the mitochondrial enzyme, into heme d-depleted enzyme. They found some oxidase activity in each of the reconstituted proteins but in no case did the specific activity approach that of the preparation reconstituted with heme d.

We have recently developed a new reconstitution procedure for Pseudomonas cytochrome oxidase which involves adding a neutral aqueous solution of heme to the protein which is dissolved in 6 M urea at neutral pH. The heme to be reconstituted is added at a concentration that is equimolar with that of the heme c already in the protein. We have found that using this procedure the protein reconstituted with heme d routinely has 90 to 100% of the original oxidase activity whereas that reconstituted according to the procedure of Yamanaka and Okunuki (1963) exhibits only 30 to 40% of the original activity, in agreement with their earlier observations. Furthermore, the absorption spectra of the oxidase reconstituted by the new procedure are identical to the original preparation at the same pH.

The heme d-depleted enzyme has been reconstituted with the heme groups shown in Table 1. As can be seen in the table only the protein reconstituted with heme d and with heme a has oxidase activity and even that containing heme a is only about 5% of the native enzyme. With the exception of heme a these results do not agree with those reported by Yamanaka and Okunuki (1963b) who found significant activity in the oxidase reconstituted with protoheme and hematoheme.

The absorption maxima of the reduced form of the oxidase reconstituted with the various heme groups are shown in Table 2. It is of interest to note that the maxima of the heme c moiety at 554, 549, 521 and 418 nm are essentially unchanged in their wavelength no matter which heme is used for reconstitution.

The EPR spectra of native Pseudomonas cytochrome oxidase and the various reconstituted forms have been compared. The EPR spectrum of the native enzyme in its oxidized state is shown in Figure 8. The signals with g-values near 2.45 and 1.71 are due to the heme _d_ while those at g = 2.93, 2.31 and 1.4 are due to the heme _c_ component of the enzyme. Those signals at g = 4.40 and g = 2.08 are attributable to nonheme iron and copper, respectively, neither of which is a functional constituent of the enzyme. When heme _d_ is removed from the enzyme the signals at g = 2.45 and 1.71 disappear completely but those due to heme _c_ are not effected. The g-values for the signals observed in the various reconstituted preparations are shown in Table III. The EPR spectrum of the protein reconstituted with heme _d_ is identical to that of the native oxidase. In those preparations reconstituted with the other heme groups signals are present near g = 6 and g = 2 which are characteristic for high-spin ferrihemoproteins and are attributed to transitions within the lowest Kramers doublet (Griffith, 1956). In addition, the preparations reconstituted with protoheme and deuteroheme exhibit an additional signal at g = 3.8. A similar EPR signal has been observed in one form of mitochondrial cytochrome _b_ by Orme-Johnson _et al_. (1971) who suggested that it is produced by low field resonances ($g_z$) of low spin ferric heme. Until further studies are carried out it is premature to conclude whether this is also the case in the oxidase. Furthermore, additional experiments are required to determine if the ferric heme at g = 3.8 is bound at a second binding site that is distinct from the high spin ferric heme characterized by a g-value at 6 or whether the ferric iron in these reconstituted heme groups exists in an equilibrium between high spin and low spin forms.

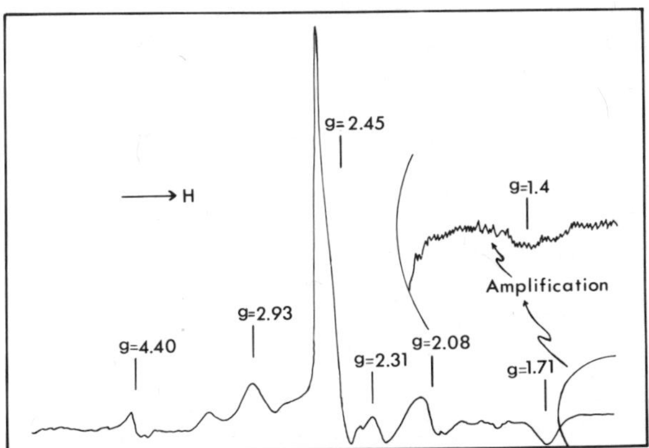


Fig. 8. EPR spectrum of oxidized Pseudomonas cytochrome oxidase at 13°K. Instrument settings were: power output, 0.3 mW; field modulation, 5G; microwave frequency, approximately 9250 MHz.

TABLE 1

OXIDASE ACTIVITY OF RECONSTITUTED *PSEUDOMONAS* CYTOCHROME OXIDASE

| Reconstituted Heme | Oxidase Activity* % of Native |
|---|---|
| Heme *d* | 90-100 |
| Heme *a* | 5 |
| Deuteroheme | 0 |
| Hematoheme | 0 |
| Mesoheme | 0 |
| Protoheme | 0 |

*Oxidase activity was measured as specific activity in terms of µmoles cytochrome *c*-551 oxidized per min per mg protein at 25° C. The heme *d*-depleted enzyme prior to reconstitution had no oxidase activity.

TABLE 2

ABSORPTION MAXIMA OF RECONSTITUTED *PSEUDOMONAS* CYTOCHROME OXIDASE IN THE REDUCED FORM*

| Reconstituted Heme | Absorption Maxima nm |
|---|---|
| Heme *d* | 625-655, 554-549, 521, 460, 418 |
| Heme *a* | 600, 555-550, 522, (440), 420 |
| Deuteroheme | 551, 525-520, 415 |
| Hematoheme | (565), 556-552, 525, 419 |
| Mesoheme | 670, 560-555, 527, 420 |
| Protoheme | 582, 555-550, 522, 418 |

*Reconstituted preparations in 0.1 M phosphate buffer, pH 7.0, were reduced by a few crumbs of $Na_2S_2O_4$. Figures in parentheses indicate major shoulders.

TABLE 3

EPR G-VALUES OF RECONSTITUTED PSEUDOMONAS CYTOCHROME OXIDASE*

| Reconstituted Heme | g-Values | | | | | | |
|---|---|---|---|---|---|---|---|
| Heme *d* | | | 2.93, | 2.45, | 2.31, | 1.71, | 1.4 |
| Heme *a* | 6.6, | | 2.9, | | 2.30 | | |
| Deuteroheme | 6.6, | 3.8, | 2.9, | | 2.31, | 2.05 | |
| Hematoheme | 6.0, | | 2.9, | | 2.30, | 2.0 | |
| Mesoheme | 6.0, | | 2.9, | | 2.31, | 2.06 | |
| Protoheme | 6.0, | 3.8, | 2.9, | | 2.30, | 2.17 | |

*EPR spectra were taken of the oxidized enzyme in a Varian Model E-4 spectrometer at $10^{o}$K. Instrument settings were: power output, lmW; field modulation, 5G; microwave frequency, approximately 9200 MHz.

As seen in Table 3 the EPR spectrum of the heme *c* component remains essentially constant with g-values at 2.93, 2.31, and 1.4. This result suggests that there is no direct heme-heme coupling between heme *c* and the second heme group since removal of the heme *d* and replacement with a different heme group in the presence of such coupling would be expected to alter the positions of the g-values of the heme *c*.

## CONCLUSIONS

The reaction between *Pseudomonas* cytochrome oxidase and azurin appears to involve the rapid formation of a complex between the two proteins followed by the rapid transfer of electrons involving the heme *c* moiety of the oxidase and the copper prosthetic group of azurin. The rapid electron transfer is followed by a slower internal electron transfer between the heme *c* group and the heme *d* moiety.

When reduced *Pseudomonas* cytochrome oxidase reacts with $O_2$ the heme *d* becomes oxidized first in a reaction that is second order with respect to the concentration of $O_2$. This reaction is followed by a slower internal electron transfer between ferroheme *c* and ferriheme *d*.

The overall reaction involving the oxidation of azurin catalyzed by *Pseudomonas* cytochrome oxidase can be expressed as:

Cu → heme c → heme d → $O_2$

where the rate-limiting step is the electron transfer between the heme c and heme d moieties within the oxidase.

Removal of the heme d group from *Pseudomonas* cytochrome oxidase results in complete loss of oxidase activity while its reconstitution yields nearly complete restoration of that activity. Reconstitution can be accomplished with other heme groups but only that with heme a restores any oxidase activity. Removal of heme d and heme reconstitution have little if any effect on the positions of the absorption maxima or EPR g-values of the protein-bound heme c suggesting a lack of intimate interaction between them.

## ACKNOWLEDGEMENTS

Supported by Grants HL10633 and GM14276 from the National Institutes of Health, USPHS.

## REFERENCES

Brunori, M., Parr, S. R., Greenwood, C., and Wilson, M. T. (1975) Biochem. J. 151, 185-188

Gibson, Q. H., Greenwood, C., Wharton, D. C., and Palmer, G. (1965) J. Biol. Chem. 240, 888-894

Griffith, J. S. (1956) Proc. Roy. Soc. Ser. A, 235, 23

Gudat, J. C., Singh, J. S., and Wharton, D. C. (1973) Biochim. Biophys. Acta 292, 376-390

Horio, T. (1958a) J. Biochem. (Tokyo) 45, 195-205

Horio, T. (1958b) J. Biochem. (Tokyo) 45, 267-279

Horio, T., Higashi, T., Matsubara, H., Kusai, K., Nakai, M., and Okunuki, K. (1958) Biochim. Biophys. Acta 29, 297-302

Horio, T., Higashi, T., Yamanaka, T., Matsubara, H., and Okunuki, K. (1961) J. Biol. Chem. 236, 944-951

Kuronen, T. and Ellfolk, N. (1972) Biochim. Biophys. Acta 275, 308-318

Kuronen, T., Saraste, M., and Ellfolk, N. (1975) Biochim. Biophys. Acta 393, 48-54

Nagata, Y., Yamanaka, T., and Okunuki, K. (1970) Biochim. Biophys. Acta 221, 668-671

Orme-Johnson, N. R., Hansen, R. E., and Beinert, H. (1971) Biochem. Biophys. Res. Communs. 45, 871-878

Parr, S. R., Wilson, M. T., and Greenwood, C. (1975) Biochem. J. 151, 51-59

Wharton, D. C., Gudat, J. C., and Gibson, Q. H. (1973) Biochim. Biophys. Acta 292, 611-620

Yamanaka, T., and Okunuki, K. (1963a) Biochim. Biophys. Acta 67, 379-393

Yamanaka, T., and Okunuki, K. (1963b) Biochem. Z. 338, 62-72

Yamanaka, T., Kijimoti, S., Okunuki, K., and Kusai, K. (1962) Nature 194, 759-760

EQUILIBRIUM STATES AND DYNAMIC REACTIONS OF IRON IN THE CAMPHOR MONOXYGENASE SYSTEM

I. C. Gunsalus and S. G. Sligar

Department of Biochemistry

University of Illinois, Urbana

## INTRODUCTION AND RATIONALE

All basic life processes include reactions centering on the control, transfer, storage, and utilization of two essential quantities, information and energy. This chapter, in dealing with molecular enzymology, focuses on the latter commodity and deals with the *oxygenase* class of enzymes which, together with the other catalytic and metabolic systems, are ubiquitous in nature and required by many living organisms for existence.

The central reaction of oxygen as a terminal oxidant involves the reductive cleavage of the O-O bond. The complete, four electron reduction of atmospheric dioxygen as carried out by the mitochondrial respiratory chain yields two molecules of water. The other intermediate states in dioxygen reduction are no less important, and in this communication we are concerned with the two electron reduction of the O-O bond, forming a single water molecule and an oxygenated substrate. It will become immediately apparent that many of the features and protein classes which are involved in the four electron process are also involved in the mixed function oxidase systems; complete pictures of both processes share the common knowledge gleaned from the study of each.

A crucial requirement for study at the level of molecular chemistry and physics is the availability of pure and homogeneous constituents. We have been able to isolate in quantity high quality protein components of a monoxygenase system from the bacterium *Pseudomonas putida* (1). This system selectively hydroxylates the substrate camphor with atmospheric oxygen as oxidant and reduced pyridine nucleotide (NADH) as reductant. The overall reaction scheme leading to cleavage of $O_2$ involves the participation of two iron proteins; cytochrome m of molecular weight 45,000 (2) which contains the oxygen and substrate binding centers, and an $Fe_2S^*_2Cys_4$ redoxin (termed putidaredoxin and abbreviated Pd) of mass 12,000 daltons (3). Reduced redoxin ($Pd^r$) is generated *in vivo* by the action of a flavoprotein reductase on NADH. Two central problems of biochemistry are manifest in the oxygenase system. These relate to (a) the physics and chemistry of the electron and proton transport from NADH through reductase and redoxin and into the cytochrome-oxygen-substrate complex, and to (b) the molecular mechanism of the O-O bond cleavage and reaction of the activated oxygen/substrate couple which results in hydroxylation.

The overall reaction cycle of the cytochrome monoxygenase under study is described in Figure 1 (4). In addition to the division of research areas as to electron transfer or oxygenation reactions, classification as to the first or second sequential input of redox equivalents can be made (5). The first electron transfer reaction is responsible for the ultimate ferric-ferrous reduction of the heme iron of cytochrome m ($m^{os} \longrightarrow m^{rs}$) while the second is intimately associated with the cleavage of the oxygen moiety and the coupled hydroxylation of the camphor skeleton. Based on the specificity and reactive rates of these two processes, we proposed some years ago that although reduced putidaredoxin ($Pd^r$) serves as electron donor for both reactions, there exist substantial differences in the molecular requirements of these events (5-8). We subsequently showed from a wide variety of physical and chemical experiments the existence of two redoxin binding sites on the surface of the cytochrome (9, 10). Putidaredoxin binding at what is labeled the reduction or "R" site (11) was found to be required for the first electron transfer reaction, while Pd association at a second distinct locus (termed the effector or "E" site) is an obligatory intermediate in the coupled second electron transfer and product formation (8, 10). This is perhaps one of the most crucial observations in our attempt to understand the molecular aspects of the hydroxylase cycle. The precise details surrounding electron transfer and mixed function oxidation involve multiprotein complexes and hence what is tentatively termed the "active site" in visions of structure/function relationships must necessarily include a complete description of the formation and breakdown of these higher aggregate complexes. In addition, two prosthetic groups are present in concert; the iron-sulfur cluster of putidaredoxin and the heme ring of the cytochrome. The possibility that these two moieties are in close proximity must be considered when one deals in the precise physical details on which we will base our understanding of biochemical oxygenation.

It would be hopeless to try to present the complete experi-

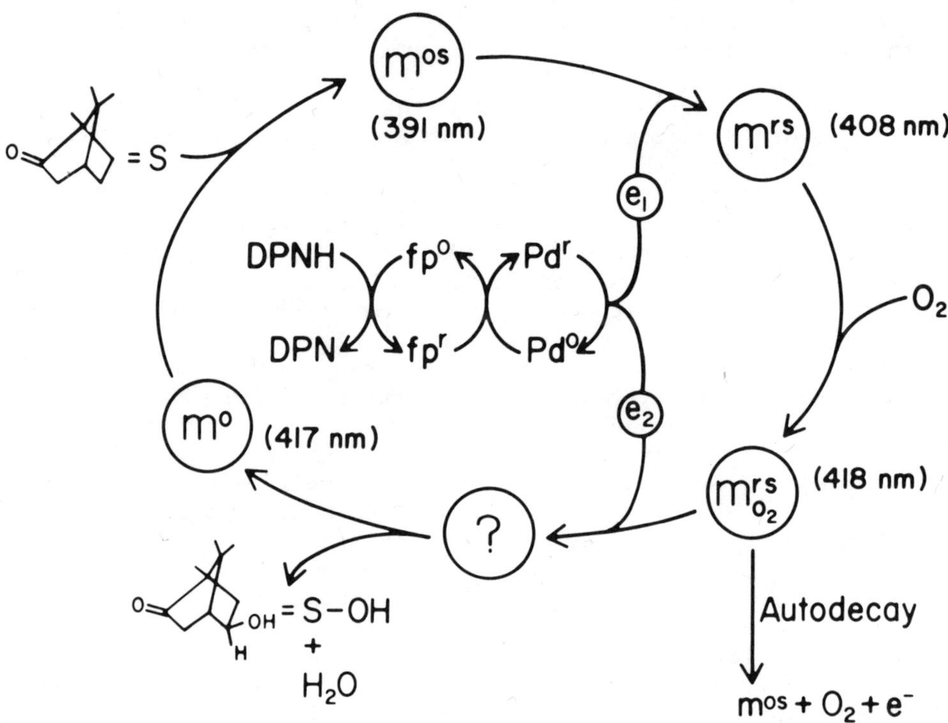
m^os
(391 nm)
= S
m^rs
(408 nm)
e1
DPNH
fp^o
Pd^r
DPN
fp^r
Pd^o
O2
e2
m^o
(417 nm)
?
m^rs O2
(418 nm)
OH = S-OH
H
+
H2O
Autodecay
m^os + O2 + e^-

Figure 1

mental accomplishments that support our mechanistic ideas in this short communication. Indeed, the space limitations would force selection and necessitate abbreviated method descriptions which are antagonistic to the spirit of scientific exchange. Hence we have chosen to reference extensively the work as it appears in the recent literature, summarize briefly our understanding to date, and focus on a generalization as to the *kinds* of information one can obtain by the newer techniques of physics and chemistry, and the *application* of the fundamental thermodynamic and kinetic principles which place our experimental measurements on a firm basis.

The wide variety of physical probes of enzyme structure fall generally into one of two classes, those that define or alter the thermodynamic *equilibrium* states of the system, and those that change or monitor the *dynamic* reactions. With this division in mind, we discuss the sequence of events in the reaction cycle (Figure 1), beginning with the cytochrome in the oxidized ferric state, $m^o$.

## SUBSTRATE BINDING ($m^o + S \leftrightarrow m^{os}$): THE STATES $m^o$ AND $m^{os}$; MODULATION OF SPIN STATE EQUILIBRIA

$m^o$ and $m^{os}$ have been extensively studied by electron paramagnetic resonance (EPR) and Mössbauer spectroscopy (4, 13-15). Resonance studies at cryogenic temperatures have shown the five d-electrons of the ferric cytochrome are paired to give a low spin (S = 1/2) ground state for $m^o$, while a mixture of approximately 60% high spin (S = 5/2) results when substrate binds. Detailed EPR measurements by Lipscomb (16) indicate that the low spin fraction of $m^{os}$ contains at least two species which respond differently to pH and temperature. These spin transitions have been correlated with changes in the optical spectrum, the low spin form being characterized by a Soret maximum at 417 nm while the high spin form has a peak at 390 nm. Quantitation of these optical changes permit the measurement of spin populations at room temperature where problems of thermal relaxation are avoided. Temperature dependent difference spectra in the region 275-315 K have shown that both $m^o$ and $m^{os}$ exhibit optical properties indicative of a spin mixture. Detailed analysis (17) shows that the spin equilibrium $m^o_{hs} \leftrightarrow m^o_{ls}$ is defined by an equilibrium constant $K_1 = [m^o_{hs}]/[m^o_{ls}] = 0.085$ at room temperature and with thermodynamic parameters $\Delta H = 10.3$ kcal and $\Delta S = 30.2$ e.u. The states $m^{os}_{hs} \leftrightarrow m^{os}_{ls}$, on the other hand, are connected by an equilibrium constant $K_2 = [m^{os}_{hs}]/[m^{os}_{ls}] = 15$, with $\Delta H = 2.47$ kcal and $\Delta S = 13.8$ e.u. The substrate modulation of this equilibrium is understood from a general thermodynamic model of biochemical regulation (18). Applied to this case, energy considerations make apparent that for the association of camphor to shift the spin equilibrium to the high spin form, the substrate must bind selectively tighter to this state of the enzyme. The camphor binding reactions are characterized by dissociation constants $K_4$ = 0.05 μM and $K_3$ = 9.0 μM for the processes $m^o_{hs}$ + S ↔ $m^{os}_{hs}$ and $m^o_{ls}$ + S ↔ $m^{os}_{ls}$, respectively. $K_1K_3 = K_2K_4$ is required by energy conservation for this four state system. In addition to these equilibria the spin state of $m^o$ and $m^{os}$ is strongly dependent on the concentration of monovalent cations (17) and thus, through direct and coupled free energy changes, the substrate association is modulated by the external salt concentrations.

A further effect of camphor binding is observed as a shift in the oxidation/reduction potential from -340 mV for $m^o$ (19) to -170 mV when substrate is bound (19, 20). This potential modulation is a direct outgrowth of the selective and assymetric interaction of substrate binding and the internal spin equilibrium (17). Thus, at fixed cation concentration, there exist three coupled equilibria: redox, spin, and camphor ligation. The complete thermodynamic description of the microreversible relationships between these reactions is made possible using the free energy concepts pioneered by Weber (21, 22) as applied to redox regulation by Sligar (18).

A selective protein modification resulting in a perturbation of these coupled reactions was discovered during an extensive and comprehensive study of sulfhydryl modification using N-ethyl maleimide (NEM) (8, 19, 23). When four of the six total sulfhydryl groups react with NEM, a protein is formed which binds substrate an order of magnitude weaker than native material and exhibits a redox potential midway between $m^o$ and $m^{os}$. The optical and resonance spectra of this derived protein show essentially a pure low spin ground state, and the application of further selective modification and thermodynamic characterization promises to reveal molecular details surrounding iron and substrate ligation reactions with the globin structure.

Large single crystals of $m^{os}$ have been grown and X-ray structure determination is in progress. A complete optical study of the single crystals has determined that the protein crystallizes with the heme plane nearly aligned (24). In addition to the strong absorption moments seen polarized in the heme plane, an additional and unusual z-polarized band is observed which resembes in strength and energy the iron-sulfur charge transfer transitions previously observed in iron-sulfur proteins. Detailed extended Hückel calculations predict this additional component, and very strongly suggest the existence of a mercaptide ligand to the heme iron.

## PUTIDAREDOXIN REDUCTION OF CYTOCHROME $m^{os}$: Pd BINDING AT THE "R" SITE

Putidaredoxin binds to cyt m in a mole ratio 2:1. The site

monitored by quenching of a fluorescent label covalently attached to an external sulfhydryl on the cytochrome surface has been shown to be involved in the transfer of reducing equivalents from redoxin to cytochrome (8, 11, 18, 25). The association of oxidized ($Pd^o$ + $m^{os} \leftrightarrow Pd^o \cdot m^{os}$) and reduced ($Pd^r + m^{rs} \leftrightarrow Pd^r \cdot m^{rs}$) protein components at this locus is assymetric, with dissociation constants 2.9 μM and 0.46 μM, respectively. This nonequivalency of binding is analogous to the simpler case of camphor binding to oxidized and reduced cytochrome and is directly related to the observed shift in the potential of the redoxin iron-sulfur center from -239 mV to -195 mV on binding to the cytochrome. The resulting potential through which the reducing equivalent is transferred from $Pd^r$ to $m^{os}$ is thus less than what would be calculated assuming the potentials of the free components. Energetically speaking, the more equi-potential transfer means a portion of the available redox energy is expended in order to selectively bind reduced redoxin at this locus. These equilibrium measurements strongly suggest that the *dynamics* of the electron transfer process require the formation of a Pd-cyt m complex. Rapid reaction experiments using a combination of stopped flow and flash photolytic techniques have supported this hypothesis (25). The preliminary kinetic constants obtained are in excellent agreement with the grouped rate parameters measured by the equilibrium techniques, and the advantages of combined thermodynamic and kinetic measurements are obvious.

Ferrous cytochrome, $m^{rs}$, has been extensively characterized with detailed Mössbauer (4, 26, 27) and magnetic susceptibility (28) measurements. Optical studies on single crystals of $m^{rs}$ show a split in-plane polarized Soret region and, perhaps most interestingly, the complete **absence** of the z-polarized iron-sulfur charge transfer band (24, 29). Together with further theoretical calculations these measurements suggest a mercaptan group as proximal ligand to the heme iron.

## OXYGEN BINDING AND THE OXYGENATED INTERMEDIATE

The rapid binding of oxygen to $m^{rs}$ yields the ferrous-oxygenated intermediate $m^{rs}_{O_2}$ which has been well characterized in terms of resonance properties. The mechanism of oxygen cleavage is cloaked in the interaction of this intermediate with an additional electron donor to reductively split dioxygen and carry out substrate oxygenation. $m^{rs}_{O_2}$ decays in the absence of an additional reducing equivalent to generate superoxide anion and $m^{os}$, with a first order rate constant at room temperature of 0.008 $sec^{-1}$ (18, 30). The oxygenated cytochrome, both with ($m^{rs}_{O_2}$) and without substrate ($m^{r}_{O_2}$), has been stabilized using low temperatures (-40 to 0 C) and mixed solvent techniques (31, 32). This is the first isolation of a completely stable oxygen complex of a hydroxylase system, and promises to be of immeasurable value in studies of the chemistry and intermediate states in the reduction of the O-O bond.

## PUTIDAREDOXIN CATALYZED DECAY OF $m^{rs}_{O_2}$: Pd BINDING AT THE "E" SITE

In the presence of putidaredoxin and an additional reducing equivalent, $m^{rs}_{O_2}$ breaks down to form hydroxycamphor, water, and $m^{o}$.

Detailed kinetic studies (8-10) have shown that an obligatory intermediate step in this reaction is the formation of a dienzyme complex between redoxin and cytochrome at a site distinct from that observed in the reduction of $m^{os}$. The kinetic constants obtained correlate with the equilibrium binding constants of redoxin and cytochrome as monitored optically by a shift in the cytochrome Soret maximum to that characteristic of the low spin form (10). Such an interaction has also been observed by EPR and Mössbauer techniques (12, 33). The overall rate-limiting step in the camphor monoxygenase cycle is the breakdown of the putidaredoxin-cytochrome complex to release oxidized proteins, water and hydroxylated substrate. Other compounds which elicit product from the oxygenated intermediate include organic dyes, hydroperoxides, cytochrome $b_5$, rubredoxin, and "lipoic acid". Other substances such as nucleophiles displace $O_2$ from the iron but do not catalyze product formation. Still a third class of compounds such as dithionite are completely inert to $m^{rs}_{O_2}$ and are not able to input the second required reducing equivalent despite extremely low potential.

Further characterization of the product-forming reaction must await the results of research in progress on selective modification of the protein and substrate components coupled with the newer methods of biochemical and physical measurement.

## SUMMARY AND FINAL NOTE

We have tried to summarize in a brief account the types of methods and analyses that are being employed to answer the fundamental questions proposed in the introduction. Two opposite patterns are emerging. As we probe in ever finer detail the molecular events surrounding the important catalytic reactions, we realize the need for more precise probes and measurements on an even more precise scale. On the other hand, as we dig deeper, we find higher levels of organization, from cytochrome binding of substrate and oxygen to the requirement of multiprotein aggregates. Evolution in both directions simultaneously is not incompatible, but is in all probability a requirement for understanding of efficient cellular catalysis and regulatory events. The fundamental knowledge gained in regard to interprotein communication and oxygen reduction is of prime biochemical significance in regard to oxygenation and respiratory processes.

This work has been supported under grants from the National Institutes of Health and the National Science Foundation. The expert and essential editorial work by Ms. Kathryn Skelton is gratefully acknowledged.

## REFERENCES

1. Hedegaard, J. and Gunsalus, I. C. (1965) *J. Biol. Chem.* 240, 4038-4043.
2. Tsai, R. L., Gunsalus, I. C. and Dus, K. (1971) *Biochem. Biophys. Res. Commun.* 45, 1300-1306.
3. Tanaka, M., Haniu, M., Yasunobu, K. T., Dus, K. and Gunsalus, I. C. (1974) *J. Biol. Chem.* 249, 3689-3701.
4. Gunsalus, I. C., Meeks, J. R., Lipscomb, J. D., Debrunner, P. G. and Münck, E. (1974) in *Molecular Mechanisms of Oxygen Activation* (O. Hayaishi, ed.), Academic Press, New York, pp. 559-613.
5. Gunsalus, I. C. and Lipscomb, J. D. (1972) in *Molecular Basis of Electron Transport* (J. Schultz and B. F. Cameron, eds.), Academic Press, New York, pp. 179-196.
6. Gunsalus, I. C., Lipscomb, J. D., Marshall, V., Frauenfelder, H., Greenbaum, E. and Münck, E. (1972) in *Biological Hydroxylation Mechanisms* (G. S. Boyd and R. M. S. Smellie, eds.), Biochem. Soc. Symp. 34, Academic Press, pp. 135-157.
7. Gunsalus, I. C., Tyson, C. A. and Lipscomb, J. D. (1973) in *Oxidases and Related Redox Systems* (T. E. King, H. S. Mason and M. Morrison, eds.), University Park Press, Baltimore, pp. 583-603.
8. Lipscomb, J. D., Sligar, S. G. Namtvedt, M. J. and Gunsalus, I. C. (1976) *J. Biol. Chem.* (in press).
9. Sligar, S. (1975) Ph.D. Thesis, University of Illinois, Urbana.
10. Sligar, S., Debrunner, P. G., Namtvedt, M. J. and Gunsalus, I. C. (1975) *Fed. Proc.* 34, 622, ABSTRACT; Lipscomb, J. D. and Gunsalus, I. C. (1973) in *Iron-Sulfur Proteins*, Vol. I (W. Lovenberg, ed.), Academic Press, New York, pp. 151-172.
11. Sligar, S. G., Debrunner, P. G., Lipscomb, J. D., Namtvedt, M. J. and Gunsalus, I. C. (1974) *Proc. Nat. Acad. Sci. USA* 71, 3906-3910.
12. Sharrock, M. (1973) Ph.D. Thesis, University of Illinois, Urbana.
13. Schlosnagle, D. C., Tsibris, J. C. M. and Gunsalus, I. C. (1974) *Fed. Proc.* 33, 1370, ABSTRACT.

14. Sharrock, M., Münck, E., Debrunner, P. G., Marshall, V. P., Lipscomb, J. D. and Gunsalus, I. C. (1973) *Biochemistry* 12, 258-265.
15. Frauenfelder, H., Gunsalus, I. C. and Münck, E. (1972) in *Mössbauer Soectroscopy and its Applications,* International Atomic Energy Agency, Vienna, Austria, pp. 231-255.
16. Lipscomb, J. D. and Gunsalus, I. C. (1976) *J. Biol. Chem.* (in press).
17. Sligar, S. G. and Gunsalus, I. C. (1976) *Biochemistry* (in press).
18. Sligar, S. G. and GUnsalus, I. C. (1976) *Proc. Nat. Acad. Sci. USA* (in press).
19. Lipscomb, J. D. (1974) Ph.D. Thesis, University of Illinois, Urbana.
20. Wilson, G. S., Tsibris, J. C. M. and Gunsalus, I. C. (1973) *J. Biol. Chem.* 248, 6059-6061.
21. Weber, G. (1972) *Biochemistry* 11, 864.
22. Weber, G. (1975) *Advances Prot. Chem.* 29, 1.
23. Lipscomb, J. D. and Gunsalus, I. C. (1976) *J. Biol. Chem.* (in press).
24. Hanson, L., Eaton, W., Sligar, S. and Gunsalus, I. C. (1976) (in preparation).
25. Pederson, T. and Gunsalus, I. C. (1976) (in preparation).
26. Champion, P. (1974) Ph.D. Thesis, University of Illinois, Urbana.
27. Champion, P. M., Lipscomb, J. D., Münck, E., Debrunner, P. and Gunsalus, I. C. (1975) *Biochemistry* 14, 4151-4158.
28. Champion, P. M., Münck, E., Debrunner, P. G., Moss, T. H., Lipscomb, J. D. and Gunsalus, I. C. (1975) *Biochim. Biophys. Acta* 376, 579-582.
29. Sligar, S. G., Debrunner, P. G., Lipscomb, J. D. and Gunsalus, I. C. (1973) 9th Int'l Congr. Biochem., Stockholm, ABSTRACT.
30. Sligar, S. G., Lipscomb, J. D., Debrunner, P. G. and Gunsalus, I. C. (1974) *Biochem. Biophys. Res. Commun.* 61, 290-296.
31. Travers, F., Douzou, P., Pederson, T. and Gunsalus, I. C. (1975) *Biochimie* 57, 43-48.
32. Debey, P., Eisenstein, L., Douzou, P. and Gunsalus, I. C. (1976) (in preparation).
33. Sharrock, M., Debrunner, P. G., Schulz, C., Lipscomb, J. D., Marshall, V. and Gunsalus, I. C. (1976) *Biochemistry* (in press).

# CURRENT STATUS OF THE SEQUENCE STUDIES OF THE *PSEUDOMONAS PUTIDA* CAMPHOR HYDROXYLASE SYSTEM

Masaru Tanaka, Scott Zeitlin, Kerry T.Yasunobu and
I.C. Gunsalus*
Dept. of Biochem-Biophysics, University of Hawaii Medical School, Honolulu, Hawaii and *Dept. of Biochemistry, University of Illinois, Urbana, Illinois

A methylene hydroxylase system from camphor induced *Pseudomonas putida* has been separated into a putidaredoxin reductase, putidaredoxin (an iron-sulfur protein), and a hydroxylase fraction known as soluble cytochrome P-450 (1). This mixed function oxidase catalyzes the hydroxylation of methylene carbon 5 of camphor with reduced diphosphopyridine nucleotide as the primary electron donor and with molecular oxygen as the acceptor. The physicochemical and kinetic properties of the hydroxylase system are discussed in review articles by Gunsalus et al (2,3 and see articles by Gunsalus and Sligar in this book).

Similar biological hydroxylating systems are found in animals which hydroxylate drugs, steroid hormones and potentially toxic molecules. Other important P-450 cytochromes, in particular from the adrenal cortical mitochondria are involved in steroid metabolism (4). However, the animal hydroxylases are membrane bound, and only recently has it been possible to isolate them in active and homogeneous forms (5). Since the *P. putida* camphor hydroxylase system is water soluble and since the protein components can be obtained in homogeneous forms relatively rapidly, it will serve as a model hydroxylase for some years to come.

The research on the *P. putida* camphor hydroxylase system has progressed to a point where the amino acid sequence data of the three protein components are very desirable. For example, crystal X-ray diffraction studies of the cytochrome P-450 are in progress and chemical studies suggest that the heme ligands are probably a cysteine residue and possibly a histidine residue (6 and see article by R. Holm in this book). However, at the present time, it is not clear which of these residues in the protein are chelated to the heme iron. Sequence studies of the methylene hydroxylase protein components have been in progress in our laboratory for several years and the objective of this report is to discuss the previous work that has been completed as well as to discuss the present status of the sequence studies.

## METHODS AND PROCEDURES

Cytochrome P-450 was isolated and purified from *P. putida*, strain PpG786 as described by Tsai et al (7) from cells grown on camphor. The purity and methods used to determine protein and heme content have been previously described (7).

The procedure of Crestfield et al (8) was used to prepare the S-β-carboxymethylcysteinyl-cytochrome P-450.

The amino acid content of the various samples were determined in the Beckman Model 120 automatic amino acid analyzer as described by Spackman et al (9).

The $NH_2$-terminal sequence of the cytochrome P-450 derivative was analyzed in the Beckman 890 Protein Sequencer. About 7 mg of protein was analyzed and the sequencer reagents were purchased from Pierce Chemical Co. The protein double cleavage program was used. The amino acid phenylthiohydantoins were identified by gas liquid chromatography as reported by Pisano and Bronzert (10), by thin layer chromatography as described by Edman and Begg (11) or by amino acid analyses of acid hydrolyzates of the amino acid hydantoins as described by Van Orden and Carpenter (12).

COOH-terminal amino acid analyses of the S-β-carboxymethyl-cysteinyl-cytochrome P-450 were determined by the hydrazinolysis procedure of Bradbury (13) or by the carboxypeptidase procedure of Ambler (14).

## RESULTS AND DISCUSSION

### Amino Acid Composition and Other Properties of the *P. putida* Camphor Hydroxylase Protein Components

Some of the properties of the putidaredoxin reductase, putidaredoxin (Fe-S protein) and cytochrome P-450 are summarized in Table I and were determined by Tsai et al. (7). The amino acid composition of the putidaredoxin has been corrected to agree with the amino acid sequence reported by Tanaka et al (15).

### Sequence Investigations of the Hydroxylase System

Thus far, the complete amino acid sequence of only putidaredoxin has been determined (15) and is shown in Fig. 1 along with

TABLE I

Amino acid composition and some properties of the P. putida Camphor Hydroxylase Components Reported by Tsai et al.(7)

| Amino acid | P-450 | Reductase | Putidaredoxin* |
|---|---|---|---|
| Asp | 27 | 25 | 9 |
| Asn | 9 | 15 | 4 |
| Thr | 19 | 20 | 5 |
| Ser | 21 | 18 | 7 |
| Glu | 42 | 16 | 6 |
| Gln | 13 | 24 | 4 |
| Pro | 27 | 18 | 4 |
| Gly | 26 | 33 | 8 |
| Ala | 34 | 48 | 9 |
| Val | 24 | 34 | 14 |
| Met | 9 | 6 | 3 |
| Ile | 24 | 24 | 6 |
| Leu | 40 | 42 | 6 |
| Tyr | 9 | 6 | 3 |
| Phe | 17 | 10 | 1 |
| His | 12 | 6 | 2 |
| Lys | 13 | 13 | 3 |
| Trp | 1 | 3 | 1 |
| Arg | 24 | 24 | 5 |
| Cys | 6 | 6 | 6 |
| Total | 397 | 393 | 106 |
| Free SH | 6 | 6 | 4 |
| N-terminus | Asx | Ser | Ser |
| C-terminus | Val | Ala | Trp |
| Mol. wt x $10^3$ | 45 | 43.5 | 11.6 |
| Prosthetic Group | Heme | 1 FAD | $(FeS)_2$ |

* Data of reference (15).

the amino acid sequence of bovine adrenodoxin. From a structure-function standpoint, the observed sequence homology of these iron-sulfur proteins are important since the homology helps to establish the cysteine residues involved in iron chelation. It seems likely that cysteine residues 39,45,48 and 85 or 86 are chelated to the iron. In addition, Sligar et al (16) have reported that the removal of the COOH-terminal tryptophan residue with carboxypeptidase greatly reduces the ability of putidaredoxin to funnel electrons to cytochrome P-450.

```
Ad (1)  Ser Ser Ser Gln Asp Lys Ile Thr Val His Phe Ile Asn Arg Asp
Pu (1)   1                      Ser Lys Val Val Tyr Val Ser His Asp
                                 1

Ad (2)  Gly Glu Thr Leu Thr Thr Lys Gly Lys Ile --- Gly Asp Ser Leu
Pu (2)  Gly --- Thr Arg Arg Gln Leu Asp Val Ala Asp Gly Val Ser Leu

Ad (3)  Leu Asp Val Val Val Gln Asn Asn Leu Asp Ile Asp Gly Phe Gly
Pu (3)  Met Gln Ala Ala Val Ser Asn Gly Ile --- Tyr Asp Ile Val Gly

             *                       *           *
Ad (4)  Ala Cys Glu Gly The Leu Ala Cys Ser Thr Cys His Leu Ile Phe
Pu (4)  Asp Cys Gly Gly Ser Ala Ser Cys Ala Thr Cys His Val Tyr Val
             *                       *           *

Ad (5)  Glu Gln His Ile Phe --- Glu Lys Leu Glu Ala Ile Thr Asn Glu
Pu (5)  --- Asn Glu Ala Phe Thr Asp Lys Val Pro Ala Ala --- Asn Glu

Ad (6)  --- Glu Asn Asn Met Leu Asp Leu Ala Tyr Gly --- Leu Thr Asp
Pu (6)  Arg Glu Ile Gly Met Leu Glu Cys Val Thr Ala Glu Leu Lys Pro

                             *
Ad (7)  Arg Ser Arg Leu Gly Cys Gln Ile Cys Leu Thr Lys Ala Met Asp
Pu (7)  Asn Ser Arg Leu Cys Cys Gln Ile Ile Met Thr Pro Glu Leu Asp
                             *

Ad (8)  Asn Met Thr Val Arg Val Pro Asp Ala Val Ser Asp Ala
Pu (8)  Gly Ile Val Val Asp Val Pro Asp Arg Gln Trp     114
                                            106
```

Fig. 1, Amino acid sequences of bovine adrenodoxin (Ad) and putidaredoxin (Pu)

The amino acid sequence investigations of the *P. putida* cytochrome P-450 are just starting in our laboratory. However, it should be mentioned that the $NH_2$- and COOH-terminal sequence of the protein have been investigated by Dus et al (6) but they differed from our results which will be presented in Table II and Table III.

The S-β-carboxymethylcysteinyl-cytochrome P-450 was analyzed in the Beckman Model 890 Protein Sequencer and the result of this analysis is summarized in Table II. Since this was a preliminary run, the yields of the various amino acid hydantoins during the Edman degradation are not reported.

When the COOH-terminal amino acid of the *P. putida* cytochrome P-450 was analyzed by the hydrazinolysis procedure, valine was found to be released quantitatively as shown in Table III. From the kinetics of the release of the COOH-terminal amino acids from cytochrome P-450 by the carboxypeptidase A, the next three amino acids to the interior of the COOH-terminal valine were shown to be alanine, threonine and a second residue of threonine.

TABLE II. $NH_2$-terminal Sequence of the P.putida cytochrome P-450.

| Step of Edman degradation | Pth-amino acid | Step of Edman degradation | Pth-amino acid |
|---|---|---|---|
| 1 | Threonine | 8 | Asparagine |
| 2 | Threonine | 9 | Alanine |
| 3 | Glutamic acid | 10 | Asparagine |
| 4 | Threonine | 11 | Leucine |
| 5 | Isoleucine | 12 | Alanine |
| 6 | Glutamine | 13 | Proline |
| 7 | Serine | 14 | Leucine * |

*The yield of Pth-leucine at the 14th step of Edman degradation was 12% as determined by gas-liquid chromatography and therefore, there is no ambiguity in the sequence data reported in the table.

TABLE III. The COOH-terminal Amino Acid Sequence of Cytochrome P-450.

| Procedure* | Amino acid released | Residues/mole of protein | | | |
|---|---|---|---|---|---|
| 1. Hydrazinolysis | Valine | 1.0 | | | |
| 2. Cpase A | | Reaction time (minutes) | | | |
| | | 10 | 30 | 60 | 120 |
| | Valine | 1.00 | 1.00 | 1.00 | 1.00 |
| | Alanine | 0.69 | 0.77 | 0.80 | 0.83 |
| | Threonine | 0.07 | 0.17 | 0.33 | 0.68 |
| 3. Cpase A for 3 hours, and Cpase B added | | Reaction time (hours) | | | |
| | | 0 | 1 | 4 | |
| | Valine | 1.00 | 1.00 | 1.00 | |
| | Alanine | 1.00 | 1.00 | 1.00 | |
| | Threonine | 1.54 | 1.86 | 1.94 | |
| | Lysine | 0.00 | 0.62 | 0.62 | |

*The experimental conditions used were: 1. Hydrazinolysis was performed at 79° for 18 hours. 2. Carboxypeptidase A digestion was carried out at 25° and an enzyme to substrate ratio of 1 to 151 at pH 8. 3. Conditions were similar to 2 except that the enzyme to substrate ratio was 1 to 116. After three hours, Cpase B was added at an enzyme tó substrate ratio of 1 to 83.

In a second experiment in which the carboxypeptidase A concentration was increased, carboxypeptidase B was added after three hours and under this condition, lysine was liberated in addition to the previous mentioned amino acids as shown in Table III. The experiments thus showed that the COOH-terminal sequence in the cytochrome P-450 was -Lys-Thr-Thr-Ala-Val-COOH.

The $NH_2$- and COOH-terminal amino acid sequences of cytochrome P-450 reported by Dus et al. (6) and sequences from the present investigation are summarized in Fig. 2.

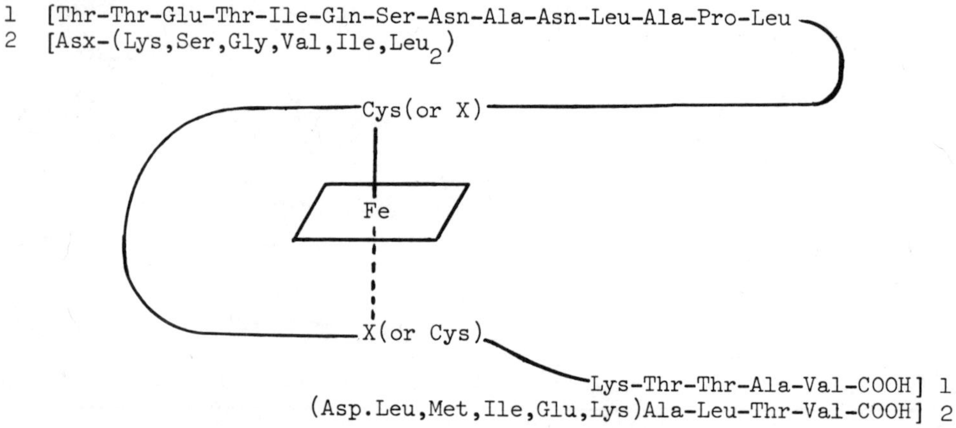

Fig. 2. The $NH_2$-terminal and COOH-terminal sequences of the *P. putida* Cytochrome P-450 found in the present study is designated 1, and the sequence reported by Dus et al. (6) is marked 2.

There are many sequence differences reported by the two groups. How can the sequence differences be rationalized? It is most likely that the dinitrophenylation technique used by Dus et al. (6) produced DNP-threonine. However, it is known that the DNP-derivatives of the hydroxyamino acids are unstable and the DNP-threonine may have decomposed during the experiment. The reported $NH_2$-terminal group, namely Asx could have come from an impurity in the P-450 preparation or may have come from other unknown sources. In our Protein Sequencer run, the $NH_2$-terminal threonine was obtained in a high yield. It is more difficult to explain the differences reported for the COOH-terminal sequence of cytochrome P-450 by the two groups. The true sequence will be known when the COOH-terminal peptide of cytochrome P-450 obtained from proteolytic digests of the protein has been isolated and sequenced. Efforts are underway in our laboratory to obtain this peptide as well as all of the peptides in order that the complete amino acid sequence of the protein can be determined. Our chief reason for presenting the preliminary sequence data of the *P. putida* cytochrome P-450 at this time is to assist the X-ray crystallographer who is currently attempting to determine the structure of the cytochrome P-450 and the correct sequence will be of interest to him in the X-ray investigation.

ACKNOWLEDGEMENTS

The research project was supported in part by research grants GM 16784, GM 16228 and AI 04865 from the National Institutes of Health and GB 43448 from the National Science Foundation.

REFERENCES

1. Katagiri, M., Ganguli, B. N., and Gunsalus, I.C. (1968). J. Biol. Chem. 243, 3543.
2. Gunsalus, I.C., Meeks, J.R., Lipscomb, J.D., Debrunner, P., and Munck, E. (1974). in Molecular Mechanisms of Oxygen Activation, ed, Hayaishi, O. (Academic Press, New York, N.Y.) pp. 559-613.
3. Gunsalus, I.C., Lipscomb, J.D., Marshall, V.P., Frauenfleder, H., Greenbaum, E., and Munck, E. (1972) in Biological Hydroxylation Mechanisms, ed. Boyd, G.S., and Smellie, R.M.S.
4. Schleyer, H., Cooper, D. Y., Levin, S.S., and Rosenthal, O. (1972), in Biological Hydroxylation Mechanisms, Boyd, G.S., and Smellie, R.M.S., ed., New York, Academic Press, p. 187.
5. Van der Hoeven, T.A., Haugen, D.A. and Coon, M.J. (1974). Biochem. Biophys. Res. Commun. 60, 569.
6. Dus, K., Katagiri, M., Yu, C-A., Erbes, D. L. and Gunsalus, I.C. (1970). Biochem. Biophys. Res. Commun. 40, 1423.
7. Tsai, R. L., Gunsalus, I.C., and Dus, K. (1971). Biochem. Biophys. Res. Commun. 45, 1300.
8. Crestfield, A. M., Moore, S., and Stein, W. H. (1963). J. Biol. Chem. 238, 622.
9. Spackman, D. H., Moore, S., and Stein, W. H. (1963). J. Biol. Chem. 238, 618.
10. Pisano, J.J., and Bronzert, T. J. (1969). J. Biol. Chem. 244, 5597.
11. Edman, P., and Begg, G. (1967). Eur. J. Biochem. 1, 80.
12. Van Orden, H. O., and Carpenter, F. H. (1964). Biochem. Biophys. Res. Commun. 14, 399.
13. Bradbury, J. H. (1958). Biochem. J. 68, 475.
14. Ambler, R. B. (1967). Methods Enzymol. 11, 436.
15. Tanaka, M., Haniu, M., Yasunobu, K.T., Dus, K. and Gunsalus, I. C. (1974). J. Biol. Chem. 249, 3689.
16. Sligar, S. G., Debrunner, P. G., Lipscomb, J. D., Namtvedt, M. J. and Gunsalus, I. C. (1974). Proc. Nat'l. Acad. Sci. U.S. 71, 3906.

# HIGHLY PURIFIED CYTOCHROME P-450 FROM LIVER MICROSOMAL MEMBRANES: RECENT STUDIES ON THE MECHANISM OF CATALYSIS

M. J. Coon, D. P. Ballou, F. P. Guengerich,† G. D. Nordblom, and R. E. White†

Department of Biological Chemistry, Medical School, The University of Michigan, Ann Arbor, Michigan, U.S.A.

## INTRODUCTION

The purpose of this paper is to summarize our knowledge of the properties of purified liver microsomal cytochrome P-450, including evidence for multiple forms, and to present data from recent experiments bearing on the mechanism of action of this cytochrome. For the sake of brevity we will confine our discussion to studies in this laboratory on cytochrome P-450 solubilized and purified from rabbit liver microsomes.

## CHARACTERIZATION OF MULTIPLE FORMS OF LIVER CYTOCHROME P-450 (P-$450_{LM}$)

As recently reviewed elsewhere, a reconstituted system containing highly purified P-$450_{LM}$, NADPH-cytochrome P-450 reductase, and phosphatidylcholine is capable of catalyzing the NADPH- and $O_2$-dependent hydroxylation of various substrates (1,2). During the purification of the phenobarbital-inducible form of P-$450_{LM}$ to apparent homogeneity (3,4), evidence was obtained for the occurrence of multiple forms of the cytochrome (5). These forms differ in their subunit molecular weights and spectral properties as well as

* This research was supported by Grant BMS71-01195 from the National Science Foundation and Grant AM-10339 from the United States Public Health Service.

†Postdoctoral Fellow, United States Public Health Service.

in their catalytic activities. We now have evidence for as many as six forms of P-$450_{LM}$; pending complete characterization of these proteins and a thorough study of their substrate specificity, it seems advisable to designate them by their electrophoretic characteristics. Neither the patterns of induction nor the chemical properties and catalytic activities so far investigated provide a simpler method of nomenclature. Even the designation "P-450" and "P-448" frequently seen in the literature is inadequate, since we now have evidence for several forms of the latter type differing in the exact λmax of the reduced CO complexes. The two forms which we have purified and studied most thoroughly will be referred to as P-$450_{LM_2}$ and $LM_4$ (5).

P-$450_{LM_2}$ is induced by phenobarbital, has subunits of molecular weight 50,000, and exhibits a spectral peak as the CO complex at 451 nm (5). The purified preparations are free of detectable amounts of cytochromes P-420 and $\underline{b}_5$ and of NADPH-cytochrome P-450 reductase (measured by activity toward cytochrome $\underline{c}$) and NADH-cytochrome $\underline{b}_5$ reductase (measured by activity toward ferricyanide). P-$450_{LM_2}$ and bacterial P-$450_{cam}$ (6) show immunological cross reaction by competitive binding and by inhibition of catalytic activity and are of similar subunit molecular weight and amino acid composition (7,8). Furthermore, upon treatment with cyanogen bromide they yield small heme-containing peptides of highly similar amino acid composition. These findings are surprising in view of the marked differences in these proteins in solubility, substrate specificity, and the requirement of P-$450_{LM}$ for a phospholipid and of P-$450_{cam}$ for an iron-sulfur protein for hydroxylation activity.

The second form of liver microsomal cytochrome P-450 which we have obtained in apparently homogeneous form, P-$450_{LM_4}$, is present at low levels in normal microsomes and is induced by β-naphthoflavone in rabbits (5). The administration of this inducer or of 3-methylcholanthrene to mice has previously been shown to stimulate hydroxylation activity toward carcinogens such as benzo[*a*]pyrene (9). P-$450_{LM_4}$ has subunits of molecular weight 54,000 and exhibits a spectral peak as the CO complex at 447 nm. The lack of reactivity of P-$450_{LM_4}$ with anti-P-$450_{LM_2}$ γ-globulin, as shown by Ouchterlony diffusion analysis and radioimmune assay (2) indicates that P-$450_{LM_2}$ and $LM_4$ are structurally different. On the other hand, the carbohydrate composition is similar; each protein contains, per polypeptide subunit, close to two molecules of mannose, one of glucosamine, a variable amount (not exceeding one molecule) of glucose, and no sialic acid (2).

These forms of P-$450_{LM}$ and others not yet completely purified appear to differ somewhat in their substrate specificity as well as in their chemical properties; for example, P-$450_{LM_2}$ is more active toward drugs such as benzphetamine and ethylmorphine. The 16α position of testosterone, and the 4 position of biphenyl, whereas $LM_{1,7}$ acts preferentially on the 6β position of testosterone (5). The position-specific oxygenation of benz[*a*]pyrene by the different forms of purified $P\text{-}450_{LM}$ has also been reported (10).

## MECHANISM OF ACTION OF $P\text{-}450_{LM}$

### Evidence for Uptake of Two Electrons Prior to Oxygenation

The availability of highly purified $P\text{-}450_{LM_2}$ has permitted us to carry out a quantitative study of electron uptake by this protein (11,12). The unexpected finding was made that the reduction of this cytochrome by dithionite under anaerobic conditions requires two electrons per molecule. Under our conditions $P\text{-}450_{cam}$ consumes a single electron, as anticipated from the work of others (13,14). Our results are summarized in Table I. $P\text{-}450_{LM}$ with CO or cyanide present, or with no ligand added, consumed two electrons from dithionite or from NADPH in the presence of catalytic amounts of the purified reductase (Expt. 1). Although the results clearly indicate

Table I

Electron Uptake by Purified $P\text{-}450_{LM}$

| Expt. No. | Cytochrome preparation | Reductant or oxidant added | Added ligand | Electron uptake per molecule of cytochrome[a] |
|---|---|---|---|---|
| 1 | $P\text{-}450_{LM}$ | Dithionite | None | 1.9 |
| | " | Dithionite | Cyanide | 1.9 |
| | " | Dithionite | CO | 2.1 |
| | " | NADPH | CO | 1.9 |
| 2 | " (reduced) | Dichlorophenol-indophenol | CO | -2.0 |
| | " (reduced) | Cytochrome *c* | CO | -2.2 |
| | " (reduced) | $O_2$ | CO | -2.1 |
| 3 | $P\text{-}420_{LM}$ | Dithionite | CO | 1.9 |

The reductive titrations and reoxidations were carried out under anaerobic conditions (11) unless indicated otherwise, and the spectra were recorded with an Aminco DW-2 spectrophotometer in the split beam mode.

[a]Electron transfer from the reduced cytochrome to an oxidant is indicated by a negative sign.

the presence of an electron acceptor in addition to the heme iron, no significant amounts of nonheme iron, other metals or cofactors, or disulfide bonds were found, and free radicals were not detected by EPR spectrometry. We are indebted to Dr. R. H. Sands of the Biophysics Research Division of The University of Michigan and Dr. J. Peisach of the Albert Einstein College of Medicine for the EPR determinations. As indicated in Expt. 2, the reduced P-450$_{LM}$ is capable of donating two electrons to a variety of acceptors, including molecular oxygen. Additional experiments with P-420$_{LM}$ prepared by treating P-450$_{LM}$ with urea or p-mercuribenzoate showed that this altered form of the cytochrome also accepts two electrons. The role of the second electron acceptor is not yet clear, but chemically or photochemically reduced P-450$_{LM}$ is capable of hydroxylating substrates in the presence of molecular oxygen as well as yielding hydrogen peroxide.

## Hydroperoxide-Dependent Substrate Hydroxylation

The ability of liver microsomal suspensions to utilize organic hydroperoxides for the hydroxylation of various substrates and the possible role of cytochrome P-450 in these reactions have been reported by several laboratories (15-20). We have recently shown that highly purified P-450$_{LM}$ catalyzes the peroxide-dependent hydroxylation of various substrates, including N-methylaniline, cyclohexane, benzphetamine, aminopyrine, and methyl benzyl ether, in the absence of NADPH and the reductase. Such experiments rule out the involvement of other microsomal electron carriers. A variety of oxygenating agents, including hydrogen peroxide, cumene hydroperoxide, sodium chlorite, and m-chloroperbenzoic acid, were found to support benzphetamine hydroxylation and demethylation. As shown in Table II, benzphetamine demethylation shows a complete dependence on P-450$_{LM}$, which cannot be replaced by P-420$_{LM}$, heme, or ferric ions at the same concentration. The reaction is also completely dependent on the presence of hydrogen peroxide and requires phosphatidylcholine for full activity. Unlike the usual reconstituted microsomal system which requires molecular oxygen when NADPH serves as the electron donor in the presence of the reductase, the peroxide-dependent demethylation is unaffected when $O_2$ is replaced by $N_2$ under highly anaerobic conditions. The failure of carbon monoxide to inhibit the reaction indicates that the ferrous form of P-450$_{LM}$ is probably not involved. In other experiments, the cumene hydroperoxide-dependent hydroxylation of cyclohexane was carried out in $^{18}O$-labeled water, and the resulting cyclohexanol was found to contain only a very low level of $^{18}O$. Since, as already shown, peroxide-dependent hydroxylations do not require molecular oxygen, it appears that the oxygen atom must have been derived from the hydroperoxide.

Table II

Requirements for Peroxide-dependent Substrate Hydroxylation Catalyzed by Purified P-$450_{LM}$

| System | Rate of benzphetamine demethylation (nmol per nmol P-$450_{LM}$ per min) |
|---|---|
| Complete | 27 |
| P-$450_{LM}$ heat-inactivated | 2 |
| No P-$450_{LM}$ | 0 |
| " " ; P-$420_{LM}$ added | 0 |
| " " ; hemin chloride added | 0 |
| " " ; $FeCl_3$ added | 0 |
| No peroxide | 0 |
| No phospholipid | 14 |
| No $O_2$ ($N_2$ present) | 27 |
| No $O_2$ (CO present) | 28 |

The complete system contained P-$450_{LM}$ (1.0 μ*M*), dilauroylglyceryl-3-phosphorylcholine (0.06 m*M*), potassium phosphate buffer, pH 7.4 (0.1 *M*), radioactive benzphetamine (1.0 m*M*), and hydrogen peroxide (50 m*M*), and was incubated at 30° in the presence of air unless indicated otherwise.

## Spectral Intermediates in the Reaction of Reduced P-$450_{LM}$ with Oxygen

Spectral evidence for the binding of molecular oxygen to ferrocytochrome P-450 in the presence of substrate to form a ternary complex has been reported by Estabrook *et al.* (21) with liver microsomal suspensions and by Gunsalus *et al.* (22) and Ishimura *et al.* (23) with the bacterial cytochrome. We have recently confirmed the formation of the oxyferro complex in steady state experiments with purified P-$450_{LM}$, as shown in Fig. 1. The difference spectrum of the steady state oxyferro complex (dashed line) formed upon the addition of oxygen to the NADPH-reduced cytochrome resembles that reported earlier when allowance is made for the contribution of cytochrome $b_5$ to the spectrum in intact microsomes (21). The spectrum is clearly different from that of the reduced protein in the absence of oxygen.

The reaction of the photochemically reduced, purified P-$450_{LM}$ with oxygen was then examined by stopped flow techniques with the results presented in Fig. 2. The spectra of two distinct intermediates are seen, as constructed from kinetic traces at various wavelengths. The intermediate observed at 10 msec decays to form a second intermediate at 0.8 sec; the latter intermediate is probably the one observed in steady state experiments. In additional studies the

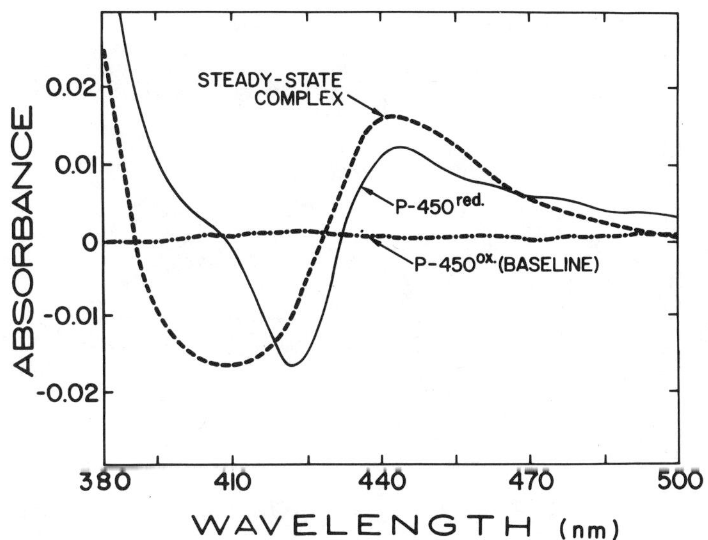


Fig. 1. Steady state difference spectra of P-$450_{LM}$ upon reduction and oxygenation. A reaction mixture containing P-$450_{LM_2}$ (1.3 μ$\underline{M}$; 0.11 mg protein per ml), NADPH-cytochrome P-450 reductase (25 μg protein per ml), dilauroylglyceryl-3-phosphorylcholine (0.03 m$\underline{M}$), benzphetamine (1.0 m$\underline{M}$), potassium HEPES buffer (pH 7.7; 0.05 $\underline{M}$), $MgCl_2$ (15 m$\underline{M}$), glucose 6-phosphate (1.0 m$\underline{M}$), and glucose 6-phosphate dehydrogenase (1.0 μg per ml) in a final volume of 2.5 ml was equilibrated with $N_2$ under anaerobic conditions. The baseline was recorded against an identical reaction mixture in an Aminco DW-2 spectrometer. NADPH (50 μ$\underline{M}$) was added anaerobically to one cuvette, and the resulting reduced difference spectrum was recorded. The reaction mixture containing reduced P-$450_{LM}$ was then aerated, and the oxyferro difference spectrum was recorded. The temperature was 20°.

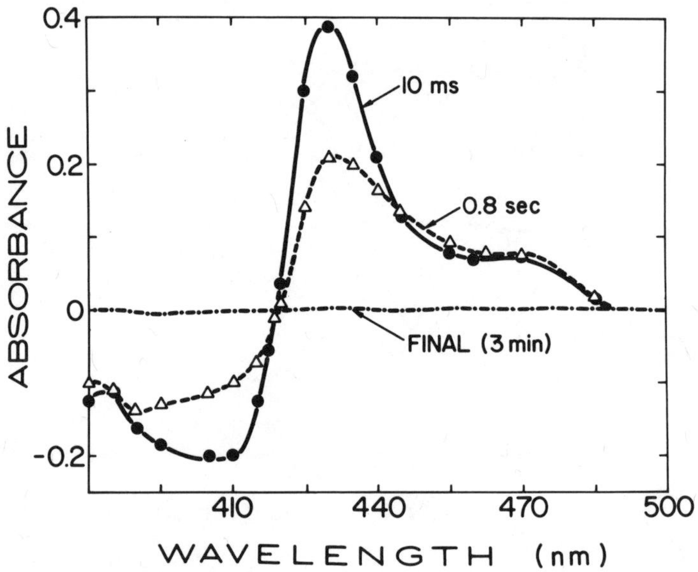


Fig. 2. Difference spectra of the complexes resulting from the reaction of reduced P-450$_{LM}$ with oxygen in the presence of benzphetamine. One syringe of the stopped flow apparatus contained P-450$_{LM_2}$ (13.5 μM; 1.0 mg protein per ml), dilauroylglyceryl-3-phosphorylcholine (0.22 mM), Tris-acetate buffer, pH 7.5 (0.06 M), EDTA (10 mM), proflavin sulfate (0.7 μM), methylviologen (0.2 μM), and benzphetamine (1.0 mM) under anaerobic nitrogen; the mixture was photoreduced at 10°. The other syringe contained an aerobic solution of benzphetamine (1.0 mM) in Tris-acetate buffer containing EDTA. The solutions were mixed, and the spectra were determined at the time intervals indicated from the kinetic traces at various wavelengths.

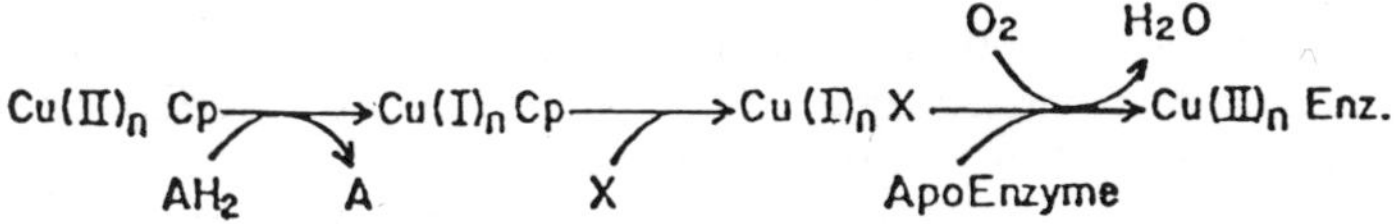


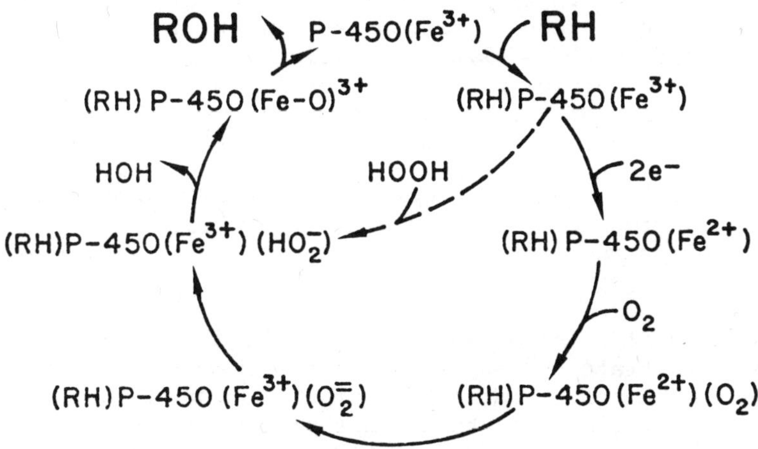


Fig. 3. Proposed scheme for mechanism of catalysis by $P\text{-}450_{LM}$. RH represents the substrate and ROH the product.

rate constants have been determined for the individual steps: reduced P-$450_{LM}$ ⟶ first oxyferro complex ⟶ second oxyferro complex ⟶ oxidized P-$450_{LM}$. The last of these steps is rate-limiting and, of particular interest, the rate is enhanced by the presence of cytochrome $\underline{b}_5$. Since in such experiments the reduced cytochrome $\underline{b}_5$ is not significantly oxidized in the time required for P-$450_{LM}$ to be reoxidized, cytochrome $\underline{b}5$ may be playing an effector role rather than functioning as an electron carrier.

## Proposed Scheme for Mechanism of Action of P-$450_{LM}$

A possible mechanism for the reactions catalyzed by liver microsomal cytochrome P-450 is given in Fig. 3. The binding of substrate to the oxidized form of the cytochrome is necessary before rapid electron transfer can occur. The uptake of two electrons is then indicated, with the formation of the ferrous protein and with a second electron bound elsewhere in this hemeprotein. Oxygen then combines with the protein and undergoes a two-electron reduction to yield $O_2^{=}$, perhaps with the intermediate formation of superoxide, $O_2^{\cdot -}$ (24). The fully reduced oxygen is then pictured as undergoing protonation to form $HO_2^-$, followed by elimination of water and formation of an Fe-O complex having an overall charge of $3^+$ and a formal oxidation state of $5^+$. Insertion of the oxygen atom into a favorably positioned C-H bond of the substrate, RH, would then yield the hydroxylated product, ROH, with regeneration of oxidized P-$450_{LM}$. The activated oxygen may be an oxenoid species (25,26), which we have shown as a ferric-bound oxygen atom, but it is recognized that other resonance structures involving higher oxidation states of the iron or the porphyrin (27) may contribute.

The scheme also indicates the manner in which hydrogen peroxide may support substrate hydroxylation in the absence of $O_2$ and an external electron donor. Postulation of $HO_2^-$ as an intermediate explains the observation, in studies not presented here, that the hydrogen peroxide-dependent hydroxylation increases in rate with increasing pH. The proposed scheme shares common features with the mechanism of action of peroxidases, as described by others (28,29). Peroxides may react with P-$450_{LM}$ to produce the same species of activated oxygen as generated from molecular oxygen and NADPH in the complete reconstituted system.

## REFERENCES

1. Coon, M. J., van der Hoeven, T. A., Haugen, D. A., Guengerich, F. P., Vermilion, J. L., and Ballou, D. P. (1975) in Cytochromes P-450 and b5: Structure, Function, and Interaction, Adv. Exp. Med. Biol. (Cooper, D. Y., Rosenthal, O., Snyder, R., and Witmer, C., eds) Vol. 58, pp. 25-46, Plenum Press, New York.
2. Coon, M. J., Haugen, D. A., Guengerich, F. P., Vermilion, J. L., and Dean, W. L. (1976) Symposium on The Structural Basis of Membrane Function, Tehran, Academic Press, in press.
3. van der Hoeven, T. A., and Coon, M. J. (1974) J. Biol. Chem. 249, 6302-6310.
4. van der Hoeven, T. A., Haugen, D. A., and Coon, M. J. (1974) Biochem. Biophys. Res. Commun. 60, 569-575.
5. Haugen, D. A., van der Hoeven, T. A., and Coon, M. J. (1975) J. Biol. Chem. 250, 3567-3570.
6. Yu, C-A., Gunsalus, I. C., Katagiri, M., Suhara, K., and Takemori, S., (1974) J. Biol. Chem. 249, 94-101.
7. Dus, K., Litchfield, W. J., Miguel, A. G., van der Hoeven, T. A., Haugen, D. A., Dean, W. L., and Coon, M. J. (1974) Biochem. Biophys. Res. Commun. 60, 15-21.
8. Dus, K., Litchfield, W. J., Miguel, A. G., van der Hoeven, T. A., Haugen, D. A., Dean, W. L., and Coon, M. J. (1975) in Cytochromes P-450 and b5: Structure, Function, and Interactions, Adv. Exp. Med. Biol. (Cooper, D. Y., Rosenthal, O., Snyder, R., and Witmer, C., eds) Vol. 58, pp. 47-53, Plenum Press, New York.
9. Nebert, D. W., Heidema, J. K., Strobel, H. W., and Coon, M. J. (1973) J. Biol. Chem. 248, 7631-7636.
10. Wiebel, F. J., Selkirk, J. K., Gelboin, H. V., Haugen, D. A., van der Hoeven, T. A., and Coon, M. J. (1975) Proc. Nat. Acad. Sci. USA 72, 3917-3920.
11. Ballou, D. P., Veeger, C., van der Hoeven, T. A., and Coon, M. J., (1974) FEBS Letters 38, 337-340.
12. Guengerich, F. P., Ballou, D. P., and Coon, M. J. (1975) J. Biol. Chem. 250, 7405-7414.
13. Peterson, J. A. (1971) Arch. Biochem. Biophys. 144, 678-693.
14. Gunsalus, I. C., Lipscomb, J. E., Marshall, V., Frauenfelder, H., Greenbaum, E., and Münck, E. (1972) in Biological Hydroxylation Mechanism (Boyd, G. S., and Smellie, R. M. S., eds) pp. 135-138, Academic Press, London.
15. Kadlubar, F. F., Morton, K. C., and Ziegler, D. M. (1973) Biochem. Biophys. Res. Commun. 54, 1255-1261.
16. Rahimtula, A. D., and O'Brien, P. J. (1974) Biochem. Biophys. Res. Commun. 60, 440-447.
17. Ellin, Å., and Orrenius, S. (1975) FEBS Letters 50, 373-381.
18. Rahimtula, A. D., and O'Brien, P. J. (1975) Biochem. Biophys. Res. Commun. 62, 268-275.

19. Hrycay, E. G., Gustafsson, J. A., Ingelman-Sundberg, M., and Ernster, L. (1975) FEBS Letters 56, 161-165.
20. Hrycay, E. G., Gustafsson, J. A., Ingelman-Sundberg, M., and Ernster, L. (1975) Biochem. Biophys. Res. Commun. 66, 209-216.
21. Estabrook, R. W., Hildebrandt, A. G., Baron, J., Netter, K. J., and Leibman, K. (1971) Biochem. Biophys. Res. Commun. 42, 132-139.
22. Gunsalus, I. C., Tyson, C. A., Tsai, R., and Lipscomb, J. D., (1971) Chem.-Biol. Interactions 4, 75-78.
23. Ishimura, Y., Ullrich, V., and Peterson, J. A. (1971) Biochem. Biophys. Res. Commun. 42, 140-146.
24. Strobel, H. W., and Coon, M. J. (1971) J. Biol. Chem. 246, 7826-7829.
25. Rahimtula, A. D., O'Brien, P. J., Seifried, H. E., and Jerina, D. M. (1975) Fed. Proc. 34, 623.
26. Coon, M. J., Nordblom, G. D., White, R. E., and Haugen, D. A., (1975) Biochem. Soc. Transactions 3, 813-817.
27. Dolphin, D., Forman, A., Borg, D. C., Fajer J., and Felton, R. H. (1971) Proc. Nat. Acad. Sci. USA 68, 614-618.
28. Schonbaum, G. R., and Lo, S. (1972) J. Biol. Chem. 247, 3353-3360.
29. Hager, L. P., Doubek, D. L., Silverstein, R. M., Hargis, J. H., and Martin, J. C. (1972) J. Am. Chem. Soc. 94, 4364-4366.

# CHARACTERIZATION OF PURIFIED CYTOCHROME P-450$_{scc}$ AND P-450$_{11\beta}$ FROM BOVINE ADRENOCORTICAL MITOCHONDRIA

Masayuki Katagiri, Shigeki Takemori, Eiji Itagaki, Katsuko Suhara, Tomoharu Gomi and Hiroshi Sato

Department of Chemistry, Faculty of Science, Kanazawa University, Ishikawa 920, Japan

## INTRODUCTION

In the adrenal cortex, NADPH-adrenal ferredoxin reductase (adrenodoxin reductase)-adrenal ferredoxin (adrenodoxin)-cytochrome P-450 system has been known to be associated with the mitochondrial steroid hydroxylase activities, viz., cholesterol side-chain cleavage and steroid 11β- and 18-hydroxylations. In order to elucidate the reaction mechanism of the hydroxylase system, it is important that each component is available in a pure form. As previously reported, adrenodoxin reductase (1) and adrenodoxin (2,3) have been obtained in a pure state. The latter has been crystallized. Details will be discussed in a separate chapter (4).

A technique for the purification of the cytochrome P-450, using an aniline-substituted Sepharose chromatography, has been developed in our laboratory for efficient separation of cytochrome P-450 from other heme proteins (5,6). Two species of adrenal P-450 cytochromes, catalyzing the cholesterol side-chain cleavage (P-450$_{scc}$) and the steroid 11β-hydroxylation (P-450$_{11\beta}$), have been highly purified in functional active forms from sonicated mitochondrial pellets (7,8). These separation and purification studies have made it possible to further investigate the structure and function of adrenal cytochrome P-450.

## PURIFICATION OF P-450

Sonicated mitochondrial pellets, prepared by the method of Omura _et al._ (9) with minor modification and stored at -60°, were thawed and suspended in 100 mM potassium phosphate buffer, pH 7.0. To the suspension, 100 μM dithiothreitol, sodium cholate (1 mg per 1.5 mg protein) and 10 μM deoxycorticosterone (DOC) were added. Purification of P-$450_{11\beta}$ was carried out in the presence of DOC, an 11β-hydroxylatable substrate, which markedly protected the protein from inactivation (Fig. 1). The procedure for separation and purification of P-$450_{scc}$ and P-$450_{11\beta}$ consists of the following three main steps (Fig. 2). 1. Ammonium sulfate fractionation to separate each P-450. 2. Dialysis against DOC-containing buffer without cholate. Cytochrome P-$450_{11\beta}$ was selectively precipitated and could be solubilized again in the cholate-containing buffer. Minor contaminant of P-$450_{11\beta}$ in P-$450_{scc}$ preparation was removed by buffer lacking cholate and DOC. 3. Aniline-Sepharose chromatography.

The yield was usually 20-30 mg of each P-450 from 5 g protein of mitochondrial pellet. P-$450_{scc}$, in the buffer containing 0.01% cholate, could be stored for weeks at -60° without substantial loss of activity. P-$450_{11\beta}$ in the substrate-complexed form at the concentration less than 5 μM remains clear on standing at 4° for a week in the presence of both 0.3% cholate and 0.3% Tween 20.

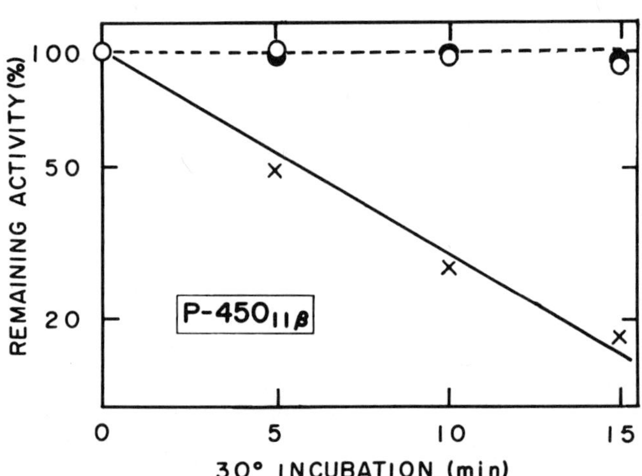


Fig. 1. Effect of 11β-hydroxylatable substrate on stability of P-$450_{11\beta}$ activity in 50 mM potassium phosphate buffer, pH 7.3, containing 100 μM EDTA and 100 μM dithiothreitol. Incubated at 30° with 50 μM DOC (o), with 50 μM testosterone (•) or without substrate (x). Crude cholate extract was used.

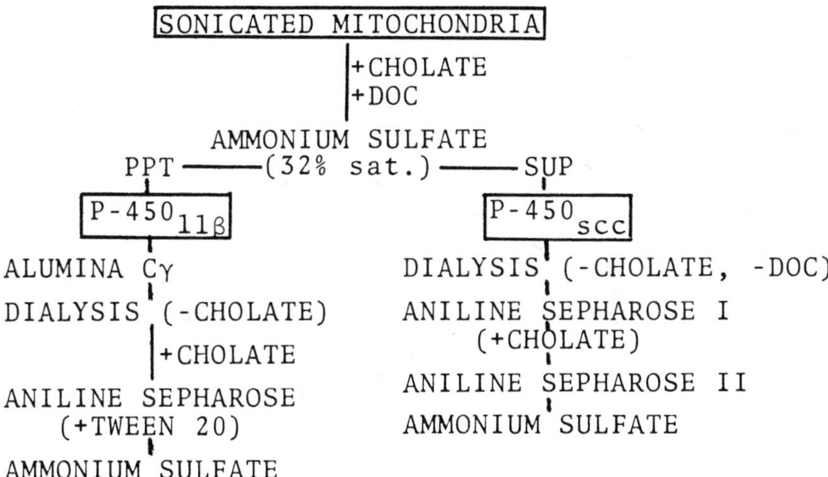


Fig. 2. Summary of the purification of $P\text{-}450_{scc}$ and $P\text{-}450_{11\beta}$.

## PURITY AND MOLECULAR WEIGHT

Sedimentation velocity experiments indicated purified $P\text{-}450_{scc}$ and $P\text{-}450_{11\beta}$ was essentially pure with $s_{20,w}$ values of 5.7 S and 3.3 S, respectively. When the purified preparations were subjected to SDS polyacrylamide gel electrophoresis, only one protein band was observed. As shown in Fig. 3, when a mixture of equal amounts of $P\text{-}450_{scc}$ and $P\text{-}450_{11\beta}$ was run, two distinct bands clearly separated From these data, the two P-450's appear to be satisfactorily pure for further analysis.

The molecular weight of native $P\text{-}450_{scc}$ was determined to be 200,000 by the method of Yphantis. Reduced and carboxymethylated preparations of $P\text{-}450_{scc}$ and $P\text{-}450_{11\beta}$ showed molecular weights of 48,000 and 45,000 respectively in 6 M guanidine-HCl. By comparing mobilities of the two cytochrome P-450's on SDS-polyacrylamide gel electrophoresis with four marker proteins, apparent molecular weights of 51,000 and 46,000 were estimated for $P\text{-}450_{scc}$ and $P\text{-}450_{11\beta}$. The optical absorbance ratio ($A_{394\ nm}/A_{280\ nm}$) and the grams of protein (biuret procedure) per mole of heme (pyridine hemochromogen method) were 0.83 and 81,000 for $P\text{-}450_{scc}$ and 0.85 and 85,000 for $P\text{-}450_{11\beta}$, respectively.

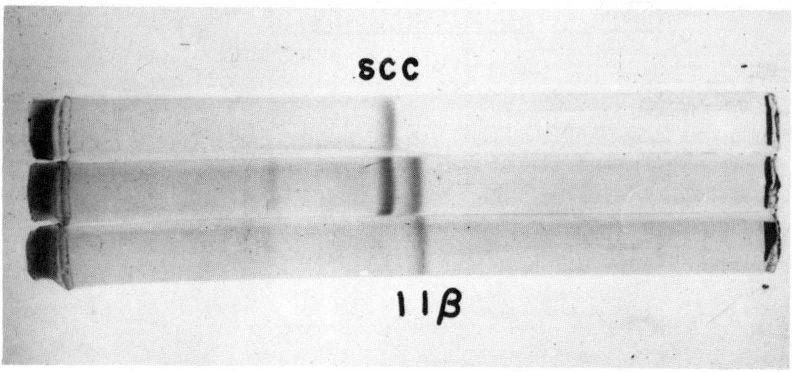


Fig. 3. SDS-polyacrylamide gel electrophoresis of P-$450_{scc}$ (upper), P-$450_{11\beta}$ (lower) and equimolar mixture of both (middle)

## SPECTRAL PROPERTIES

The absorption spectra of the two purified P-450's are illustrated in Fig. 4. and Fig. 5, respectively. Either purified preparation, in its ferric state, showed a high spin type absorption spectrum having maxima at 394, 510 and 645 nm. The CO spectrum of the dithionite reduced form gave maxima at 448 and 550 nm. The purified P-$450_{scc}$ preparation contained endogenous cholesterol (0.6 to 1.0 mole per mole of heme). P-$450_{11\beta}$ contained DOC which had been added as a stabilizing agent during the course of purification and did not have significant amount of cholesterol..When either P-450 was treated with the adrenodoxin dependent electron transfer system, the respective absorption maxima of the preparation shifted completely to 418, 539 and 570 nm. By the addition of the substrate (cholesterol of DOC), the high spin type species was recovered. The rate of conversion of P-$450_{scc}$ induced by cholesterol was slow with a half-time of 40 sec (Fig. 6). Upon the addition of 11β-hydroxylatable substrate, DOC, 11-deoxycortisol, or testosterone, absorption peaks of the low-spin type species of oxidized P-$450_{11\beta}$ were shifted instantly to those of the typical high-spin or substrate-bound type spectrum. Under the same conditions, cholesterol, progesterone, and 11β-hydroxycorticosteroids did not induce a change in the spectrum of purified P-$450_{11\beta}$.

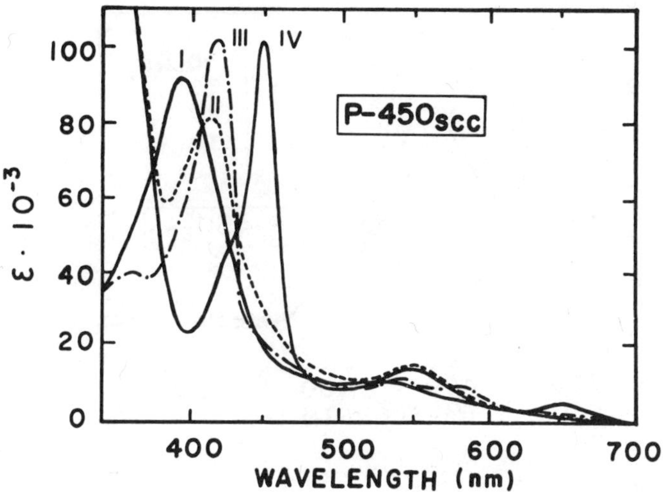


Fig. 4. Absorption spectra of purified $P\text{-}450_{scc}$. The buffer system was 50 mM potassium phosphate, pH 7.4, containing 100 μM EDTA, 100 μM dithiothreitol and 0.01% sodium cholate. I. Oxidized (cholesterol-complexed), II. dithionite reduced, III. after treatment with the adrenal electron transfer system and IV. dithionite-reduced CO complex.

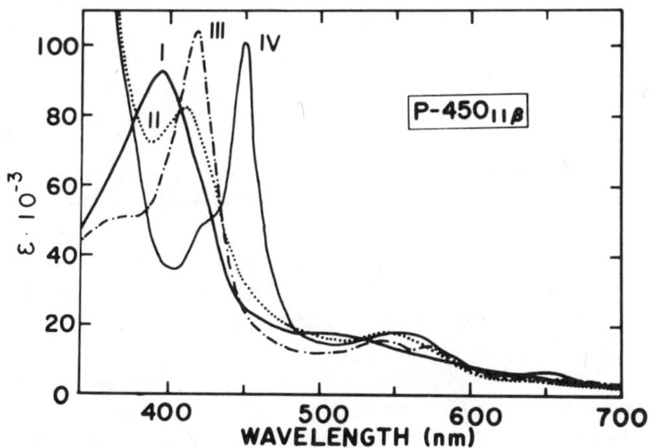


Fig. 5. Absorption spectra of purified $P\text{-}450_{11\beta}$. The buffer system was the same as described in the legend of Fig. 4, except 0.3% cholate and 0.3% Tween 20 were present. I. Oxidized (DOC-complexed), II. dithionite reduced, III. after treatment with the electron transfer system and IV. dithionite-reduced CO complex.

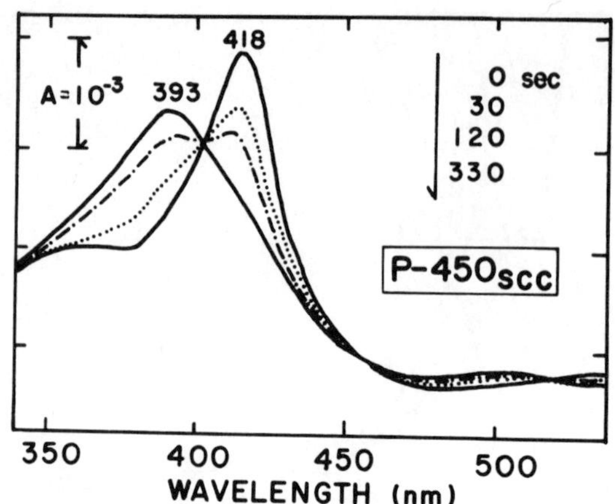


Fig. 6. Effect of cholesterol (lecithin-emulsion) upon the formation of cholesterol complex of P-450$_{scc}$, before (0) and after the indicated times. For the buffer used, see the legend for Fig. 4.

## ACTIVITY AND SPECIFICITY

The cholesterol side-chain cleavage activity was measured by incubating P-450 in potassium phosphate buffer, containing NADP$^+$, glucose-6-phosphate, glucose-6-phosphate dehydrogenase, $MgCl_2$, adrenodoxin reductase, adrenodoxin and cholesterol as a sonicated lecithin-emulsion. Following the incubation, the medium was extracted with chloroform and the trimethylsililated derivatives of the extract were assayed by the gas chromatographic method of Ando and Horie (10).

When this assay was used, P-450$_{scc}$ catalyzed the conversion of cholesterol to pregnenolone at a rate of 7 moles/min/mole of heme.

Assay of 11β-hydroxylase activity was performed in a similar manner but with DOC as the substrate. The product was quantitated by measuring fluorimetrically the rate of corticosterone formation(11).

P-450$_{11\beta}$ converted DOC into the corresponding 11β-hydroxylated product, corticosterone (130 moles/min/mole of heme). The reaction mixture was extracted with dichloromethane and then subjected to silica gel thin layer chromatography (Eastman 13181) with chloroform-methanol (97:3) or benzene-acetone-water (75:50:0.2) as solvents. The Rf values of the main spot from the reaction mixture and the corresponding authentic 11β-hydroxysteroid were identical.

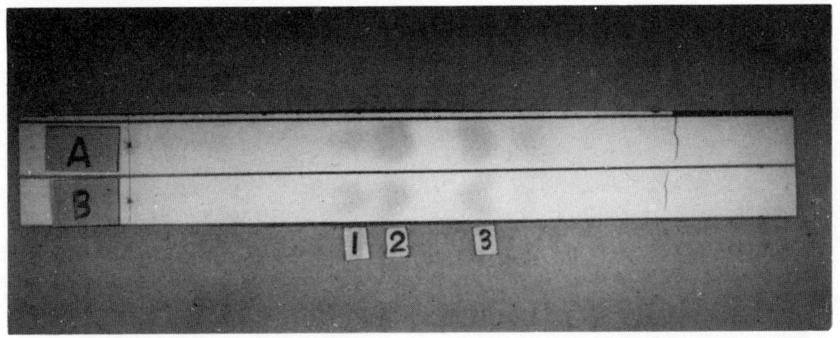


Fig. 7. Thin layer chromatography of dichloromethane extracted product from the $P\text{-}450_{11\beta}$ - DOC reaction mixture. A: Dichloromethane extracts. B: Authentic samples of; 18-hydroxy DOC, 2; corticosterone and 3; DOC. Photographed while illuminating with a UV lamp.

Björkhem and Karlmer (12) reported recently that unfractionated adrenal mitochondrial extracts produced 11β-hydroxylated- and concomitantly a lower level of 18-hydroxylated DOC. Our purified enzyme preparation produced an additional minor product. The Rf value of the second product corresponded to the Rf of authentic 18-hydroxy DOC. The ratio between intensities of the 11β- and 18- hydroxy DOC was estimated densitometrically at 250 nm to be about 6:1 (Fig. 7). At the present time, it is not clear whether the 18-hydroxylation can be attributed to $P\text{-}450_{11\beta}$ itself or to a co-purified "$P\text{-}450_{18}$".

11-Deoxycortisol was an equally active substrate of $P\text{-}450_{11\beta}$ as DOC, while testosterone was less active. Progesterone was inactive. This agrees with the observation that progesterone did not induce a spectral change of $P\text{-}450_{11\beta}$.

In a completly reconstituted hydroxylase system composed of NADPH, adrenodoxin reductase, adrenodoxin and $P\text{-}450_{11\beta}$, DOC dependent NADPH oxidation was observed with the formation of corticosterone.

## ANALYSES

Table I lists the amino acid compositions of the cytochrome P-450 proteins which were determined, on acid hydrolyzates of the protein, in the automatic amino acid analyzer. Half-cystine and tryptophan values have been determined by specific methods (13,14). The data inidcate some common features between $P\text{-}450_{scc}$ and $P\text{-}450_{11\beta}$: these are the high percentage (36%) of hydrophobic amino acid residues and similar contents of the sulfur amino acids.

| | $P\text{-}450_{scc}$ | $P\text{-}450_{11\beta}$ | | $P\text{-}450_{scc}$ | $P\text{-}450_{11\beta}$ |
|---|---|---|---|---|---|
| Lys | 25.9 | 12.2 | Ala | 18.4 | 28.3 |
| His | 12.4 | 10.8 | Cys* | 3.3 | 3.4 |
| Arg | 21.2 | 30.3 | Met | 11.5 | 11.2 |
| Asx | 36.5 | 28.4 | Val | 23.0 | 24.5 |
| Thr | 21.9 | 21.4 | Ile | 23.4 | 14.5 |
| Ser | 21.4 | 20.2 | Leu | 39.1 | 45.8 |
| Glx | 48.1 | 47.7 | Tyr | 13.4 | 9.6 |
| Pro | 25.9 | 22.5 | Phe | 23.5 | 16.4 |
| Gly | 21.2 | 21.1 | Try† | 4.5 | 6.4 |

Table I. The amino acid compositions. Values are calculated on the basis of molecular weight of 48,000 and 45,000 for $P\text{-}450_{scc}$ and for $P\text{-}450_{11\beta}$, respectively.
* After performic acid oxidation (13); † determined spectrophotometrically (14).

Experiments aimed to discover copper by bathocuproine method (15) have yielded negative results in both proteins. A similar result has been obtained in the case of highly purified liver microsomal cytochrome P-450 ($P\text{-}450_{LM}$) by Ballor et al. (16). Analysis of total iron content by bathophenanthroline method (15) and heme contents by pyridine hemochromogen method (17) of $P\text{-}450_{scc}$ and $P\text{-}450_{11\beta}$ indicated the absence of non-heme iron in the purified hemoprotein preparations.

## CONCLUSION

Although our study is still in progress, the results obtained thus far provide direct evidence that two cytochrome P-450 species exist in the adrenal cortex. These P-450's have important differences in the specificities with regard to the compounds which produce spectral shifts and compounds which they hydroxylate.

$P\text{-}450_{scc}$ and $P\text{-}450_{11\beta}$ possess many similar analytical characteristics, optical properties and molecular size per heme but also some differences in solubility. $P\text{-}450_{11\beta}$ is less soluble than $P\text{-}450_{scc}$ and is more hydrophobic and requires the presence of a non-ionic detergent, viz. Tweens, Triton X-100 or Emulgen 220 is required during the purification and storage of the enzyme.

ACKNOWLEDGEMENTS

We wish to thank Ms. Sachiko Tohoda and Messrs. Tsuneo Takadera and Kiyotaka Yamakawa for the valuable technical assistances, and Dr. Ryo Sato for supplying adrenal glands.

REFERENCES

1. Suhara, K., Ikeda, Y., Takemori, S., and Katagiri, M., Fed. Eur. Biochem. Soc. Lett., 28, 45 (1972).
2. Suhara, K., Takemori, S., and Katagiri, M., Biochim. Biophys. Acta, 263, 272 (1972).
3. Suhara, K., Kanayama, K., Takemori, S., and Katagiri, M., Biochim. Biophys. Acta, 336, 309 (1974).
4. Takemori, S., Suhara, K., and Katagiri, M., This book, p.36
5. Hashimoto, S., Suhara, K., Takemori, S., and Katagiri, M., Seikagaku, 43, 535 (1971).
6. Katagiri, M., and Takemori, S., Abstract Book, Ninth International Congress of Biochemistry, Stockholm, p. 327 (1973).
7. Takemori, S., Suhara, K., Hashimoto, S., Hashimoto, M., Sato, H., Gomi, T., and Katagiri, M., Biochem. Biophys. Res. Commun., 63, 588 (1975).
8. Takemori, S., Sato, H., Gomi, T., Suhara, K., and Katagiri, M., Biochem. Biophys. Res. Commun., 67, 1151 (1975).
9. Omura, T., Sanders, E., Estabrook, R. W., Cooper, D. Y., and Rosenthal, O., Arch. Biochem. Biophys., 117, 660 (1966).
10. Ando, N., and Horie, S., J. Biochem., 65, 269 (1969).
11. Mattingly, D., J. Clin. Pathol., 15, 374 (1962).
12. Björkhem, I., and Karlmar, K-E., Eur. J. Biochem., 51, 145 (1975).
13. Hirs, C. H. W., J. Biol. Chem., 219, 611 (1956).
14. Goodwin, T. W., and Morton, R. A., Biochem. J., 40, 628 (1946).
15. Van de Bogart, M., and Beinert, H., Anal. Biochem., 20, 325 (1967).
16. Ballou, D. R., Veeger, C., Van der Hoeven, T. A., and Coon, M. J., Fed. Eur. Biochem. Soc. Lett., 38, 337 (1974).
17. Appleby, C. A., and Morton, R. K., Biochem. J., 73, 539 (1959).

# PURIFICATION AND PROPERTIES OF CYTOCHROME P-450 FROM ADRENOCORTICAL MITOCHONDRIA AND ITS INTERACTION WITH ADRENODOXIN

Toshihiro Sugiyama, Retsu Miura and Toshio Yamano

Department of Biochemistry, Osaka University Medical School, 33 Joancho, Kitaku, Osaka, Japan

Cytochrome P-450s from mammalian sources have been purified to a homogeneous state by several groups, and highly purified cytochrome P-450 from bacterial origin has been obtained in a crystalline form. Among the mammalian P-450s, the one from adrenal mitochondria (Harding et al., 1964) is the easiest to solubilize and the hydroxylating system can be reconstituted with purified components.

There have been some uncertainties in the enzyme activity measurements (Horie and Watanabe, 1975; Wang et al., 1974) and in the criteria of enzyme purity. Concerning activity measurements, the effect of the medium in the assay medium should be carefully controlled. This includes the effect of buffers, ionic strengths, detergents and organic solvents in which the substrates are dissolved. It should also be noted that substrates do not easily dissolve in water and also that cytochrome P-450 polymerizes with ease during the purification procedure. The state of polymerization (Shikita and Hall, 1973) may in turn affect the enzyme activity. While the spin states of the two P-450s, one with a high spin type and the other with mixed type spectrum, have been assigned for the side chain cleavage and 11β-hydroxylation, respectively (Jefcoate et al., 1970; Ando and Horie, 1972), it is still desirable to further clarify the effect of additives on the activity of P-450 and the spin states. The purity of cytochrome P-450 has not yet been clearly established. For example, what is the correct subunit structure of cytochrome P-450 and how are the subunits linked to form the functional polymer(s)?

We have developed a new method for purifying adrenal P-450 and have attempted to analyse the environmental effects on its mode of action and on its physical properties both intra- and intermolecularly. Our success in crystallizing NADPH-adrenodoxin reductase (Sugiyama and Yamano, 1975) led us to use adrenodoxin affinity chromatography for the purification of P-450 which donates electrons to cytochrome P-450.

## MATERIALS AND METHODS

Bovine adrenals were purchased frozen from a slaughterhouse. Crystalline adrenodoxin was prepared by the method of Suhara et al. (1972). Immobilized adrenodoxin-Sepharose (Ad-Sepharose) and NADPH-adrenodoxin reductase were prepared according to the method previously reported (Sugiyama and Yamano, 1975). The mitochondria were suspended in 0.1 M potassium phosphate, pH 7.4, disrupted with a blender for 10 minutes and centrifuged at 55,000 x g for 90 minutes. The precipitate was used for the preparation of P-450 and the supernatant was reserved for the preparation of adrenodoxin and adrenodoxin reductase. All manipulations were carried out at 4°.

The concentration of P-450 was determined by the method of Omura and Sato (1964), using an extinction coefficient of 91 $cm^{-1}$ $mM^{-1}$. The heme content was evaluated spectrophotometrically after converting the heme into pyridine-hemochromogen. Protein content was estimated by the method of Lowry et al. (1951). SDS acrylamide gel electrophoresis was carried out according to the method of Weber and Osborn (1969) and bovine serum albumin, catalase, glutamic dehydrogenase, ovalbumin, alcohol dehydrogenase, α-chymotrypsin and cytochrome c were used as the standards for the molecular weight determination.

The enzyme activities of 11β-hydroxylation and side chain cleavage of steroids were measured gas-chromatographically and/or fluorometrically (Mattingly, 1962). The assay mixture (2.0 ml) contained substrate (400 nmoles in 20 μl isopropyl alcohol), adrenodoxin (20 nmoles), adrenodoxin reductase (2 nmoles), P-450, NADPH (2 μmoles) and 50 mM potassium phosphate, pH 7.4. The reaction was initiated by the addition of NADPH and after the reaction proceeded at 37°, the mixture was immediately extracted with an equal volume of dichloromethane. The organic layer was separated and evaporated to dryness. The residue was subjected to analysis. In gas chromatographic analysis pregnenolone acetate was used as the internal standard.

Spectrophotometric measurements were carried out with High-Sens Spectrophotometer SM-401 equipped with Spectral Data Processor SM-4012, Union Giken Co. Ltd. EPR measurements were performed in the Varian E12 Spectrometer.

All the reagents were from commercial sources and of the highest grade available.

## RESULTS AND DISCUSSION

The outline of the preparation procedure is presented in Fig. 1. Mitochondrial pellet prepared as described above from 500 g of bovine adrenal cortex was suspended in 10 mM phosphate, pH 7.4 to a final protein concentration of 20 mg/ml. To this suspension was added an equal volume of the same buffer containing 2% (w/v) Triton X-100 and 10% (w/v) polyethylene glycol (#6000). After stirring for an hour, the mixture was centrifuged at 55,000 x g for 90 minutes. The supernatant, without further manipulation, was directly applied to the adrenodoxin affinity column (2 x 10 cm) previously equilibrated with 10 mM phosphate, pH 7.4. The column was washed with 200 ml of 10 mM phosphate, pH 7.4, containing 0.1% Triton X-100 and 1 mM EDTA, then with 1000 ml of 10 mM phosphate, pH 7.4 was eluted with 0.6 M sodium chloride in 50 mM phosphate, pH 7.4. Fig. 2 shows the elution pattern. Fractions containing P-450 with absorbance ratio (394 nm/275 nm) of 0.4 to 0.5 were collected and subjected to ammonium sulfate fractionation with 45% saturation. The precipitate was dialyzed against 0.6 M sodium chloride in 50 mM phosphate, pH 7.4. The sample contained a main component with a molecular weight of about 54,000 and a minor component of molecular weight 49,000 when the sample was analyzed by the SDS acrylamide gel electrophoresis. The side chain cleavage activity of the sample was 5 nmol pregnenolone/nmol P-450/min when 400 nmoles of 20α-hydroxy-cholesterol was used. The 11β-hydroxylation activity was 5-15 nmol deoxy-corticosterone (DOC) consumed/nmol P-450/min when 400 nmoles of DOC was used. The heme content was estimated as 12.4 nmol/mg protein (80,600 g protein/mol heme). Most of the endogenous cholesterol had been washed off and the content of cholesterol was estimated by gas-chromatography to be less than 0.5 nmol/nmol P-450. For further purification the sample was applied to a Sephadex G-200 column, which was eluted with 0.6 M sodium chloride in 50 mM phosphate, pH 7.4. The elution profile is shown in Fig. 3. Two peaks were obtained and the component eluted first contained the so-called mixed type P-450 spectrum. The latter shows a high spin type of absorption spectrum with A(394 nm)/A(275 nm) ratio of 0.81. SDS acrylamide gel electrophoresis analysis showed a single component whose molecular weight was about 54,000. The details of the purification steps are summarized in Table I.

The Sephadex G-200 gel filtration step showed that P-450 is in a polymeric state as previously reported by Shikita and Hall (1973). The minor component found formerly was eluted in the earlier fractions, which suggested that it exists in a more highly polymerized state than the major component.

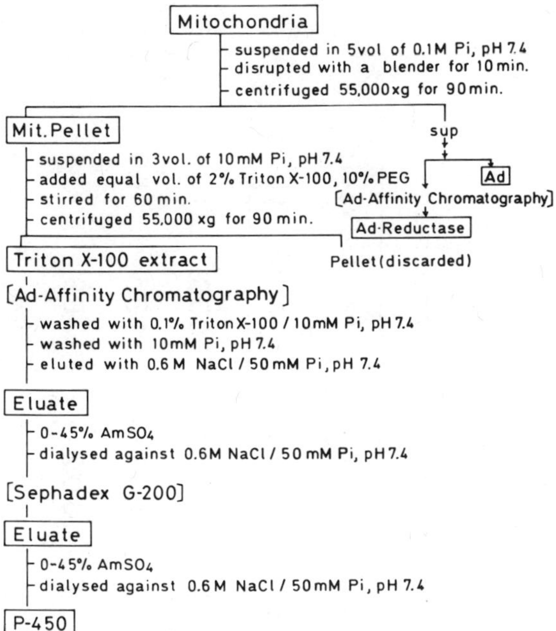


Figure 1. Flow diagram of purification of P-450 from adrenocortical mitochondria.

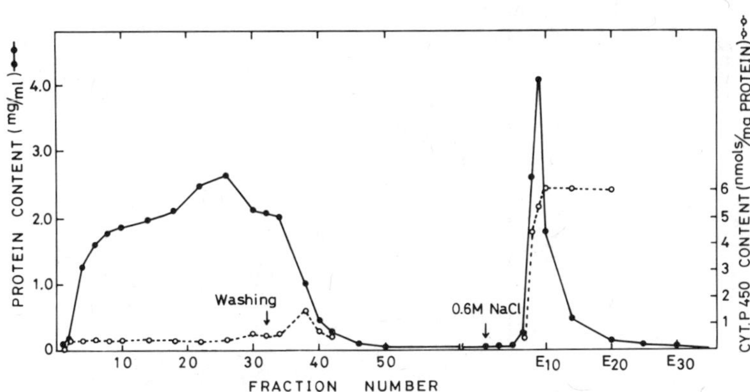


Figure 2. Chromatography of P-450 from bovine adrenocortical mitochondria on the Ad-Sepharose column. To the column (2 x 10 cm), equilibrated with 10 mM phosphate, pH 7.4, was applied 440 ml of the Triton X-100 extract (2.63 mg protein/ml). The arrows indicate where the washing and the increase in ionic strength occurred. The details are described in the text.

TABLE I

Recovery of Cytochrome P-450 from Bovine Adrenocortical Mitochondria

| Fractions | Total volume (ml) | Total Protein (mg) | P-450 (nmoles/mg P.) | Total P-450 (nmoles) | Yield (%) |
|---|---|---|---|---|---|
| 1) Mit. pellet suspension | 250 | 5350 | 0.37 | 2000 | 100 |
| 2) Triton X-100 extract | 440 | 1157 | 1.03 | 1190 | 60 |
| 3) Affinity Chromatogr. eluate | 85 | 61.3 | 6.02 | 369 | 18 |
| 4) Sephadex G-200 eluate | 93 | 15.4 | 10.8 | 166 | 9 |

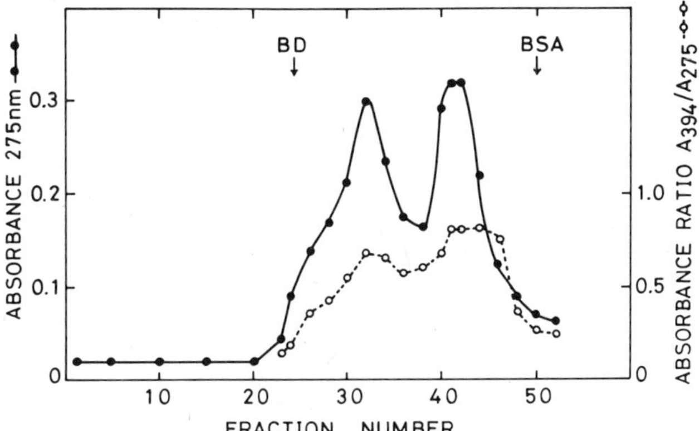


Figure 3. Gel filtration pattern of P-450 after the affinity chromotography step. To the Sephadex G-200 column (1.5 x 100 cm) equilibrated with 10 mM phosphate, pH 7.4, containing 0.6 M NaCl, was applied 0.5 ml of P-450 solution (23 mg protein/ml). BD and BSA designate the positions where blue dextran and bovine serum albumin are eluted, respectively.

The purified P-450 showed a high spin type absorption spectrum in the Soret region, namely a peak at 394 nm with a shoulder at 414 nm. This spectrum was converted to that of the low spin type by the addition of cholesterol oxidase from *Nocardia erythropois*. As shown in Fig. 4, the addition of Triton X-100 to the P-450 sample (spectrum 1) caused a decrease in absorbance at 394 nm with a concomitant increase at 418 nm. When the final Triton X-100 concentration was 0.1% or higher, the peak at 418 nm reached its maximum, hence the absorbance at 394 nm its minimum. Replot of the spectral conversion (Fig. 5) shows that the conversion *vs.* the detergent concentration is sigmoidal with the maximum slopes occurring near the critical micelle concentration (Robinson and Tanford, 1975). This observation may suggest that the conversion is closely related to the micelle formation of P-450 with the detergent. Most of the cholesterol bound to the enzyme came off when the P-450 adsorbed to Ad-Sepharose was washed with Triton X-100, although it was not removed completely even when the detergent concentration was as high as 0.1%. The increase in the 418 nm absorption peak (low spin spectrum) was observed on the addition of Triton X-100 even after the treatment of the sample with cholesterol oxidase. The observed absorbance changes may thus give a further insight into the cholesterol binding site and clarify whether the same mechanism is responsible for the spectral changes induced by the detergent and by the cholesterol oxidase. The same spectral conversion was found to occur when the enzyme was treated with a series of alcohols and acetone as will be described later.

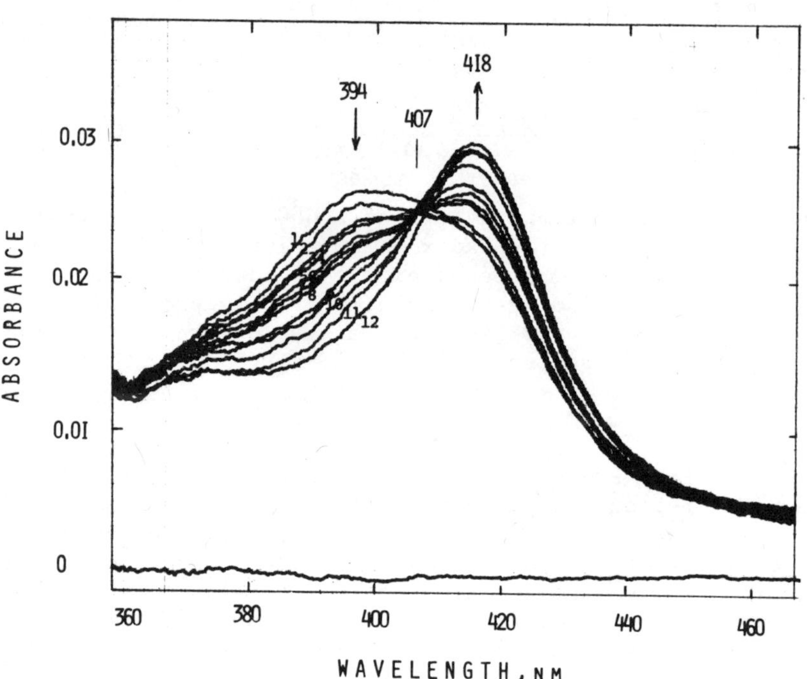


Figure 4. Spectral titration with Triton X-100 in the Soret region of P-450. Triton X-100 (0.1 or 1.0 w/v %) was added to 2.2 ml of the sample, 0.77 nmole P-450 in 10 mM phosphate, pH 7.4 containing 0.4 M NaCl. Prior to the recording of each spectrum the mixture was left to stand for 5 min. at 33°. The final Triton X-100 concentrations are: 1, none; 2, 0.91 x $10^{-4}$ % (w/v); 3, 2.7 x $10^{-4}$; 4, 7.2 x $10^{-4}$; 5. 1.17 x $10^{-3}$; 6, 1.62 x $10^{-3}$; 7, 2.07 x $10^{-3}$; 8, 4.27 x $10^{-3}$; 9, 6.54 x $10^{-3}$; 10, 8.81 x $10^{-3}$; 11, 1.34 x $10^{-2}$; 12, 1.79 x $10^{-2}$ M.

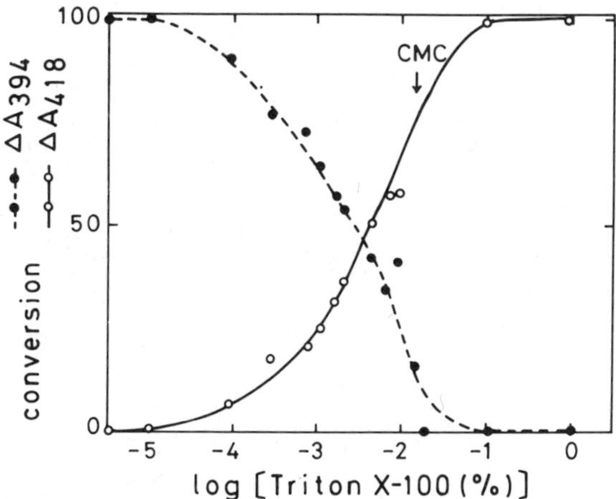


Figure 5. Replot of the spectral changes in the Soret region against Triton X-100 concentration. The absorbances were corrected for the change in the total volume. One hundred per cent represents the maximum difference in absorbance at 394 nm (418 nm).

The spectral changes caused by the addition of Triton as well as alcohols or acetone were found to correlate with the reduction rate of P-450 by dithionite in the presence of these reagents. The rate was measured by the appearance of the 450 nm peak in the presence of carbon monoxide. The rate was greatly accelerated by the addition of Triton X-100, alcohols or acetone. The time courses of the appearance of 450 nm band are shown in Fig. 6 in the presence of various concentrations of Triton X-100. Increasing the concentration of Triton caused a decrease in the peak height, once it had attained its maximum value, possibly due to a degradation of the native conformation of P-450. The pseudo-first order reaction rate constants were calculated, in the presence of saturating concentration of carbon monoxide and excess dithionite, from the plots of logarithm of ($\Delta A_\infty$ - $\Delta A_t$) against time, $\Delta A$ being the difference in absorbances at 450 nm and 490 nm. The rate constants are listed in Table II. Similar increases in rate similar to that presented in Fig. 6 were observed with alcohols or acetone and the rate constants were likewise calculated (Table III). The blanks in the figures occur when the rates were unobtainable due to a secondary degradative reaction of the P-450. It seems unlikely that the effect of alcohols or acetone is to increase the solubility of carbon monoxide and thereby accelerate the reduction rates of P-450. Indeed, the acceleration of the rates of the reactions were not proportional to the concentrations of these reagents and the $K_m$ for carbon monoxide is much lower than the $K_m$ for saturating concentrations of CO in the

TABLE II

Rate Constants of Reduction of P-450

| TRITON X-100 (% w/v) | k ($min^{-1}$) |
|---|---|
| none | 0.09 |
| $1 \times 10^{-3}$ | 0.12 |
| $5 \times 10^{-3}$ | 0.13 |
| $1 \times 10^{-2}$ | 0.24 |
| $2 \times 10^{-2}$ | 0.77 |
| $3 \times 10^{-2}$ | 1.39 |

TABLE III

Rate Constants ($min^{-1}$) of Reduction of P-450

| % v/v | MeOH | EtOH | i-PrOH | n-PrOH | t-BuOH | n-BuOH | Propylene Glycol | Glycerol | Acetone |
|---|---|---|---|---|---|---|---|---|---|
| 2.5 | 0.11 | 0.11 | 0.13 | 0.20 | 0.28 | 0.58 | 0.18 | 0.25 | 0.22 |
| 5.0 | 0.13 | 0.15 | 0.20 | 0.30 | 0.35 | - | 0.22 | 0.27 | 0.35 |
| 7.5 | 0.14 | 0.19 | 0.30 | 0.58 | 1.16 | - | 0.24 | 0.28 | 1.39 |
| 10.0 | 0.23 | 0.41 | - | - | - | - | 0.27 | 0.27 | 1.65 |
| 20.0 | 1.87 | 1.80 | - | - | - | - | 0.73 | 0.30 | - |

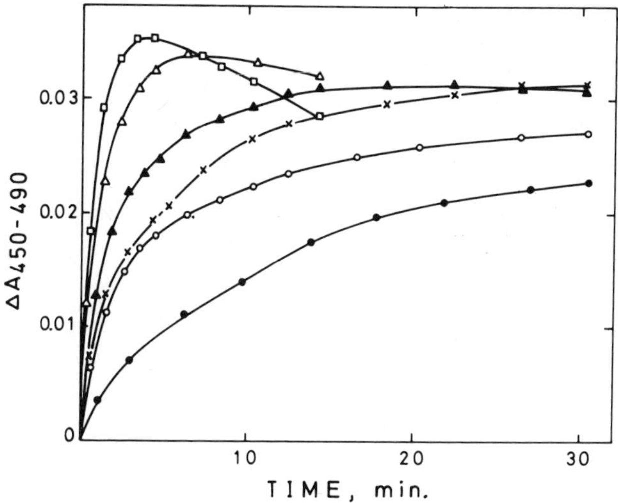


Figure 6. Effect of Triton X-100 on the reduction rate of P-450. Both the sample and reference cuvettes contained 0.75 nmol P-450 and Triton X-100 in 2.0 ml of 50 mM phosphate, pH 7.4. Difference spectra were taken as reported by Omura and Sato (1964). The temperature was kept at 20°. Final concentrations of Triton X-100 were: (●—●) none; (○—○) 1 x $10^{-3}$% (w/v); (x—x) 5 x $10^{-3}$, (▲—▲) 1 x $10^{-2}$; (△—△) 2 x $10^{-2}$; (□—□) 3 x $10^{-2}$ M.

absence of the reagents. Their effect on the acceleration of P-450 reduction, however, can be explained by their hydrophobic character and their detergenic power. This acceleration of the reduction rate by the above-mentioned components can be explained in terms of a disruption of intra- and intermolecular hydrophobic interactions, as have been reported when lysozyme (Kurono and Hamaguchi, 1964) and cytochrome P-450 (Ichikawa et al., 1968) are denatured by alcohols.

Electron paramagnetic resonance (EPR) spectra of P-450 were measured in the presence and absence of Triton X-100 or isopropyl alcohol. As seen in Fig. 7, distinct shift of g values and narrowing of the width of the signal were observed upon addition of Triton X-100 or isopropyl alcohol. Different set of g values (2.42, 1.91; 2.39, 1.92) in the low magnetic field region and those in the higher field region (Mitani et al., 1973) had been ascribed to the presence of different P-450s or different substrate binding sites (Ando and Horie, 1972; Jefcoate et al., 1973; Schleyer et al., 1972). However, additional studies are necessary to correlate the shift in g values of P-450 in the low magnetic field region caused by

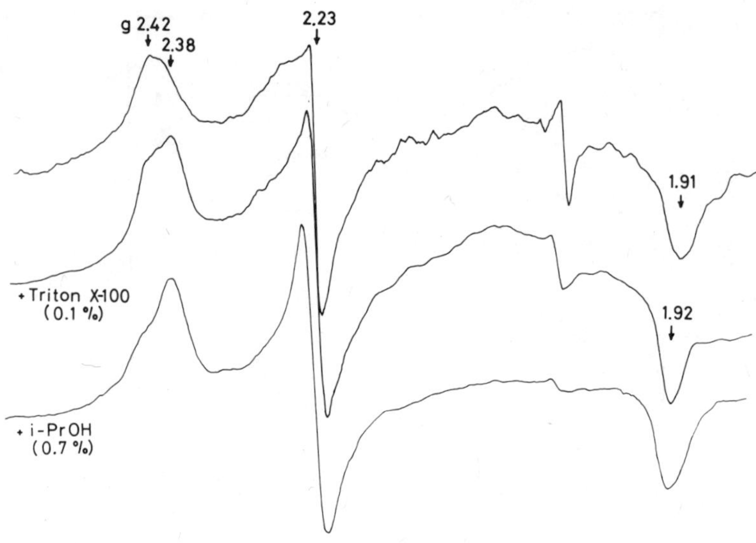


Figure 7. Effect of Triton X-100 and isopropyl alcohol on EPT spectrum of P-450. The spectrum on top is the control with P-450 concentration of 52 μM in 50 mM phosphate, pH 7.4. The final concentrations of Triton X-100 and isopropyl alcohol are given in the figure. EPR conditions: temperature 77°K, microwave power 100 mW, modulation amplitude 10 Gauss and microwave frequency 9.08 GHz.

Triton X-100 or isopropyl alcohol with the changes mentioned earlier which are caused by the addition of Triton X-100, alcohols or acetone.

When P-450 was treated with DOC, one of the substrates of P-450, an acceleration in the reduction rate was observed (Fig. 8) similar to those observed with Triton X-100, alcohols and acetone (Fig. 6, Tables II and III). This is also comparable to the marked stimulation of the reduction of the P-450 seen in P-450 from acetone extracted mitochondria (Cheng and Harding, 1973). However, P-450 was altered by DOC concentrations of 0.5 mM or higher. This effect is probably due to DOC itself, since P-450 stays intact in the presence of 1% isopropyl alcohol.

Side chain cleavage activity of P-450 was greatly enhanced by the addition of tert-butyl alcohol. Studies on the effect of other reagents on the enzyme activity as well as on the effect of these reagents in increasing the rate of reduction of P-450 are now in progress and will be correlated when the studies are completed.

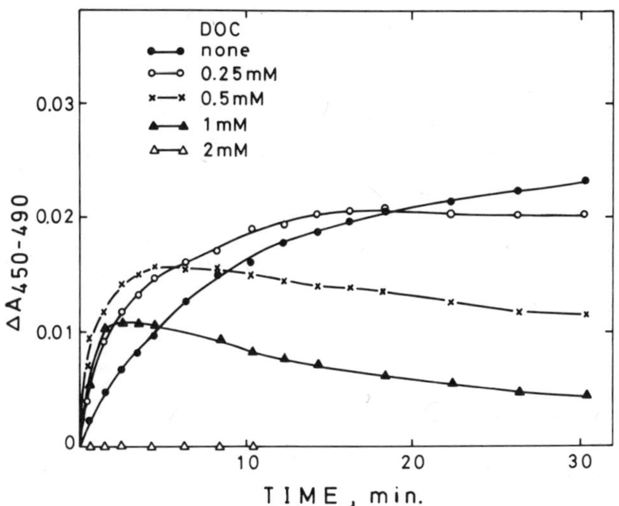


Figure 8. Effect of DOC on the reduction rate of P-450. Assay conditions are the same as those used for the Triton X-100 experiments (Fig. 6). DOC was dissolved in 20 µl of isopropyl alcohol and was introduced into the mixture. The final concentrations of DOC are given in the figure.

## REFERENCES

Ando, N., and Horie, S. (1972), J. Biochem. 72, 583-597.

Cheng, S.C., and Harding, B.W. (1973), J. Biol. Chem. 248, 7263-7271.

Harding, B.W., Wong, S.H., and Nelson, D.H. (1964), Biochim. Biophys. Acta 92, 415-417.

Horie, S., and Watanabe, T. (1975), J. Steroid Biochem. 6, 401-409.

Ichikawa, Y., Uemura, T., and Yamano, T. (1968), Structure and Function of Cytochromes (ed. Okunuki, K., Kamen, M.D., and Sekuzu, I.) University Tokyo Press and University Park Press, 634-644.

Jefcoate, C.R., Hume, R., and Boyd, G.S. (1970), FEBS Lett. 9, 41-44.

Jefcoate, C.R., Simpson, E.R., and Boyd, G.S. (1973), Ann. New York Acad. Sci. 212, 243-261.

Kurono, A., and Hamaguchi, K. (1964), J. Biochem. 56, 432-440.

Lowry, O.H., Rosebrough, N.J., Farr, A.L., and Randall, R.J. (1951), J. Biol. Chem. 193, 265-275.

Mattingly, D. (1962), J. Clin. Path. 15, 374-379.

Mitani, F., Ando, N., and Horie, S. (1973), Ann. New York Acad. Sci. 212, 208-224.

Omura, T., and Sato, R. (1964), J. Biol. Chem. 239, 2370-2378.

Robinson, N.C., and Tanford, C. (1975), Biochemistry 14, 369-378.

Schleyer, H., Cooper, D.Y., Levin, S.S., and Rosenthal, O. (1972), Biological Hydroxylation Mechanisms (ed. Boyd, G.S., and Smellie, R.M.S.) Academic Press, 187-206.

Shikita, M., and Hall, P.F. (1973), J. Biol. Chem. 248, 5598-5604; 5605-5609.

Sugiyama, T., and Yamano, T. (1975), FEBS Lett. 52, 145-148.

Suhara, K., Takemori, S., and Katagiri, M. (1972), Biochim. Biophys. Acta 263, 272-278.

Wang, H.-P., Pfeiffer, D.R., Kimura, T., and Tchen, T.T. (1974), Biochem. Biophys. Res. Commun. 57, 93-99.

Weber, K., and Osborn, M. (1969), J. Biol. Chem. 224, 4406-4412.

# THE ROLE OF CYTOCHROME P-450 IN THE REGULATION OF STEROID BIOSYNTHESIS

Peter F. Hall

Department of Physiology, University of California

Irvine, California 92717

A mitochondrial cytochrome P-450 catalyzes the conversion of cholesterol to pregnenolone in the various steroid-forming organs: pregnenolone is converted to the steroid hormones secreted by these organs. The conversion of cholesterol to pregnenolone is commonly called side-chain cleavage (Figure 1). Side-chain cleavage appears to determine the rate of steroid biosynthesis so that the role of P-450 in that reaction is important in considering the regulation steroid production in the whole animal (1-3).

We have recently purified the side-chain cleavage enzyme from bovine adrenocortical mitochondria in a form which does not catalyze 11β- or 18-hydroxylation of steroids which use other mitochondrial cytochromes P-450. Moreover, the enzyme appears as a single peak in electrophoresis on polyacrylamide gel containing sodium dodecyl sulfate. The enzyme can be prepared in three stable forms which are named according to the number of subunits present: protein 16, protein 8 and protein 4. This P-450, together with the highly purified electron carriers required for enzyme activity (adrenodoxin and adrenodoxin reductase), have been used to study the side-chain cleavage reaction. This paper presents the results of three approaches to the investigation of the role of P-450 in the regulation of steroidogenesis, using these purified proteins.

## METHODS

The preparation and properties of the enzyme, together with methods for assaying its activity, have been published (4,5). The same is true for methods of measuring oxidation of TPNH, consumption

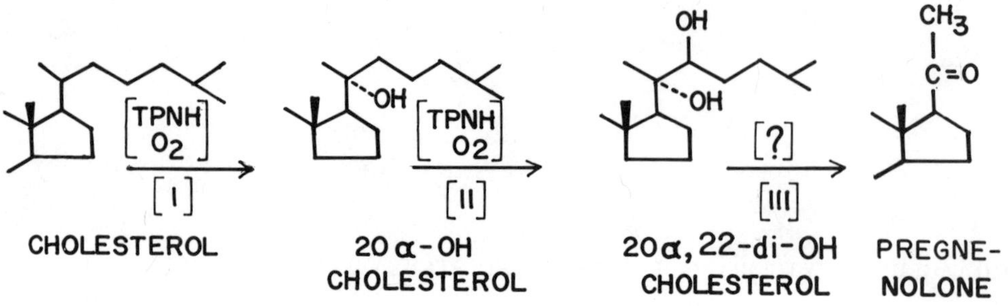

Figure 1. Side-chain cleavage of cholesterol: a proposed mechanism.

of oxygen (6), determination of photochemical action spectra (7) and the preparation and handling of sucrose density gradients (8).

## RESULTS

### Stoichiometry of Side-Chain Cleavage

The consumption of TPNH by cytochrome P-450 with adrenodoxin, adrenodoxin reductase and TPNH was measured with and without added substrate by recording $A_{340}$. Production of pregnenolone was determined at the end of 20 minutes incubation at $37^{o}$. Without substrate, background oxidation of TPNH was observed and no pregnenolone was formed since the enzyme is virtually free of endogenous cholesterol. Figure 2 shows that consumption of TPNH increased when substrate (cholesterol) was present and that the ratio of additional utilization of TPNH to pregnenolone formed (n moles/20 minutes in each case) was approximately 3.0 at each of the various time intervals examined.

Similar studies were performed with a Clark electrode to determine oxygen consumption of the enzyme system with and without added substrate. It is clear that whatever reaction(s) occur in the absence of substrate, TPNH and oxygen are consumed mole for mole (Figure 2). Moreover, when cholesterol was added, the ratio of the additional utilization of oxygen to pregnenolone produced was approximately 3.0 (Figure 2).

Similar studies were performed as a function of enzyme concentration; the results were the same (3 TPNH:3 $O_2$: 1 pregnenolone. Consumption of hydrogen ions was also measured by means of a pH meter (6). The results of these experiments are summarized in Table I.

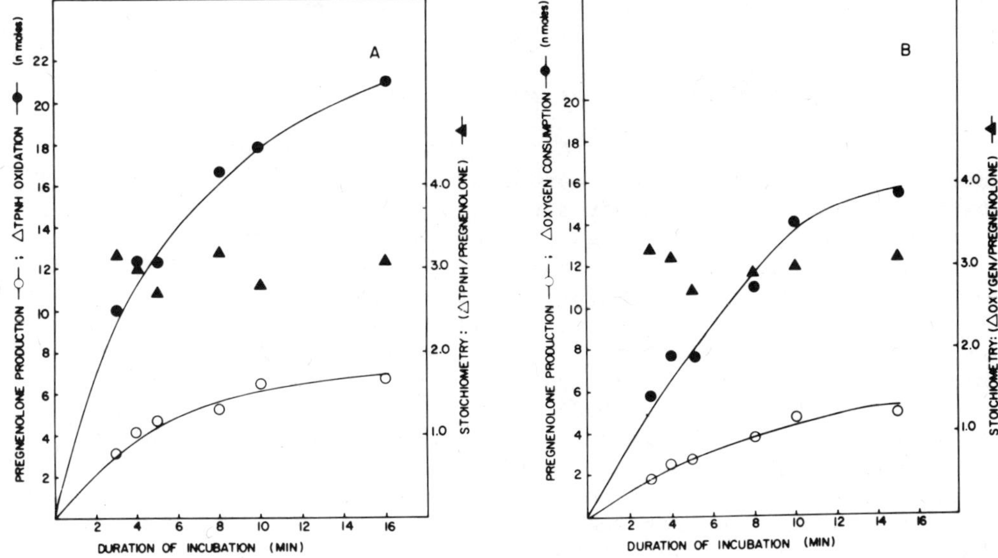


Figure 2A. Oxidation of TPNH and conversion of cholesterol to pregnenolone by P-450, adrenodoxin and adrenodoxin reductase.

Figure 2B. Oxygen consumption and conversion of cholesterol to pregnenolone by the same enzyme as that used in A.

Consumption of TPNH and oxygen without substrate (background reaction) has been substrated.

●: Additional consumption of TPNH and oxygen with substrate
O: Pregnenolone production
A▲: Δ TPNH used/pregnenolone formed
B▲: Δ $O_2$ used/pregnenolone formed

Table I

Stoichiometry of Side-Chain Cleavage

| Substrate | ΔTPNH / Pregnenolone | $\Delta O_2$ / Pregnenolone | $\Delta H^+$ / Pregnenolone |
|---|---|---|---|
| Cholesterol (9) | 3.0 ± 0.2 | 3.1 ± 0.3 | 3.0 ± 0.2 |
| 20α-OH-Cholesterol (6) | 1.9 ± 0.3 | 2.1 ± 0.3 | 2.1 ± 0.1 |
| 20α,22-$(OH)_2$-Cholesterol (6) | 1.0 ± 0.3 | 0.9 ± 0.4 | 1.1 ± 0.1 |

Numbers in parentheses indicate the number of different enzyme preparations examined. Values are means and ranges.

## Side-Chain Cleavage of 20S,22R-Dihydroxycholesterol

Figure 3 shows that CO (4%) inhibits side-chain cleavage of 20S,22R-dihydroxycholesterol and that this inhibition is reversed by white light. Preliminary studies were performed to measure the partition coefficient K (the ratio $CO:O_2$ required to produce 50% inhibition of cleavage) and to determine the influence of intensity of light on reversal of inhibition; these data were used to correct values observed with monochromatic light of various wavelengths using interference filters. White light from a Xenon lamp was passed through three filters as shown in Figure 4; the neutral density filter was used for studies of the effect of light intensity on reversal of CO inhibition. The photochemical action spectrum was obtained without neutral density filters (15 $mW/cm^2$) and the intensity at the sample (i.e. after passing through the interference filter) was measured for each wavelength. Values for enzyme activity at various wavelengths were corrected for intensity. It was observed that light of the wavelengths and intensities used in these studies was without effect on enzyme activity (only on inhibition by CO).

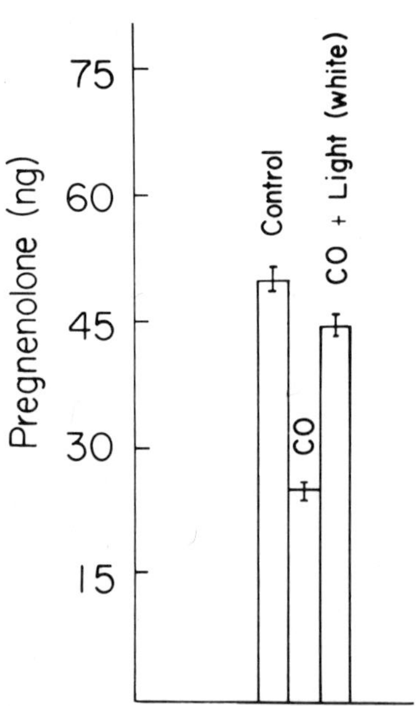


Figure 3. Conversion of 20S,22R-dihydroxycholesterol to pregnenolone: effect of CO with and without white light.

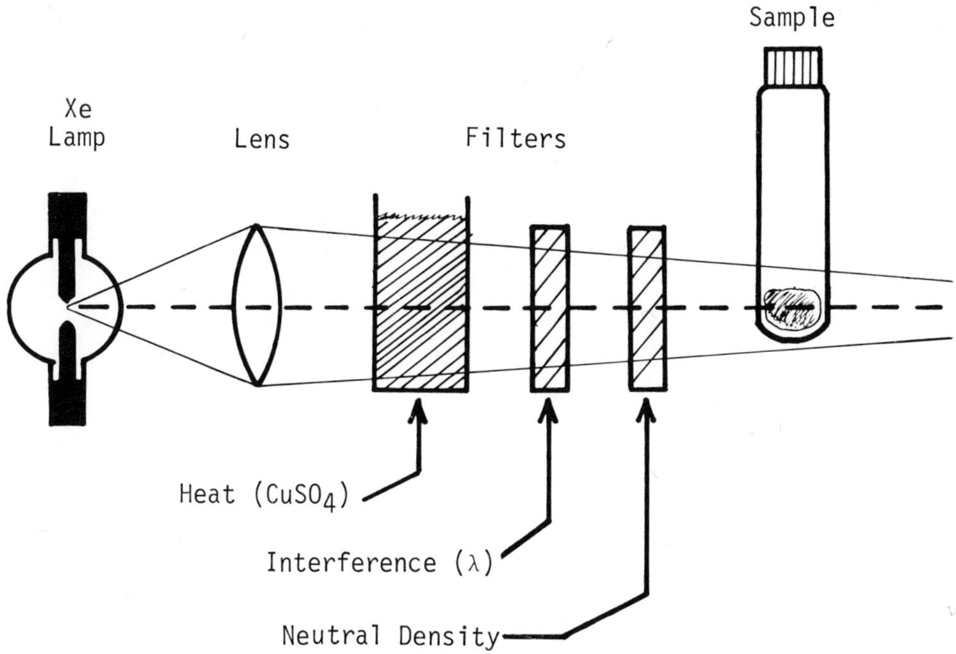


Figure 4. Apparatus used to determine photochemical action spectra of side-chain cleavage.

The effect of light of six different wavelengths upon reversal of CO inhibition of the cleavage of 20S,22R-dihydroxycholesterol was measured and the photochemical action spectrum was calculated. Similar studies were performed with cholesterol. Figure 5 shows the action spectra with the two substrates; L is the light sensitivity factor, i.e. the reciprocal of the quantal energy required to double the partition coefficient (K) (9). The ordinate of Figure 5 shows the so-called relative light sensitivity, i.e. the ratio of L at various wavelengths to L at 451 nm. The action spectra show typical bell-shaped curves indicating that maximal reversal of CO inhibition is seen at the wavelength of maximal light absorbance of the P-450-CO complex (451 nm). The curves do not show shoulders seen with crude enzyme preparations and are similar to those observed with other steroid hydroxylations (10).

## The Active Form of Side-Chain Cleavage P-450

The side-chain cleavage P-450 is stable and enzymatically active in forms consisting of 16, 8 and 4 subunits called respectively

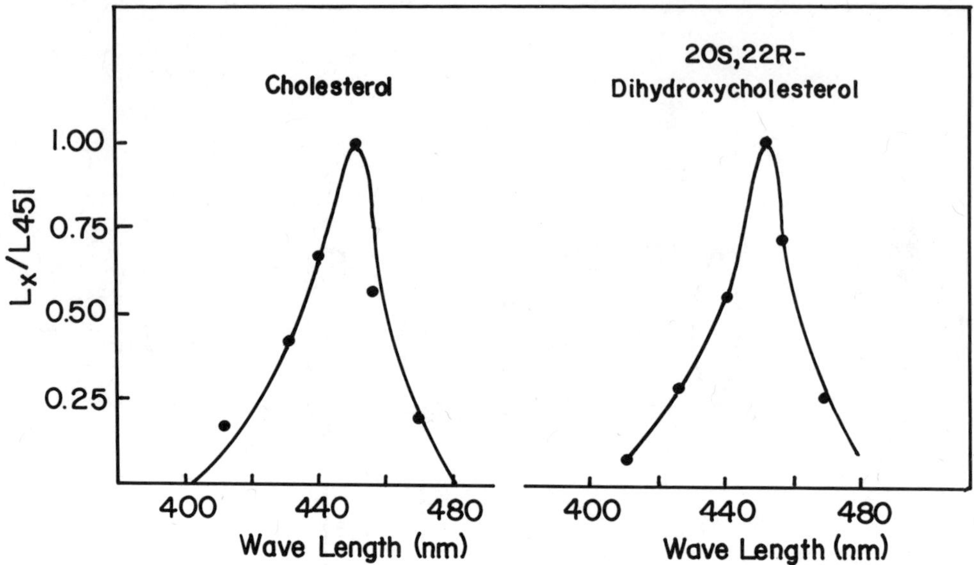


Figure 5. Photochemical action spectra for CO-inhibition of side-chain cleavage of cholesterol and 20S,22R-dihydroxycholesterol. $L_x/L_{451}$: see text.

protein 16, protein 8 and protein 4. To determine whether these three forms are all active as such or are interconverted to some subunit combination which is the "active form", we prepared sucrose density gradients (5 ml; 5 to 20%) containing uniform concentrations of phosphate buffer (0.1M, pH 7.0), adrenodoxin (2μM), adrenodoxin reductase (2μM), TPNH (1 mM) and substrate. Samples of the three forms of P-450 were layered onto separate gradients and centrifuged at 37° for 100 minutes at 225,000 x $g_{av}$ (49,000 rpm) in the SW50.1 Beckman rotor. Following centrifugation, 0.2 ml fractions were collected and examined for pregnenolone. Some studies were performed at 21° and in certain experiments, one of the essential ingredients was omitted from the gradient or $TPN^+$ was substituted for TPNH. The three substrates used were [4-$^{14}$C] cholesterol, 20S-hydroxycholesterol and 20S,22R-dihydroxycholesterol.

Figure 6 shows that whether protein 4, protein 8 or protein 16 was layered onto the gradient, the maximal amount of [$^{14}$C] pregnenolone was found in fractions of sedimentation coefficient 20-22; this value corresponds to protein 16. The locations of proteins 4, 8 and 16 when centrifuged with buffered sucrose only,

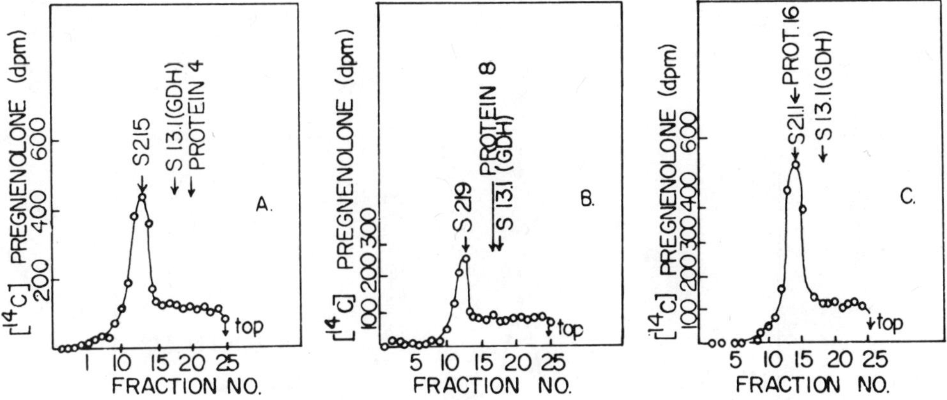


Figure 6. Sedimentation of P-450 through sucrose density gradients containing adrenodoxin, adrenodoxin reductase, TPNH, and [$^{14}$C] cholesterol at 37°. Fractions were collected from the gradient, [$^{14}$C] pregnenolone was isolated and $^{14}$C content was measured. P-450 was layered onto gradients as follows: A:protein 4, B:protein 8; C:protein 16. The location of each of these proteins when sedimented in buffered sucrose only is indicated by a labeled arrow: GDH: glutamic dehydrogenase; EDG: ethanol dehydrogenase; S: sedimentation constant, calculated as described by Martin and Ames (14).

are indicated on the figure. Figure 7 shows that at 21° no enzyme activity is observed (open circles) unless the fractions are incubated at 37° for 20 minutes after centrifugation. When incubation was performed, peak activity again corresponded to protein 16. The figure also shows that protein 4 sediments as protein 16 with TPN$^+$ instead of TPNH - in this case it was necessary to incubate the fractions collected after centrifugation, with TPNH to reveal the location of enzyme activity. These results are summarized in Table II.

## DISCUSSION

Side-chain cleavage cytochrome P-450 has been purified from bovine adrenocortical mitochondria to near homogeneity (4,5). The protein can be prepared in oligomeric species from subunits of molecular weight 53,000 - the following forms are stable and enzymatically active: protein 16, protein 8 and protein 4, named

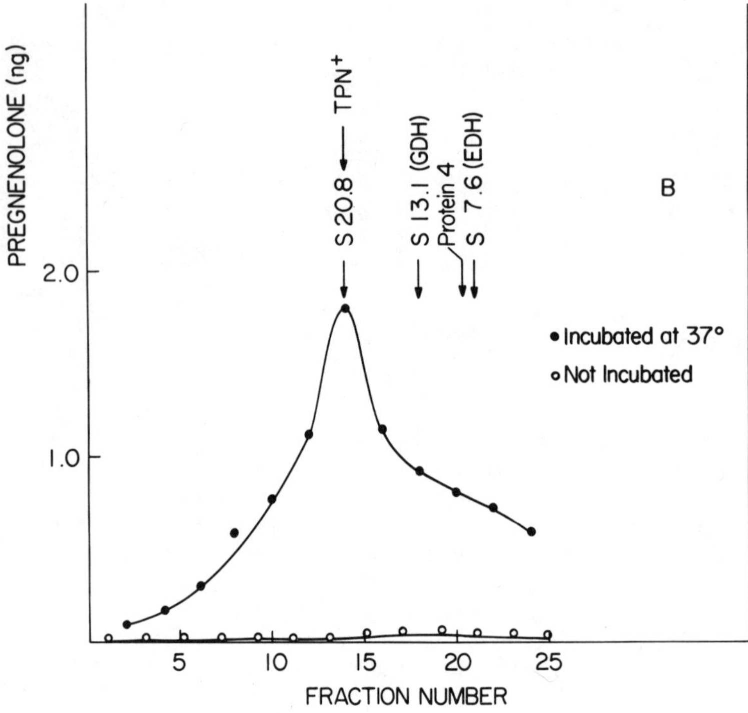


Figure 7. Sedimentation of protein 4 at 21°. O: no incubation after centrifugation, ●: incubated at 37° for 20 minutes after centrifugation and $TPN^+$: $TPN^+$ replaced TPNH throughout the gradient. See legend to Figure 6.

Table II

| P-450 layered on gradient | S value | T° of Centrifugation | Other (Conditions) | S value of Enzyme activity |
|---|---|---|---|---|
| Protein 4 | 8.8 | 37 | Standard | 21.5 |
| Protein 8 | 13.8 | 37 | Standard | 21.9 |
| Protein 16 | 20-22 | 37 | Standard | 21.1 |
| Protein 4 | 8.8 | 37 | $TPN^+$ | 20.8 |
| Protein 4 | 8.8 | 21 | Standard | 20.8 |
| Protein 4 | 8.8 | 37 | 20S-hydroxy-cholesterol | 20.8 |
| Protein 4 | 8.8 | 37 | 20S,22R-dihy-droxycholesterol | 20.4 |

according to the number of subunits present; these forms show the following molecular weights (850,000, 420,000 and 210,000, respectively) (5). Preliminary evidence suggests that there are two species of subunits of similar molecular weight. Each form contains half as many heme groups as subunits (unpublished observations).

This enzyme was used to determine the stoichiometry of side-chain cleavage which, for cholesterol, is as follows:

$$C_{27}H_{46}O + 3\ O_2 + 3\ TPNH + 3H^+ \rightarrow C_{21}H_{32}O_2 + C_6H_{12}O + 3\ TPN^+ + 4\ H_2O$$

Apparently one oxygen, one TPNH and one $H^+$ is required for each of the two hydroxylation reactions and one of each for the cleavage of the diol. The stoichiometry observed with the purified enzyme is compatible with the "classical" mechanism (Figure 1). It is also consistent with a recent proposal that the diol is synthesized via an epoxide (11,12).

The mechanism of cleavage of the diol has been examined with purified P-450. The reaction is inhibited by CO and inhibition is specifically reversed by light of wavelength 450 nm. Cleavage of the diol evidently involves heme and both the photochemical action spectrum and the stoichiometry are consistent with typical monoxygenation or mixed function oxidation (13). The photochemical action spectra with cholesterol and 20S,22R-dihydroxycholesterol are very similar to those of other steroid hydroxylation reactions (10).

Although protein 16, protein 8 and protein 4 are all stable at 4°C and are enzymatically active, it is evident from the studies reported here that they do not all act as such. In the presence of the appropriate cofactor, substrate and electron carrier proteins, protein 4 and protein 8 are converted to protein 16 in the process of establishing enzyme activity. Protein 16 is active as such. These results were seen when enzyme activity was permitted to occur in the gradient at 37° or in fractions collected from the gradient after centrifugation and subsequently incubated at 37°. In this process of forming protein 16, $TPN^+$ can substitute for TPNH; however, the latter must be added, after centrifugation, to demonstrate enzyme activity. All the factors needed for enzymatic activity (substrate, TPNH, adrenodoxin and adrenodoxin reductase) must be present for conversion of proteins 4 and 8 to protein 16. It would appear that, at least under the conditions examined _in vitro_, protein 16 is the active form of the enzyme.

The information provided by these studies forms the basis of our present efforts to explore the regulation of this complex reaction in further detail.

ABSTRACT

A cytochrome P-450 from bovine adrenocortical mitochondria has been purified to near homogeneity. The protein catalyzes side-chain cleavage of cholesterol (cholesterol → pregnenolone) but neither 11β- nor 18-hydroxylation. It consists of 16 subunits of two species (MW 52,000) and contains 8 heme groups. The enzyme has been used to determine the stoichiometry of side-chain cleavage with the following results: (TPNH and $O_2$ consumed/mole of cleavage), cholesterol 3:3:1, 20S-hydroxycholesterol 2:2:1 and 20S,22R-dihydroxycholesterol 1:1:1. These findings support the occurrence of the proposed pathway for the side-chain cleavage of cholesterol. Cleavage of the diol is inhibited by CO and shows a characteristic P-450 photochemical action spectrum. Evidently the diol is cleaved in a typical monoxygenase reaction. The active form of the enzyme contains 16 subunits (protein 16); forms consisting of 8 (protein 8) and 4 (protein 4) subunits can be isolated and are enzymatically active only by prior conversion to protein 16.

REFERENCES

1. Karaboyas, G.C. and Koritz, S.B. Biochemistry 4:462, 1965.

2. Hall, P.F. and Young, D.G. Endocrinology 82:559, 1968.

3. Koritz, S.B. and Hall, P.F. Biochemistry 4:2740, 1965.

4. Shikita, M. and Hall, P.F. J. Biol. Chem. 248:5598, 1973.

5. Shikita, M. and Hall, P.F. J. Biol. Chem. 248:5605, 1973.

6. Shikita, M. and Hall, P.F. Proc. Nat'l. Acad. Sci. 71:1441, 1974.

7. Hall, P.F., Lee Lewes, J. and Lipson, E.D. J. Biol. Chem. 250:2283, 1975.

8. Takagi, Y., Shikita, M. and Hall, P.F. J. Biol. Chem. 250: 8445, 1975.

9. Warburg, O. Heavy Metal Prosthetic Groups and Enzyme Action. Clarendon Press, Oxford, 1949.

10. Conney, A. H., Levin, W., Ikeda, M., Kuntzman, R., Cooper, D.Y. and Rosenthal, O. J. Biol. Chem. 243:3912, 1968.

11. Kraaipoel, R. J., Degenhart, H. J., Leferink, J. G., van Beek, V., De Leeuw-Boon, H. and Visser, H. K. A. FEBS Letters 50:204, 1975.

12. Kraaipoel, R.J., Degenhart, H.J. and Leferink, J.G. FEBS Letters 57:294, 1975.

13. Mason, H.S. Advan. Enzymol. 19:79, 1959.

14. Martin, R.G. and Ames, B.N. J. Biol. Chem. 236:1372, 1961.

# ON THE PARTICIPATION OF CYTOCHROME P-450 IN THE MECHANISM OF PREVENTION OF HEPATIC CARCINOGENESIS

A. N. Saprin *

Cancer Center of Hawaii

1997 East-West Road, Honolulu, Hawaii 96822

The microsomal mixed-function oxidase system metabolizes many compounds with different chemical structures and biological activity i.e. carcinogens, drugs, poisons, etc. (1-4). The terminal oxidase of the microsomal electron transfer system - cytochrome P-450 - plays an important role in these processes as has been shown by many investigators over the past few years (3,4).

The question of the specific participation of the microsomal mixed-function oxidase system in carcinogenesis is well discussed in the literature (1,3,5-7). The necessity for enzymatic conversion of certain carcinogens by microsomes to their active intermediates is well established (8). There have been numerous studies on the induction of microsomal mixed-function oxidases in animal liver by carcinogenic polycyclic aromatic hydrocarbons and the effect of these inducers on hepatic carconigensis (1,3,5,7,9). However, the alterations in the microsomal mixed-function oxidases themselves during the action of hepatic carcinogens have not been extensively studied, particularly the kinetic aspects of these alterations as a function of carcinogenesis.

In the work presented here we have studied the kinetics of cytochrome P-450 levels in rat liver during dimethylamino-azobenzene (DAB) and diethylnitrosoamine (DENA) induced carcinogenesis. We have used Wistar rats with both carcinogens.

---

* Present address: Department of Kinetics of Chemical and Biological Processes, Institute of Chemical Physics, Academy of Sciences, U.S.S.R., MOSCOW, U.S.S.R.

DAB was administered by feeding in a special diet as was DENA for 5 days a week for a total period of 4 months. Such procedures generally produce hepatic tumors in about 5 months and 4 months, respectively. ESR absorption measurements at low temperatures (-180°C) were used for the determination of the level of cytochrome P-450 in the whole normal and malignant liver tissue and in the microsomes, isolated from these tissues by the method Omura and Sato (10). It is well known that in the oxidized state cytochrome P-450 gives a characteristic ESR signal at g-value 2.42, 2.25 and 1.91, which is due to the low spin state of the heme iron in an asymmetric environment (11-13).

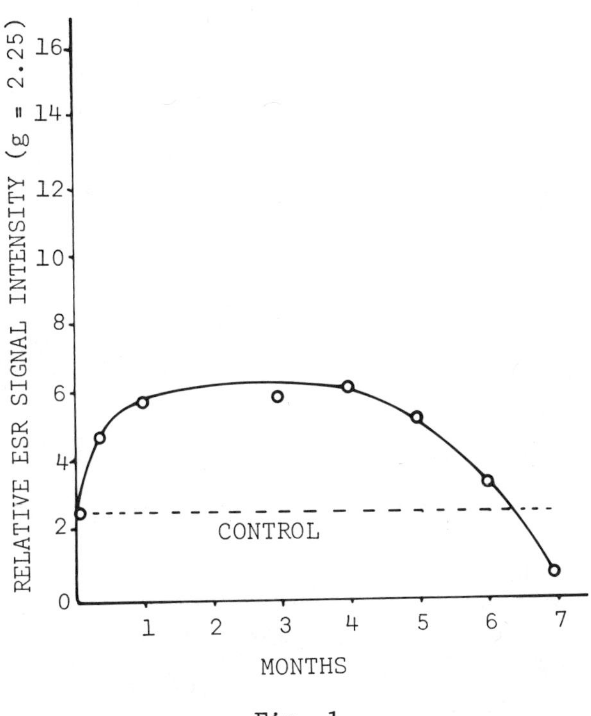


Fig. 1

Figure 1 shows the changes observed in cytochrome P-450 levels in the liver and in the liver tumors during the various stages of DAB carcinogenesis (14). One can see that during the pre-cancer stages the level of cytochrome P-450 in normal liver or hyperplastic tissue increases 2 and one-half times. At the time of tumor appearance and subsequent growth, the level of cytochrome P-450 drops.

In the actual tumors during the terminal stages of growth, the level of cytochrome P-450 decreases as much as four-fold from normal liver.

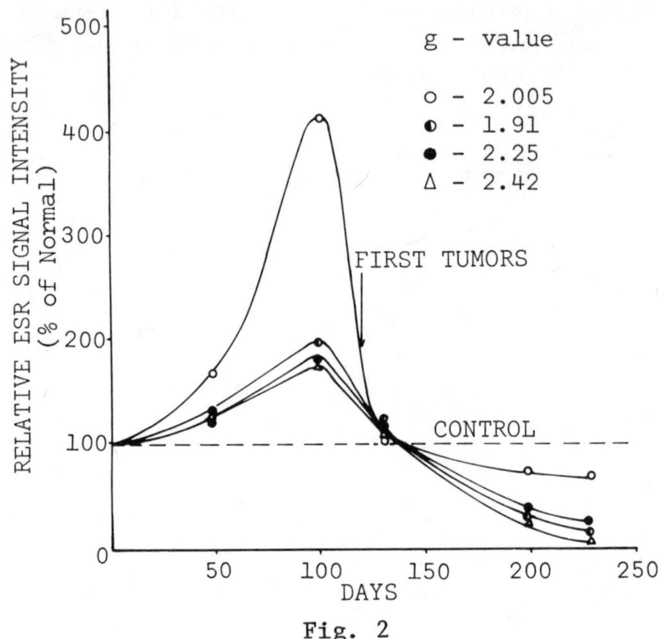


Fig. 2

Figure 2 shows the alterations in the level of cytochrome P-450 in microsomal fractions of both liver and tumors at the different stages of carcinogenesis induced by DENA (15). (The same alterations observed here in microsomes are also observed for the whole tissue). One can see that in the pre-cancer stages the level of cytochrome P-450 increases 1.8 times and then decreases in the actual tumors 4 to 5 fold. Low levels of cytochrome P-450 have been reported for some transplantable tumors by a number of other workers.

It seems reasonable to suggest that the increase in the amount of cytochrome P-450, which we have observed for both carcinogens here, may be a consequence of the activation requirement for both carcinogens (3,8). It remains, however, to account for the subsequent decrease in the amount of cytochrome P-450 during tumor formation and growth. One of several possible mechanisms seem plausible based on the following additional results. In Figure 2 we show the changes that occur in the concentration of free radicals in microsomes as measured by ESR at a g-value of 2.005 at the various stages of DENA - carcinogenesis. One can see that the free radical levels in microsomes increase more than 4 fold during the development of the pre-cancerous stages as compared to normal liver. Moreover, this significant increase in the free radical concentration observed prior to the appearance of tumors precedes the onset of a decrease in the level of cytochrome P-450 in the microsomes.

It is well known that free radicals in general are very reactive intermediates. Therefore, such an increase in concentration of free radicals in the microsomes, the particles at which the first interaction with the carcinogen probably takes place, suggest the possibility that the free radicals may be causing damage of the cytochrome P-450 and the decreased levels observed. It is important to differentiate between at least two types of free radicals generally observed in biological systems: 1) Very reactive radicals which appear possible during lipid peroxidation and 2) the more stable free radicals such as the semiquinone intermediates formed in the steady state during enzymatic redox reactions which are easily detected by ESR. The radicals formed during lipid peroxidation are virtually impossible to detect by ESR because of their short life times. It is possible, however, that the free radicals observed here are produced by the lipid peroxide intermediates and are electron transfer products resulting from electron transfer to proteins or other molecules with subsequent formation of more stable semiquinone radicals. The results in Figure 3 seem to support this later mechanism. One can see that

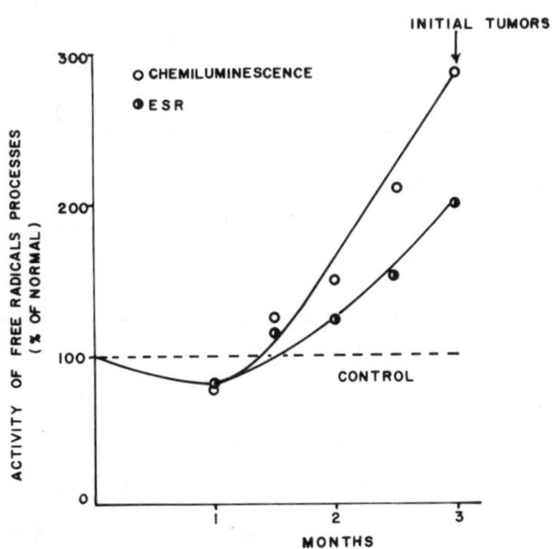


Fig. 3

there appears to be a correlation between the changes in free radical concentration as measured by ESR in whole epidermal tissue in pre-cancer stages during carcinogenesis with benzo (a) pyrene, and the changes in the intensity of chemiluminescence of lipids which have been isolated from this same tissue. It has been suggested by several authors that chemiluminescence, such as seen here, results from the recombination of lipid peroxide radicals (16).

Finally, the suggestion made here of inactivation of cytochrome P-450 as a result of the activation of lipid peroxidation has been confirmed by Archakov and Levin et al. (16,17). These authors observed a decrease in optical absorption of cytochrome P-450 in normal rat liver microsomes after the activation of lipid peroxidation. We have also confirmed these results in our laboratory.

Following the suggestions of the possible role of free radicals as initiators of molecular alterations which can lead to malignant growth, N. M. Emanuel and co-workers in 1958 proposed the use of free radical inhibitors as a means of inhibiting carcinogenesis (18). They showed by simultaneous feeding the non-toxic inhibitor of free radical processes, butylatedhydroxytoluene (BHT) and the carcinogen DAB, that BHT was able to inhibit the development of carcinogenesis in rats (19). This effect of the inhibition can be monitored by ESR as seen in Figure 4 (14).

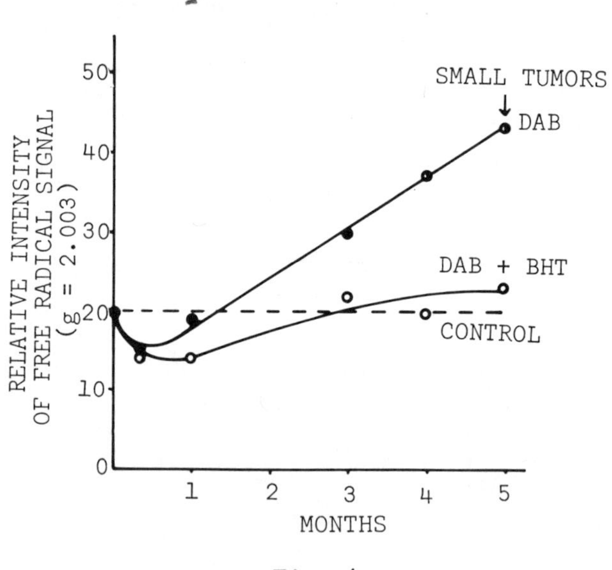


Fig. 4

Liver tissue from rats treated with DAB alone show, in the pre-cancer stages, an increase in free radical production both in the liver and primary tumors. In contrast, liver tissue from rats fed both DAB and BHT showed no significant change in free radical concentration over normal controls. No tumors formed in this group after one year. These results would suggest that the anti-carcinogenic properties of BHT are due its anti-radical properties. However, the simultaneous observations in these experiments of changes in cytochrome P-450 levels suggest the possibility of yet another mechanism of anti-carcinogenic action of BHT (Fig. 5).

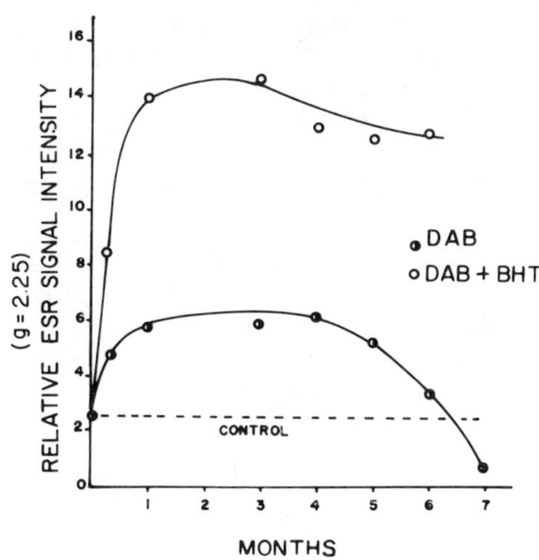


Fig. 5

It can be seen in this figure that already in the first 30 days, in animals fed both DAB and BHT the levels of cytochrome P-450 increased approximately 6 fold over the level in normal liver and 2.3 fold higher than in rats fed DAB alone. Such high levels of cytochrome P-450 remained for a long period of time (6-months). It should be noted that this increase in cytochrome P-450 level began at a much earlier time than the time of increase in the free radical concentration during the pre-cancer stage as one can see in Fig. 4. This finding seems to indicate that the anti-radical mechanism as a means of explaining the anti-carcinogenic action of BHT is unlikely, although it cannot be excluded completely. The properties of BHT as an inducer of cytochrome P-450 seems to be more important. The induction of high levels of cytochrome P-450 in the rat liver by the simultaneous action of DAB and BHT apparently promotes the detoxification of the carcinogen DAB. It should be pointed out that it has been shown in the same system that BHT sharply decreased the amount of DAB binding to proteins (19). Finally, it has also been shown that some phenolic antioxidants, including BHT, induce the synthesis of micromosal mixed-function oxidase system and promote an increase in the metabolism of several substrates (20-23). The results presented in this paper seem to indicate that cytochrome P-450 plans an important role in the prevention of hepatic carcinogenesis.

## REFERENCES


1. Conney, A.K., Miller, E.C., Miller, J.A., Cancer Res., 1956, 16, 450.
2. Gillette, J.R. In: Advances in Pharmacology, v. 4. Academic Press, New York, 1966, pp. 219-261.
3. Gelboin, H.V. In: Advances in Cancer Research, v. 10, Academic Pr-ss Inc., New York and London, 1967, pp. 9-91.
4. Maslova, Y.M., Reichman, L.M., Skulachev, V.P. Uspechi sovremennoi biologii, U.S.S.R., 1969, 67, 400.
5. Miller, E.C., Miller, J.A., Brown, R.R., McDonald, Y.C., Cancer Res., 1958, 18, 469.
6. Cramer, J.W., Miller, J.A., Miller, E.C. J. Biol. Chem., 1960, 235, 250.
7. Gelboin, H.V., Kinoshita, N., Wiebel, F.J. In: Environment and Cancer (24 Annual Symposium on Fundamental Cancer Research, 1971), The Williams and Wilkins Como. 1972, pp. 476.
8. Magee, P.N., Barnes, J.M. In: Advances in Cancer Research, v. 10, Academic Press, New York - London, 1967, pp. 242-323.
9. Gelboin, H.V., Kinoshita, N., Wiebel, F.J. Fed. Proc., 1972, 31, 1298.
10. Omura, T., Sato, T., J. Biol. Chem., 1964, 239, 2370.
11. Miyake, Y., Mason, H.S., Landgraf, W., J. Biol. Chem., 1967, 242, 393.
12. Miyake, Y., Gayler, J.L., Mason, M.S., J. Biol. Chem., 1968, 243, 5788.
13. Ichikawa, J., Yamano, T. Biochim. at Biophys. Acta 1968, 153, 753.
14. Emanuel, N.M., Saprin, A.N., Shuljakovskaja, T.S., Kozlova, L.E., Bunto, T.V., Dokladi Academii Nauk U.S.S.R., 1973, 209, 1449.
15. Saprin, A.N., Gurevitch, S.M., Zvegintseva, E.Y., Dokladi Academii Nauk, U.S.S.R., 1973, 208, 973.
16. Vladimirov, Ju.A., Archakov, A.I. In: The Peroxide Oxidation of Lipids in Biological Membranes, Moscow, 1972.
17. Levin, W., Lu, A.Y.M., Jacobson, M. et al. Archives of Biochem. and Biophys., 1973, 158, 842.
18. Emanuel, N.M., Lipchina, L.P. Dokl. Acad. Nauk, U.S.S.R., 1958, 121, 141.
19. Frankfurt, O.S., Lipchina, L.P., Bunto, T.V., Emanuel, N.M., Bull. Exp. Biol. Med. (U.S.S.R.), 1967, 8, 163.
20. Gilbert, D., Goldberg, L. Food Cosmet. Toxicol. 1965, 3, 417.
21. Creaven, P.J., Davies, W.H., Williams, R.T., J. Pharmacol., 1966, 18, 485.
22. Martin, A.D., Gilbert, D. Biochem. J., 1968, 106, 22.
23. Gilbert, D., Martin, A.D., Gargolli, S.D. et al. Food Cosmet. Toxicol., 1969, 7, 603.

# MODEL SYSTEM STUDIES OF AXIAL LIGATION IN THE OXIDIZED REACTION STATES OF CYTOCHROME P-450 ENZYMES

R.H. Holm,[a] S.C. Tang,[b] S. Koch,[b] G.C. Papaefthymiou,[d] S. Foner,[d] R.B. Frankel,[d] and J.A. Ibers[c]
Department of Chemistry, [a]Stanford University, Stanford, CA 94305, [b]Massachusetts Institute of Technology, Cambridge, MA 02139, [c]Northwestern University, Evanston, ILL 60201, [d]Francis Bitter National Magnet Laboratory, M.I.T.

## I. INTRODUCTION

Cytochrome P-450 enzymes have been isolated from bacteria, microsomes, and mitochondria and catalyze the insertion of one atom of dioxygen into substrate while the other is reduced to water. The biological and many of the physical properties of these enzymes are summarized elsewhere (Hill *et al*., 1970a; Ullrich, 1972; Gunsalus *et al*., 1973; Tomaszewski *et al*., 1974). Substantial clarification of the nature of P-450 dependent oxygenase reactions has been afforded by isolation of the soluble cytochrome (P-$450_{cam}$) from *Pseudomonas putida* grown on camphor. *In vitro* assembly of the of the enzyme system, which also includes an electron transfer chain comprised of a reductase and the 2Fe-2S protein putidaredoxin, has led to deduction of reaction sequence (1) (Tyson *et al*., 1972; Gunsalus *et al*.,1973). In this scheme S = substrate, ox = Fe(III), red = Fe(II), and s-red is the one-electron reduction product of

$$\begin{array}{ccc} & \xrightarrow{S} & \\ \text{ox-P-450} & \longrightarrow & \text{ox-P-450·S} \\ \uparrow \searrow \text{S-OH} + H_2O & & \downarrow e^- \\ \text{s-red-P-450·S·}O_2 & & \text{red-P-450·S} \\ \nwarrow e^- & & \swarrow O_2 \\ & \text{red-P-450·S·}O_2 & \end{array} \qquad (1)$$

the Fe(II)·$O_2$ form. More recently microsomal P-450 cytochromes have been purified and enzyme systems assembled (Imai and Sato, 1974; van der Hoeven and Coon, 1974; Haugen *et al*., 1975; Guengerich et al., 1975). While scheme (1) appears to apply to these systems as well, the details of substrate binding, oxygen activation, and the hydroxylation step remain to be established. This matter is exascerbated by the lack of structural information for the heme active site. While all P-450 enzymes are b-type (protoporphyrin IX) cytochromes, the axial ligand(s) to iron, on which structural, electronic, and reactivity properties are significantly dependent in both natural and synthetic Fe(II,III) porphyrin systems, have not been securely identified in any of the five reaction states. Considerable speculation has centered on cysteinate sulfur (Cys-S) ligation, a possibility first suggested in the mid-1960's (Mason et al., 1965; Murikami and Mason, 1967) and subsequently supported by the findings that thiol addition to met-Mb and Hb and hemin chloride (usually in the presence of nitrogenous bases) afforded epr g-values close to those of the low-spin (S=1/2) ox-P-450 state (Jefcoate and Gaylor, 1969; Bayer et al., 1969; Röder and Bayer, 1969). The substrate-bound state ox-P-450·S of at least the P. putida enzyme is predominantly high-spin (S=5/2) (Peterson, 1971; Tsai et al., 1970; Sharrock et al., 1973).

The possibility of axial sulfur ligation in one or more of the enzyme reaction states has proven difficult to assess in the abscence of known properties of fully characterized sulfur-bound iron porphyrins. Indeed, the [Fe(III)$N_4$SR] coordination unit, potentially unstable to intramolecular electron transfer, has only recently been obtained in isolable porphyrin (Koch et al., 1975c; Collman et al., 1975; Ogoshi et al., 1975; Tang et al., 1976) and other tetraazamacrocyclic complexes (Koch et al., 1975ab). In this investigation we have attempted to develop certain empirical criteria for identification of ligands L and L' in porphyrin complexes containing the high-spin [Fe(III)$N_4$L] and low-spin [Fe(III)$N_4$LL'] coordination units, in which L and L' are O-, S-, and N-donor ligands intended to simulate binding by protein side-chains. Two basic assumptions underlie this approach: (i) high-spin ox-P-450·S and low-spin ox-P-450 states are effectively five- and six-coordinate, respectively, there being no exceptions to these spin state-structure correlations in natural and synthetic porphyrin systems (Hoard, 1971, 1975); and (ii) at nominal parity of axial ligand(s) properties of synthetic porphyrins and enzyme sites will be sufficiently similar to allow deduction of axial ligation modes in the latter.

## II. RESULTS AND DISCUSSION

### A. Synthetic Fe(III) Porphyrins

Five-coordinate complexes of protoporphyrin IX dimethyl ester dianion (PPIXDME) were obtained by cleavage of the μ-oxo dimer with HL and six-coordinate species were formed *in situ* by reaction with ligands L', as shown in reactions (2) and (3). Five-coordinate

$$[Fe(PPIXDME)]_2O + HL \longrightarrow Fe(PPIXDME)L + H_2O \quad (2)$$

$$\downarrow L'$$

$$Fe(PPIXDME)LL' \quad (3)$$

thiolate complexes sufficiently stable for isolation were obtained only from arylthiols; alkylthiols caused reduction to Fe(II). Fe-(PPIXDME)L species served as precursors to six-coordinate species containing a variety of L/L' axial ligand combinations. If the initial cleavage was performed in the presence of excess L', it was possible in some cases to trap at low temperatures six-coordinate complexes with L = alkylthiolate. Low-spin complexes containing acetylcysteinate-N-methylamide (S-Cys(Ac)NHMe) were obtained in this way. The following ligands were used as simulators of protein side-chain coordination: Cys, $ArS^-$, MeNH(Ac)Cys-$S^-$; Met, tetrahydrothiophene (THT); His, N-methylimidazole (N-MeIm); Tyr, $ArO^-$; Asp, Glu, $OAc^-$; Ser, Thr, $RO^-$; Lys, Arg, $RNH_2$; Asn, Gln, dimethylformamide (DMF). Stable five-coordinate complexes were isolated with L=p-$ClC_6H_4S$, p-$O_2NC_6H_4S$, p-$O_2NC_6H_4O$, OEt, and OAc. Analogous complexes were obtained in the octaethylporphyrinate (OEP) series.

Amino acid analyses of two P-450 enzymes (Tsai *et al.*, 1971; Dus *et al.*, 1974) reveal the presence of all residues potentially capable of coordinating to metal ions. Not all of the large number of possible enzyme ligation modes could be tested in synthetic five- and six-coordinate Fe(III) porphyrins owing to the failure to isolate or generate in solution all desired species. Magnetic and spectroscopic features of accessible complexes were determined and, together with available data for the synthetically inaccessible ligation modes in met-Hb and Mb (His-Fe) and cytochromes c (Met-Fe-His), were compared to corresponding properties determined for ox-P-450•S and ox-P-450. Such comparisons are listed in Table I, to which reference should be made in the following discussion, and are limited to those synthetic species which most closely approach the enzyme properties. Full details concerning preparation of compounds, ligation modes examined, and necessary theory, as well as the complete body of physical data, are given elsewhere (Tang *et al.*, 1976).

## B. High-Spin Forms

All Fe(PPIXDME)L complexes are high-spin with magnetic moments near the theoretical spin-only value of 5.92 BM for the $d^5$ configuration. In non-coordinating solvents these complexes exhibit diagnostic high-spin absorption spectra (Caughey *et al.*, 1973; Smith and Williams, 1970). Of various species with anionic sulfur and

TABLE I

COMPARISON OF PHYSICAL PROPERTIES OF SYNTHETIC Fe(III) PORPHYRIN THIOLATES AND OXIDIZED CYTOCHROME P-450 REACTION STATES

| | High-Spin Forms | | | |
|---|---|---|---|---|
| Magnetism and Epr Data | $\mu$(BM) | g-values[o] | | |
| $Fe(PPIXDME)(SC_6H_4NO_2)$ | 5.90[a] | --- | | |
| $Fe(PPIXDME)(SC_6H_4Cl)$ | 5.87[a] | 7.2, 4.8, 1.9 | | |
| ox-P-$450_{cam}$•S | 5.2[b] | 8.0, 4.0, 1.8[c] | | |
| microsomal ox-P-450•S | -- | 8.1, 3.7, 1.7[d] | | |
| *R. japonicum* ox-P-450•S | -- | 7.9, 3.8, 1.8[e] | | |
| Electronic Spectra | $\lambda_{max}$, nm ($\varepsilon_{mM}$) | | | |
| | $\alpha$ | $\beta$ | | $\gamma$(Soret) |
| $Fe(PPIXDME)(SC_6H_4NO_2)$ | 646(4.8) | 540(sh,10.4), 517(11.2) | | 391(86.8) |
| $Fe(PPIXDME)(SC_6H_4Cl)$ | 643(4.4) | 571(10.4), 533(11.5) | | 396(83.3) |
| ox-P-$450_{cam}$•S[f] | 646(4.5) | 540(sh,10), ∿520 | | 391(87) |
| microsomal ox-P-450•S[g] | 647 | 540(sh), 517 | | 394 |
| Mössbauer Data[h] | $\delta$(mm/sec)[i] | $\Delta E_Q$(mm/sec) | $H^{\circ}_{hf}$(kOe) | D($cm^{-1}$) |
| $Fe(PPIXDME)(SC_6H_4NO_2)$[j] | 0.33 | 0.76 | $-476\pm10$ | $10\leq D\leq12$ |
| ($CH_2Cl_2$ soln.) | 0.32 | 0.88 | -- | -- |
| ox-P-$450_{cam}$•S[k] | 0.35 | 0.79 | $-448\pm10$ | 3.8[c] |

TABLE I (Cont'd.)

Low-Spin Forms

| Electronic Spectra | $\lambda_{max}$, nm ($\varepsilon_{mM}$) α | β | γ(Soret) |
|---|---|---|---|
| $Fe(PPIXDME)(SC_6H_4NO_2)(2\text{-}MeTHF)$[l] | 566(1.1) | 535(1.0) | 418(14.9) |
| $Fe(PPIXDME)(SC_6H_4Cl)(2\text{-}MeTHF)$[l] | 566(1.0) | 536(1.0) | 422(14.2) |
| $Fe(PPIXDME)(SC_6H_4NO_2)(DMF)$[l] | 567(1.1) | 533(1.0) | 420(14.1) |
| ox-P-$450_{cam}$[f] | 571(10.5) | 535(10) | 417(105) |
| microsomal ox-P-450[m] | 568(14.2) | 534(13.6) | 417(124) |
| mitochondrial ox-P-450[n] | 567 | 532 | 416 |

| Epr Data | Ligand(L') | g-values[o] | | |
|---|---|---|---|---|
| $Fe(PPIXDME)(SC_6H_4NO_2)$ | N-MeIm[p] | 2.42 | 2.26 | 1.91 |
| | n-$PrNH_2$ | 2.48 | 2.26 | 1.90 |
| | THT[q] | 2.40 | 2.28 | 1.92 |
| | PhSH[s] | 2.43 | 2.28 | 1.92 |
| | DMF[r] | 2.46 | 2.28 | 1.90 |
| Fe(PPIXDME)(S-Cys(Ac)NHMe) | N-MeIm[p] | 2.37 | 2.24 | 1.94 |
| | n-$PrNH_2$ | 2.40 | 2.23 | 1.93 |
| | THT[q] | 2.35 | 2.24 | 1.94 |
| | DMF[r] | 2.37 | 2.24 | 1.95 |
| ox-P-$450_{cam}$[c] | -- | 2.45 | 2.26 | 1.91 |
| microsomal ox-P-450[t] | -- | 2.41 | 2.25 | 1.91 |
| mitochondrial ox-P-450[u] | -- | 2.42 | 2.26 | 1.91 |

[a]295°K. [b]94-253°; Peterson, 1971. [c]Tsai _et al_., 1970. [d]Peisach and Blumberg, 1970. [e]Peisach _et al_., 1972. [f]Gunsalus _et al_., 1973. [g]Estabrook _et al_., 1973. [h]Values at 4.2°K unless otherwise noted. [i]Relative to Fe metal. [j]Solid state. [k]200°K, Sharrock _et al_., 1973. [l]77°K, values in parentheses are the absorbance ratios $A/A_\beta$. [m]van der Hoeven and Coon, 1974. [n]Mitani and Horie, 1969. [o]∼90°K. [p]N-methylimidazole. [q]Tetrahydrothiophene. [r]N,N-dimethylformamide. [s]Low-spin form is minority component. [t]Miyake _et al_., 1969. [u]Cheng and Harding, 1973.

oxygen ligands and with His(Im) coordination (met-Hb, Mb, Antonini and Brunori, 1971), only thiolate complexes showed a high degree of correspondence with the substrate-bound reaction state in terms of $\alpha$ and $\gamma$ band positions and intensities. Further comparisons are afforded by MCD spectra, a conspicuous feature of which for ox-P-$450_{cam}$•S is substantial negative ellipticity in the Soret region (Dolinger et al., 1974; Vickery et al., 1975). Among all synthetic porphyrin complexes examined this feature is observed only with thiolate species; all others exhibit positive ellipticity of their Soret bands (Dawson et al., 1976).

The epr spectra of all ox-P-450•S states are characterized by large rhombic splittings manifested by three g-values, two of which are symmetrically split about g = 6 and the other is <2. For strict axial symmetry $g_\perp = 6$ and $g_\parallel = 2$. From the observed g-values the spin Hamiltonian parameter ratio E/D = 0.083-0.092, corresponding to degrees of rhombicity of 25-28% (Peisach and Blumberg, 1971). The epr spectra of Fe(PPIXDME)($SC_6H_4Cl$) and Fe(OEP)(SPh) reveal substantial rhombic splittings with E/D $\cong$ 0.05 or 15% rhombicity. In contrast, complexes with O- and N-donor ligands have axial or near-axial symmetry, with the largest splitting observed for Fe(PPIXDME)-(OEt) (E/D = 0.02). For Mbs and normal intact Hbs the degree of rhombicity is usually $\lesssim$3% (Peisach and Blumberg, 1971; Peisach et al., 1969; Kotani and Watari, 1971). Consequently, substantial rhombicity in the present group of enzymes and synthetic complexes appears to correlate with thiolate ligation.

The Mössbauer spectral parameters of isomer shift ($\delta$) and quadrupole splitting ($\Delta E_Q$) are typical of high-spin Fe(III) porphyrins and are in good agreement with results for ox-P-$450_{cam}$•S. However, these parameters are rather insensitive to the nature of axial ligands (Tang et al., 1976; Maricondi et al., 1972; Torrens et al, 1972) and do not serve to identify enzyme ligation. Considerably more useful in this respect is the quantity $H^{\circ}_{hf}$, the saturation magnetic field at the $^{57}$Fe nucleus, which is expected to be sensitive to changes in axial ligation. This has proven to be the case. Values of $H^{\circ}_{hf}$ for synthetic porphyrins were evaluated from low magnetization and Mössbauer measurements in large applied magnetic fields and the preceding epr data. Experimental methods and necessary theory are available elsewhere (Tang et al., 1976). For typical Fe(III) salts, met-Mb (Lang, 1970), and oxygen-ligated porphyrins $-H^{\circ}_{hf} \gtrsim$ 500 kOe, substantially larger than the value of -448 kOe for ox-P-$450_{cam}$•S. Only in the thiolate complex Fe(PPIXDME)-($SC_6H_4NO_2$), for which $-H^{\circ}_{hf}$ = 476±10 kOe, is the enzyme value reasonably closely approached.

### C. Low-Spin Forms

Combination of reactions (2) and (3) has led to the formation of six-coordinate low-spin Fe(III) porphyrin complexes in solution. Attempts to isolate thiolate species proved unsuccessful. Epr monitoring of reaction mixtures with a variety of ligands L' has shown that, after the initial low temperature quench, further thaw-quench cycles were accompanied by rapid decrease in the intensity of low-spin Fe(III) signals with no other resonances detectable at 80-95°K. This behavior is interpreted as arising from the redox reaction (4). Because of the instability of six-coordinate species and in some cases the lack of complexation by excess L', the scope of physical

$$2Fe(III)(PPIXDME)(SR)L' + 2L' \longrightarrow 2Fe(II)(PPIXDME)L'_2 + RSSR \quad (4)$$

studies of these complexes is less extensive than for five-coordinate complexes.

Satisfactory electronic spectra of Fe(PPIXDME)LL' complexes were obtained only in glass media at ∿77°K and only with certain combinations of L and L'. Spectral data for 2-MeTHF and DMF adducts are those of low-spin species (Smith and Williams, 1970) and are in fairly good agreement with results for the ox-P-450 state. Further spectral examination has revealed that insufficient differences exist between the N-MeIm/N-MeIm and $SC_6H_4NO_2$/DMF cases to permit secure ligation criteria to be established. However, epr results (vide infra) eliminate the first mode of ligation. Because of the inability to survey a wide range of Fe(PPIXDME)LL' species, it can be noted only that spectral similarities are consistent with, but do not require, thiolate coordination in the resting enzyme state.

Prior to this study ample evidence has been presented, and usefully quantitated in "truth" diagrams (Peisach and Blumberg, 1971; Peisach et al., 1973a), that g-values of the rhombic epr spectra of low-spin ferrihemes are sensitive to the nature of axial ligands. Using both PPIXDME and OEP complexes the spectra of species containing some 25 L/L' ligand combinations were examined. Together with data for cytochromes b and c and $Fe(PPIXDME)(N\text{-}MeIm)_2^+$, this information indicated that in the ox-P-450 state one ligand is cysteinate and the other corresponds to the biological counterparts of ligands L' in Table I, which in combination with L = ArS produced satisfactory agreement with the g-values of the enzymes.

## D. Conclusions

Noting the modes of axial ligation which could not be directly tested with synthetic porphyrin complexes and the limitations of such complexes as models for biological heme coordination, matters which are dealt with more fully elsewhere (Tang et al., 1976), the following principal conclusions are drawn: (i) On the basis of collective comparative results from electronic absorption, MCD, epr, and magnetically perturbed Mössbauer spectra, the most probable axial ligation mode in the high-spin ox-P-450•S state is Cys-S-Fe. (ii) From comparative epr and, marginally, electronic absorption spectra, the Cys-S-Fe coordination in ox-P-450•S is retained in the resting ox-P-450 state with axial ligation Cys-S-Fe-L'; no choice can be made among the possibilities L' = His, Lys(Arg), Cys-SH, Met, Asn(Gln), and $H_2O$. All but the last have been directly tested. From these conclusions the simplified representation of substrate binding, reaction (5), follows for those cases in which a spin-state change occurs. It is consistent with certain explicit earlier proposals concerning one or both reaction states (Hill <u>et al</u>, 1970ab; Peisach <u>et al</u>., 1973b) except that a weakly bound sixth ligand is


(5)

ox-P-450 (low-spin) ox-P-450•S (high-spin)

not included in the high-spin state. Removal of L' upon substrate binding with retention of thiolate coordination conflicts with an earlier proposal of sulfur ligand displacement (Tsai <u>et al</u>., 1970), but is in accord with another proposal (Estabrook <u>et al</u>., 1973) that L' = His is detached from the coordination sphere. More recently it has been suggested that in ox-P-450 L' = $H_2O$, which is displaced when substrate is bound (Griffin and Peterson, 1975). We have been unable to detect $H_2O$ binding to Fe(III) porphyrin thiolates by epr measurements. If 2-MeTHF is considered a rough simulator of axial $H_2O$ coordination, spectral data for Fe(PPIXDME)-$(SC_6H_4NO_2)$(2-MeTHF) widen the choice of L' ligands to include $H_2O$. In assigning the ligation modes above, the resemblance between certain electronic features of the oxidized P-450 reaction states and chloroperoxidase is noted (Chiang <u>et al</u>., 1975). In the latter the two cysteinyl residues are described as existing as a disulfide in both the Fe(III) and Fe(II) forms of the enzyme and thus are unavailable as thiolate ligands. Possibly in at least this one case thiolate coordination may be a sufficient but not necessary condition for development of properties similar to those of oxidized P-450 reaction states. However, this situation is not consistent with the collective body of data for model Fe(III) PPIXDME complexes presented here and elsewhere (Tang et al., 1976).

### E. Structure of Fe(PPIXDME)($SC_6H_4NO_2$)

In order to define in detail the stereochemistry of a presumably representative Fe(III) porphyrin thiolate and thus provide the

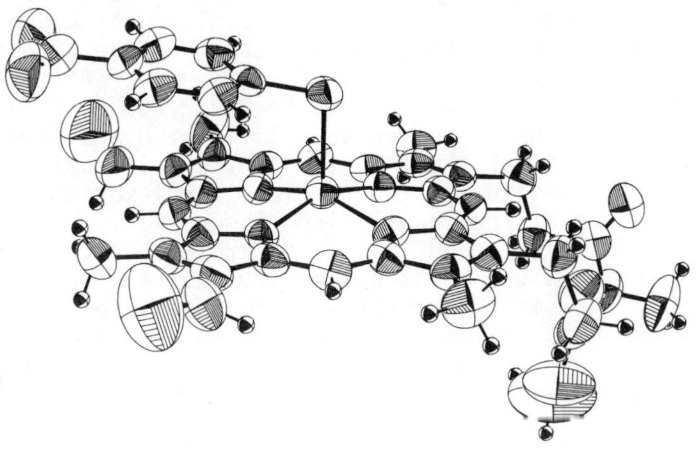

Figure 1. Overall molecular structure of Fe(PPIXDME)($SC_6H_4NO_2$).

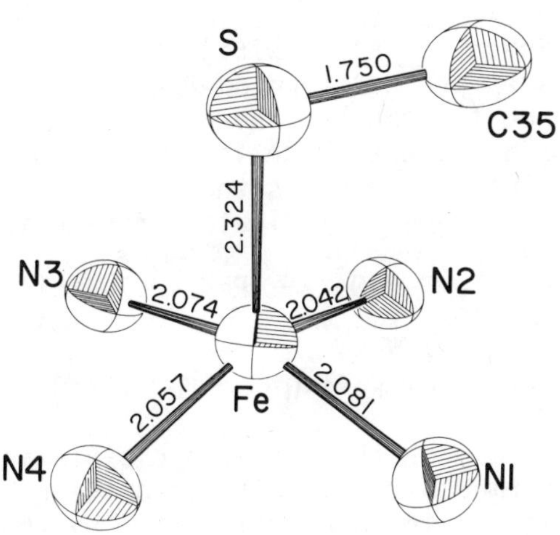


Figure 2. Coordination sphere of Fe(PPIXDME)($SC_6H_4NO_2$).

first structural model for the active site of the ox-P-450•S reaction state, the molecular structure of Fe(PPIXDME)($SC_6H_4NO_2$) was determined by X-ray diffraction. A view of the entire molecule is presented in Figure 1 and the coordination sphere geometry with bonded distances is given in Figure 2. The Fe(III)$N_4$S coordination unit has the pyramidal arrangement found in all high-spin Fe(III) porphyrins but with marginally significant differences in Fe-N distances, which average to 2.064 Å. The Fe atom lies 0.448 Å out of the mean porphyrin plane and 0.434 Å out of the $N_4$ plane toward the axial sulfur. The porphyrin ring is "domed" toward the metal by 0.014 Å. The average Ct-N distance is 2.017 Å, although definition of the $N_4$ center (Ct) is somewhat arbitrary in the absence of axial symmetry. These values agree closely with representative parameters for high-spin Fe(III) porphyrins of Fe-N = 2.065 Å, Ct•••N = 2.015 Å, and Ct•••Fe = Fe•••$N_4$ plane = 0.45 Å given by Hoard (1971, 1975). The Fe-S distance is at the long end of the 2.21-2.31 Å range for non-porphyrin high-spin Fe(III) thiolate complexes (Tang et al., 1976). The locus of the phenyl ring (Figure 1) is apparently dictated by steric effects and, with the marginally different Fe-N distances, degrades the Fe site symmetry in the coordination sphere from axial to rhombic. This situation may contribute to the substantial rhombic splittings observed in the epr spectra.

Provided the conclusion that Fe-S-Cys ligation occurs in the oxidized enzymes is correct, the utility of well-defined synthetic Fe(III) porphyrin thiolate complexes may extend beyond providing electronic and structural models for active sites. It has recently been shown that P-450 enzymes can function as peroxidases (Hrycay et al., 1975b), that certain organic hydroperoxides can replace NADH and $O_2$ in enzymic substrate hydroxylation (Rahmitula and O'Brien, 1975), and that hydroperoxide, $NaClO_2$, $NaIO_4$, or $H_2O_2$ support substrate hydroxylation with the oxidized enzyme in the absence of molecular oxygen (Hrycay et al., 1975a). The latter finding implies that the three step process from ox-P-450•S to s-red-P-450•S•$O_2$ in the reaction sequence (1) can be "short-circuited" by supplying to the oxidized enzyme a reagent which in combination with Fe(III) results in the same oxidation level as the s-red state before hydroxylation. A formalistic scheme is indicated in reaction (6), where

$$\begin{array}{c} Fe^{3+} + O_2^{2-} \longrightarrow [Fe^{3+}\text{-}O\text{-}O^{2-} \longleftrightarrow Fe^{2+}\text{-}O_2^{-}] \\ \downarrow 2H^+ \\ [Fe^{3+}\text{-}O \longleftrightarrow Fe^{4+}\text{-}O^{-} \longleftrightarrow Fe^{5+}\text{-}O^{2-}] + H_2O \end{array} \quad (6)$$

the ferryl (FeO) entity may be the active hydroxylating agent and conceivably could be formed directly with chlorite or periodate. Employment of such reagents in combination with synthetic Fe(III) porphyrin thiolates and potential substrate could afford useful model hydroxylation systems for mechanistic studies, which would be more easily manipulated than others requiring formation and reduction of Fe(II)·$O_2$ species prior to hydroxylation. Aromatic hydroxylation has been accomplished in model systems composed of hemin chloride, aqueous base, excess thiol, $O_2$, and substrate aniline (Sakurai and Ogawa, 1975). In this case presumably thiolate reduces Fe(III) to Fe(II), and subsequent reactions may proceed in a manner similar to scheme (1).

## ACKNOWLEDGMENTS

This research was supported at the Department of Chemistry, M.I.T., by National Science Foundation Grant GP-40089X, at Northwestern University by NIH Grant HL-13157, and at the Francis Bitter National Magnet Laboratory by the National Science Foundation.

## REFERENCES

Antonini, E., and Brunori, M. (1971). In "Hemoglobin and Myoglobin in their Reactions with Ligands", North Holland Publishing Co., Amsterdam, Chapter 3.

Bayer, E., Hill, H. A. O., Röder, A., and Williams, R. J. P. (1969). Chem. Commun., 109.

Caughey, W. S. (1973). In "Inorganic Biochemistry", Vol. 2, G. Eichhorn, Ed., Elsevier Publishing Co., Amsterdam, Chapter 24.

Cheng, S. C., and Harding, B. W. (1973). J. Biol. Chem., 248, 7263.

Chiang, R., Makino, R., Spomer, W. E., and Hager, L. P. (1975). Biochemistry, 14, 4166.

Collman, J. P., Sorrell, T. N., and Hoffman, B. M. (1975). J. Amer. Chem. Soc., 97, 913.

Dawson, J. H., Tang, S. C., Holm, R. H., Bunnenberg, E., and Djerassi, C. (1976). results to be published.

Dolinger, P. M., Kielczewski, M., Trudell, J. R., Barth, G., Lindner, R. E., Bunnenberg, E., and Djerassi, C. (1974). Proc. Nat. Acad. Sci. U.S., 71, 399.

Dus, K., Litchfield, W. J., Miguel, A. G., van der Hoeven, T. A., Haugen, D. A., Dean, W. L., and Coon, M. J. (1974). Biochem. Biophys. Res. Commun., 60, 15.

Estabrook, R. W., Baron, J., Peterson, J. and Ishimura, Y. (1973). In "Biological Hydroxylase Mechanisms", Biochem. Soc. Symp. 34, R. M. S. Smellie, Ed., pp. 159-185.

Griffin, B. W., and Peterson, J. A. (1975). J. Biol. Chem., 250, 6445.

Guengerich, F. P., Ballou, D. P., and Coon, M. J. (1975). J. Biol. Chem., 250, 7405.

Gunsalus, I. C., Meeks, J. R., Lipscomb, J. D., Debrunner, P., and Münck, E. (1973). In "Molecular Mechanisms of Oxygen Activation", O. Hayaishi, Ed., Academic Press, New York, Chapter 14.

Haugen, D. A., van der Hoeven, T. A., and Coon, M. J. (1975). J. Biol. Chem., 250, 3567.

Hill, H. A. O., Röder, A., and Williams, R. J. P. (1970a). Struct. Bonding (Berlin), 8, 123.

Hill, H. A. O., Röder, A., and Williams, R. J. P. (1970b). Naturwiss., 57, 69.

Hoard, J. L. (1971). Science, 174, 1295.

Hoard, J. L. (1975). In "Porphyrins and Metalloporphyrins", K. M. Smith, Ed., Elsevier Publishing Company, Amsterdam, Chapter 8.

Hrycay, E. G., Gustafsson, J., Ingelman-Sundberg, M., and Ernster, L. (1975a). Biochem. Biophys. Res. Commun., 66, 209.

Hrycay, E. G., Jonen, H. G., Lu, A. Y. H., and Levin, W. (1975b). Arch. Biochem. Biophys., 166, 145.

Imai, Y., and Sato, R. (1974). Biochem. Biophys. Res. Commun., 60, 8.

Jefcoate, C. R. E., and Gaylor, J. L. (1969). Biochemistry, 8, 3464.

Koch, S., Holm, R. H., and Frankel, R. B. (1975a). J. Amer. Chem. Soc., 97, 6714.

Koch, S., Tang, S. C., Holm, R. H., and Frankel, R. B. (1975b). J. Amer. Chem. Soc., 97, 914.

Koch, S., Tang, S. C., Holm, R. H., Frankel, R. B., and Ibers, J. A. (1975c). J. Amer. Chem. Soc., 97, 916.

Kotani, M., and Watari, H. (1971). In "Magnetic Resonance in Biological Research", C. Franconi, Ed., Gordon and Breach, New York, pp. 75-96.

Lang, G. (1970). Quart. Rev. Biophys., 3, 1.

Maricondi, C., Straub, D. K., and Epstein, L. M. (1972). J. Amer. Chem. Soc., 94, 4157.

Mason, H. S., North, J. C., and Vanneste, M. (1965). Fed. Proc., 24, 1172.

Mitani, F., and Horie, S. (1969). J. Biochem. (Tokyo), 66, 139.

Miyake, Y., Mori, K., and Yamano, T. (1969). Arch. Biochem. Biophys., 133, 318.

Murikami, K., and Mason, H. S. (1967). J. Biol. Chem., 242, 1102.

Ogoshi, H., Sugimoto, H., and Yoshida, Z. (1975). Tetrahedron Letters, 2289.

Peisach, J., Appleby, C. A., and Blumberg, W. E. (1972). Arch. Biochem. Biophys., 150, 725.

Peisach, J., and Blumberg, W. E. (1970). Proc. Nat. Acad. Sci. U.S., 67, 172.

Peisach, J., and Blumberg, W. E. (1971). In "Probes of Structure and Function of Macromolecules and Enzymes", Vol. II, B. Chance, T. Yonetani, and A. S. Mildvan, Ed., Academic Press, New York, pp. 215-229, 231-239.

Peisach, J., Blumberg, W. E., and Adler, A. (1973a). Ann. N. Y. Acad. Sci., 206, 310.

Peisach, J., Blumberg, W. E., Wittenberg, B. A., and Kampa, L. (1969). Proc. Nat. Acad. Sci. U.S., 63, 934.

Peisach, J., Stern, J. O., and Blumberg, W. E. (1973b). Drug. Metab. Disp., 1, 45.

Peterson, J. A. (1971). Arch. Biochem. Biophys., 144, 678.

Rahmitula, A. D., and O'Brien, P. J. (1975). Biochem. Biophys. Res. Commun., 62, 268.

Röder, A., and Bayer, E. (1969). Eur. J. Biochem., 11, 89.

Sakurai, H., and Ogawa, S. (1975). Biochem. Pharmacol., 24, 1257.

Sharrock, M., Münck, E., Debrunner, P. G., Marshall, V., Lipscomb, J. D., and Gunsalus, I. C. (1973). Biochemistry, 12, 258.

Smith, D. W., and Williams, R. J. P. (1970). Struct. Bonding (Berlin), 7, 1.

Tang, S. C., Koch, S., Papaefthymiou, G. C., Foner, S., Frankel, R. B., Ibers, J. A., and Holm, R. H. (1976). J. Amer. Chem. Soc., in press.

Tomaszewski, J. E., Jerina, D. M., and Daly, J. W. (1974). Ann. Reports Med. Chem., 9, 290.

Torréns, M. A., Straub, D. K., and Epstein, L. M. (1972). J. Amer. Chem. Soc., 94, 4162.

Tsai, R., Yu, C. A., Gunsalus, I. C., Peisach, J., Blumberg, W., Orme-Johnson, W. H., and Beinert, H. (1970). Proc. Nat. Acad. Sci. U.S., 66, 1157.

Tsai, R. L., Gunsalus, I. C., and Dus, K. (1971). Biochem. Biophys. Res. Commun., 45, 1300.

Tyson, C. A., Lipscomb, J. D., and Gunsalus, I. C. (1972). J. Biol. Chem., 247, 5777.

Ullrich, V. (1972). _Angew. Chem. Int. Ed._, _11_, 701.

van der Hoeven, T. A., and Coon, M. J. (1974). _J. Biol. Chem._, _249_, 6302.

Vickery, L., Salmon, A., and Sauer, K. (1975). _Biochim. Biophys. Acta_, _386_, 87.

# A NEW ASSAY PROCEDURE FOR INDOLEAMINE 2,3-DIOXYGENASE*

O. Hayaishi, F. Hirata, T. Ohnishi, R. Yoshida and T. Shimizu

Department of Medical Chemistry, Kyoto University Faculty of Medicine, Sakyo-ku, Kyoto 606 JAPAN

Indoleamine 2,3-dioxygenase is a heme-containing dioxygenase that catalyzes the oxygenative ring cleavage of various substituted and unsubstituted indoleamines, such as tryptophan, 5-hydroxytryptophan, tryptamine, serotonin and so forth. The product of the reaction was identified to be the corresponding formylanthraniloylalkylamine (Fig. 1). This enzyme is ubiquitously distributed in various organs of mammals and appears to play an important role in the metabolism of these biogenic indoleamines.

In previous reports from our laboratory (1, 2), we presented some evidence indicating that this enzyme utilizes superoxide anion rather than molecular oxygen as an oxidizing agent. This conclusion was based mainly upon the following two lines of evidence. First, the purified enzyme by itself is completely inactive unless superoxide anion is generated in the reaction mixture either chemically by ascorbic acid plus methylene blue or enzymatically by the xanthine oxidase or glutathione reductase systems. Secondly, scavengers for superoxide anion, such as superoxide dismutase or Tiron, inhibited the dioxygenase-catalyzed reaction under these conditions.

In order to further establish the participation of superoxide anion in the catalytic process and to clarify its role in this reaction, we developed a new assay procedure for this enzyme, in

---

* This investigation was supported in part by research grants from the Sakamoto Foundation, Iyakushigen Kenkyu Shinkohkai, the Mitsubishi Foundation, Nippon Shinyaku Co., Ltd., the Science and Technology Agency, Japan, and a Grant-in-Aid for Scientific Research from the Ministry of Education, Science and Culture, Japan.

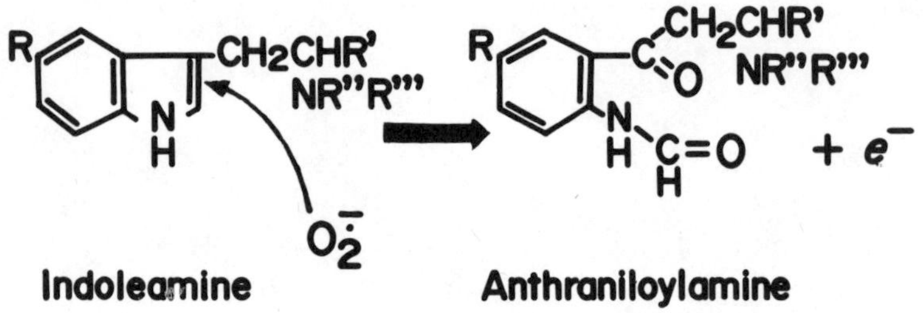

Fig. 1. Indoleamine 2,3-dioxygenase

which superoxide anion *per se* was directly infused into the reaction mixture. The principle of this new assay procedure is based upon the use of [ring-2-$^{14}$C]-labelled substrate, as originally described by Beverly Peterkofsky in 1968 for the tryptophan dioxygenase assay (3). As can be seen in Fig. 2, when the pyrrole ring of the substrate, such as tryptophan, is cleaved by the action of indoleamine dioxygenase, formylkynurenine is produced. When formamidase, an enzyme which catalyzes the hydrolysis of the formyl group, is included in the reaction mixture, the radioactive carbon is released as formate, which can be isolated either by distillation or by column chromatography and then quantitatively determined. The details of the new assay procedure are illustrated in Fig. 3.

The reaction mixture contains D,L-[2-$^{14}$C]tryptophan as substrate, formate as a carrier, catalase to decompose hydrogen peroxide which is formed from superoxide anion by spontaneous decomposition, a highly purified formamidase preparation from rat liver, indoleamine dioxygenase which was purified from rabbit intestine to apparent homogeneity and buffer, in a total volume of 0.2 ml. Potassium superoxide was dissolved in an aprotic solvent, such as dimethylsulfoxide, and was directly but slowly infused into the

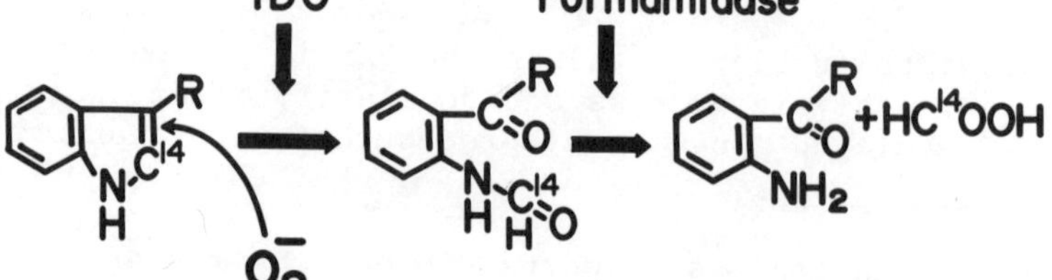

Fig. 2. Principle of enzyme assay

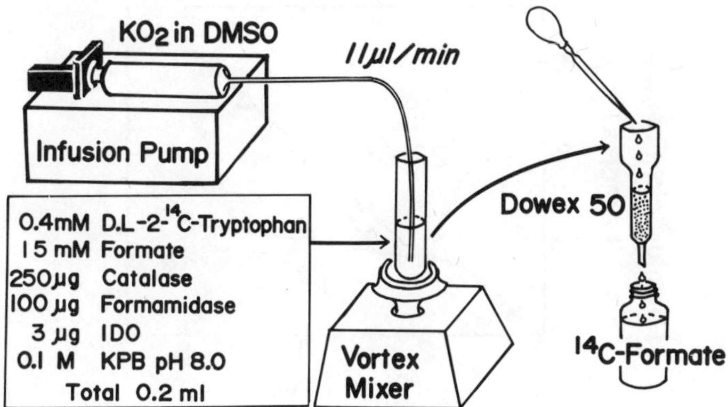


Fig. 3. New assay method of IDO

reaction mixture, which was being vigorously mixed by means of a vortex mixer. The reaction was terminated by 5% TCA and the acidified reaction mixture was passed through a short column of Dowex-50. The radioactive formate, which was eluted with water, was quantitated with a scintillation counter. The time course of the reaction is illustrated in Fig. 4. When methylene blue was not included in the reaction mixture, the rate of the reaction quickly decreased after about 1 minute. But, thereafter, the reaction seemed to proceed almost linearly for at least another several minutes. With increasing concentrations of methylene blue, the rate of the reaction increased, and in the presence of 25 μM methylene blue, the reaction proceeded linearly for at least 6 minutes. These results seem to indicate that the initial velocity was not significantly affected by methylene blue, but that methylene blue was

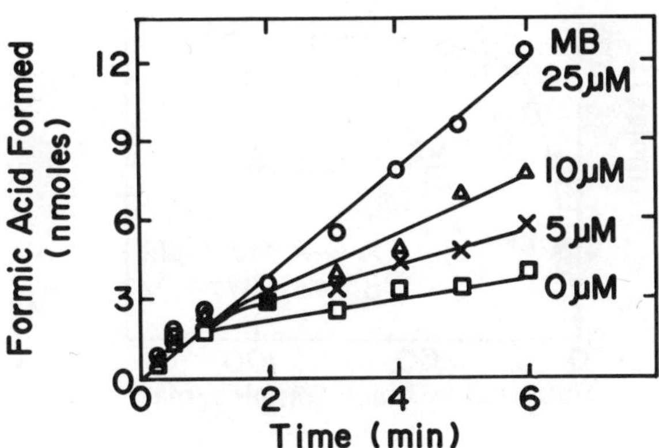


Fig. 4. Time course at various concentrations of MB

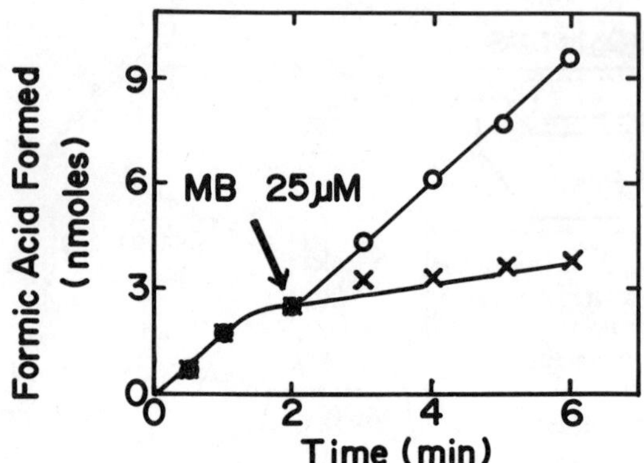


Fig. 5. Effect of MB during course of reaction

essential for maintaining the steady state of the reaction even when superoxide anion was continuously supplied to the incubation mixture. In order to see whether methylene blue was necessary to protect the enzyme from inactivation, we then added methylene blue during the course of the reaction (Fig. 5).

The lower curve represents the time course of the reaction in the absence of methylene blue. 25 µM methylene blue was added to the reaction mixture 2 minutes after the initiation of the reaction, when the rate of the reaction started to decline. As soon as methylene blue was added to the reaction mixture, the rate of the reaction was greatly accelerated and became essentially identical to that observed when 25 µM methylene blue was added at the

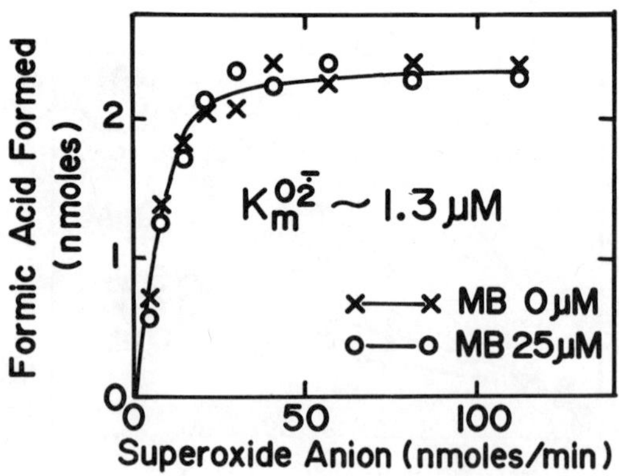


Fig. 6. Km for $O_2^-$ in presence and absence of MB

onset of the reaction. These results seem to indicate that methylene blue is required not for protection of the enzyme from inactivation but as an essential component of the catalytic process. In order to see if methylene blue acted as an allosteric effector, we determined the Km values for superoxide anion and tryptophan in the presence and absence of methylene blue. The results are presented in Fig. 6.

The velocity-substrate concentration curves were essentially identical either in the presence or absence of methylene blue. Km values for superoxide anion were close to 1.3 μM in both cases.

Km values for the substrate, tryptophan, were also identical both in the presence and absence of methylene blue (Fig. 7). This value, 0.03 mM, was also identical with the Km value determined by the conventional assay procedure in the presence of ascorbate and methylene blue. In this case, there was no activity when methylene blue was omitted, presumably because methylene blue was necessary for the generation of superoxide anion in the presence of ascorbate.

The pH optimum of the initial velocity lies in the neighbourhood of 8.5 with the new assay procedure both in the presence and absence of methylene blue, whereas in the previous report, the pH optimum was found to be about 7.0 in the presence of ascorbic acid and methylene blue (Fig. 8). The reason for this discrepancy is not yet completely understood, but it may be attributable to the pH dependency of the rate of generation and the stability of superoxide anion.

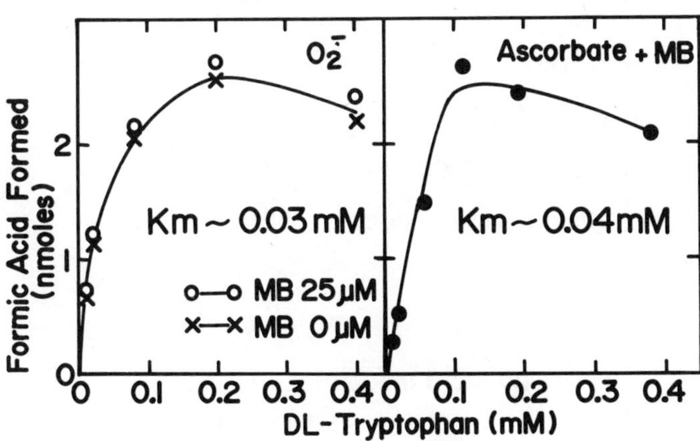


Fig. 7. Km for D,L-tryptophan

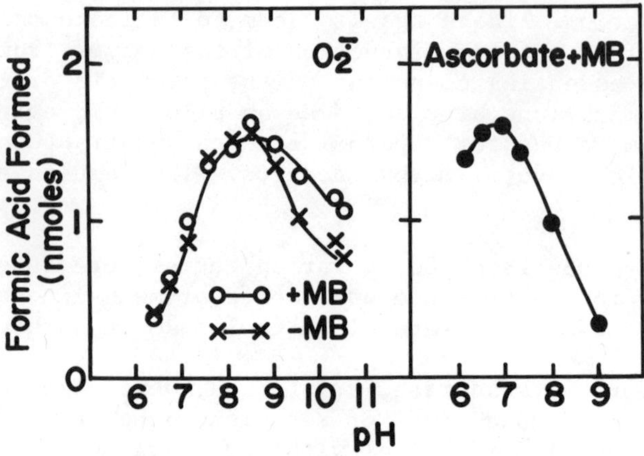


Fig. 8. Optimum pH

Utilizing this new assay procedure, we then reexamined the requirement for superoxide anion and the inhibition by superoxide dismutase. When the highly purified indoleamine dioxygenase preparation was incubated with the substrate under aerobic conditions, it was completely inactive unless superoxide anion was present in

Table I. Effect of $O_2^-$ or $H_2O_2$ generating systems

| Addition | Product formation |
|---|---|
| | nmoles |
| None | 0 |
| Ascorbate | 28.0 |
| Xanthine oxidase | 30.0 |
| Glutathione reductase | 25.4 |
| $H_2O_2$ | 0 |
| Glucose oxidase | 0 |
| L-Amino acid oxidase | 0 |
| $KO_2$ | 24.5 |

the reaction mixture. As can be seen in Table I, there was no activity unless ascorbic acid plus methylene blue, or xanthine oxidase or glutathione reductase with their respective substrates were added to the reaction mixture. Ascorbate, the xanthine oxidase system and the glutathione reductase system are all known to generate superoxide anion as well as hydrogen peroxide. However, neither hydrogen peroxide as such nor the glucose oxidase or amino acid oxidase systems that were known to generate hydrogen peroxide but not superoxide anion, were able to support the enzyme activity. When superoxide anion was infused into the reaction mixture by this new assay procedure, it was able to replace superoxide anion which was generated chemically or enzymatically. Furthermore, the rates of the reaction, Vmax values, observed under these conditions were almost identical. These results provided unequivocal evidence that superoxide anion rather than oxygen or hydrogen peroxide is required for this reaction.

Fig. 9 illustrates the percentage inhibition of the indoleamine dioxygenase-catalyzed reaction and the reduction of cytochrome c by superoxide anion as plotted against varying amounts of a highly purified preparation of superoxide dismutase from bovine erythrocytes. Superoxide dismutase is usually assayed by its ability to trap superoxide anion and to inhibit the reduction of cytochrome c in the presence of superoxide anion. In fact, the amount of superoxide dismutase which produces 50% inhibition of cytochrome c reduction is defined as one unit of enzyme as illustrated in Fig. 9. The reaction catalyzed by indoleamine dioxygenase was also inhibited by superoxide dismutase and the degree of inhibition was almost identical to that observed with the reduction of cytochrome c. These results, taken together, further substantiate our previous conclusion that this enzyme utilizes superoxide anion as an oxidizing agent.

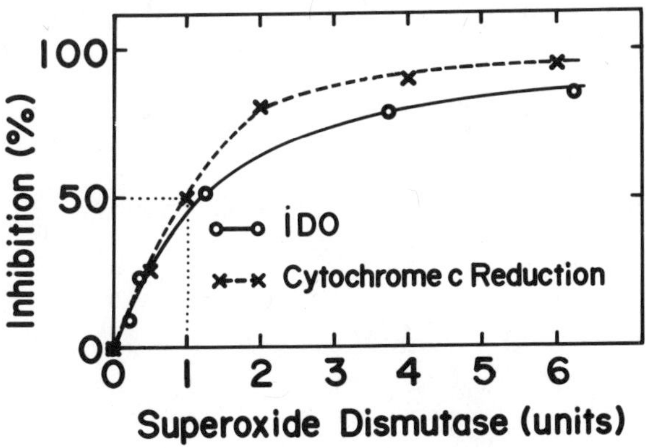


Fig. 9. Inhibition by superoxide dismutase

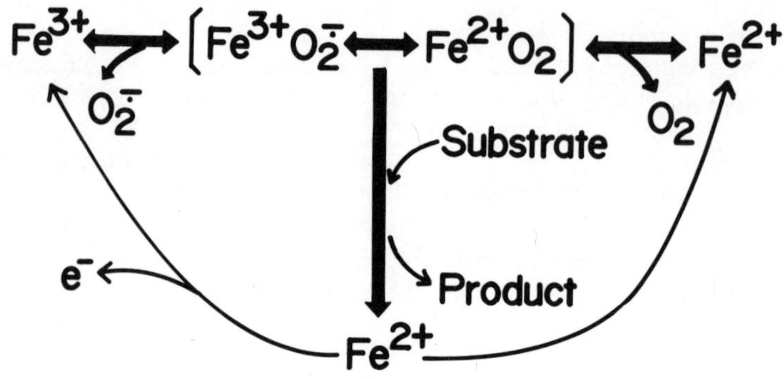


Fig. 10. Plausible reaction sequence

These results are consistent with, but do not necessarily prove, the plausible reaction sequence, illustrated in Fig. 10, which was previously proposed as a working hypothesis (2). The native, ferric form of the enzyme reacts with superoxide anion to form the oxygenated enzyme, which then reacts with substrate to produce the endproduct of the reaction. The resulting ferrous form of the enzyme may react with oxygen to form the oxygenated intermediate but this reaction appears to be sluggish. In the normal course of events, the ferrous enzyme is quickly oxidized to the native ferric enzyme to complete the catalytic cycle and methylene blue may play a role in this final step of the catalytic cycle.

## References

1. Hirata, F., and Hayaishi, O. (1971) J. Biol. Chem. 246, 7825-7826

2. Hirata, F., and Hayaishi, O. (1975) J. Biol. Chem. 250, 5960-5966

3. Peterkofsky, B. (1968) Arch. Biochem. Biophys. 128, 637-645

# IS INDOLEAMINE 2,3-DIOXYGENASE ANOTHER HEME AND COPPER CONTAINING ENZYME?

Frank O. Brady and Albert Udom
Department of Biochemistry
School of Medicine
The University of South Dakota
Vermillion, South Dakota 57069

Indoleamine 2,3-dioxygenase catalyzes a reaction similar to that catalyzed by L-tryptophan 2,3-dioxygenases except that the former shows little substrate specificity while the latter shows absolute specificity for L-tryptophan. Originally referred to as D-tryptophan oxygenase, indoleamine 2,3-dioxygenase has been detected in and/or purified from rabbit small intestine (1-7), rabbit brain (8), rat intestine (9), and rat brain (10-11). Serotonin, melatonin, D- and L-5-hydroxytryptophan, tryptamine, and D- and L-tryptophan have been shown to be substrates for indoleamine 2,3-dioxygenase. An unusual characteristic of this enzyme is the requirement for methylene blue and ascorbate in eliciting catalytic activity *in vitro* (1,2,5), although a recent report suggests that tetrahydrobiopterin can also serve in this process(12). Hirata and Hayaishi (3,5) have implicated the involvement of $O_2^-$ in the reaction catalyzed by indoleamine 2,3-dioxygenase, based mainly on the ability of superoxide dismutase to inhibit the ongoing reaction. Based on my previous work with L-tryptophan 2,3-dioxygenases (13, 14), we became interested in the possibility that indoleamine 2,3-dioxygenase might also be a heme and copper containing enzyme. Reported herein are the results of our efforts to purify the enzyme and analyses for its heme and copper content.

Indoleamine 2,3-dioxygenase was purified essentially as before (2), with some modifications, the details of which will be published elsewhere.[1] Our scheme of purification and the results of analyses are shown in Table I. The enzyme as purified through the disc acrylamide gel stage of purification contains nearly 17 ng-

[1]Albert Udom and Frank O. Brady, manuscript in preparation.

Table I

PURIFICATION OF INDOLEAMINE 2,3-DIOXYGENASE AND ANALYSIS FOR COPPER

| Stage of Purification[1] | Protein[2] (mg/ml) | Enzyme Activity[3] (eu/ml) | Specific Activity (eu/mg) | Total Enzyme Activity (eu) | Copper Content[4] (ng-atoms/ml) | Copper Content[4] (ng-atoms/mg protein) |
|---|---|---|---|---|---|---|
| High Speed Supernatant | 2.21 | 0.0048 | 0.0022 | 3.79 | 9.48 | 4.29 |
| Streptomycin-$SO_4$ Supernatant | 1.29 | 0.0058 | 0.0045 | 4.38 | 5.46 | 4.23 |
| $(NH_4)_2SO_4$ Precipitate | 9.00 | 0.0159 | 0.0018 | 0.86 | 59.36 | 6.60 |
| DEAE-Cellulose | 3.60 | 0.0217 | 0.0060 | 0.37 | 38.62 | 10.73 |
| Sephadex G-75 | 3.52 | 0.0834 | 0.0240 | 0.75 | 18.68 | 5.31 |
| Disc Acrylamide Gel | 0.38 | 0.0445 | 0.1170 | 0.067 | 6.44 | 16.96[5] |
| Isoelectric Focusing | 0.075 | 0.0233 | 0.3110 | 0.049 | --[6] | --[6] |

[1]Rabbit small intestines (Pel-Freez), 252 g from 10 intestines, were used as starting material. Details of purification will be published elsewhere.
[2]Determined turbidometrically (15).
[3]Assayed according to Hayaishi (1-5,8), using D-tryptophan.
[4]Determined by atomic absorption spectroscopy after ashing of samples in a muffle furnace for 24 hrs at 500°.
[5]Assuming a molecular weight of 58,000 Daltons, 1 g-atom Copper/mole protein is equal to 17.2 ng-atoms copper per mg protein.
[6]Not determined, insufficient sample.

atoms copper per mg protein and has a specific activity of 0.117 when assayed at pH 7.0 and 37°, using D-tryptophan as substrate. The enzyme is not homogeneous at this stage of purification, but it seems to be so after purification by isoelectric focusing (specific activity, 0.311). However, we have not yet obtained sufficient sample of enzyme from this step in purification to do copper analyses. Table II summarizes the heme and copper analyses of the enzyme as purified through the disc acrylamide gel stage. As can be seen, a near equivalence between heme and copper is found, the ratio of copper to heme being 0.79 ± 0.08. Calculating minimal molecular weights based on cofactor contents yields estimates quite close to the molecular weight of 58,000 ± 2,000 for the enzyme, as determined by molecular sieve chromatography (Figure 1). Analytical disc acrylamide gel electrophoresis of the enzyme (Figure 2) indicates that there is still some contamination remaining in the preparation at this stage of purification. Further purification can be obtained by isoelectric focusing as shown in Figure 3, the enzyme having a specific activity of 0.311 after this stage.

Table II

HEME AND COPPER ANALYSES OF INDOLEAMINE 2,3-DIOXYGENASE PURIFIED THROUGH THE DISC ACRYLAMIDE GEL STAGE

| | |
|---|---|
| Protein[1] | 0.95 mg |
| Specific Activity[2] | 0.117 eu/mg protein |
| Heme[3] | 21.6 ± 0.3 nmoles/mg protein |
| Copper[4] | 16.96 ± 0.9 ng-atoms/mg protein |
| Ratio, Copper/Heme | 0.79 ± 0.08 |
| Minimal Molecular Weight | |
| Based on Copper | 58,400 |
| Based on Heme | 46,300 |

[1]Determined turbidometrically (15).
[2]Assayed according to Hayaishi (1-5,8), using D-tryptophan.
[3]Dipyridine hemochromogen (16,17). $\varepsilon_M^{406}$ = 168,000.

[4]Atomic absorption spectroscopy on ashed samples, using internal standards to ascertain recovery (method of additions).

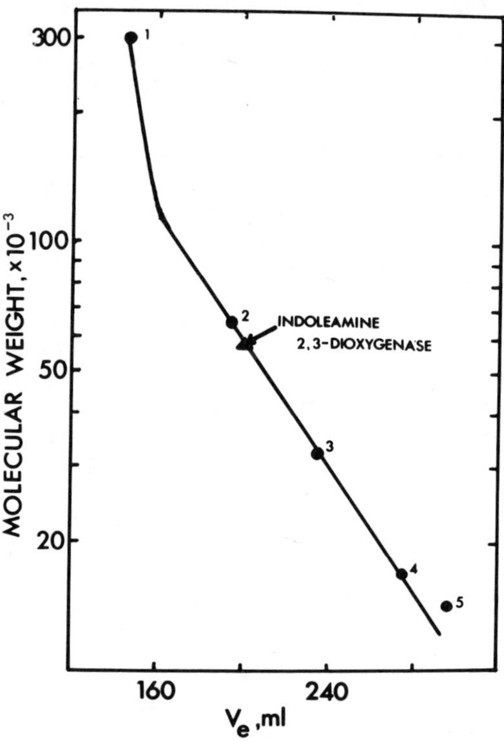


Figure 1. Sephadex G-75 column chromatography of indoleamine 2,3-dioxygenase. The column size was 2.5 x 95 cm. Marker proteins (●) were run simultaneously with the enzyme: 1, bovine milk xanthine oxidase, 300,000; 2, hemoglobin, 64,000; 3, Cu-Zn superoxide dismutase, 32,000; 4, myoglobin, 17,500; 5, ribonuclease, 14,000. The position of indoleamine oxygenase (▲) was determined by assaying enzymatic activity.

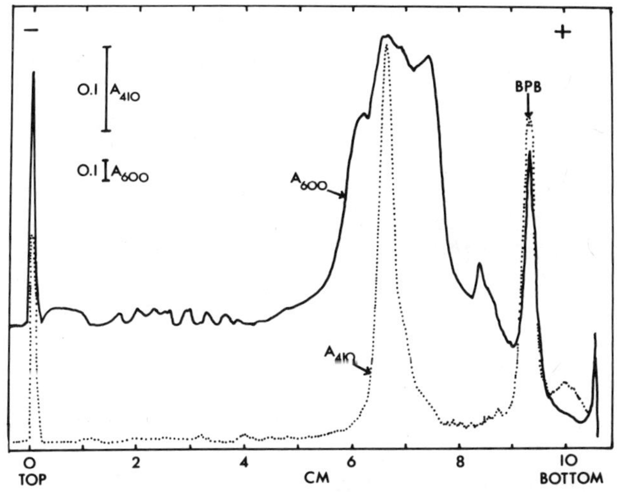


Figure 2. Analytical disc acrylamide gel electrophoresis of indoleamine 2,3-dioxygenase, purified through the preparative disc acrylamide gel stage. Electrophoresis was according to the method of Davis (18), using 7.5% gels, 0.6 x 10 cm. After electrophoresis the gels were scanned for heme ($A_{410}$, ············), stained with Coomassie blue, and rescanned for protein ($A_{600}$,————). BPB, bromphenol blue tracking dye.

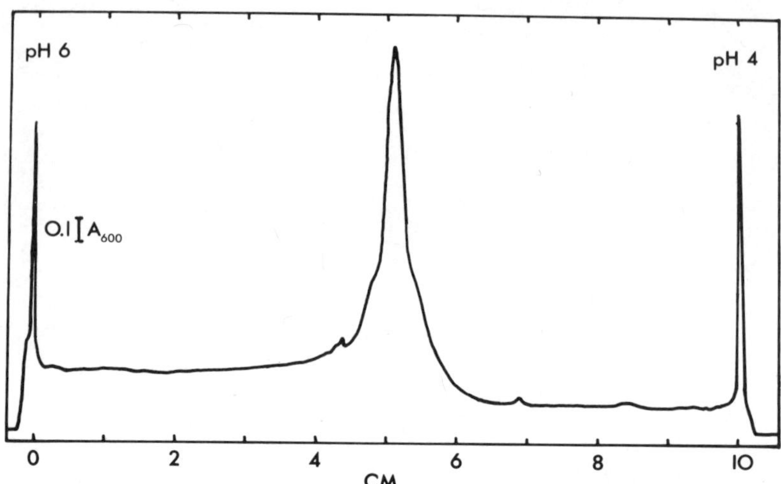


Figure 3. Polyacrylamide isoelectric focusing gel of indoleamine 2,3-dioxygenase, purified through the preparative disc acrylamide gel stage. Electrophoresis was according to the method of Swanson and Sanders (19) using 7.5% gels, 0.6 x 10 cm, with a pH 6 buffer in the upper reservoir and a pH 4 buffer in the lower reservoir. The gel was stained with Coomassie blue and scanned for protein ($A_{600}$, ———).

The visible absorption spectrum of the enzyme is shown in Figure 4. The oxidized enzyme has a Soret peak at 406 nm with a molar extinction coefficient of 168,000 per heme and a nondescript $\alpha,\beta$ region. In the presence of dithionite the Soret exhibits a hypochromic shift to 428 nm and a prominent peak at 556 nm emerges.

Observation of the enzyme during steady state catalysis revealed the following (Figure 5). When the enzyme was assayed in the presence of catalase, it exhibited the same absorption spectrum as that of the isolated ferriheme enzyme (Spectra 1 and 2). If catalase was omitted from the assay mixture, a shift of the Soret from 406 nm to 418 nm occurred (Spectrum 3). A slow, progressive inactivation of the enzyme also occurred during assay, which is prevented by catalase. The addition of cyanide to the assay containing catalase resulted in the cessation of catalysis and the formation of a cyanoferriheme enzyme complex (Soret, 421 nm) (2 + NaCN). These results are reminiscent of steady state catalysis by tryptophan oxygenase in which the half reduced enzyme, $[\text{ferriheme}]_2[\text{Cu(I)}]_2$, was shown to have an oxidized heme during catalysis (20). The possibility that indoleamine 2,3-dioxygenase functions as

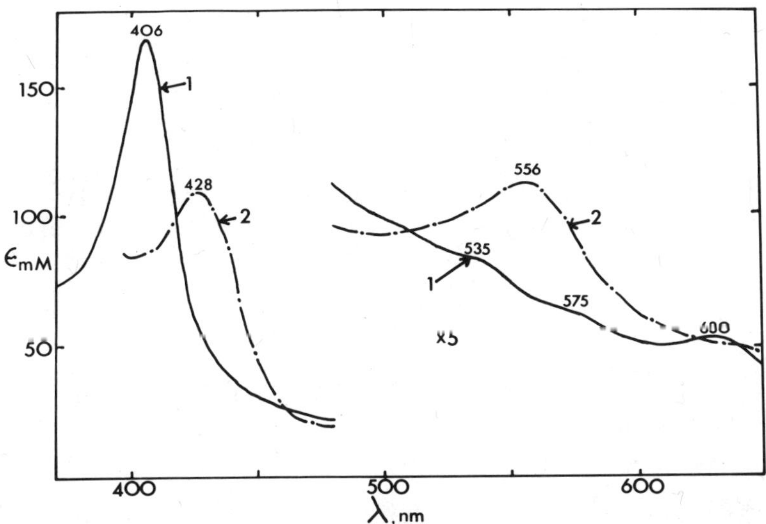


Figure 4. Visible absorption spectra of indoleamine 2,3-dioxygenase. Spectra were recorded on an Aminco DW-2 spectrophotometer at pH 7.0 and 20°. Oxidized enzyme, ——— ; enzyme reduced with dithionite, —·—·— .

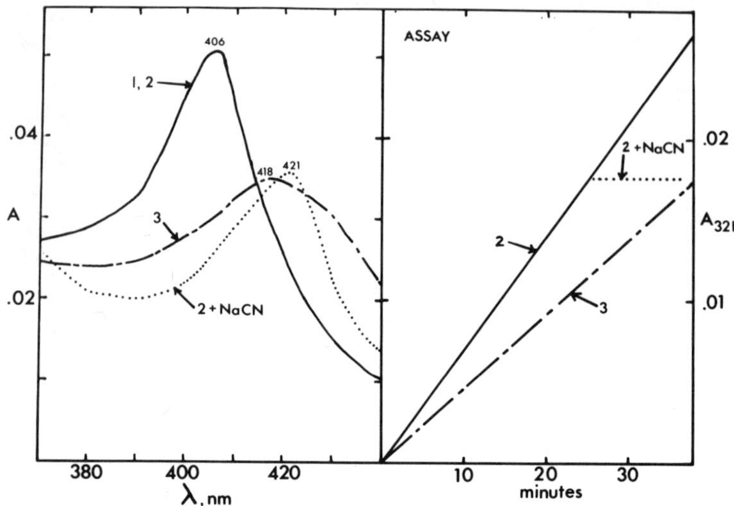


Figure 5. Steady state catalysis by indoleamine 2,3-dioxygenase. Spectra were recorded with an Aminco DW-2 spectrophotometer. Spectrum 1, oxidized enzyme in 0.1 M $KP_i$, pH 7.0. Spectrum 2 and assay 2, the same amount of enzyme as in 1, but in an assay mixture containing catalase (40 µg/ml); the reference cuvette contained an assay mixture minus enzyme. At the indicated time, NaCN (6mM) was added to assay 2, and its absorption spectrum was recorded (2 + NaCN). Spectrum 3 and assay 3 were the same as in 2, but catalase was omitted.

[ferriheme] [Cu(I)] during steady state catalysis is further suggested by inhibition studies using Cu(I) specific chelators (21). Both bathocuproinesulfonate and bathophenanthrolinesulfonate are uncompetitive inhibitors with respect to D-tryptophan of catalysis by indoleamine oxygenase ($K_i^{BCS}$ = 220 µM, $K_i^{BPS}$ = 54 µM, $K_i^{D-Trp}$ = 8.9 µM), indicating the presence of a functioning Cu(I) cofactor (Figure 6). That the inhibition by bathocuproinesulfonate is due to a specific complexing with Cu(I) is supported by the results shown in Figure 7. In this experiment the absorption spectrum of the enzyme was recorded during steady state catalysis in the presence of catalase. A second assay was run in which 100 µM BCS was included, and a spectrum was recorded. A difference spectrum was then calculated [(enzyme + BCS) minus enzyme alone], and as can be seen, it is that of a BCS-Cu(I) complex with a maximum near 475 nm.

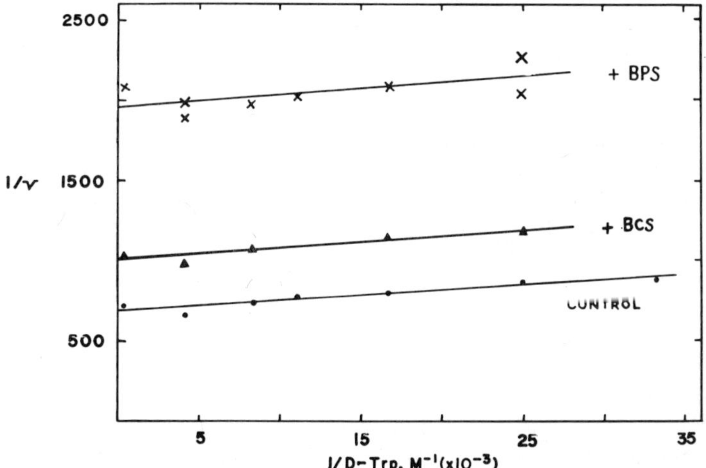


Figure 6. Lineweaver-Burk plot of the inhibition of catalysis by indoleamine 2,3-dioxygenase by bathocuproinesulfonate (BCS) and by bathophenanthrolinesulfonate (21). v, nmoles D-N-formylkynurenine synthesized $min^{-1}$. Control, ●; + BCS, ▲; + BPS, X .

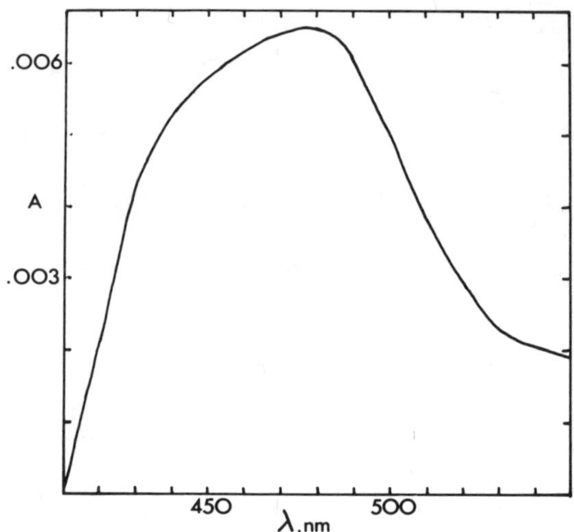


Figure 7. Calculated difference spectrum for indoleamine 2,3-dioxygenase and its complex with bathocuproinesulfonate. Spectra during steady state catalysis in the absence and presence of 100 μM BCS were recorded with an Aminco DW-2 spectrophotometer. Enzyme concentration was 0.35 μM based on heme. Copper-BCS complex concentration was 0.53 μM, based on $\varepsilon_M^{475}$ of 12,250 (22), although this may be an erroneous calculation due to the extinction coefficient being based on two molecules of BCS bound to one copper.

In summary we have obtained a nearly homogeneous preparation of rabbit intestinal indoleamine 2,3-dioxygenase. Analysis of our preparations for heme and copper suggests an equivalence in stoichiometry of the two cofactors, although we cannot unequivocally declare them to be cofactors until we obtain and analyze a homogeneous enzyme. Supporting the possibility that copper may be a cofactor of indoleamine oxygenase are the experiments with the cuprous chelators, bathocuproinesulfonate and bathophenanthrolinesulfonate, which inhibit an ongoing catalytic reaction. In the case of BCS the inhibition seems to be a specific one in which the inhibitor is indeed bound to Cu(I). Since the heme apparently remains in the ferriheme state during catalysis, our conclusion is that the enzyme functions as [ferriheme] [Cu(I)] during steady state catalysis, a state reminiscent of the half reduced form of L-tryptophan 2,3-dioxygenases (20). Future work must focus on the involvement of $O_2^-$ in catalysis and its binding site on the enzyme. The intriguing possibility presents itself that copper, possibly as Cu(II), may be the binding site for $O_2^-$, making it similar in nature to the eucaryotic Cu-Zn superoxide dismutases.

Further studies must also be devoted to an elucidation of the function of methylene blue in the *in vitro* catalytic reaction before any firm conclusions can be made as to a physiologically important catalytic mechanism.

Acknowledgements. We would like to thank the Research Committee of The University of South Dakota School of Medicine for the support of this research, the money for which came from the Parson's Fund for Medical Research and a General Research Support Grant from NIH.

References

1. Higuchi, K., and Hayaishi, O. (1967) *Arch. Biochem. Biophys.*, 120, 397-403.
2. Yamamoto, S., and Hayaishi, O. (1967) *J. Biol. Chem.*, 242, 5260-5266.
3. Hirata, F., and Hayaishi, O. (1971) *J. Biol. Chem.*, 246, 7825-7826.
4. Hirata, F., and Hayaishi, O. (1972) *Biochem. Biophys. Res. Commun.*, 47, 1112-1119.
5. Hirata, F., and Hayaishi, O. (1975) *J. Biol. Chem.*, 250, 5960-5966.
6. Loh, H.H., and Berg, C.P. (1971) *J. Nutr.*, 101, 465-475.
7. Loh, H.H., and Berg, C.P. (1973) *J. Nutr.*, 103, 397-406.
8. Hirata, F., Hayaishi, O., Tokuyama, T., and Senoh, S. (1971) *J. Biol. Chem.*, 249, 1311-1313.
9. Loh, H.H., and Berg, C.P. (1972) *J. Nutr.*, 102, 1331-1340.
10. Gal, E.M., Armstrong, J.D., and Ginsburg, B. (1966) *J. Neurochem.*, 13, 643-654.
11. Tsuda, H., Noguchi, T., and Kido, R. (1972) *J. Neurochem.*, 19, 887-889.
12. Nishikimi, M. (1975) *Biochem. Biophys. Res. Commun.*, 63, 92-98.
13. Feigelson, P., and Brady, F.O., in *Molecular Mechanisms of Oxygen Activation* (O. Hayaishi, Ed.), Academic Press, New York, 1974, pp. 87-133.
14. Brady, F.O. (1976) *Bioinorganic Chem.*, 5, in press.
15. Layne, E. (1957) *Methods Enzymol.*, 3, 447-454.
16. Paul, K.G., Theorell, H., and Akeson, A. (1953) *Acta Chem. Scand.*, 7, 1284-1287.
17. Falk, J.E., *Porphyrins and Metalloporphyrins*, American Elsevier, New York, 1964, pp. 181-182, 240-241.
18. Davis, B.J. (1964) *Ann. N. Y. Acad. Sci.*, 121, 404-427.
19. Swanson, M.J.,and Sanders, B.E. (1975) *Anal. Biochem.*, 67, 520-524.
20. Brady, F.O., and Feigelson, P. (1975) *J. Biol. Chem.*, 250, 5041-5048.
21. Brady, F.O. (1975) *FEBS Letters*, 57, 237-240.
22. Diehl, H., and Smith, G.F. in *The Copper Reagents: Cuproine, Neocuproine, Bathocuproine*, G.F. Smith Chemical Co., Columbus, Ohio, 1958, pp. 39-44.

COPPER CONTENT OF INDOLEAMINE 2,3-DIOXYGENASE

F. Hirata, T. Shimizu, R. Yoshida, T. Ohnishi,
M. Fujiwara and O. Hayaishi

Department of Medical Chemistry, Kyoto University
Faculty of Medicine, Kyoto, Japan

Indoleamine 2,3-dioxygenase is a hemoprotein which catalyzes the oxygenative ring cleavage of substituted and unsubstituted indoleamines such as tryptophan, 5-hydroxytryptophan, tryptamine and serotonin (1). This enzyme utilizes superoxide anion ($O_2^-$) as an oxidizing agent (2).

When indoleamine 2,3-dioxygenase was assayed in the presence of methylene blue and ascorbic acid at low concentrations, bathocuproinesulfonate, a chelating agent for Cu(I), inhibited the enzyme

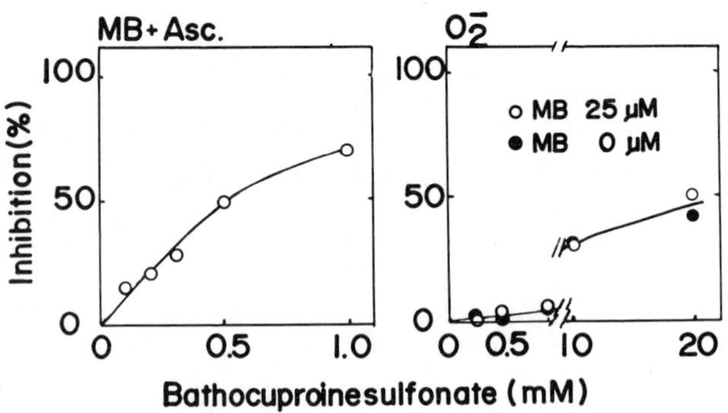


Fig. 1. Inhibition of enzyme by bathocuproinesulfonate

as described by Brady (3) (Fig. 1, left). On the other hand, when the new assay method described by Hayaishi et al. (4) was employed, the enzyme was not inhibited significantly by 1 mM bathocuproinesulfonate and a 40-fold higher concentration was required for 50% inhibition (Fig. 1, right). These results led us to suspect that the inhibition observed with bathocuproinesulfonate might not necessarily be due to specific chelation of Cu(I) contained in the dioxygenase.

The analysis of pure indoleamine 2,3-dioxygenase for copper would resolve the conflicting data. Therefore, we purified the enzyme from rabbit intestine. The final preparation of the enzyme was essentially homogeneous upon ultracentrifugation (Fig. 2). Polyacrylamide gel electrophoresis of the enzyme revealed a single broad protein band which was found to correspond to the enzyme activity. When the enzyme was treated with sodium dodecylsulfate, two protein bands appeared upon polyacrylamide gel electrophoresis, suggesting that the enzyme consists of subunits.

Using this homogeneous indoleamine 2,3-dioxygenase, copper was analyzed by the use of a carbon rod atomizer (flameless atomic absorption spectrophotometer, Varian Techtron model 63). The assay of the standard copper was linear over a range of 0.02 to 0.6 ppm (Fig. 3, left). When crude and pure preparations of enzyme, $E_1$ and $E_2$ respectively, were included in the standard copper solution, the absorbance was increased. However, the slope of plots of absorbance against the concentration of the standard copper was not affected. These results suggest that the inclusion of the protein did not interfere with the analysis of standard copper solutions, although Brady et al. have reported that low values were obtained if the protein was not wet-ashed first (5). The copper contents of a series of

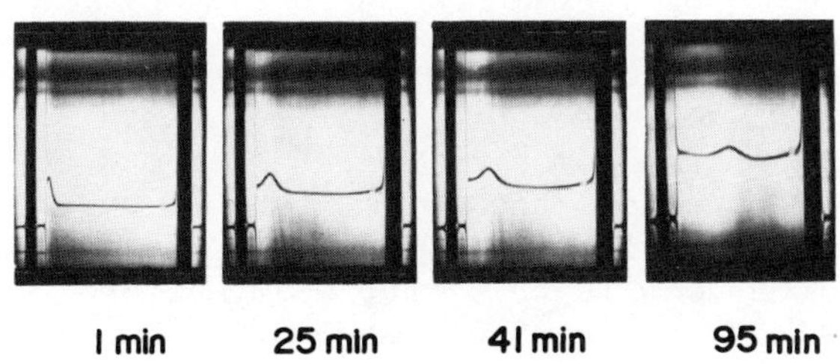


Fig. 2. Sedimentation pattern of enzyme

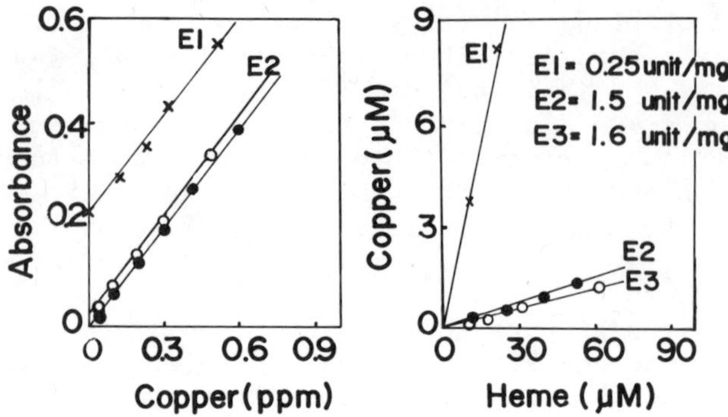


Fig. 3. Quantitative estimation of copper in enzyme preparation

concentrations of indoleamine dioxygenase were, then, measured, and the results were plotted against the enzyme-heme content obtained by determination of the absorbance at the Soret band ($\varepsilon_{\underline{mM}}^{406\ nm}$ = 142) (Fig. 3, right). (The enzyme was found to contain approximately 2 moles of protoheme IX per mole of enzyme (molecular weight = 105,000)). This value for the heme content was also confirmed by the determination of iron with a carbon rod atomizer. The ratios of copper to heme in two different preparations of pure enzyme with specific activities of 1.5 and 1.6 units per mg of protein, $E_2$ and $E_3$ respectively, were found to be about 0.03. On the other hand, a crude enzyme preparation with a low specific activity, $E_1$ (0.2 unit per mg of protein), gave a ratio of about 0.38.

Therefore, the levels of copper associated with indoleamine 2,3-dioxygenase were examined at each step of the enzyme purification (Table I). During the first four column chromatographies (P-cellulose, QAE-Sephadex, Sephadex G-200 and hydroxyapatite), the increase in specific activity of the enzyme paralleled an enrichment of copper in the protein. Consequently, the ratio of copper to iron progressively increased. The enzyme preparation from the hydroxyapatite chromatography step, which appeared to be more than 90% pure upon polyacrylamide gel electrophoresis, was found to contain both nonheme iron and heme iron, in a 1:1 ratio. However, the last step in the purification, the second chromatography on Sephadex G-200, reduced the copper content by about 95%, and the heme content became identical with the iron content. Most of the copper and iron lost in this step was recovered in the fractions which were eluted in front of indoleamine 2,3-dioxygenase, suggesting that a copper-rich protein was separated from indoleamine 2,3-dioxygenase.

In summary, we have purified indoleamine 2,3-dioxygenase to apparent homogeneity from rabbit intestine. Although the data on its molecular properties were not presented here, the enzyme was

Table I. Purification and copper content of enzyme

| | Specific Activity (munit/mg protein) | Copper (ng atom/mg protein) | Ratio of Cu to Fe |
|---|---|---|---|
| Crude Extracts | 11 | 0.70 | |
| Streptomycin | 18 | 0.68 | |
| Amm. Sulfate | 38 | 0.56 | |
| P-Cellulose | 45 | 0.35 | 0.06 |
| QAE-Sephadex | 139 | 0.87 | 0.16 |
| Sephadex G-200 | 857 | 5.0 | 0.19 |
| Hydroxyapatite | 1458 | 8.8 | 0.22 |
| Sephadex G-200 | 1700 | 0.57 | 0.038 |

found to have a molecular weight of about 105,000 and to consist of two each of two nonidentical subunits. The specific activity was about 2 katals per mole of enzyme. The dioxygenase contained approximately 2 moles of protoheme IX per mole of enzyme, but no significant amount of copper.

This investigation was supported in part by research grants from the Sakamoto Foundation, Iyakushigen Kenkyu Shinkohkai, the Mitsubishi Foundation, Nippon Shinyaku Co. Ltd., the Science and Technology Agency, Japan, and a Grant-in-Aid for Scientific Research from the Ministry of Education, Science and Culture, Japan.

## References

1. Hirata, F., and Hayaishi, O. (1972) Biochem. Biophys. Res. Commun. 47, 1112-1119

2. Hirata, F., and Hayaishi, O. (1975) J. Biol. Chem. 250, 5960-5966

3. Brady, O. F. (1975) FEBS Letters 57, 237-240

4. Hayaishi, O., Hirata, F., Ohnishi, T., Yoshida, R., and Shimizu, T., presented in this meeting

5. Brady, O. F., Manaco, E. M., Forman, J. H., Schutz, G., and Feigelson, P. (1972) J. Biol. Chem. 247, 7915-7922

# PSEUDOMONAD AND HEPATIC L-TRYPTOPHAN 2, 3-DIOXYGENASE

Philip Feigelson

The Institute of Cancer Research and the Department Biochemistry, College of Physicians and Surgeons, Columbia University, New York, New York 10032

Kotake and collaborators discovered the ability of liver preparations to convert L-tryptophan to L-kynurenine (1). Knox and Mehler found that this overall reaction was the resultant of the activities of two distinct enzymes the first of which was tryptophan oxygenase which catalyzed the conversion of L-tryptophan to L-formylkynurenine (2). A similar enzyme was identified by Hayaishi and Stanier within extracts of Pseudomonas grown in L-tryptophan (3). Tanaka and Knox found this enzyme to be a heme protein (4). Subsequently Feigelson and Greengard prepared the apoenzyme and demonstrated restoration of tryptophan oxygenase catalytic activity by the addition of heme (ferriprotoporphyrin IX) thus confirming the catalytically essential role of heme in the functioning of this enzyme (5). We have since purified tryptophan oxygenase to homogeneity from both rat liver and induced pseudomonas (6, 7). Our laboratory also demonstrated the presence of two moles of copper and two moles of heme per tetremer of enzyme and have produced other evidence suggesting a catalyic function for the copper as well as the heme moiety (8, 9, 10). In contrast, reports from the laboratories of Hayaishi and Ishimura indicate that they have purified, from induced pseudomonas, preparations of L-tryptophan oxygenase of high specific activity which they report to be devoid of stoichiometric levels of copper (11). The presence and possible catalytic role of copper within this enzyme molecule is therefore uncertain.

L-tryptophan oxygenase is inducible in pseudomonas grown in media containing L-tryptophan as essentially the sole carbon source (3, 6, 8). In mammals the hepatic level and catalytic efficiency of this enzyme is subject to a variety of interesting control processes. Parenteral administration to rats of either glucocorticoidal steroid hormones or high levels of its substrate, L-tryptophan, results within a few hours in elevations in the hepatic level of the catalytic activity of this enzyme (8, 12, 13). The functional level of catalytic activity of this enzyme is thus subject to both hormonal and substrate induction in mammals. Immunochemical studies with anti-tryptophan oxygenase demonstrated that this increase in catalytic activity is accompanied by corresponding elevation in the level of the enzyme protein during both substrate or hormonal induction (14). Radioactive amino acid incorporation studies have shown that substrate induction is due to enzymic stabilization, i.e., a decreased rate of intracellular degradation of L-tryptophan oxygenase (15). Hormonal enzyme induction is due to a selective hormonally enhanced synthesis of this enzyme protein (16). Recent studies have demonstrated that hormonal enzyme induction is accompanied by a proportional elevation in the tissue level of the specific messenger RNA coding for tryptophan oxygenase whereas during substrate induction no alteration in mRNA for tryptophan oxygenase occurs (16, 17, 18). In addition to the aforementioned mechanisms which regulate the cellular level of tryptophan oxygenase additional controls exist which regulate the catalytic efficiency by which this enzyme molecule functions. This allosteric control is evidenced by sigmoidal saturation of psuedomonad and hepatic tryptophan oxygenases by the substrate, tryptophan (8, 19, 20, 21), feedback negative modulation of the hepatic enzyme by the tryptophan metabolite, 3 hydroxyanthranylic acid (21), and the existence of α-methyl-tryptophan which serves as a positive allosteric modulator of this enzymic activity (8). Recent evidence indicates that this allosteric behavior is dependent upon the ionization state of the enzyme (22); lasar stopped flow studies indicate that saturation of the allosteric site increases the the affinity of the catalytic site for the substrate tryptophan which in turn determines the rate of interaction of enzymic heme at the catalytic site with carbon monoxide and oxygen (23).

L-tryptophan dioxygenase has been purified to homogenity from rat liver (7) and pseudomonad sources (6). In each instance the enzyme is a tetramer. The rat liver enzyme has been shown to have the protomeric composition $\alpha 2 \beta 2$ where the α and β subunits each have molecular weights of approximately 43,000 (7). As mentioned above, in our laboratory we find two moles of heme and two moles of copper present per tetramer of purified enzyme from either microbial or hepatic origins (8).

Tryptophan oxygenase exists in more than one catalytically active oxidation state. When the enzyme is fully reduced, with its heme prosthetic group as ferroheme, it is catalytically active without reducing agents being added to the enzymic assay. The enzyme also exists in a fully oxidized state wherein the heme exists as ferriheme and requires addition of exogenous reductants to the assay medium for catalysis to proceed. We have recently demonstrated the existence of another form of the enzyme apparently of an intermediate oxidation state (24, 25). This latter enzymic form is generated by separation of tryptophan from the ferroheme enzyme by Sephadex G25 chromatography. The tryptophan-free ferroheme tryptophan oxygenase then undergoes rapid autooxidation generating a half reduced form of this enzyme in which the heme is in the oxidized ferriheme state but with 80% of the catalytic activity being expressed in the absence of exogenous reductants.

The oxygenated complexes of the two catalytically active forms of pseudomonad and rat liver L-tryptophan-2, 3-dioxygenase have been studied. In confirmation of Ishimura et al. (26) we observe that the fully reduced form of pseudomonad tryptophan oxygenase during steady state catalysis exists predominatly as the L-tryptophan ferroheme-$O_2$ enzyme complex ($\lambda$ max - 415 nm, 540 nm, 570 nm). However, during steady state catalysis by the half-reduced form of both the pseudomonad and hepatic enzymes, the predominant species present manifest absorption spectra indicative of ternary complexes in which all the heme exists as ferriheme (Soret, 407 nm), there being no trace of an enzyme ferroheme-$O_2$ complex.

Carbon monoxide is a competitive inhibitor with respect to molecular oxygen of catalysis by either the half-reduced or fully-reduced forms of pseudomonad tryptophan oxygenase. During steady state catalysis in the presence of an $O_2$:CO mixture, the fully reduced form of the enzyme exists as a mixture of the oxyferroheme (Soret 415 nm) and carboxyferroheme (Soret - 421 nm) enzyme complexes. However, if the same experiment is repeated with the half-reduced form of the pseudomonad enzyme, all of the enzyme is in the ferriheme state, even though CO is inhibiting this form of the enzyme to the same degre as it does the fully reduced form. We conclude that for the half-reduced form of pseudomonad tryptophan oxygenase the substrate, $O_2$, and the inhibitor, CO, are not binding to the heme moieties, but are bound elsewhere on the enzyme.

On the basis of the information available to us to date we

are willing to consider the possibility that in the partially reduced form of the enzyme the enzymic moiety which is binding and activating the oxygen is the enzymic copper which we find to be present in our enzyme preparations. We therefore consider the possibility that oxygen may serve as a bridging ligand between the enzymic heme and copper moieties. The valence state of the heme would determine its proximity to and the degree to which it interacts with oxygen.

Parallel steady state and inhibitor studies have apparently not been undertaken to date in the laboratory posessing tryptophan oxygenase preparations devoid of copper. Therefore at the present time it remains unknown whether similar ferriheme forms of tryptophan oxygenase which do not contain copper would or would not be capable of catalyzing the oxygenation of L-tryptophan to L-formylkynurenine while the heme remains in its oxidized state. Collaborative experiments between the American and Japanese laboratories are being planned which hopefully should resolve these questions and illuminate the nature of the underlying catalytic events.

## REFERENCES


1. Kotake, Y., and Masayama, T. (1936) Z. Physiol. Chem. 243, 237-244.
2. Knox, W.E., and Mehler, A.H. (1950) J. Biol. Chem. 187, 419-430.
3. Hayaishi, O., and Stanier, R.Y. (1951) J. Bacteriol. 62, 691-709.
4. Tanaka, T., and Knox, W.E. (1959) J. Biol. Chem. 234, 1162-1170.
5. Feigelson, P., and Greengard, O. (1961) J. Biol. Chem. 236, 153-157.
6. Poillon, W.N., Maeno, H., Koike, K., and Feigelson, P., (1969) J. Biol. Chem. 244, 3447-3456.
7. Schutz, G., and Feigelson, P. (1972) J. Biol. Chem. 247, 5327-5332.
8. Feigelson, P. and Brady, F.O. (1974) in Molecular Mechanisms of Oxygen Activation, O. Hayaishi (Ed), Academic Press, 87-133.
9. Brady, F.O., Feigelson, P., and Rajagopalan, K.V. (1973) Arch. Biochem. Biophys. 157, 63-72.
10. Brady, F.O., Monaco, M.E., Forman, H. J., Schutz, G., and Feigelson, P. (1972) J. Biol. Chem. 247, 7915-7922.
11. Ishimura, Y., and Hayaishi, O. (1973) J. Biol. Chem. 248, 8610-8612.
12. Knox, W.E., and Mehler, A.H. (1951) Science 113, 237-238.
13. Feigelson, P., Feigelson, M., and Greengard, O. (1962) Recent Progr. Horm. Res. 18, 491-512.

14. Feigelson, P. and Greengard, O. (1962) J. Biol. Chem. 237, 3714-3717.
15. Schimke, R.T., Sweeney, E.W., and Berlin, C.M. (1965) J. Biol. Chem. 240, 322-331.
16. Feigelson, P., Beato, M., Colman, P., Kalimi, M., Killewich, L., and Schutz, G. (1975) Recent Progr. Horm. Res. 31, 231-242.
17. Schutz, G., Killewich, L., Chen, G., and Feigelson, P. (1975) Proc. Nat'l. Acad. Sci. USA 72, 1017-1020.
18. Killewich, L., Schutz, G., and Feigelson, P. (1975) Proc. Nat. Acad. Sci. 72, 4285-4287.
19. Feigelson, P., and Maeno, H. (1967) Biochem. & Biophys. Res. Commun. 28, 289-293.
20. Forman, H.J., and Feigelson, P. (1972) J. Biol. Chem. 247, 256-259.
21. Schutz, G., Chow, E., and Feigelson, P. (1972) J. Biol. Chem. 247, 5333-5337.
22. Colman, P., Blanchet, S., Chow, E., and Feigelson, P. (1975) J. Biol. Chem., 250, 6208-6213.
23. Feigelson, P., Brady, F.O., and McCray, J.A. (1973) J. Biol. Chem. 248, 5267-5271.
24. Maeno, H., and Feigelson, P. (1967) J. Biol. Chem., 242, 596-601.
25. Brady, F.O., and Feigelson, P. (1975) J. Biol. Chem., 250, 5041-5048.
26. Ishimura, Y., Nazaki, M., Hayaishi, O., Tamura, M., and Yamazaki, I. (1970) J. Biol. Chem. 245, 3593-3602.

# ON THE PROSTHETIC GROUPS OF L-TRYPTOPHAN 2,3-DIOXYGENASE FROM PSEUDOMONAS: EVIDENCE FOR NONINVOLVEMENT OF COPPER IN THE REACTION

Yuzuru Ishimura and Osamu Hayaishi

From the Department of Biochemistry Keio University School of Medicine, Shinjuku-ku, Tokyo 160 and the Department of Medical Chemistry, Kyoto University Faculty of Medicine, Sakyo-ku, Kyoto 606, Japan

## Summary

The amounts of copper present in highly purified preparations of L-tryptophan 2,3-dioxygenase from *Pseudomonas fluorescens* have been shown to be negligible by six different methods of copper determination. It has also been demonstrated that, during the purification, the heme content of enzyme preparations increased in parallel with the specific enzyme activity, whereas that of copper decreased.

These results, together with the finding that the inhibitory effects of copper chelators on the enzyme could be attributable to some other action of these chemicals rather than to their chelating properties, indicate that copper is not an essential component of L-tryptophan 2,3-dioxygenase.

## Introduction

L-tryptophan 2,3-dioxygenase, commonly known as L-tryptophan pyrrolase, is a protohemoprotein which catalyzes the incorporation of molecular oxygen into L-tryptophan yielding L-formylkynurenine as the product (1-3). Since the enzyme is a unique dioxygenase containing protoheme as an essential component, attention has long been focused on the nature, properties and the role of the heme in the catalysis. It has been demonstrated that the enzyme with its heme in a ferrous state (Fe ) is catalytically active (1-3) and that the ferrous enzyme-L-tryptophan complex reacts with oxygen to form a ternary complex of oxygen, L-tryptophan and the ferrous enzyme (the oxygenated form), the decomposition of which is the rate determining step of the overall reaction (2-6). Thus, the role of heme in the catalysis has been unambiguously established (7). Attempts to find out other metal cofactors have been reported to be unsuccessful by Poillon et al. (10) and by us (8, 9).

On the other hand, Brady and Feigelson, and their coworkers recently reported that highly purified enzyme preparations both from Pseudomonas and rat liver contained significant amounts of copper (11-14). They also described that the enzyme activity could be inhibited by certain copper chelating agents such as diethyldithiocarbamate and bathocuproine sulfonate (11). More recently, they emphasized that the ferric form of the enzyme was also catalytically active and that no oxygenated form of the enzyme was observed during the steady state of the reaction. According to their view, the catalytically active enzyme contains 2 cuprous copper ($Cu^{+}$) and 2 ferric protoheme IX ($Fe^{3+}$) and the copper could be the site of oxygen binding (14, 15).

The purpose of this paper is to review our investigations on the prosthetic groups of L-tryptophan 2,3-dioxygenase in which particular emphasis is laid on the analyses of copper in highly purified and fully active enzyme preparations. Pertinent portions of this work have been reported previously (8, 9).

## Materials and Methods

Preparation of enzyme --- The enzyme was extracted and purified from cells of Pseudomonas fluorescens (ATCC 11299B) grown on L-tryptophan as the major carbon source. The details were already reported elsewhere (3, 16). The specific activity of the final preparation was usually above 18 units per mg protein. One unit of enzyme activity is defined as that amount which catalyzes the conversion of 1 μmole of L-tryptophan to L-formylkynurenine per min under the standard assay conditions (3, 16).

Determinations --- The standard assay system of the enzyme activity is a modification of that originally described by Tanaka and Knox (1) and contained 25 μmoles of L-tryptophan, 10 μmoles of L-ascorbate, 250 μmoles of potassium phosphate buffer (pH 7.0), and the enzyme in a final volume of 2.5 ml (3, 16). The assays were carried out at 24° under normal atmospheric conditions in a cuvette with a 1-cm light path. The reaction was initiated by the addition of the ferric enzyme and the increase in optical density at 321 nm due to the formation of formylkynurenine ($\varepsilon$=3750) was followed continuously by a spectrophotometer. When the ferrous form of the enzyme was used for the assay instead of the ferric form, ascorbate was omitted from the system. The methods for the preparation of the ferrous enzyme were previously described (3).

Protein was measured either by the biuret method (17) with crude enzyme preparations, by a turbidimetric method (18) or by the method of Warburg and Christian (19).

Copper was determined by either of the following methods: (a) the micromethod of Van De Bogart and Beinert (20) which involves wet ashing of the sample and solvent extraction; (b) the method of Poillon and Dawson (21) using cuproine and glacial acetic acid; (c) the direct addition of bathocuproine sulfonate after the reduction of the enzyme by the addition of sodium dithionite (11); the methods described by Sandell with dithizone after dry combustion (22); and atomic absorption (9). The hematin content of the enzyme preparation was determined as the pyridine-ferrohemochrome according to the method of Rieske (23).

Instruments --- Usual spectrophotometric measurements were carried out either with a Cary model 15 recording spectrophotometer, with a Union-Giken model SM-401 high sens.-spectrophotometer equipped with a digital data processor or with a Shimadzu model MPS 500 spectrophotometer. Atomic absorption spectrophotometers used were a Nippon Jarrel-Ash atomic absorption spectrometer model AA-70 with a total combustion hydrogen-air burner and a Perkin-Elmer Atomic Absorption Spectrophotometer model 403 with a heated graphite atomizer model HGA 2000.

Reagents --- For the metal analyses as well as the enzyme purification, water distilled with a stainless steel still and then deionized by passage through a column of Amberlite IR-120B and IRA 410 was further distilled with a glass still. A standard copper solution ($CuCl_2$ in 0.1 N HCl) was obtained from Wako Pure Chemical Industries, Ltd., Japan. Sodium diethyldithiocarbamate and sodium bathocuproine sulfonate were products of Nakarai Chemicals, Ltd., and Dojindo Co., Ltd., Japan, respectively. All other chemicals were of analytical grade.

## Results

Purification and copper contents of enzyme --- In Table 1 is shown the copper content of L-tryptophan 2,3-dioxygenase along with its purification. The starting enzyme preparation for the experiment was obtained by aluminum grinding of cells of *Pseudomonas fluorescens* grown on L-tryptophan, followed by the extraction with phosphate buffer, streptomycin treatment and the first ammonium sulfate fractionation. The enzyme sample thus prepared was subjected to further purification as depicted in the table and the contents of both copper and heme in each step were estimated by atomic absorption spectrometry and pyridine-ferrohemochrome method, respectively. The details of the purification method have been described elsewhere (3, 9).

Table 1. Purification and copper content of enzyme[a)]

| Purification step | Specific activity[b)] | Copper content[c)] | | Copper/Heme |
|---|---|---|---|---|
| | e.u./mg | ng atom/e.u. | ng atom/mg | |
| $(NH_4)_2SO_4$·Ppt. | 0.44 | 2.2 | 0.97 | --- |
| Heat treatment | 0.89 | 0.7 | 0.62 | --- |
| Second $(NH_4)SO_4$ fractionation | 2.77 | 0.3 | 0.83 | 0.52 |
| DEAE-cellulose | 12.2 | 0.14 | 1.71 | 0.23 |
| Third $(NH_4)_2SO_4$ fractionation | 11.3 | 0.13 | 1.46 | 0.18 |
| Sephadex G200 | 20.8 | 0.033 | 0.7 | 0.04 |

a) Data are taken from reference (9) by Ishimura and Hayaishi.
b) Protein was determined by a turbidimetric method (18).
c) Copper was determined by atomic absorption spectrometry.

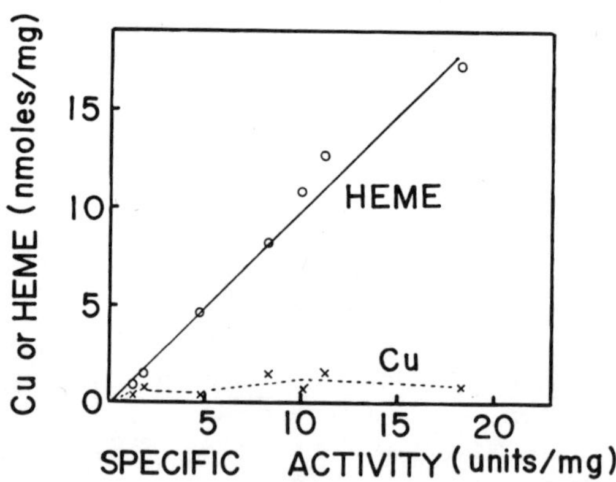


Fig. 1. Relationship between specific catalytic activity of L-tryptophan 2,3-dioxygenase and its copper or heme content. Copper and heme was determined by atomic absorption spectrometry and the method of Rieske (23), respectively. Protein was determined by the biuret method (17).

As seen in Table 1, the copper content of enzyme preparations did not increase in parallel with the specific enzyme activity. Although the amount of copper as expressed per mg protein increased slightly as purification proceeded, it decreased in the final step of the purification. Moreover, the amount of copper found decreased along with purification when expressed by either per unit of enzyme activity or per hematin molecule. On the other hand, the amount of heme in these preparations increased progressively in parallel with the specific enzyme activity. This relationship was also illustrated in Fig. 1.

Copper content in the purified anzyme preparations --- While the above determinations were performed exclusively by means of atomic absorption, the experiment in Table 2 were carried out with five different methods to show that the result was reproducible and independent of the methods employed. It can be seen that the values for copper in these preparations were about one-tenth or less of the heme content which was found to be approximately 2 moles per mole of enzyme. Somewhat higher values obtained by the Routine procedure after digestion may have been due to insufficient wet ashing, since the ashed samples which appeared colorless, became faint yellow upon the addition of sodium acetate to neutralize the acidic ashed solution. The yellow color was extractable with 1-hexanol and therefore gave erroneous optical absorption at 479 nm where the amounts of copper-bathocuproine complex were colorimetrically determined (20). When the time for ashing was increased from 3 to 10 min, no such coloration was observed and the values for copper were very small, as depicted in the Table. The insufficient ashing may have been due to the large amounts of organic materials applied in this experiment, since the sample contained a few mgs of Tris acetate and L-tryptophan in addition to enzyme protein. The micro method of Van De Bogart and Beinert was originally designed to estimate copper in organic materials of less than 2 mg (20). However, the lower values obtained by the prolonged ashing do not appear to be due to the evaporation of copper as recently rebutted by Brady (13). In order to avoid such a loss of copper during the prolonged ashing, hydrogen peroxide was used as oxidant in place of perchloric acid and the recovery of exogenously added copper either in a form of free ion or in a protein bound form was always found to be over 80% by control experiments. Superoxide dismutase from bovine erythrocyte and a *Pseudomonas* blue protein were used as authentic copper proteins.

The experiments with the routine procedure, namely 3 min for ashing as specified in the original paper by Van De Bogart and Beinert (20), were executed to compare our results directly with those of Brady et al. (11) who found 16.1 ng atoms of copper per mg of protein with a specific enzyme activity of 13.6 μmoles per mg of protein by this method. Colorimetric determination of copper by the direct addition of bathocuproine sulfonate to the reduced enzyme solution was carried out as described by Brady et al. (11) and the

Table 2. Copper content of highly purified L-tryptophan 2,3-dioxygenase as assayed by different methods

| Enzyme preparations | | | Amounts of copper found | | | | |
|---|---|---|---|---|---|---|---|
| | | | | Chemical, After digestion[a] | | Chemical, Without digestion | |
| Specific activity | Protein | Protoheme | Atomic absorption | Routine ashing | Prolonged ashing | BCS[b] | Cuproine[c] |
| e.u./mg | mgs | moles/mole of enzyme | g atoms/mole of enzyme[d] | | | | |
| 13.8 | 0.68 | 2.26 | 0.28 | 0.28 | --- | --- | <0.35 |
| 13.8 | 1.36 | 2.26 | 0.14 | 0.35 | 0.07 | --- | <0.18 |
| 13.8 | 2.72 | 2.26 | --- | 0.28 | --- | --- | --- |
| 20.8 | 1.45 | 2.08 | 0.08 | --- | --- | <0.16 | --- |
| 19.8 | 0.58 | 1.68 | --- | 0.25 | 0.06 | <0.46 | --- |
| 19.8 | 1.16 | 1.68 | --- | 0.16 | --- | <0.21 | --- |

a) Method a in the text (20).
b) BCS denotes bathocuproine sulfonate. Method c in the text (11).
c) Method b in the text (21).
d) The amount of enzyme was calculated assuming the molecular weight to be 122,000 (10).

method using cuproine in glacial acetic acid without digesting the protein is the same to that employed by Maeno and Feigelson (24) when they first discovered copper in their L-tryptophan 2,3-dioxygenase preparations. In any event, however, no significant amount of copper could be detected in our purified enzyme preparations.

It should be mentioned here that we have also tried to detect significant amounts of copper employing rather crude enzyme preparations by the methods other than those described heretofore. The amounts of copper found by the dithizone method after dry combustion (22) were approximately 0.2 and 0.7 ng atoms per enzyme unit with enzymes having specificactivity of 4.9 and 2.9, respectively. Here again, the amounts of copper were found to decrease as purification proceeded.

<u>The effect of copper-specific chelators on the enzyme activity</u> --- The inhibition of <u>Pseudomonas</u> L-tryptophan 2,3-dioxygenase by sodium diethyldithiocarbamate was first reported by Maeno and Feigelson (24) and subsequently by Brady et al. (11). Their results were found to be reproducible even with our copper deficient preparations and hence the mode of inhibition was examined as described below.

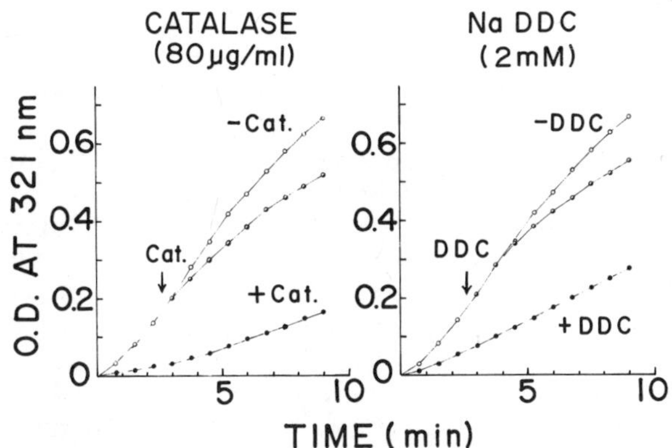


Fig. 2. The inhibition of L-tryptophan 2,3-dioxygenase by sodium diethyldithiocarbamate (DDC) and catalase (Cat). The reaction was carried out under the standard assay conditions. DDC and catalase were added to the reaction mixture either before the reaction was started by the addition of enzyme (+Cat and +DDC) or at the time indicated by arrows.

We found that the mode of inhibition by sodium diethyldithiocarbamate was identical with that of catalase which prevents reductive activation of the inactive ferric enzyme by removing hydrogen peroxide from the medium but has no effect on the enzyme activity of the ferrous enzyme. Both catalase and sodium diethyldithiocarbamate inhibited the reaction when added before the start of reaction but had little effect on the enzyme activity, if added during the steady state of reaction. The results were shown in Fig. 2. The effect of ascorbate, which is always included in the standard assay medium as an activator, is attributable to its ability to generate hydrogen peroxide through autooxidation (1). Thus, it was anticipated that sodium diethyldithiocarbamate might inhibit the enzyme by removing hydrogen peroxide and this possibility was affirmed by the following experiment (9).

When sodium diethyldithiocarbamate was incubated with hydrogen peroxide at a slightly alkaline pH, its ultraviolet absorption spectrum ($\lambda$max; 257 and 282 nm) changed to that of a new compound ($\lambda$max; 263 and 325 nm) with an isosbestic point at 295 nm. Kinetic analysis of this spectral change revealed that the reaction is first order with respect to either sodium diethyldithiocarbamate or $H_2O_2$, and that the second order rate constant at pH 7.2 and 24° is 130 $M^{-1}min^{-1}$. From the latter value, the half-life of $H_2O_2$ in the presence of 5 mM sodium diethyldithiocarbamate, the concentration required for the half-maximal inhibition of the enzyme activity, was expected to be approximately 1 min. Thus, the copper-specific chelator, sodium diethyldithiocarbamate, was found to be an effective trapping agent for $H_2O_2$ and the inhibition of L-tryptophan 2,3-dioxygenase can be attributable to this action of diethyldithiocarbamate.

The inhibition by another copper-chelating agent, bathocuproine sulfonate, was also described by Brady et al. (11) who reported that the inhibition was competitive with respect to L-tryptophan and that the addition of bathocuproine sulfonate to their reduced enzyme preparation resulted in spectral changes of the enzyme attributable to the formation of a bathocuproine-cuprous ion complex. Although their findings on the inhibition of enzyme activity were reproducibl we failed to confirm the latter finding on the spectral changes of the enzyme induced by the addition of bathocuproine sulfonate. Thus, the inhibition by bathocuproine sulfonate appeared to occur regardles of the formation of the copper-chelator complex and this led us to suspect the chaotropic action of the reagent. Along with this line of concept, we examined the effects of some other sulfonated aromatic compounds such as toluidinyl naphthalene sulfonate and anilinonaphthalene sulfonate, which are unlikely to be strong chelators and are known as hydrophobic probes and found that they were potent inhibitors of L-tryptophan 2,3-dioxygenase. Chemical structures of these sulfonated aromatic compounds as well as their concentrations required for the half maximal inhibition of the enzyme

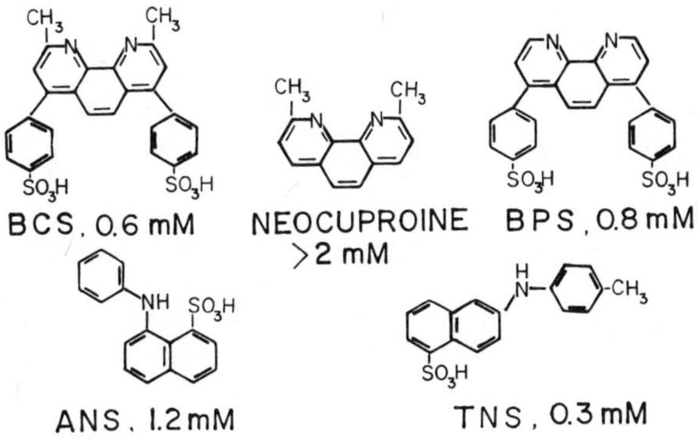

BCS, 0.6 mM NEOCUPROINE >2 mM BPS, 0.8 mM

ANS, 1.2 mM TNS, 0.3 mM

Fig. 3. Some sulfonated aromatic compounds and their concentrations required for the half-maximal inhibition of L-tryptophan 2,3-dioxygenase. Abbreviation used are: BCS, Bathocuproine sulfonate; BPS, Bathophenanthroline sulfonate; ANS, Anilinonaphthalene sulfonate; TNS, Toluidinylnaphthalene sulfonate.

activity under the standard assay conditions were illustrated in Fig. 3. Thus, the inhibition by bathocuproine sulfonate may not be necessarily due to its chelating action but can be due to some hydrophobic interaction with the enzyme protein.

## Discussion

It is well known that the quantitative determination of heavy metals, especially that of copper in the protein, meets several difficulties which often lead to erroneous results. The possible cause for such misleadings can be summarized as follows:

(1) Insufficient ashing of the organic materials.
(2) Loss of copper during the procedure, especially evaporation on ashing.
(3) Interferences caused either by copper chelating substances remaining in the system or by other heavy metals.
(4) Errors in standardization.
(5) Special state of copper in the sample.

However, none of them seems to be the case in our determinations. We have employed six different methods of copper estimation using either native enzyme, dry ashed or wet ashed samples in which a set of control run was always carried out to check the loss of copper and reagent blanks. It should be emphasized that the recovery of authentic copper either added as a free ion or a protein bound form to the enzyme was always satisfactory. Thus, the results presented in this paper shows that copper is not a constituent of our L-tryptophan 2,3-dioxygenase preparations.

On the other hand, the data reported by Feigelson, Brady and their coworkers (11-15) clearly indicate that their preparations of L-tryptophan 2,3-dioxygenase contain significant amounts of copper. Two possible explanations seem to exist for the above discrepancy: (1) Their purified preparations contain still contaminant copper proteins, the molecular properties of which are very similar to those of L-tryptophan 2,3-dioxygenase. (2) Copper is necessary to keep the structural integrity of the enzyme but is not essential to the catalytic activity. There may be some differences between the stability of their enzyme preparation and ours. In any event, the final conclusion as to whether L-tryptophan 2,3-dioxygenase is a copper-hemoprotein or not, must be deferred until either the resolution of copper from their preparation and the reconstitution with restoration of the activity is achieved or the catalytic significance of copper is established.

Acknowledgements --- This investigation has been supported in part by the Scientific Research Fund of the Ministry of Education of Japan, and grants from the Waksman Foundation of Japan, and the Naito Foundation. One of the authors (Y. I.) is grateful to the Toray Research Foundation for a travel grant to attend this symposium

## References

1. Tanaka, T., and Knox, W. E., J. Biol. Chem., 234, 1162 (1959)

2. Hayaishi, O., Proceedings of the Plenary Sessions, Sixth International Congress of Biochemistry 1964, IUB Vol. 33, p.31

3. Ishimura, Y., Nozaki, M., Hayaishi, O., Tamura, M., Nakamura, T. and Yamazaki, I., J. Biol. Chem., 245, 3593 (1970)

4. Ishimura, Y., Nozaki, M., Hayaishi, O., Tamura, M., and Yamazaki I., J. Biol. Chem., 242, 2574 (1967)

5. Ishimura, Y., Nozaki, M., Hayaishi, O., Tamura, M., and Yamazaki I., Advan. Chem. Ser., 77, 235 (1968)

6. Ishimura, Y., Nozaki, M., Hayaishi, O., Tamura, M., and Yamazaki I., in K. Okunuki, M. D. Kamen, and I. Sekuzu (Editors), Structure and function of cytochromes, University of Tokyo Press Tokyo, 1968, p.188

7. Hayaishi, O., Ishimura, Y., Fujisawa, H., and Nozaki, M., On the oxygenated intermediate in enzymic oxygenation. In T. E. King, H. S. Mason and M. Morrison (Editors), Oxidases and Related Redox Systems, University Park Press, Baltimore Vol. I, p.125 (1971)

8. Ishimura, Y., Okazaki, T., Nakazawa, T., Ono, K., Nozaki, M., and Hayaishi, O., in K. Bloch and O. Hayaishi (Editors), Biological and chemical aspects of oxygenases, Maruzen Company, Ltd., Tokyo, 1966, p.416

9. Ishimura, Y., and Hayaishi, O., J. Biol. Chem., 248, 8610 (1973)

10. Poillon, W. N., Maeno, H., Koike, K., and Feigelson, P., J. Biol. Chem., 244, 3447 (1969)

11. Brady, F. O., Monaco, M. E., Forman, H. J., Schutz, G., and Feigelson, P., J. Biol. Chem., 247, 7915 (1972)

12. Brady, F. O., Feigelson, P., and Rajagopalan, K.V., Arch. Biochem. Biophys., 157, 63 (1973)

13. Brady, F. O., J. Biol. Chem., 250, 344 (1975)

14. Feigelson, P., and Brady F. O., in O. Hayaishi (Editor), Molecular mechanisms of oxygen activation, Academic press, New York, 1973, p.87

15. Brady, F. O., and Feigelson, P., J. Biol. Chem., 250, 5041 (1975)

16. Ishimura, Y., Methods Enzymol., 17A, 429 (1970)

17. Layne, E., Methods Enzymol., 3, 450 (1957)

18. Feigelson, P., Ishimura, Y., and Hayaishi, O., Biochim. Biophys. Acta, 96, 283 (1965)

19. Warburg, O., and Christian, W., Biochem. Z., 298, 150 (1938)

20. Van De Bogart, M.,and Beinert, H., Anal. Biochem., 20, 325 (1967)

21. Poillon, W. N., and Dawson, C. R., Biochim. Biophys. Acta., 77, 1195 (1955)

22. Sandell, B. E., Colorimetric determination of traces of metals, Interscience Publishers, Inc., New York, 2nd Ed., 1950, p.295

23. Rieske, J. S., Methods Enzymol., 10, 488 (1967)

24. Maeno, H., and Feigelson, P., Biochem. Biophys. Res. Commun., 21, 297 (1965)

# THE SEARCH FOR COPPER IN L-TRYPTOPHAN 2,3-DIOXYGENASES

Frank O. Brady
Department of Biochemistry
School of Medicine
The University of South Dakota
Vermillion, South Dakota 57069

L-Tryptophan 2,3-dioxygenase (EC 1.13.11.11) catalyzes the formation of N-formylkynurenine from L-tryptophan and $O_2$ (Figure 1) (see references 1, 2 for reviews). The enzymes from L-tryptophan induced *Pseudomonas acidovorans* (ATCC 11299b) and from L-tryptophan and cortisone induced rat liver have been purified to homogeneity (3,4). Both proteins are allosteric enzymes composed of tetramers of subunits of equal size, having a total molecular weight of 122,000 for the bacterial enzyme and 167,000 for the hepatic enzyme. Before tryptophan oxygenase was purified to homogeneity it was established as a heme protein, based on inhibitor and spectral studies (5) and the demonstration of a restoration of catalytic activity to hepatic apo-tryptophan oxygenase by the addition of ferriprotoporphyrin IX (6-8). Both the bacterial and hepatic enzymes contain two g-atoms copper per mole enzyme, as functioning cofactors. The possible presence of copper in these enzymes has been a source of dispute amongst several laboratories for many years. In this paper I will present historically the evidence for copper being a cofactor of tryptophan oxygenase and will attempt to rebut the evidence for copper not being a cofactor.

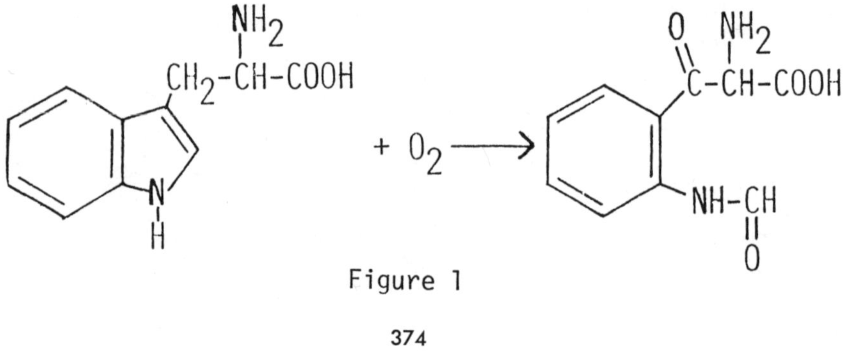

Figure 1

Maeno and Feigelson in 1965 were the first to indicate that copper was a component of pseudomonad tryptophan oxygenase(9). During purification of the enzyme they showed a progressive enrichment of copper with respect to protein. In addition the effects of various metal chelators on the catalytic activity of tryptophan oxygenase were surveyed. Diethyldithiocarbamate, salicylaldoxime, cuprizone, and bathocuproinesulfonate, all markedly inhibited catalysis, whereas 8-hydroxyquinoline, o-phenanthroline, α,α'-dipyridyl, and EDTA did not inhibit under similar conditions. They proposed,"At this stage of our knowledge one may suspect that the substrates of tryptophan pyrrolase, oxygen and tryptophan, are bound respectively to the cuprous and hematin components of the enzyme."

In the same year Ishimura, *et al.* reported that they could not find copper in pseudomonad tryptophan oxygenase in quantities stoichiometric with the heme content (10). Their preparation of enzyme was not yet pure, but examination of their data reveals a variability in copper content from sample to sample. In our experience the copper in pseudomonad tryptophan oxygenase can be quite labile, especially when the enzyme is not purified in the presence of high levels of L-tryptophan (circa 0.01 M), as Ishimura, *et al.* used to do.

The first homogeneous preparation of tryptophan oxygenase was obtained from *P. acidovorans* in 1969 as reported by Poillon, *et al.* (3). This enzyme was shown to be a tetrameric protein, containing less copper than heme. Using an extinction coefficient from the literature, they mistakenly calculated a stoichiometry of one heme per mole enzyme. They obtained low readings in their copper analyses, because copper was determined by atomic absorption spectroscopy on unashed protein samples. We now know that the amount of protein present in such samples resulted in severe interference and low determinations, when assayed by conventional flame atomic absorption spectroscopy. Because of this problem Poillon, *et al.* concluded that copper was not a cofactor of pseudomonad tryptophan oxygenase.

Our interest in copper in tryptophan oxygenase was resurrected because of two different observations. The first was concerned with the problem of reductive activation of tryptophan oxygenase. Although no electrons are lost or gained in the formation of N-formylkynurenine from L-tryptophan and $O_2$, certain enzyme preparations manifested a requirement for a reducing agent in order to elicit catalytic activity. Hayaishi's group and Knox's group thought this was because only a reduced heme on the enzyme could bind $O_2$, and indeed an L-tryptophan-oxyferroheme enzyme complex could be demonstrated in the presence of a reductant (11, 12). Maeno and Feigelson were the first to show that another interpretation of reductive activation could be made (13). They treated a catalytically inactive ferriheme form of pseudomonad tryptophan oxygenase with L-tryptophan and ascorbate, generating a fully active ferroheme enzyme.

Upon removal of ascorbate and L-tryptophan by chromatography on Sephadex G-25, the enzyme rapidly reoxidized to a ferriheme state, but it still retained full catalytic activity. It was concluded that another component of the enzyme, presumably copper, was undergoing oxidation-reduction, resulting in a requirement for reductive activation *in vitro*. Further work by me on the processes of reductive activation of tryptophan oxygenase confirmed the artifactual nature of reductive activation (14). The second observation (15) which led us back to copper was the analysis for metals in a homogeneous preparation of rat liver tryptophan oxygenase (4). Initially this was done spectrographically and by atomic absorption, courtesy of Dr. Bert Vallee, and revealed an equivalence of iron and copper. After heme analysis the stoichiometry for this enzyme was established to be two hemes and two coppers per mole enzyme. Also, in this paper analyses for metals in both enzymes by several techniques was reported. Copper was detected in tryptophan oxygenase as the bathocuproine complex after wet ashing of the protein, using $H_2O_2$ as an oxidant, by atomic absorption spectroscopy after ashing, as the bathocuproinesulfonate and bathophenanthrolinesulfonate complexes of the holoenzyme, and by inhibition of catalytic activity and of reductive activation by copper chelators (15).

An EPR study in collaboration with K.V. Rajagopalan at Duke University examined pseudomonad tryptophan oxygenase (16). The enzyme was revealed to contain a high spin ferriheme and Cu(II), if it had been oxidized by treatment with ferricyanide. As the enzyme was reductively activated, the Cu(II) signal disappeared, presumably because of the formation of Cu(I) in active enzyme molecules. The EPR spectrum of Cu(II) in inactive pseudomonad tryptophan oxygenase is shown in Figure 2. Using the analysis of Blumberg and Peisach

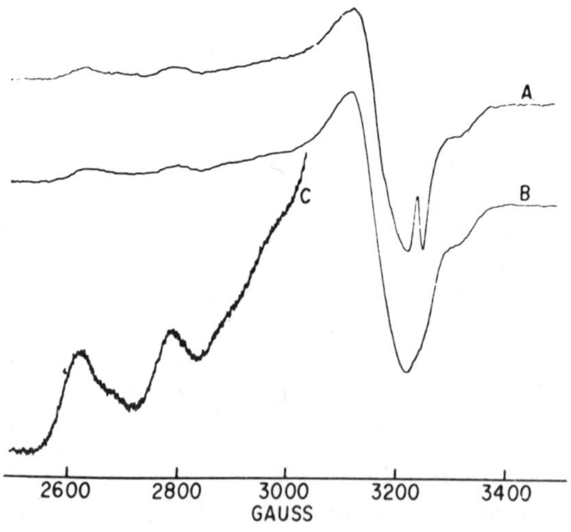


Figure 2. EPR spectrum of Cu(II) in pseudomonad tryptophan oxygenase (16). $g_{\perp} = 2.065$, $g_{\parallel} = 2.265$, hyperfine splitting const. 170 G.

(17), this copper is seen to be similar in nature to the "non-blue" coppers in dopamine β-hydroxylase and in ceruloplasmin in which the copper is ligated by two to four nitrogens and zero to two oxygens. In Figure 3 is shown an EPR spectrum from 0 to 5000 gauss of pseudomonad tryptophan oxygenase after the addition of 2.67 mM L-tryptophan. A new spectrum with $g_{1,2,3}$ = 2.66, 2.20, 1.81, ascribable to low spin ferriheme, appears with a concomitant disappearance of about half of the high spin ferriheme. This low spin ferriheme signal appears in a nonhyperbolic or sigmoidal manner during a titration with L-tryptophan, as is shown in Figure 4, with a Hill coefficient comparable to that seen when catalytic activity is determined as a function of L-tryptophan concentration. The position of the g-values of the low spin ferriheme signal is consistent with the possibility that the indole nitrogen of the substrate is ligated to the sixth position of the heme iron (18).

In 1973 Ishimura and Hayaishi reported that they were only able to detect trace amounts of copper in their nearly homogeneous preparations of pseudomonad tryptophan oxygenase, although they found the correct amount of heme (19). They seem to have repeated some earlier errors in the analysis of copper previously encountered and overcome in Feigelson' laboratory, namely, (a) conventional atomic absorption analysis of unashed samples containing interfering amounts of protein, (b) analysis of wet ashed samples by the bathocuproine method, using $HClO_4$ as an oxidant, leading to volatilization of copper perchlorate ($H_2O_2$ is used by us instead), and (c) use of the neocuproine-acetic acid method which does not release all of the copper from tryptophan oxygenase, and with which heme

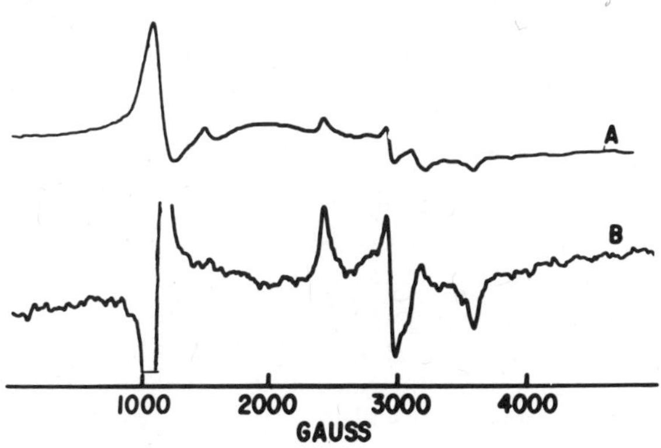


Figure 3. Effect of L-tryptophan (2.67 mM) on the EPR spectrum of pseudomonad tryptophan oxygenase (16). A, derivative spectrum. B, difference spectrum. Receiver gain, 800. Modulation amplitude, 10 G. Temperature, -180°.

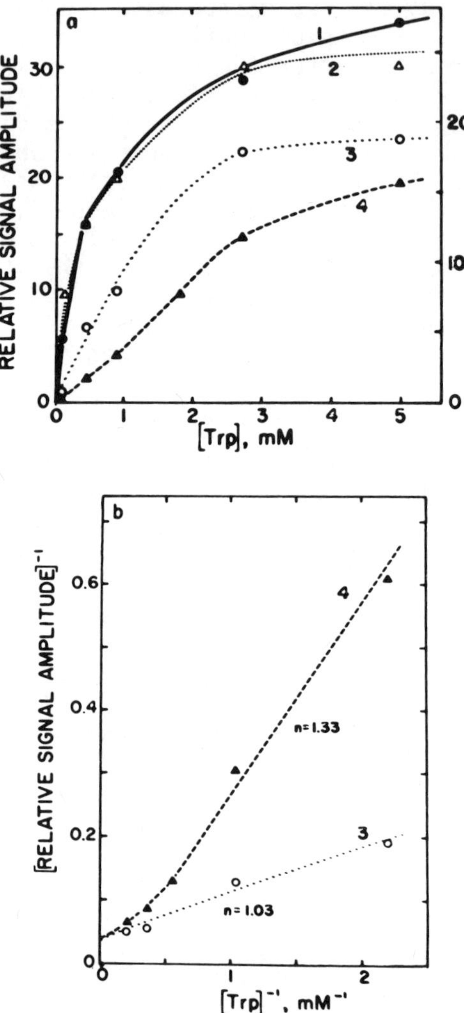


Figure 4. Titration with L-tryptophan of the disappearance of the high spin ferriheme and the appearance of the low spin ferriheme EPR signals in pseudomonad tryptophan oxygenase (16). In *a* curves 1 and 2 (left ordinate) are the high spin ferriheme signal; curves 3 and 4 (right ordinate) are the low spin ferriheme signal. D,L-$\alpha$-methyltryptophan (1.0 mM), an allosteric effector, was present in curves 2 and 3. In *b* Lineweaver-Burk plots of the low spin ferriheme signals are shown (curves 3 and 4), with n signifying the respective Hill coefficients.

interferes in the spectrophotometric determination of the neocuproine-copper complex. The only results of ours which they could reproduce was the inhibition of catalysis of tryptophan oxygenase by diethyldithiocarbamate and bathocuproinesulfonate.

As a result of this report, I was prompted in 1974 to do an experiment in which I grew *Pseudomonas acidovorans* in the presence of $^{64}Cu(NO_3)_2$ (20). Under appropriate conditions of induction by L-tryptophan, tryptophan oxygenase was synthesized, and following purification was shown to contain radiocopper. In Figure 5 are shown disc acrylamide gel profiles of tryptophan oxygenase at several stages of purification. As can be seen, a considerable amount of radiocopper is present coincident with the enzyme. Also, this peak of radiocopper could be diminished by prior treatment of the sample with anti-tryptophan oxygenase γ-globulins, but it could not be diminished by treatment with preimmune γ-globulins, indicating that the radiocopper present in this peak was specifically associated with the enzyme.

In summary although analysis for copper in pseudomonad tryptophan oxygenase has been full of detours and pitfalls, copper indeed seems to be a cofactor of this enzyme. The copper in rat liver tryptophan oxygenase is much less labile, and it has been easier to

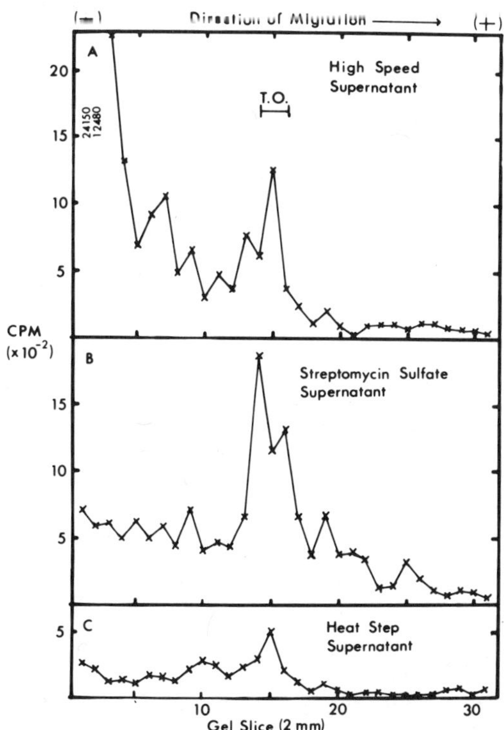


Figure 5. Disc acrylamide gel profiles of the $^{64}Cu$ in pseudomonad tryptophan oxygenase during several steps of purification (20).

analyze this protein. The recent discovery of three different oxidation states in tryptophan oxygenases has clarified the reductive activation question and shown the need for copper as a functioning cofactor of these enzymes (21, see paper by Feigelson in this meeting). Thus, we are presently able to prepare three different forms of tryptophan oxygenase: oxidized, $[\text{ferriheme}]_2[Cu(II)]_2$; half reduced, $[\text{ferriheme}]_2[Cu(I)]_2$; and fully reduced, $[\text{ferroheme}]_2[Cu(I)]_2$. The oxidized enzyme is catalytically inactive and the other two are catalytically active, but with differing mechanistic properties.

Acknowledgements. I would like to thank Dr. Philip Feigelson in whose laboratory I did a great deal of this work while I was a postdoctoral fellow. Support for this research has come from USPHS Grants CA 02332 and HL 17205, from a Training Grant CRTY-05011, and a Postdoctoral Fellowship F02-CA 45807.

References.

1. Feigelson, P., and Brady, F.O., in *Molecular Mechanisms of Oxygen Activation* (O. Hayaishi, Ed.), Academic Press, New York, 1974, pp. 87-133.
2. Brady, F.O. (1976) *Bioinorganic Chem.*, 5, in press.
3. Poillon, W.N., Maeno, H., Koike, K., and Feigelson, P. (1969) *J. Biol. Chem.*, 244, 3447-3456.
4. Schutz, G., and Feigelson, P. (1972) *J. Biol. Chem.*, 247, 5327-5332.
5. Tanaka, T., and Knox, W.E. (1959) *J. Biol. Chem.*, 234, 1162-1170.
6. Feigelson, P., and Greengard, O. (1961) *J. Biol. Chem.*, 236, 153-157.
7. Feigelson, P., and Greengard, O. (1961) *Biochim. Biophys. Acta*, 50, 200-202.
8. Greengard, O., and Feigelson, P. (1962) *J. Biol. Chem.*, 237, 1903-1907.
9. Maeno, H., and Feigelson, P. (1965) *Biochem. Biophys. Res. Commun.*, 21, 297-302.
10. Ishimura, Y., Okazaki, T., Nakazawa, T., Ono, K., Nozaki, M., and Hayaishi, O., in *Biological and Chemical Aspects of Oxygenases* (K. Bloch and O. Hayaishi, Eds.), Maruzen Press, Tokyo, 1966, pp. 416-422.
11. Ishimura, Y., Nozaki, M., Hayaishi, O., Nakamura, T., and Yamazaki, I. (1967) *J. Biol. Chem.*, 242, 2574-2576.
12. Ishimura, Y., Nozaki, M., Hayaishi, O., Nakamura, T., Tamura, M., and Yamazaki, I. (1970) *J. Biol. Chem.*, 245, 3593-3602.
13. Maeno, H., and Feigelson, P. (1967) *J. Biol. Chem.*, 242, 596-601.
14. Brady, F.O., Forman, H.J., and Feigelson, P. (1971) *J. Biol. Chem.*, 246, 7119-7124.
15. Brady, F.O., Monaco, M.E., Forman, H.J., Schutz, G., and Feigelson, P. (1972) *J. Biol. Chem.*, 247, 7915-7922.
16. Brady, F.O., Feigelson, P., and Rajagopalan, K.V. (1973) *Arch. Biochem. Biophys.*, 157, 63-72.

17. Peisach, J., and Blumberg, W.E. (1974) *Arch. Biochem. Biophys.*, 165, 691-708.
18. Blumberg, W.E., and Peisach, J., in *Probes of Structure and Function of Macromolecules and Membranes* (B. Chance, T. Yonetani, and A.S. Mildvan, Eds.), Academic Press, New York, 1971, Vol. 2, pp. 215-229.
19. Ishimura, Y., and Hayaishi, O. (1973) *J. Biol. Chem.*, 248, 8610-8612.
20. Brady, F.O. (1975) *J. Biol. Chem.*, 250, 344-347.
21. Brady, F.O., and Feigelson, P. (1975) *J. Biol. Chem.*, 250, 5041-5048.

# COMPARISON OF FUNCTION OF THE DISTAL BASE BETWEEN MYOGLOBIN AND PEROXIDASE

I. Yamazaki, Y. Hayashi, R. Makino, and H. Yamada

Biophysics Division, Research Institute of Applied Electricity, Hokkaido University, Sapporo 060

From the reactions of heme-substituted hemoproteins and their heme-linked protonation, the following two possibilities are suggested. 1. There is a proton-dissociable amino acid residue in the vicinity to the sixth ligand of horseradish peroxidase. 2. Some of the functional differences between myoglobin and peroxidase can be explained in terms of the differences in the interaction of the distal base with the sixth ligand.

## I. HEME-LINKED PROTONATION IN HORSERADISH PEROXIDASE A AND C

The presence of a base with a $pK_a$ = ca. 7 in reduced horseradish peroxidase has been suggested by Theorell[1)] and Harbury[2)]. We have confirmed this result by four different experimental methods and the results are summarized in Table I. The results indicate that the same proton-dissociable group is involved in the reactions of each peroxidase isoenzyme and also that the proton dissociation constants of the two isoenzymes are different. Assuming that the proton dissociable group is located in the vicinity of the sixth ligand, the pH dependence of proton release in the reaction from the ferrous to ferric enzyme and proton uptake in the reaction from the ferrous to its oxygenated enzyme can be explained by the simple model shown in Scheme I[4)]. Participation of the distal base in reactions of catalase or peroxidase has been suggested by several workers[5-8)]. The results shown in Table I, however, can not necessarily exclude the possibility that the proton dissociation occurs at the proximal amino acid residue. The model in Scheme I in which the proton dissociation is ascribed to the distal amino acid residue will be further substantiated by the following two experimental results.

Table I. Heme-linked Protonation Constant of Horseradish Ferroperoxidases A and C[a)]

| Method | Isoenzyme A | Isoenzyme C | Reference |
|---|---|---|---|
| | | 7.38 | Harbury[2)] |
| pH dependence of redox potential | 5.8 | 7.32 | Yamada *et al.*[3)] |
| Proton release in the reaction from the ferrous to ferric form | 5.86 | 7.17 | Yamada *et al.*[3)] |
| Proton uptake in the reaction from the ferrous to its oxy-form | ---b) | 7.17 | Yamada *et al.*[3)] |
| Spectrophotometric titration | 5.87 | 7.07[c)]<br>7.25[c)] | Yamada and Yamazaki[4)] |

a Peroxidase A is a mixture of $A_1$ and $A_2$. Our peroxidase C preparation contains a small amount of isoenzyme B which is chemically very similar to isoenzyme C[21)].

b The experiment becomes difficult as the pH decreases.

c The basicity increases in the presence of 0.1 M KCl.

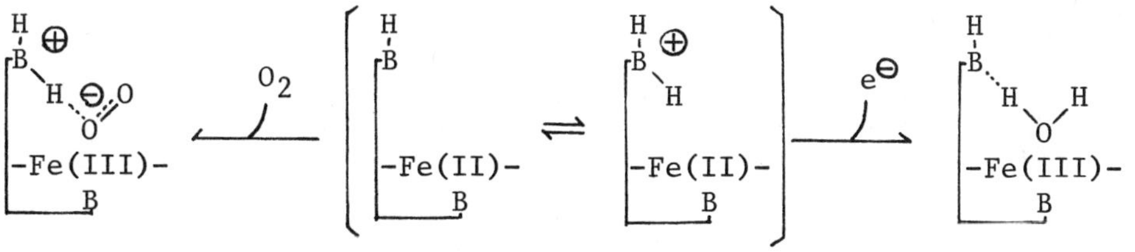

Scheme I

1. Effect of $pK_3$ on the Proton Dissociation Constant

The basicity of pyrrole nitrogen atoms ($pK_3$) can be controlled by substituents at the 2 and 4 positions of deuteroporphyrin IX. Hemes used in this experiment are shown in Table II. The proton dissociation constant of ferroperoxidases listed in Table I is termed here $K_r$. It is well known that a proton of a ferriperoxidase dissociates at certain pH forming its alkaline form. The dissociation constant is termed $K_o$. There proton dissociation reactions may be formulated as follows:

$$H^{+}PrFe(II) \underset{}{\overset{K_r}{\rightleftharpoons}} H^{+} + PrFe(II) \quad (1)$$

$$PrFe(III)OH_2 \underset{}{\overset{K_o}{\rightleftharpoons}} H^{+} + PrFe(III)OH^{-} \quad (2)$$

It has been shown[3,12] that the ratio $\Delta pK_r/\Delta pk_3$ equals 0.1 and $\Delta pK_o/\Delta pK_3$ equals 1.0. The result that $\Delta pK_o/\Delta pK_3 = 1.0$ seems reasonable because the proton dissociation from water molecule in Reaction (2) must be closely affected by the basicity of the pyrrole nitrogen atoms through the heme iron. On the other hand, the fact that $pK_r$ is almost independent of $pK_3$ will exclude the possibility that $K_r$ is ascribed to the proximal amino acid residue. It should be noted here that differences of $pK_r$ and $pK_o$ between horseradish peroxidases A and C are both about 1.5 pH units[3]. This close correlation can be also explained by the following model.

Scheme II

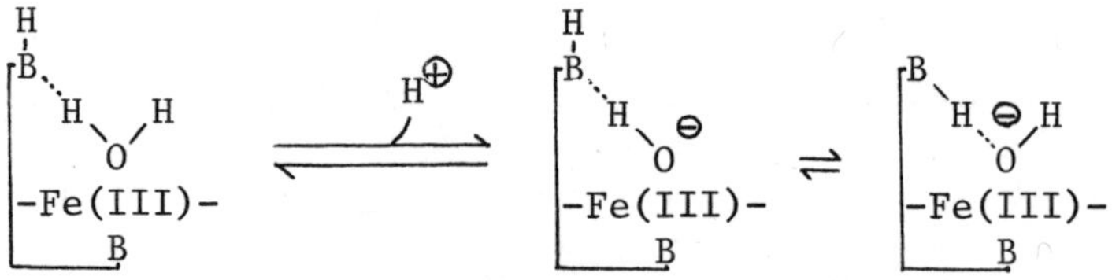

H–B···H–O(–H)–Fe(III)–B ⇌ (H⁺) H–B···H–O⁻–Fe(III)–B ⇌ B···H–O⁻–H–Fe(III)–B

2. Dependence of $pK_r$ upon the π* Energy Level of Diatomic Ligands[13]

For this particular experiment diacetyldeuteroperoxidase A is used because diacetyldeuteroperoxidases form the stable oxygenated forms[14] and $pK_r$ values of isoenzyme A derivatives are expected to lie in a measurable pH range. There are pH-dependent reversible changes in the absorption spectra of CO, NO and oxygenated

Table II

| Heme | Substituents at 2 and 4 positions | | $pK_3$[a] |
|---|---|---|---|
| Meso | 2,4 | $-CH_2CH_3$ | 5.8 |
| Deutero | 2,4 | -H | 5.5 |
| Proto | 2,4 | $-CH{=}CH_2$ | 4.8 |
| Chlorocruoro | 2<br>4 | -CHO<br>$-CH{=}CH_2$ | 3.75 |
| Diacetyldeutero | 2,4 | $-COCH_3$ | 3.3 |

a $K_3$ is a constant of proton dissociation from monocationic species of metal-free porphyrin[9,10].

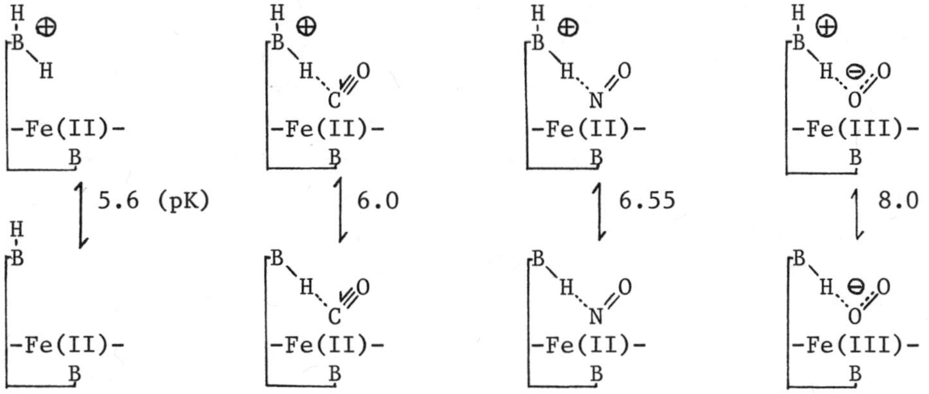

Scheme III

complexes of ferrous diacetyldeuteroperoxidase A. The proton dissociation constants can then be measured from spectrophotometric titration experiments. The results are summarized in Scheme III with possible structural models of the complexes. The value of 8.0 for the $pK_r(O_2)$ is also obtained from the pH dependence of proton uptake in the reaction of the ferrous enzyme with $O_2$. From the fact that the basicity is increased with a decrease of the $\pi^*$ energy level of these diatomic ligands[15)], it can be safely said that the proton dissociation does occur at the group located near the sixth ligand but not at the proximal group. It would be also suggested that the formal oxidation state of the iron atom is increased from +2 to +3 in the order of CO, NO and $O_2$ complexes. In Scheme III the formalism $Fe(III)O_2^-$ is tentatively used for the $O_2$ complex.

## II HEME-LINKED PROTONATION IN MYOGLOBIN

The spectrophotometric titration with myoglobin shows the presence of a heme-linked proton dissociable group[16)]. If the pH dependent changes of visible spectra of myoglobin and its CO complex are explained in the same way as for peroxidase, the data in Table III might indicate that in the CO-myoglobin complex, the interaction between the distal histidine and CO is too weak to increase the basicity of the imidazole group.

Table III. Comparison of Heme-linked Proton Dissociation between Myoglobin and Ferroperoxidases (Spectrophotometric titration)[16]

| | $pK_r$ | $pK_r$(CO) |
|---|---|---|
| Peroxidase A | 5.8 | 6.72 |
| Peroxidase C | 7.25 | 8.25 |
| Myoglobin | 5.57 | 5.67 |

## III. COMPARISON OF FUNCTION BETWEEN MYOGLOBIN AND PEROXIDASE

Differences of the reactivities between myoglobin and peroxidase listed in Table IV can be explained simply by assuming that the interaction between the distal base and the sixth ligand is much stronger in peroxidases than in myoglobin.

It is demonstrated in Scheme I that proton uptake occurs upon combination with $O_2$ or CO when the basicity of the distal base is increased by the ligand bound to the heme iron. The dependence of the $O_2$ affinity of myoglobin upon $pK_3$ clearly indicates that the binding is stabilized by the electron migration from the heme iron to $O_2$[17]. But, it seems likely that in the case of myoglobin the interaction between the distal imidazole and the negatively charged ligand is so weak that there is no measurable increase in the basicity of the imidazole group.

It has been accepted that autoxidation of the heme iron is prevented by the hydrophobic character of the polypeptide chain surrounding the iron. Though the problem is not wholly solved, it may be concluded that in the case of oxyperoxidases the distal base, by inducing charge separation in the $O_2$-iron bonding, makes the $O_2$ dissociation difficult and the autoxidation easy.

Table IV Functional Differences Between Myoglobin and Peroxidase

| | Myoglobin | Peroxidase |
|---|---|---|
| Bohr proton | no | yes |
| $O_2$ binding | reversible | irreversible |
| Autoxidation | slow | fast |
| Redox potential (at neutral pH) | high | low |
| Ligand bound | ionized form | acid form |

The redox potentials of horseradish peroxidases are known to be markedly low in comparison with other protoheme proteins. The cause may be due to the stabilization of the ferric form through hydrogen bonding between the distal base and the oxygen atom of the water molecule[3]. If one assumes that the bond is broken at extremely acidic pH, the redox potential would become close to the normal value of the protoheme complexes, such as myoglobin, cytochrome $b_5$ and pyridine protoheme[3].

From the pH dependence of the affinities of ferric hemoproteins for various ligands, high-spin protoheme proteins have been classified into two groups, one group which includes catalase and peroxidase which bind the acid form and the other group which includes hemeglobin and myoglobin which bind the anion form of the ligand[5,18,19] The difference has been interpreted in terms of the crevice structure[19] or the net charge on the heme iron[5]. For the time being, it is difficult to give a full explanation for the large number of existing experimental data. However, one may safely predict that the most important factor that differenciates the two types of ligand reactions is the nature of interaction between the distal base and the sixth ligand.

## IV CONCLUSION

The heme-linked protonation of a ferrous horseradish peroxidase is assigned to a distal amino acid residue. The conclusion is drawn from analyses of reactions that involve the protonation. Unfortunately it is difficult to appy it to the myoglobin case because no reaction has been found which is coupled with the protonation of the distal histidine. It is therefore of special interest to note that the pK value of 5.7 has been assigned to the distal histidine of metmyoglobin from binding kinetics with ligands[20].

It is well known that the reaction with hydrogen peroxide is quite different for the two types of hemoproteins. The distal base may be associated with the stabilization of the primary compound with hydrogen peroxide and also another amino acid residue may serve as a nucleophile for the stabilization of the $\pi$-cation radical of porphyrin in the case if peroxidases.

Numerous papers have dealt with heme substitution, heme linked protonation and reactions of hemoproteins related to the present subject. In this short paper, however, the discussion has mostly centered around the data obtained recently in our laboratory.

## REFERENCES


1. Theorell, H. (1947) Advan. Enzymol. 7, 265.
2. Harbury, H.A. (1957) J. Biol. Chem. 225, 1009.
3. Yamada, H., Makino, R., and Yamazaki, I. (1975) Arch. Biochem. Biophys. 169, 344.
4. Yamada, H., and Yamazaki, I. (1974) Arch. Biochem. Biophys. 165, 728.
5. Nicholls, P. (1962) Biochim. Biophys. Acta 60, 217.
6. Nichols, P., and Schonbaum, G.R. (1963) in The Enzymes (Boyre, P.D., Lardy, H.A., and Myrback, K. eds.) 2nd ed, Vol. 8, p.147. Academic Press, New York.
7. Ricard, J., Massa, G., and Williams, R.J.P. (1972) Eur. J. Biochem. 28, 566.
8. Schonbaum, G.R. (1973) J. Biol. Chem. 248, 502.
9. Caughey, W.S., Fujimoto, W.Y., and Johnson, B.P. (1966) Biochemistry 5, 3830.
10. Falk, J.E., Phillips, J.N., and Magnusson, E.A. (1966) Nature 212, 1531.
11. Makino, R., and Yamazaki, I. (1972) J. Biochem. (Tokyo) 72, 655.
12. Makino, R., and Yamazaki, I. (1973) Arch. Biochem. Biophys. 157, 356.
13. Yamada, H., and Yamazaki, I. (1975) Arch. Biochem. Biophys. 171, xxx.
14. Makino, R., Yamada, H., and Yamazaki, I. (1976) Arch. Biochem. Biophys. 172, xxx.
15. Wayland, B.B., Minkiewitz, J.V., and Abd-Elmageed, M.E. (1974) J. Am. Chem. Soc. 96, 2795.
16. Hayashi, Y., Yamada, H., and Yamazaki, I. Biochim. Biophys. Acta, in press.
17. Makino, R., and Yamazaki, I. (1974) Arch. Biochem. Biophys. 165, 485.
18. Chance, B. (1951) in The Enzymes (Sumner, J.B., and Myrback, K., eds.) Vol. II., Part I, p. 428. Academic Press, New Yorl.
19. George, P., and Lyster, R.L.J. (1958) Proc. Natl. Acad. Sci. U.S. 44, 1013.
20. Goldsack, D.E., Eberlein, W.S., and Alberty, R.A. (1966) J. Biol. Chem. 241, 2653.
21. Shin, J.H.C., Shannon, L.M., Kay, E., and Lew, J.Y. (1971) J. Biol. Chem. 246, 4546.

# X-RAY ABSORPTION SPECTROSCOPY:

# PROBING THE CHEMICAL AND ELECTRONIC STRUCTURE OF METALLOPROTEINS

W. E. Blumberg, P. Eisenberger, J. Peisach*,
and R. G. Shulman

Bell Laboratories, Murray Hill, New Jersey 07974 and
*Albert Einstein College of Medicine, New York,
New York, 10461

Recently high fluxes of X-rays have become available in the form of synchrotron radiation from electron storage rings such as at the Stanford Synchrotron Radiation Project (SSRP). At SSRP one has $10^4$ to $10^5$ times more intensity of X-rays with a continuous spectrum than that available from standard X-ray tubes. This increase in intensity has allowed X-ray absorption to be observed successfully in dilute materials such as metalloproteins and metal complexes in solution. While X-ray absorption measurements have been made sporadically since 1931 (1), the measurements, their interpretation and their applications to dilute systems have only recently been extensively developed due to the work of Sayers, Stern, and Lytle (2) and the use of synchrotron sources (3).

It is convenient in X-ray absorption to distinguish between the absorption of energies less than the ionization potential, where the transition is to an empty bound orbital, and of energies greater than the ionization potential where the transition is to an unbound final state. Transitions to empty bound states can be interpreted in terms of the formal charge on the ion and of the site symmetry and covalency (4). On the other hand the higher energy electrons are back scattered by neighboring atoms leading to an energy dependent modulation of the absorbance, $\Omega$, which gives rise to the extended X-ray absorption fine structure (EXAFS) (2,3).

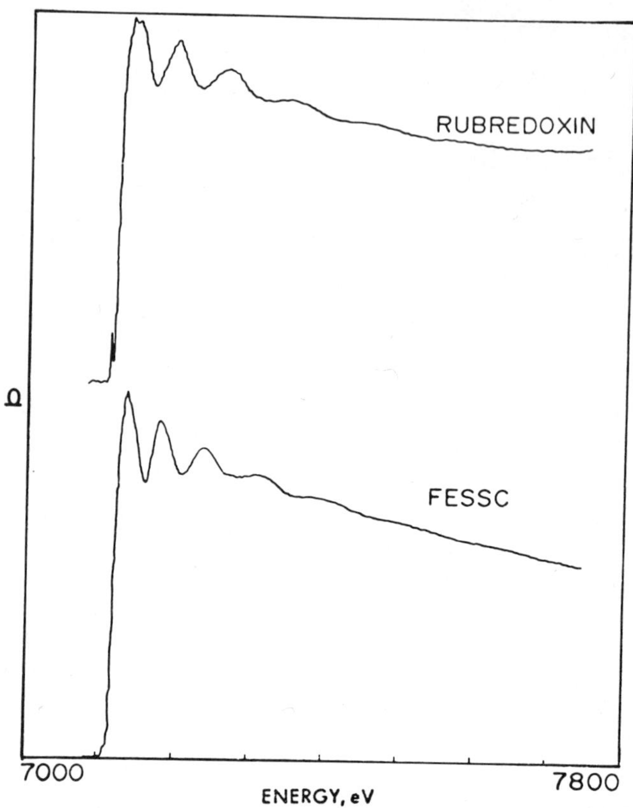


Fig. 1. X-ray absorbance Ω of solid samples of rubredoxin and FESSC as a function of energy. The data were collected in 401 channels with a dwell time of 2 sec in both cases. These and all the other spectra have been converted to energy by calibrating the spectrometer with metallic copper, where the first absorption peak on the low energy side of the edge is assumed to be at 8979 eV.

The absorption spectrum vs. energy of lyophilized rubredoxin Peptococcus aerogenes is shown by the upper curve in Fig. 1. There is a small peak before the edge, which arises from the 1s → 3d transition, followed by the intense absorption near 7100 eV corresponding to the 1s → 4p transition (4). At still higher energies the absorption enters the EXAFS region whose gentle modulations are detectable almost out to 7800 eV. The lower curve in Fig. 1 shows the analogous absorption spectrum for tris(pyrrolidine carbodithiolate-S,S′)Fe(III) (FESSC) over the same energy region.

## ELECTRONIC EXCITATIONS

First let us examine the absorptions arising from electronic transitions 1s → nl, where n = 3,4,5... and l = s,p,d, and focus our attention on the lower energy end of such an absorption spectrum. A detailed comparison of $FeF_2$ and $FeF_3$ is given in Fig. 2, where the lowest energy absorptions have been enlarged several fold. In both compounds the first absorption is weak and contains structure. About 15 volts to higher energy in the ferrous compound and 23 volts to higher energy in the ferric compound there is an intense peak whose position depends upon the oxidation state of the iron. At somewhat lower energies there are shoulders of varying degrees of resolution on these intense peaks. Calculations have shown that the iron atoms in these two compounds differ by about 0.9 of an electron, and this difference is reflected in the differing positions of each of the transitions.

Figure 3 shows the absorption spectra of a series of cubic metal fluorides having the perovskite crystal structure where the divalent metal ion changes from $Mn^{+2}$ to $Zn^{+2}$. The interpretation of the fluoride spectra is that the first weak ab-

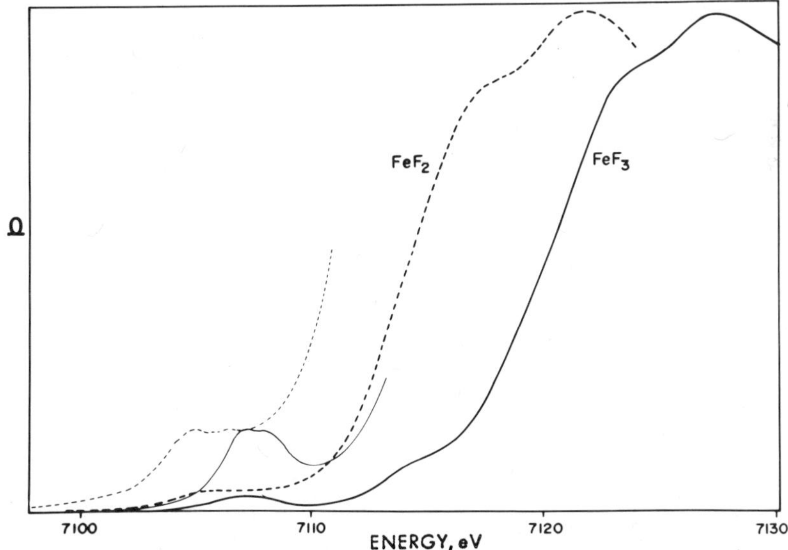


Fig. 2. Expanded view of the low energy region of the $FeF_2$ and $FeF_3$ absorption spectra. The incomplete dashed and solid curves are the $FeF_2$ and $FeF_3$ absorptions multiplied by 4 and 5 respectively.

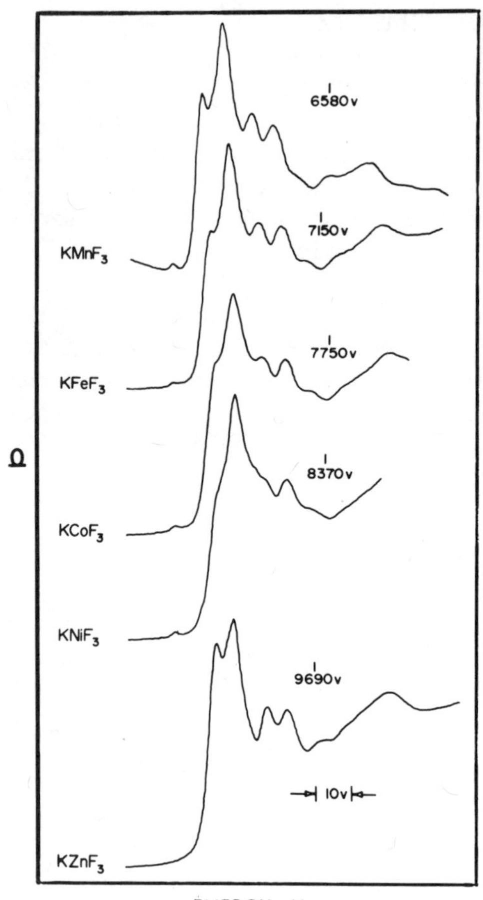


Fig. 3. X-ray absorption spectra of the divalent metal atom X in the $KXF_3$ series, all of which are cubic and have the perovskite crystal structure. Each spectrum covers the energy range from below the first absorption feature to well above the ionization potential of each compound, and thus each spectrum is in a different energy range, but the 10 V marker applies to each range. There is no isomorphous compound in the series for X = Cu. The most intense peak in each spectrum is the 1s → 4p transition, which is flanked to its left by the 1s → 4s. The weak absorption to lowest energy is the 1s → 3d transition, which occurs in all but the zinc compound.

sorption, observed in all but $KZnF_3$, which has no empty 3d orbital, is the 1s → 3d transition. The second absorption, present either as a shoulder or a peak ~9 volts higher is the 1s → 4s transition, and the intense absorption ~16 volts above the first absorption is the 1s → 4p. These assignments are based upon the agreement between measured values of the energies from spectroscopic tables (5) and the selection rules and intensities as discussed by Shulman et al. (4).

Figure 4 shows the absorption near the edge of the two iron-sulfur complexes of Fig. 1. In the protein rubredoxin, the ferric iron is known from X-ray crystallographic studies (6) to be tetrahedrally coordinated to four sulfur atoms terminal to the side chain of cysteine. In the model compound FESSC the ferric iron is octahedrally coordinated to six sulfur atoms. The inten-

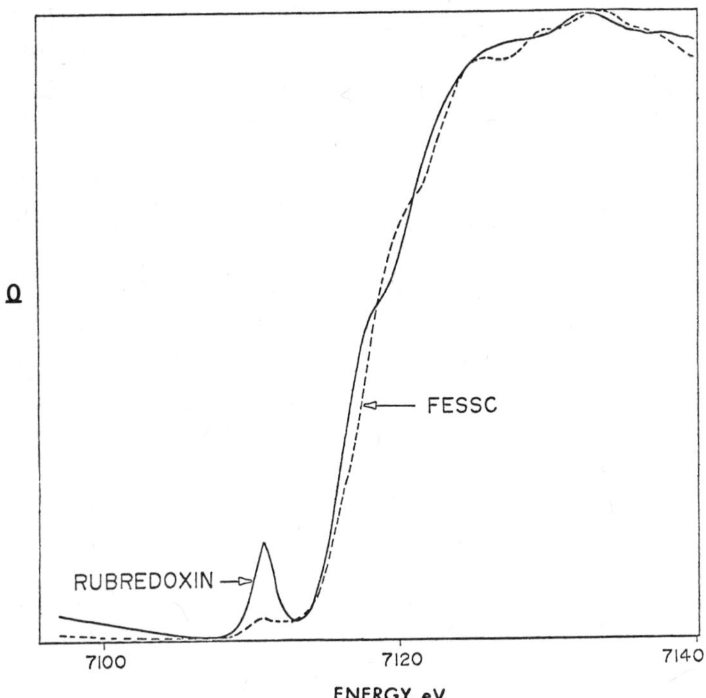


Fig. 4. Absorption of the tetrahedrally coordinated iron atom in rubredoxin from Peptococcus aerogenes and the octahedrally coordinated iron atom in FESSC showing that the energies of the lowest energy absorptions agree. The major difference is in the intensity of the lowest energy absorption (the 1s → 3d transition) which is about seven times more intense in rubredoxin.

sities of the 1s → 3d transitions are particularly interesting as the ratio of the 1s → 3d intensity to the 1s → 4p intensity is seven times larger in the tetrahedrally coordinated rubredoxin than in the octahedrally coordinated FESSC. The explanation of these intensities given by Shulman et al. (4) shows that the intensity of the 1s → 3d transition of iron group ions can be used as an indicator of the structure of the complex. Since there is no center of inversion in tetrahedral coordination, the 3d-4p mixing is stronger than in octahedral symmetry, as is well known from optical spectroscopy, and experimental results have shown that the forbidden transitions should be 10 to 100 times stronger than in octahedral symmetry. The observed intensity ratio here is about seven, which, while somewhat smaller, is qualitatively in agreement.

Since the $d^5$ ferric compounds do not have allowed transitions to excited d state multiplets it is clear that the observed splitting of the 1s → 3d transition of $FeF_3$ (~1.5 eV, see Fig. 2) must be caused by crystal field splitting of the ground state. This splitting agrees rather well with the 1.4-1.6 eV crystal field splitting of cobaltic fluorides observed by optical spectroscopy. It is clear, furthermore, that the crystal field splitting would not be observable in the divalent fluorides because optical crystal field splittings are ~1 eV and are obscured by the natural linewidths of the X-ray absorption spectra, which are ~1.5 eV.

Two final states can be reached in the case of $FeF_2$ namely, $^4F$ and $^4P$. The F to P energy separation is ~2 eV, the F level being the lowest. Shulman <u>et al.</u> (4) have calculated the ratio of the expected intensities of these two forbidden transitions. The results are that the intensities are proportional to (2L+1), i.e., in the ratio 7 to 3 for the F and P states, respectively. We expect, then, a weaker absorption ~2 eV above the lowest 1s → 3d transition in the case of $FeF_2$ but not in the case of $FeF_3$.

Superoxide dismutase is an enzyme containing active sites comprising one copper and one zinc atom bridged by the imidazole ring of a histidine side chain (7). This center can be reduced by a one-electron process, and this reduction affects the structure of the site as monitored by various optical and magnetic techniques. As electron spin density is facilely transferred across the imidazole bridge between metal atoms (8), it is of interest to determine how much charge density is transferred upon reduction. The X-ray absorption spectrum of oxidized human superoxide dismutase (1 mM in solution) is shown in Fig. 5, where the region of copper absorption has been scanned. The region of the 1s → 3d absorption is expanded five fold. Also shown are the spectra of the enzyme completely reduced with sodium dithionite and partially reduced with $H_2O_2$. Figure 6 shows the analogous spectra for the oxidized and completely reduced enzyme in the region of the zinc absorption. Using the $FeF_2$-$FeF_3$ structures as a guide, one concludes that the copper atoms in the oxidized and reduced enzymes differ by about a third of an electron while the zinc atoms differ by no more than 0.05 of an electron. Thus, the reduction proceeds entirely at the copper atom and its ligands and is not communicated to the zinc atom. In addition, one can see that the 1s → 4s transition of the copper atom increases markedly

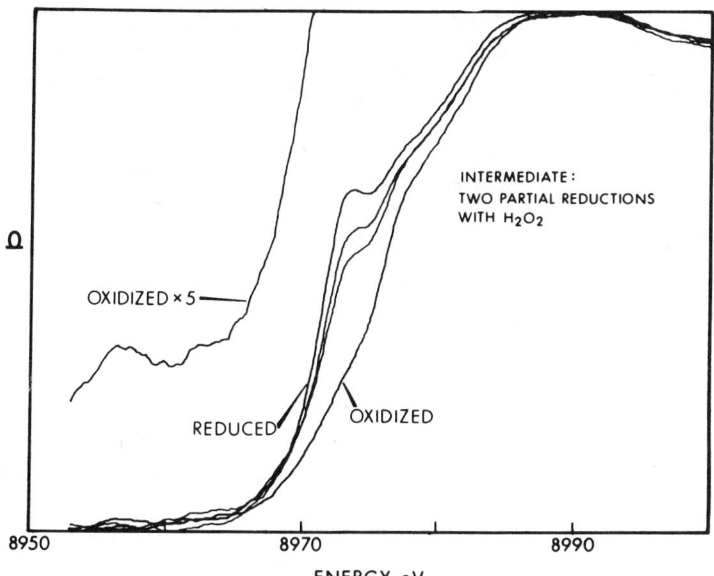


Fig. 5. X-ray absorption spectra of human superoxide dismutase near the ionization potential for copper. The region of the 1s → 3d absorption is expanded fivefold for the oxidized protein. Three levels of reduction of the enzyme are shown, two partial reductions with $H_2O_2$ and one complete reduction with dithionite.

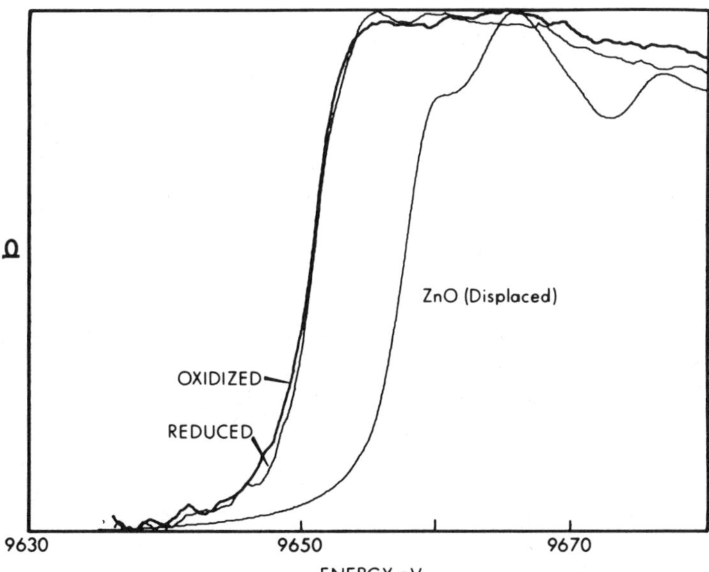


Fig. 6. X-ray absorption spectrum of human superoxide dismutase in the region of zinc absorption for the oxidized enzyme and the enzyme completely reduced with sodium dithionite. The spectrum of ZnO is shown for comparison, displaced a few volts to the right.

in intensity upon reduction. This indicates a large conformational change (probably from nearly octahedral toward tetrahedral). No such change is observed at the zinc atom.

## EXTENDED FINE STRUCTURE

Let us return to Fig. 1 and consider the origin of the oscillations observed in the X-ray absorption spectra above the absorption edge. For compounds where the electron back scattering is from N neighbors at distance R the absorbance is given by (3)

$$\Omega = x \frac{N}{kR^2} f(k,\pi) \sin[2kR + \alpha(k)] e^{-2\sigma^2 k^2}, \qquad (1)$$

where x is the sample thickness, $f(k,\pi)$ is the back scattering amplitude from the N neighbors, $\alpha(k)$ is the phase shift function, which in the present case describes the propagation of the ejected electron from the absorbing atom to the scattered atom and back, $\sigma^2$ is the mean squared fluctuation in R as in the Debye-Waller factor for a crystal. It is clear that if $\alpha(k)$ is known it is possible to interpret the measured dependence of $\Omega$ upon k so as to determine R.

In this case the absorption of the model compound FESSC can be used to determine $\alpha(k)$ as a function of k and, in conjunction with the absorption spectrum of rubredoxin, this in turn is used to determine its iron-sulfur distances. The crystal structure of FESSC, has been reported by Healey and White (9). The six Fe-S bonds range from 2.38 to 2.44 Å with a mean of 2.41 Å. This distance is 0.02Å shorter than the value used by Shulman et al. (10) so the distances reported by them should all be decreased by that amount.

Since Eq. (1) gives $\Omega$ in terms of a sinusoidal function of the free electron wave vector k, we have converted these spectra from energy, E, to k by assuming a ionization potential of 7130 eV and using

$$k = \left[\frac{2m}{h^2} (E-7130)\right]^{1/2}, \qquad (2)$$

where m = the electron mass and h = Planck's constant. Compensating for the 1/k term in Eq. (1) by multiplying by k and subtracting a best fit to a polynomial of order three in order to

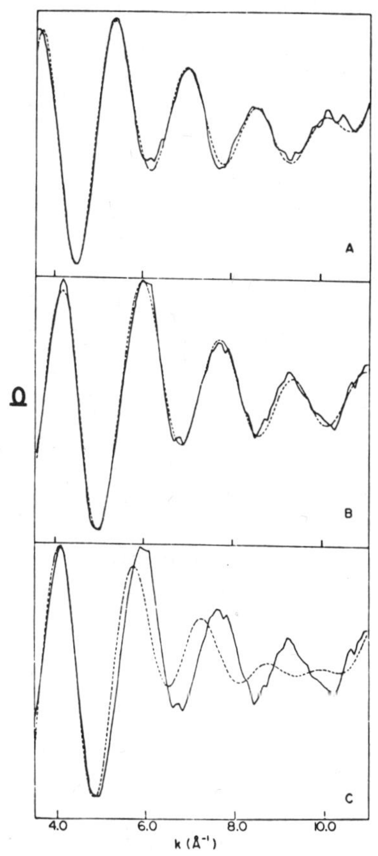


Fig. 7. Absorbance of FESSC (a) and rubredoxin (b and c) as a function of k from 3.5 to 11 $Å^{-1}$. The ionization potential was taken to be 7130 eV for both samples, and the data were smoothed by an exponential convolution. After smoothing, the best fitting third order polynomial was subtracted. (a) The dashed calculated curve was adjusted to fit the FESSC data keeping all six values of R constant at 2.41 Å and thereby determining $\alpha(k)$ for use in fitting the rubredoxin data. Relaxing the constraint that all values of R be identical did not appreciably reduce the rms error, which was 2.7% of the maximum amplitude swing, which was 0.276 in $\Omega$. (b) A comparison of the fit obtained with $R_3$ = 2.226 and $R_1$ = 2.197 Å against the rubredoxin data. This example, one of several equally close fits, had an average value of R = 2.219 and an rms error of 2.7% of the maximum amplitude swing of these data, which was 0.061 in $\Omega$. (c) A comparison of the data with a curve calculated using the X-ray crystal structure distances (11), showing that the agreement is poor, even though the average value of R = 2.238 agrees rather well.

flatten the baseline, we obtain the experimental plots of $\Omega(k)$ vs. k shown in Fig. 7(a) and 7(b)-(c) for FESSC and rubredoxin respectively. In the former spectrum six full cycles of a damped wave are resolved while in the latter five are observed.

For fitting purposes Shulman et al. (10) approximated the factors $f(k,\pi)$ and $e^{-\sigma^2 k^2}$ of Eq. (1) by a ratio of polynomials in k, and all six values of R were set to 2.430 Å for FESSC. A least squares fit was then used to determine $\alpha(k)$. The resulting computed curve is compared to the experimental EXAFS spectrum in Fig. 7(a).

Using this value of $\alpha(k)$ one approaches the experimental EXAFS spectrum of rubredoxin. A least squares fitting algorithm was used to determine the values of two radial variables: $R_3$, the distance of three of the sulfur atoms, assumed to be equivalent, and $R_1$ the distance of the fourth sulfur atom, assumed to be unique. Over a certain range of $R_1$ and $R_3$ these distances were correlated so that it was not possible to determine a unique set. In all cases the average value of R was 2.239 Å. The confidence limit for $R_1$ was ±0.16 Å from $R_3$, with no bias toward longer or shorter distances. One solution is given in Fig. 7(b).

The X-ray diffraction determination of the crystal structure of Clostridium pasteurianum rubredoxin was originally interpreted (6) as indicating that the four iron to sulfur distances were 2.31, 2.33, 2.39 and 1.97 Å, which were revised in the more recent version (11) to 2.24, 2.32, 2.34 and 2.05 Å. This poses the question as to whether our results on Peptococcus aerogenes rubredoxin agree with those of the X-ray crystal structure of Clostridium pasteurianum rubredoxin. A direct comparison with the latter results is given in Fig. 7(c) where all four crystal structure values have been used to simulate the back scattering. The fit in this case is obviously unacceptable. The heterogeneity of distances reported by Watenpaugh et al. (11) is certainly not observed in this case.

## FURTHER APPLICATIONS

In addition to the specific results reported here it should be stressed that the techniques used can be applied to a variety of systems. At the moment one is only limited to absorbing atoms with atomic number greater than 20. Interpretable EXAFS spectra have been obtained from rubredoxin in solution at concentrations of ~$6 \times 10^{-3}$M. Both these limitations should improve with time. To the extent that quantitative information about specific metal bonds in proteins is desired, we feel the general approach described here will not only improve the accuracy of the crystallographic work but will also be able to obtain results on systems such as solutions or membrane-bound proteins, which could not otherwise be studied.

Generally, in proteins, absorption peaks arising from transitions to bound states are one to two orders of magnitude more intense than are the modulations of the absorptions due to EXAFS. Although the absolute concentrations necessary for each of these measurements should decrease in the future, the transitions to bound states of metals in metalloproteins will always be more intense. Hence X-ray absorption edge spectroscopy will be more widely applicable to biological systems than EXAFS spectroscopy, especially when the metal atom's neighbors are varied and at different distances.

## REFERENCES

1. Kronig, R. de L. (1931). Physik 70, 317-323.

2. Sayers, D. E., Lytle, F. W. and Stern, E. A. "Advances in X-Ray Analysis", Vol. 13, Plenum Press, 1970, p. 248.

3. Kincaid, B. M., Eisenberger, P. (1975). Phys. Rev. Letts. 34, 1361-1364.

4. Shulman, R. G., Eisenberger, P., Yafet, Y., and Blumberg, W. E. (1976). Proc. Nat. Acad. Sci. (U.S.A.) (in press).

5. Moore, C. E. "Atomic Energy Levels" Vol. II. Circular of the National Bureau of Standards 467, 1952.

6. Watenpaugh, K. D., Sieker, L. C., Herriott, J. R. and Jensen, L. H. (1971). Cold Spring Harbor Symposium on Quantitative Biology 36, 359-367.

7. Richardson, J. S., Thomas, K. A., Rubin, B. H., and Richardson, D. C. (1975). Proc. Nat. Acad. Sci. 72 1349.

8. Rotilio, G., Calabrese, L., Mondovì, B., and Blumberg, W. E. (1974). J. Biol. Chem. 249, 3157-3160.

9. Healy, P. C. and White, A. H. (1972). J. Chem. Soc., Dalton Trans., 1163-1171.

10. Shulman, R. G., Eisenberger, P., Blumberg, W. E., and Stombaugh, N. A. (1975). Proc. Nat. Acad. Sci. (U.S.A.) 72, 4003-4007.

11. Watenpaugh, K. D., Sieker, L. C., Herriott, J. R. and Jensen, L. H. (1973). Acta Crystallogr. Sect. B 29, 943-956.

*The portion of this investigation carried out at Albert Einstein College of Medicine was supported in part by U.S. Public Health Service Research Grant HL-13399 from the Heart and Lung Institute and as such this is Comm. No. 343 from the Joan and Lester Avnet Institute of Molecular Biology. Recipient of Public Health Service Career Development Award 2-K3-GM-31,156 from the National Institute of General Medical Sciences.

Figs. 1 and 7 have been adapted from reference 10; Figs. 2-4 have been adapted from reference 4; Figs. 4 and 5 are taken from W. E. Blumberg, P. Eisenberger and J. Peisach, manuscript in preparation.

CARBONATE: KEY TO TRANSFERRIN CHEMISTRY

George W. Bates and Gary Graham

Department of Biochemistry and Biophysics

Texas Agricultural Experiment Station

Texas A&M University

College Station, Texas 77843

## INTRODUCTION

Transferrin, the serum iron transport protein, has three important functions. First, to sequester iron; second, to transport it in a form that is nontoxic and unavailable to the nonspecific complexing agents of serum; and third, to release iron to specific receptor sites on cell membranes. The chemistry of transferrin is a reflection of these physiological functions. Current topics in transferrin research are presented in the proceedings of the recent European Biology Organization Workshop on Iron Proteins (1).

An interesting aspect in the history of transferrin was the debate over the formation of a specific carbonate free $Fe^{3+}$-transferrin complex. As many positive as negative results had been published. We approached this problem through the development of a variety of synthetic routes aimed at formation of such a specific binary complex (2). The products were tested by spectrophotometric and chemical reactivity means. In all cases the carbonate free complexes exhibited the spectrum and sluggish chemical reactivity of nonspecifically bound iron. In contrast, the metal binding sites of the protein exhibited the high metal ion sequestering reactivity of apotransferrin. The conclusion was virtually inescapable: transferrin does not bind $Fe^{3+}$ in the absence of carbonate or other synergistic anions (2).

The binding of $Cu^{2+}$ to the specific metal binding site of transferrin in the absence of synergistic anions is more plausable since $Cu^{2+}$ is not as susceptible to hydrolytic polymerization as is $Fe^{3+}$

and could, therefore, be bound at a weaker site. Reports of specific $Cu^{2+}$-transferrin binary complexes have appeared in the literature (3). We approached this question in a fashion parallel to our studies with iron and found that in the absence of suitable anions $Cu^{2+}$ is nonspecifically bound to transferrin while the specific binding sites are vacant (4). These findings suggest that the metal binding sites may have virtually no affinity, or even repel metal ions, in the absence of synergistic anions. This, of course, would be of considerable value as a means of iron release to membrane receptor sites.

Our examination of forty candidate carbonate substitutes (synergistic anions) allowed us to develop a tentative rule for the chemical and structural requirements for carbonate substitutes (5). The synergistic anion must possess a carboxylic acid group and a proximal functional group within about 6.5 Å, capable of donating a coordinate bond to $Fe^{3+}$. The molecule must also have a stereochemistry such that the carboxyl group and a proximal ligand can be accommodated in a site at least 3 Å deep, approximately 6 Å wide, and between 4 and 6.5 Å in length. From this work, we suggested that the carboxylic acid group interacts with the protein while the proximal functional group is ligated to the iron (5). In a parallel fashion we envision carbonate as donating two oxygens to the protein and one to the $Fe^{3+}$ ion. Current research in our laboratory is focusing on the relationships of anion and metal ion binding.

## MATERIALS AND METHODS

Human apotransferrin was obtained from Behring Diagnostics. The protein was further purified by SP Sephadex C-50 chromatography (6) and rendered chelate free by techniques adapted from Price and Gibson (7). Chemicals used were of reagent grade and not further purified. Slow kinetic data was obtained using a Cary Model 15 spectrophotometer. Rapid kinetics were monitered on a Durrum Model 110 stopped flow spectrophotometer. $CO_2$-free techniques and preparation of $Fe^{3+}$-transferrin-anion complexes have been described (2, 5). The reduction chelation solution used in the iron removal experiments consisted of 50 m$\underline{M}$ ascorbic acid and 14 m$\underline{M}$ bathophenanthroline sulfonate (BPS). The transferrin solutions were in 5 m$\underline{M}$ Tris at pH 7.5 at a concentration of 2.7 x $10^{-5}$ $\underline{N}$. $Fe^{3+}$-transferrin-$CO_3^{2-}$ or an $Fe^{3+}$-transferrin-anion complex was placed in one drive syringe of the stopped flow while the desired reducing agents, chelate, anion mixture was placed in the other syringe. Spectra and pH measurement on the products were made by pooling material obtained from the collection syringe.

## RESULTS AND DISCUSSION

The finding that transferrin does not bind metal ions in the absence of synergistic anions (2) certainly suggests that it is the apotransferrin-$CO_3^{2-}$ complex that would be active in sequestering the metal ion. The possibility then arises that the $CO_2$ level in blood might affect transferrin reactivity. We investigated this possibility by examining the reaction of apotransferrin with metal ions as a function of $NaHCO_3$ concentration. In studies with $Fe^{3+}$-nitrilotriacetic acid (NTA), $Fe^{3+}$-fructose, $Fe^{3+}$-citrate, $Fe^{2+}$-salts, and $Cu^{2+}$-salts a hyperbolic dependence on the bicarbonate concentration was observed (4, 8). The apparent dissociation constant was 3 to 5 m*M*. The fact that this behavior is observed with such a wide variety of metal complexes suggests that it is an interaction of bicarbonate with transferrin that accounts for this phenomenon. We suggest the reaction sequence presented below is operative.

$$\text{Apotransferrin} + HCO_3^{2-} \longleftrightarrow \text{Apotransferrin-}CO_3^{2-} + H^+$$

$$\text{Apotransferrin-}CO_3^{2-} + Fe^{3+}\text{-Complex} \longleftrightarrow$$

$$Fe^{3+}\text{-transferrin-}CO_3^{2-} + \text{Complexing Agent}$$

Apotransferrin is depicted as binding carbonate to form an apotransferrin-carbonate complex. We suggest that it is this binary complex that is the neucleophilic protein specie. Since the bicarbonate ion concentration of plasma is approximately ten times higher than the apparent dissociation constant, it is apparent that under physiological conditions unsaturated transferrin exists almost entirely as the apotransferrin-$CO_3^{2-}$ complex and that carbonate cannot play a regulatory role in iron assimilation.

The rapid and effective removal of iron from $Fe^{3+}$-transferrin-$CO_3^{2-}$ by reticulocytes (9) in comparison to the relatively slow removal by chelating agents (10) clearly suggests that special mechanisms may be employed by the cell in the metal exchange reaction. While several hypotheses have been put forward no definitive evidence exists with regard to the mechanism utilized by the iron receptor site. Five possible factors promoting iron release can be suggested:

1. Chelation
2. Reduction of $Fe^{3+}$ to $Fe^{2+}$
3. Weakening of the anion-iron thermodynamic linkage
4. Disruption of $Fe^{3+}$-Amino Acid Residue Bonds
5. Allosteric (conformational) alterations

It is possible that the membrane iron receptor site utilizes a combination of these factors and perhaps others.

The removal of iron as $Fe^{3+}$ from $Fe^{3+}$-transferrin-$CO_3^{2-}$ by chelating agents has been examined in some detail (9). Because of the high affinity of transferrin for iron very high concentrations of chelating agent must be used. EDTA, citrate, and NTA removed iron at a rate that approximated the relative affinities of these agents for the metal. The reaction involving EDTA was found to be first order with regard to chelating agents and first order with regard to $Fe^{3+}$ ransferrin-$CO_3^{2-}$. While 0.1 M EDTA was capable of removing 100% of the iron from 3.6 x $10^{-4}$ N $Fe^{3+}$-transferrin-$CO_3^{2-}$, a full twelve hours were required for completion of the reaction. The ferric removal reaction appears to be a relatively simple reaction which depends upon the binding affinity of the chelating agent. It is apparent that chelation alone will not yield the rapid iron exchange observed with reticulocytes.

If, however, the $Fe^{3+}$-anion linkage is weakened by the use of a carbonate substitute, greatly enhanced exchange reactivities can be observed. In figure 2 of reference 5 the iron removal rates by 9 mM citrate are observed for a variety of $Fe^{3+}$-transferrin-anion complexes. The rate of iron removal is found to be inversely proportional to the stability of the ternary complex. In the case of the relatively weak $Fe^{3+}$-transferrin-malonate the reaction approaches completion after only twenty minutes. Certainly weakening of the carbonate-iron linkage by the membrane receptor site could allow more rapid removal of the iron.

Since transferrin binds $Fe^{2+}$ only very weakly (3), an obvious mechanism for iron release from the protein is reduction. Johnston and Cottingham (11) examined a system in which iron is released to

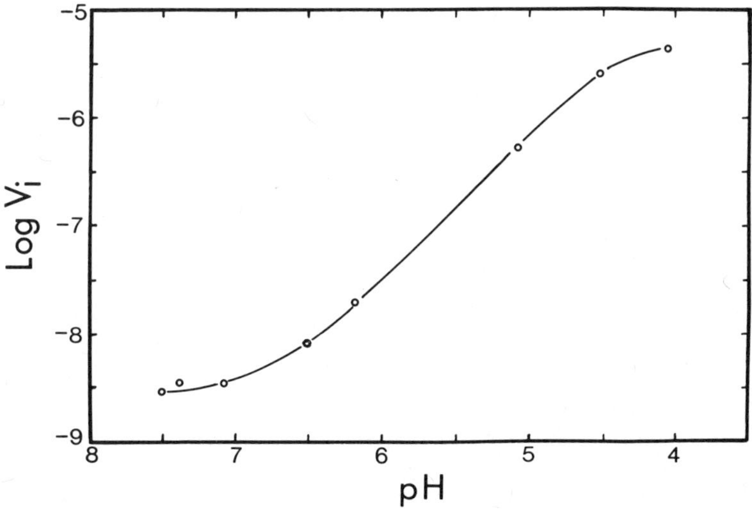


Figure 1. The effect of pH on the rate of the reaction in which iron is removed from $Fe^{3+}$-transferrin-$CO_3^{2-}$ by a mixture consisting of ascorbic acid and bathophenanthroline sulfonate. The initial velocity $V_i$, is obtained in terms of moles liters$^{-1}$ sec$^{-1}$. Experimental details are provided in the text.

TABLE I

EFFECT OF SYNERGISTIC ANIONS ON THE RELATIVE RATES OF REDUCTION

$$Fe^{3+}\text{-TRF-Anion} \xrightarrow{\text{BPS + ASC}} Fe^{2+}\text{-BPS + APO TRF + Anion}$$

| Complex | Relative Rate | Amount of Iron Removed by Citrate* |
|---|---|---|
| $Fe^{3+}$-TRF-$CO_3^{2-}$ | 1.0 | 7% |
| $Fe^{3+}$-TRF-Oxalate | 2.9 | 1% |
| $Fe^{3+}$-TRF-NTA | 116 | 70% |
| $Fe^{3+}$-TRF-Salicylate | 164 | 85% |

*Reference 5.

1,10 phenantholine after reduction by ascorbic acid. Even in the presence of 0.1 M 1,10 phenantholine a half life in excess of one year was determined. We have examined a parallel reaction in the presence of bathophenanthroline sulfonate (BPS) and find a half life on the order of fourteen hours, a great deal shorter yet still out of the range of the reticulocyte reaction.

Using the reaction system containing $Fe^{3+}$-transferrin-$CO_3^{2-}$, and the ascorbic acid and BPS reduction-chelation solution, we examined the rate of iron release from transferrin as a function of pH. $Fe^{3+}$-transferrin-$CO_3^{2-}$ at pH 7.5 was in one syringe of the stop flow and the reduction-chelation mixture at various pH values in the other. Following the reaction the contents of the acceptor syringe were emptied into a vial and the final pH determined. In Figure 1 is seen the rate of the reaction as a function of pH. It will be seen that there is a marked enhancement in the reactivity of the iron as the pH is lowered from 7.5 to 4. While pH may have a wide range of effects, this suggests that proton donation either to amino acid ligands or to the carbonate binding site may aid in the release of the iron. It might be noted that a similar profile is not exhibited in the release of iron from the $Fe^{3+}$-transferrin-oxalate complex, suggesting that the action of the proton may indeed center at carbonate. The half life of the reaction at pH 7.5 is fourteen hours, which is far too long for physiological consideration. At pH 4.1, however, the half life is reduced to two seconds. It is apparent that chemical conditions may be found which can enhance iron release to the magnitude of that found in the reticulocyte.

The reduction-chelation system was examined with several $Fe^{3+}$-transferrin-anion complexes. Since the redox potential of the bound iron is related to the stability of the $Fe^{3+}$-transferrin-anion ternary complex it appeared likely that variable occupancy of the anion binding site could affect the reaction rate. In Table I the relative rates of the reduction-chelation reaction are shown relative to $Fe^{3+}$-transferrin-$CO_3^{2-}$. As an index of the stability of the corresponding ternary complex, the amount of iron removed by citrate in a $CO_2$ free chelate competition reaction (5) is also presented. It will be seen that the rate of iron removal from the $Fe^{3+}$-transferrin-anion complex is inversely proportional to the stability of the complex. The greatest rate enhancement effect in this series is found with $Fe^{3+}$-transferrin-salicylate which reacts 164 times faster than the corresponding physiological complex.

In order to better understand the meaning of this finding we carried out experiments in which this reaction was run as a function of BPS concentration, ascorbate concentration, and the presence or absence of citrate and $NaHCO_3$. The data indicated that the reaction is zero order in BPS and that there is little or no effect of added citrate, suggesting that the chelation step is not rate limiting. Saturation kinetics were observed with regard to ascorbate concentration emphasizing the important role of reduction in the reaction. This information coupled with the observed high dependence of the reaction rate on the nature of the synergistic anion suggests that electron transfer to the iron may well prove to be an important aspect of the physiological iron removal reaction. While carbonate will always be the synergistic anion at the anion binding site, enzyme type reactions which could weaken the iron-anion linkage would have effects similar to that observed by using weaker synergistic anions. A possible insight into the mechanism of weakening this linkage was gained from the reaction in which 50 m<u>M</u> $NaHCO_3$ was added to the BPS ascorbate mixture prior to reaction with $Fe^{3+}$-transferrin-NTA. Carbonate rapidly displaces NTA from the anion binding site, and we anticipated a greatly decreased reaction rate since the more stable $Fe^{3+}$-transferrin-$CO_3^{2-}$ would form. Instead, an enhancement factor of 3000 was observed resulting in an iron removal reaction with a half life of less than one second. The possibility exists that during attack by carbonate on the anion binding site an intermediate exists in which the anion binding site is not fully occupied. A rapid reduction of the iron to the ferrous form during this intermediate state could account for the rapid kinetics observed.

In summary, occupancy of the anion binding site of transferrin has been shown to be essential with regard to binding $Fe^{3+}$, $Fe^{2+}$, and $Cu^{2+}$. In the absence of carbonate or other synergistic anions these and other metals bind nonspecifically to the protein while the metal binding site remains vacant. An examination of forty candidate synergistic anions revealed that an anion must possess a carboxylic acid group and a proximal functional group capable of forming a coordinate bond with the $Fe^{3+}$. We have suggested that the carbonate and iron binding sites of the transferrin molecule are in fact interlocking and that carbonate acts as a ligand to the protein as well as to the metal ion. Experiments described here and elsewhere emphasize the very important role of the anion in the chemical reactivity of apotransferrin and $Fe^{3+}$-transferrin-$CO_3^{2-}$. Evidence has been presented that apotransferrin binds $CO_3^{2-}$ prior to binding metal ions and it is the apotransferrin-$CO_3^{2-}$ that is the reactive metal ion sequestering specie. The removal of iron from $Fe^{3+}$-transferrin-$CO_3^{2-}$ is not well understood in chemical terms at this time, however, it is apparent that membrane bound receptor sites on developing red blood cells are quite effective in carrying out this reaction. The use of carbonate substitution shows that the iron protein bond can become much more labile to both chelation and reduction when the anion-iron linkage is weakened. It is apparent that extensive chemical studies are required to elucidate the possible mechanisms of iron removal from transferrin in order that the possibilities for cell mediated reaction can be understood.

## BIBLIOGRAPHY

1. Crichton, R. R., Ed. (1974) Proteins of Iron Storage and Transport; Proceedings of the EMBO Symposium, ASP Biological and Medical Press, Amsterdam

2. Bates, G. W., and Schlabach, M. R. (1975) J. Biol. Chem. 250, 2177-2181

3. Aasa, R., and Aisen, P. (1968) J. Biol. Chem. 243, 2399-2404

4. Bates, G. W., and Schlabach, M. R. (1974) The Synergistic Binding of Metal Ions and Anions by Transferrin in Proteins of Iron Storage and Transport; Proceedings of the EMBO Symposium, R. R. Crichton, Ed., ASP Biological and Medical Press, Amsterdam

5. Schlabach, M. R., and Bates, G. W. (1975) J. Biol. Chem. 250, 2182-2188

6. Martinez-Medellin, J. and Schulman, H. M. (1972) Biochim. Biophys. Acta 264, 272-284

7. Price, E. M. and Gibson, J. F. (1972) Biochem. Biophys. Res. Commun. 46, 646-651

8. Phelps, C. F. and Antonini, E. (1975) Biochem. J. 147, 385-391

9. Jandl, J., Inman, J., Simmons, R., and Allen, D. (1959) J. Clin. Invest. 38, 161-185

10. Bates, G. W., Billups, C., and Saltman, P. (1967) *J. Biol. Chem.* 242, 2816-2821

11. Johnston, D. O., and Cottingham, A. B. (1969) *Biochim. Biophys. Acta* 177, 113-117

Acknowledgements

This work was supported by Research Grant A-430 from the Robert A. Welch Foundation of Houston, Texas.

# OXIDATION AND REDUCTION OF COPPER IONS IN CATALYTIC REACTIONS OF *RHUS* LACCASE

Takao Nakamura

Department of Biology, Faculty of Science

Osaka University, Toyonaka, Osaka 560

## INTRODUCTION

The latex of lacquer tree, urushi, contains 50-80% urushiol which is a derivative of catechol (1-3). Urushiol in the latex undergoes rapid oxidation when exposed to air and the resultant quinone polymerizes to form a dark and lusterous solid film in combination with polysaccharides and other substances in the latex. The urushi film is superior in mechanical strength and stability against most of the chemicals and aging: thus it has been used in oriental countries in manufacturing lacquer wares in the past thousands of years. The aerobic oxidation of urushiol in the latex is catalyzed by laccase. In 1883 Yoshida (4) obtained an alcohol-soluble material and an alcohol-insoluble but cold water-soluble material from the latex. The former (urushiol) did not by itself undergo characteristic drying, as the original latex, on contact with air. The latter contained a heat-unstable, nitrogenous substance having "diastatic" properties. Yoshida found that when the former was acted by the latter substance, a dry film of urushi was formed at 20°C and the drying process required moisture and atmospheric oxygen. This is the first establishment of the characteristics of the catalyst in the latex at the time when biochemistry was only in its very early stage of development. A decade later Bertrand (5) independently performed similar investigations and gave the name of laccase to the enzyme.

## *RHUS* LACCASE

Lacquer tree laccase is a blue-colored copper protein (6,7) and can conveniently be extracted from the acetone powder of the latex. Fresh latex obtained from the lacquer tree *Rhus vernicifera* or *R. succedanea* was treated with chilled acetone and the acetone powder was extracted with water immediately after the acetone treatment. Although belonging to the same species *R. vernicifera*, the lacquer trees grown in Japan and in different districts of China (Ken Shii, Chu Shi and Mao Ba) produce latex of different quality as a paint. Laccase obtained from these latexes of different origin exhibit some differences in their enzymatic characteristics, as will be described later.

The purification process of laccase (8,9) includes ammonium sulfate fractionation, column chromatography on a cation exchange resin CG-50 and gel filtration on Sephadex G-100, and the enzyme was obtained as a homogeneous protein. The molecular weight of *R. vernicifera* laccase has been determined to be 104,000 (9). Laccase contains 4.0 gram atoms of copper per mole protein. Laccase catalyzes *in vitro* the aerobic oxidation of *p*-phenylenediamine, hydroquinone, pyrogallol, catechol or ascorbic acid. All the measurements described in this paper were performed at 25° in 0.1 M phosphate buffer pH 7.0 unless otherwise specified.

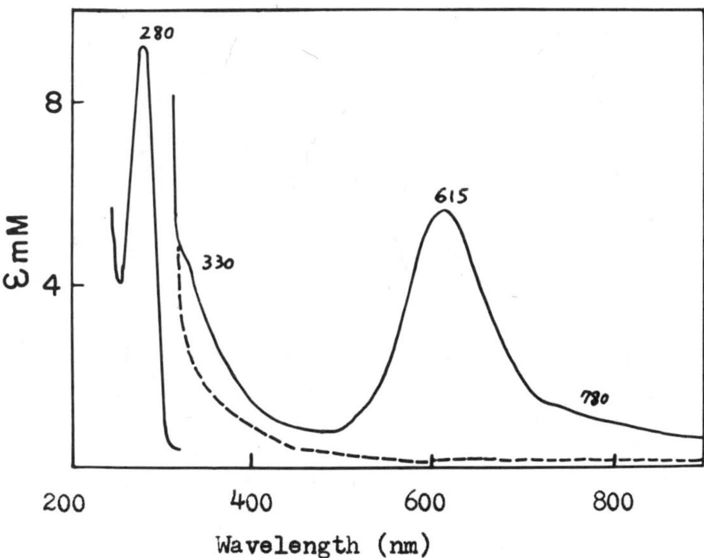


Fig. 1. Absorption spectrum of *Rhus vernicifera* laccase in phosphate buffer, pH 7.0. ———, Oxidized form; -----, Reduced form (by ascorbate). The extinction coefficient at 615 nm peak used was 5.4/mM protein (8).

## SPECTROSCOPIC CHARACTERISTICS AND VALENCE STATE OF COPPER IONS IN RHUS LACCASE

The absorption spectrum of laccase is shown in Fig. 1. There are two absorption maxima at 280 and 615 nm with a shoulder at 330 nm. All preparations of *R. vernicifera* laccase obtained from latexes of different origin were found to have the absorption maxima at the same wavelengths. The absorption bands at 615 and 330 nm disappeared on reduction by substrate or reducing agents such as dithionite, and reappeared on oxidation with $O_2$, $H_2O_2$ or ferricyanide (8,9). The bands did not disappear by simple evacuation of a laccase solution The oxidized minus reduced difference absorption spectra of laccases, ascorbate oxidase and *R. vernicifera* blue protein (stellacyanin) are presented in Fig. 2. It is apparent in the figure that copper proteins with oxidase activity are characterized by an absorption band at 330 nm in addition to the visible band at 607-615 nm in the difference spectra while those without oxidase activity (stellacyanin) lack the 330 nm band. The $Cu^+$ and total Cu contents of *R. vernicifera* laccase samples in the aerobic, resting state were determined colorimetrically by the cuproine method (11), and the $Cu^{2+}$ contents were estimated as the difference between the amount of total copper and $Cu^+$ copper. The ratio of $Cu^{2+}$ to total copper was obtained as a reproducible value in repeated experiments with the same enzyme sample, but the values differed 42-90% with different samples of latex. The laccase samples of different origin were found to show essentially the same EPR spectra as exemplified in Fig. 3. The spectrum indicated that laccase contained a single type of $Cu^{2+}$, whose EPR parameters were $g_m = 2.07$, $g_{\parallel} = 2.30$ and $A_{\parallel} = 0.004$ $cm^{-1}$. The spectrum was accompanied by a small and sharp signal at $g = 2.00$ which is due to trace amount of contaminating free radical. $Cu^{2+}$ contents of the laccase samples can also be estimated from their EPR spectra, and the ratios of $Cu^{2+}$ to total copper thus obtained were in good agreement with those obtained by the cuproine method. Laccase sample obtained from Ken Shii latex, which gave the highest value (90%) for the $Cu^{2+}$ to total Cu ratio, has previously been examined by magnetic susceptibility measurement (12) using a Gouy magnetic balance. Copper in the enzyme sample was found to be almost all in cupric state by the experiment. All preparations of *R. vernicifera* laccase obtained from the latexes of different origin were found to have the same value of $A_{280}/A_{615}$ of 19.2 (8). From the data of copper analysis, which was carried out by the cuproine method, the values of the extinction coefficient of the laccase preparations at 615 nm were calculated on the basis of the contents of total Cu and $Cu^{2+}$. All these results are summarized in Table I. It was found that the absorption at 615 nm is proportional to the content of total copper, but not to the estimated amount of $Cu^{2+}$. On the other hand, the oxidized minus difference extinction coefficient at the absorption shoulder at 330 nm was found to be proportional to the $Cu^{2+}$ content (Table II).

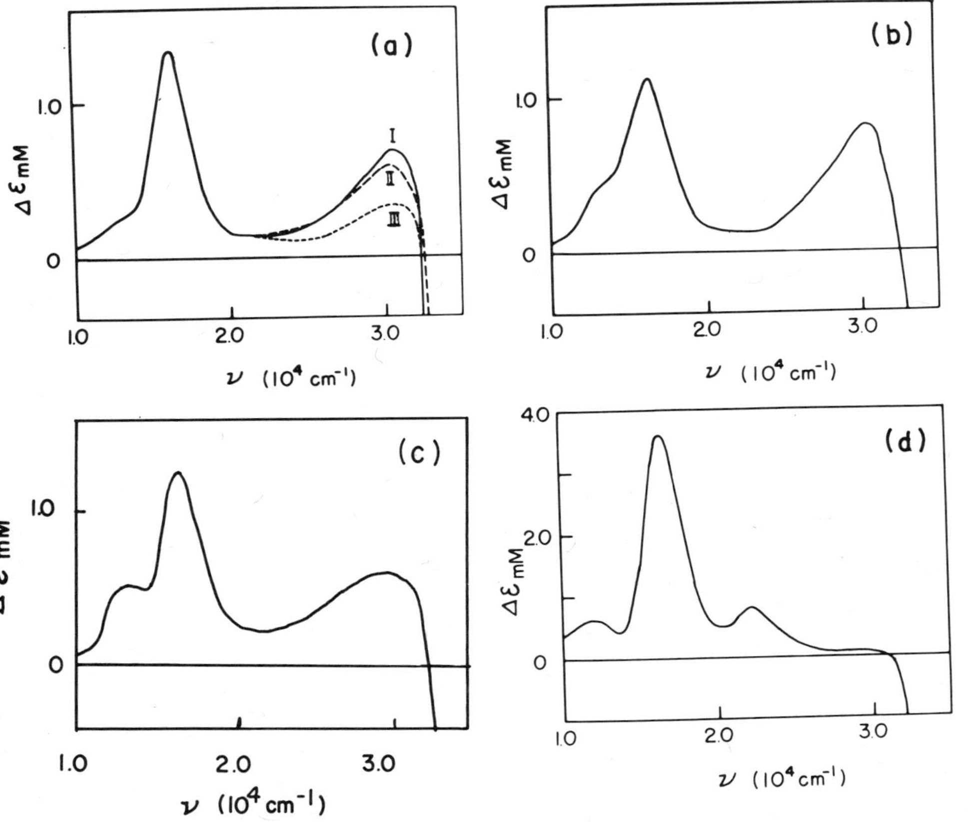


Figure 2. Oxidized minus reduced difference absorption spectra of copper proteins at pH 7.0. (a) *Rhus vernicifera* laccase (I, Ken Shii-I; II, Chu Shi; III, Mao Ba-III). (b) *Rhus succedanea* laccase. (c) Cucumber ascorbate oxidase. (d) *Rhus vernicifera* blue protein (stellacyanin). $\Delta\varepsilon_{mM}$ represented on the basis of total Cu. (10)

Table I. Copper Analysis, Spectrometry and Activity of Laccase Preparations (11)

| Laccase | Ken Shii-I | Ken Shii-II | Chu Shi | Mao Ba-I | Mao Ba-II | Mao Ba-III | Ibaraki (Japanese) |
|---|---|---|---|---|---|---|---|
| $Cu^{2+}$(%) by cuproine method | 89 | 90 | 73 | 71 | 47 | 42 | 80 |
| $Cu^{2+}$(%) by EPR | 91 | - | 75 | 80 | 57 | - | 72 |
| $A_{615}$/mM Total Cu | 1.44 | 1.46 | 1.41 | 1.59 | 1.49 | 1.43 | 1.44 |
| $A_{615}$/mM $Cu^{2+}$ | 1.62 | 1.66 | 1.92 | 2.24 | 3.20 | 3.41 | 1.80 |
| Activity/total Cu ($sec^{-1}$) | 4.2 | 4.1 | 3.2 | 2.6 | 2.2 | 1.9 | (6.5) |
| Activity/$Cu^{2+}$ ($sec^{-1}$) | 4.8 | 4.6 | 4.4 | 3.7 | 4.8 | 4.5 | (8.2) |

Activity = Turnover number of laccase copper reacting at pH 7.0 with hydroquinone as substrate.

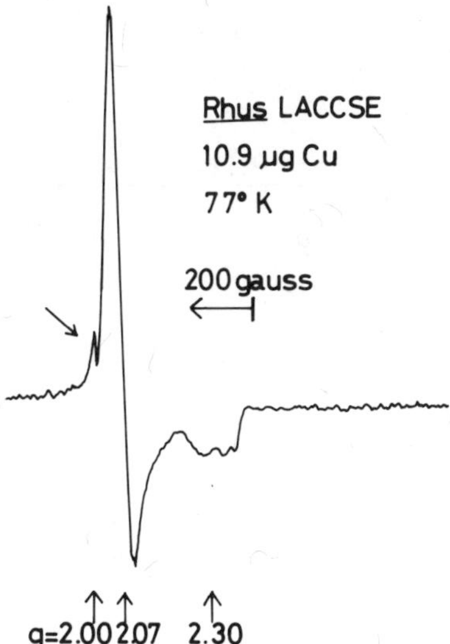


Fig. 3. Derivative curve of the EPR spectrum of Rhus vernicifera laccase in phosphate buffer, pH 7.0. Field modulation, 616 gauss. (11)

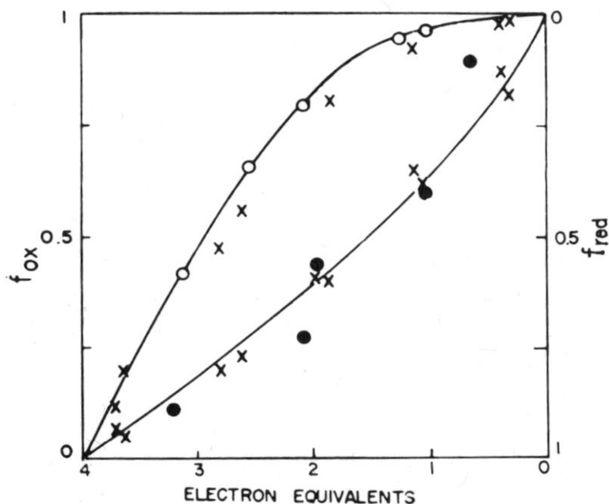


Fig. 4. Reductive titration of laccase with ascorbate under anaerobic conditions (quoted from Makino and Ogura (9)). Laccase was pretreated with $H_2O_2$ to the fully oxidized state. Absorbancé change was followed at 615 nm (○) or 330 nm (●). ×, Values obtained by oxidative titration with $H_2O_2$, pH 7.0, 25°.

Table II. $Cu^{2+}$ Content and Oxidized Minus Reduced Difference Extinction Coefficient of Rhus vernicifera Laccase at 330 nm (10)

| Laccase | Ken Shii-I | Chu Shi | Mao Ba-III |
|---|---|---|---|
| $Cu^{2+}$% by cuproine method | 89 | 73 | 42 |
| $A_{330}$/mM $Cu^{2+}$ | 0.77 | 0.79 | 0.78 |

It was found by Makino and Ogura (9) that $Cu^{+}$ in laccase in the aerobic resting state can be stoichiometrically oxidized by $H_2O_2$ with concomitant increase of the absorption at 330 nm while the absorption at 615 nm was constant during the titrimetric process (9). Reductive titration of laccase copper with substrate was carried out by the same authors by following the decrease of absorbances at 615 and 330 nm under anaerobic conditions (9). It was observed that laccase consumes 4 electrons/molecule for complete reduction (9,13). The titration curves measured at these two wavelengths followed different profiles and it was concluded that these absorption bands were due to different species of copper in laccase (Fig. 4). Makino and Ogura estimated that there was one Cu ion, Cu(615) (absorption at 615 nm) and three Cu ions, Cu(330) (absorption at 330 nm) in one molecule which were in oxidation-reduction equilibrium. The oxidation-reduction potential ($E_o'$) of Cu(615) was reported to be lower than that of Cu(330) by 45 mV at pH 7.0, 25°. Results from oxidative titration of reduced laccase with $H_2O_2$ were also consistent with the same conclusion. The $E_o'$ of Cu(615) has previously been determined spectrophotometrically to be +415 mV (8) in the presence of the ferrocyanide-ferricyanide system and by assuming $E_o'$ of ferrocyanide-ferricyanide to be +409 mV. Similar titrimetric measurements have also been carried out by Reinhammar and Vanngard (14).

## KINETICS OF OXIDATION AND REDUCTION OF COPPER IONS IN RHUS VERNICIFERA LACCASE

A laccase solution was placed in a cuvette of recording spectrophotometer, and a small volume of concentrated solution of ascorbate was rapidly mixed with the enzyme (Fig. 5). As seen in the figure, the absorbance of laccase at 615 nm quickly decreased to a level which roughly corresponded to the $Cu^{2+}$ content (estimated in the resting state) of the enzyme sample, and after a period of steady state, the enzyme was totally reduced due to the exhaustion of $O_2$ in the solution. When excess amount of

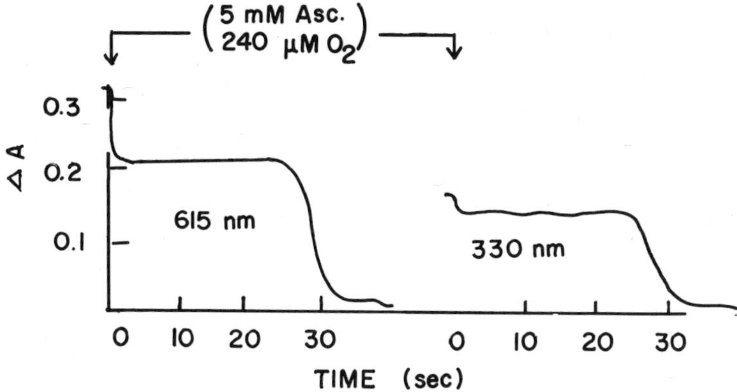


Fig. 5. Kinetics of oxidation-reduction of laccase copper at pH 7.0, 25°, followed by the change of absorbance at 615 and 330 nm. About 2.9 ml of the Rhus laccase (Chu Shi, 73% $Cu^{2+}$) was mixed with 0.1 ml of 0.15 M ascorbate under an aerobic condition. Enzyme concentration was 60 μM.

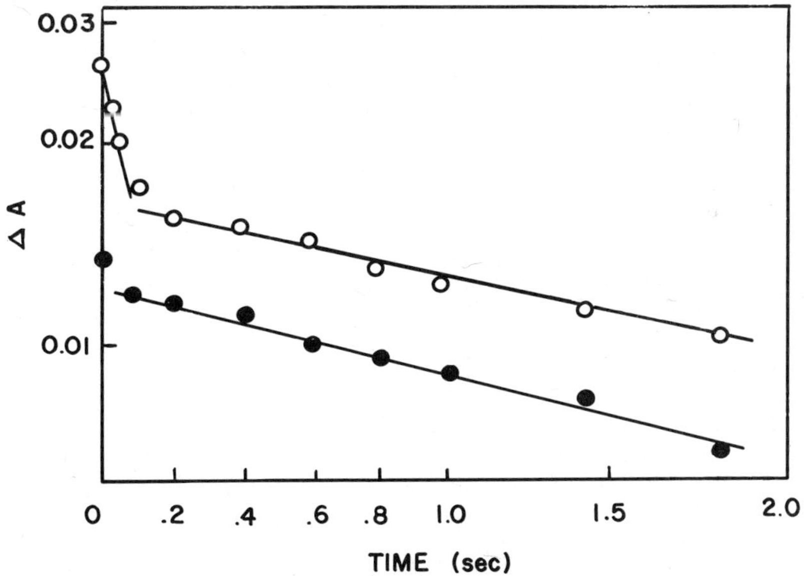


Fig. 6. Semi-log plot of the stopped flow records of the time course of reduction of laccase copper with ascorbate at pH 7.0, 25° under anaerobic conditions. 5 μM (protein) laccase was mixed with 10 mM ascorbate and the change of absorbance at 615 (○) or 330 nm (●) was recorded. Chu Shi laccase, 73% $Cu^{2+}$ content.

ferricyanide was added to the reduced enzyme the original absorbance at 615 nm was fully restored. The restoration was incomplete when $O_2$ was used as an oxidant. When the same experiment was followed at 330 nm, the initial rapid decrease of absorbance was much less than that observed at 615 nm. After a period of steady state whose duration was almost the same as that observed at 615 nm, the enzyme was totally reduced. The turnover number (k) of the enzyme in the experiment of Fig. 5 at 615 nm was calculated from the equation

$$k = [O_2]_{initial} / \int_o^{\infty} [\text{oxidized laccase}] \cdot dt$$

to be 0.2 $sec^{-1}$. This value is in good agreement with that obtained by measurement of the rate of $O_2$ consumption (0.2 $sec^{-1}$) which was performed by using an oxygen electrode under the same experimental conditions. In the latter calculation the enzyme concentration was taken as [total laccase] x ($Cu^{2+}$ to total Cu ratio, estimated by the cuproine method). Identical results have previously been reported on Rhus laccase obtained from latexes of different origin, using hydroquinone as substrate (10). These results indicate that in the laccase sample there are two kinds of $Cu^{2+}$ with 615 nm absorption, one is rapidly reducible by substrate but not oxidizable by $O_2$, i.e. catalytically inactive Cu(615), and the other, reducible by substrate and rapidly oxidizable by $O_2$, i.e. catalytically active Cu(615). The latter Cu was almost completely in the oxidized form during the dynamic equilibrium conditions of the steady state in these experiments.

The rates of reduction by various substrates and the rate of oxidation by $O_2$ of 615 nm- and 330 nm chromophores of laccase were determined by the stopped flow method. The laccase solution was mixed with the substrate solution under anaerobic conditions and the change of absorbance of the mixture at 615 or 330 nm was recorded. In the experiment shown in Fig. 6, a laccase sample of Chu Shi origin (73% $Cu^{2+}$) was mixed with 10 mM ascorbate and the absorbance at 615 or 330 nm was recorded. It is apparent in the figure that the time course of reduction at 615 nm followed a biphasic change, as expected from the previous aerobic experiment (Fig. 5). The rapid initial decrease of the absorbance (inactive Cu(615)) corresponded to 35% of total absorbance change and it was followed by a process of slower reduction (active Cu(615)). The experiment was repeated with different concentrations of ascorbate. The rate of the rapid reduction was proportional to [ascorbate], with a second order rate constant $k_{rapid}$ = 630 $M^{-1}$ $sec^{-1}$ while that of the slow rate was found to follow a Michaelis-Menten type kinetics with $k_{slow}^{max}$ = 0.8 $sec^{-1}$ (extrapolated to 1/[ascorbate] = 0). The time course of the absorbance change at 330 nm occurred at a slow rate and the rate constant was in agreement with the value obtained at 615 nm

Table III. Comparison of the Rates of Oxidation and Reduction and Spectroscopic Parameter of Copper Proteins (15)

| | | $k_{ox}$ ($M^{-1}sec^{-1}$) | Ascorbate | Hydroquinone | p-phenylene diamine | Dimethyl p-phenylene diamine | pH | Temp (°C) | $\Delta\varepsilon_{mM}$ 330 nm |
|---|---|---|---|---|---|---|---|---|---|
| | | | $k_{red}^{max}$ ($sec^{-1}$) | | | | | | |
| Copper enzymes | Ascorbate oxidase | $4.8 \times 10^6$ | 430 | - | - | - | 5.8 | 1.7 | 0.56 |
| | *R. succedanea* laccase | $7 \times 10^5$ | 37 | 85 | 50 | - | 6.0 | 25 | 0.82 |
| | *R. vernicifera* laccase | $2.8 \times 10^6$ ($2 \times 10^6$) | 0.8 (1.0) | 1.2 | 1.4 | 1.0 | 7.0 | 25 | 0.78 |
| | Porcine ceruloplasmin | $1 \times 10^5$ | 1.6 | 0.6 | 4.0 | 10.0 | 7.0 | 25 | 0.44 |
| | | | $k_{red}$ ($M^{-1}sec^{-1}$) | | | | | | |
| Blue proteins | *R.* blue protein (stellacyanin) | 1.8 | 2,100 | very fast | v.f. | v.f. | 7.0 | 25 | 0.1 |
| | Cucumber plastocyanin | 0 | 280 | 88 | 32,000 | 17,000 | 7.0 | 25 | 0 |

$k_{ox}$: second order velocity constant of the oxidation of the protein-bound cuprous ion by $O_2$.
$k_{red}$: rate constant of the reduction of the protein-bound cupric ion by the substrate.
Data in parentheses were obtained by following the absorbance change at 330 nm. Others were obtained at 615 nm.

(Table III). The obtained rates of aerobic oxidation of the 615 and 330 nm chromophores were also measured by the stopped flow method by mixing reduced enzyme (prepared by adding a minimum amount of ascorbate under anaerobic conditions) with a solution of $O_2$. The second order rate constant for the process of aerobic oxidation was determined and the results obtained, as well as the values of the rate constant which have already been determined on other copper oxidases and blue copper proteins, are summarized in Table III.

From the data in Table III, it seems that laccase and other enzymatically active copper proteins are characterized by a high rate of aerobic oxidation of the copper ion in the enzyme while rapid reduction of copper takes place both in enzymatically active and nonactive copper proteins.

## A MODEL OF THE LACCASE MOLECULE

All these results obtained may be consistent with the following model of Rhus laccase molecule (Fig. 7). It is a slightly modified model of Makino and Ogura (4) which has already been proposed. In the model it was assumed that (1) the molecule contains 3 Cu(330) ions, exhibiting absorption at 330 nm and 1 Cu(615-330) ion exhibiting absorption both at 615 and 330 nm; the latter having a lower redox potential by 45 mV than Cu(330). These absorption bands disappear when reduced to $Cu^+$. No distinction of Cu(615-330) and Cu(330) was possible from EPR spectral measurements. (2) The latex sample originally contains active and inactive laccase molecules in a variable ratio. So-called reaction-inactivation, i.e. inactivation of an enzyme occurring during its catalytic reaction, is not uncommon among copper enzymes (16). Since the rate of secretion of urushi-

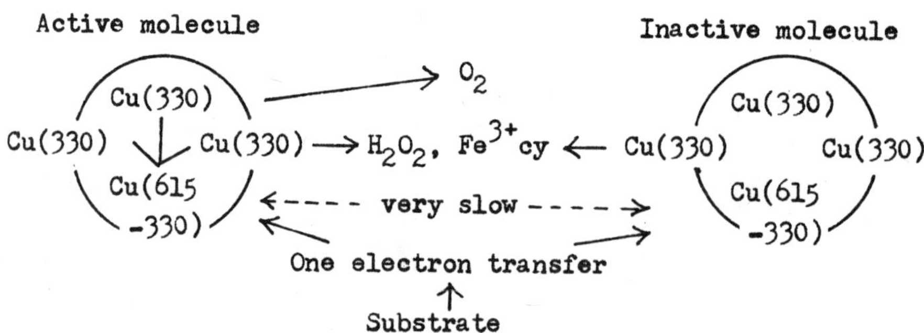


Fig. 7. A model of Rhus laccase molecule.

latex is so slow (few grams/day/tree), it seems that the laccase in the latex may undergo various degrees of inactivation during the process of collection of the latex in the field, i.e. it depends on the method used for the collection of the latex in the local districts. (3) There is an intramolecular interaction between Cu(615-330) and 3 Cu(330) ions in active laccase molecule. In the inactive molecule, the interaction is blocked. Active molecule reacts with $O_2$ only when all the Cu ions are in $Cu^+$ state. Inactive molecule does not react with $O_2$. Both active and inactive molecules are reactive with oxidants such as ferricyanide or $H_2O_2$. Inactive laccase molecule can only very slowly be oxidized by active molecule by an intermolecular electron transfer. In this case, however, only Cu(615-330) is oxidized, Cu(330) may not be oxidized due to its high redox potential. Substrate molecules are oxidized by $Cu^{2+}$ ions in laccase by a process of one electron transfer.

According to the model, experimental observations as presented in this paper can be interpreted in the following manner.

(a) The absorbance of laccase at 615 nm was (in the aerobic, resting state) proportional to total Cu, but not to the $Cu^{2+}$ for the enzyme samples whose $Cu^{2+}$/Cu ratio range from 42-90%, since in aerobic, resting state all the molecules contain 1 Cu(615-330) in cupric state.

(b) The absorption of a laccase sample at 330 nm was (in aerobic, resting state) proportional to $Cu^{2+}$ since in the active molecule Cu(615-330) and Cu(330) are all in cupric state and in the inactive molecule only Cu(615-330) is in cupric state.

(c) Enzymatic activity must be proportional to the amount of active molecule. Based on the model, the relationship between the ratios of active molecule/total molecule (relative activity/mg protein) and $Cu^{2+}$/total Cu ($Cu^{2+}$%) could be calculated. The activity vs. $Cu^{2+}$ content data in Table I was replotted in Fig. 8, and good agreement between the theoretical and the experimental results was obtained.

(d) Each $Cu^{2+}$ ion in laccase accepts one electron from the substrate, independently, to form the substrate semiquinone (17). There is no indication of existence of a $Cu^{2+}$-$Cu^{2+}$ pair in the molecule as far as the process of electron transfer from substrate to laccase molecule is concerned.

(e) Cu(615-330) in the inactive molecule is rapidly reduced by substrate but is not reoxidized by $O_2$ (rapid initial reduction in Fig. 5). $Cu^+$ ions in the active molecule is rapidly oxidized by $O_2$ and its reduction to $Cu^+$ by substrate is the rate limiting step in the catalytic process. In the dynamic equilibrium

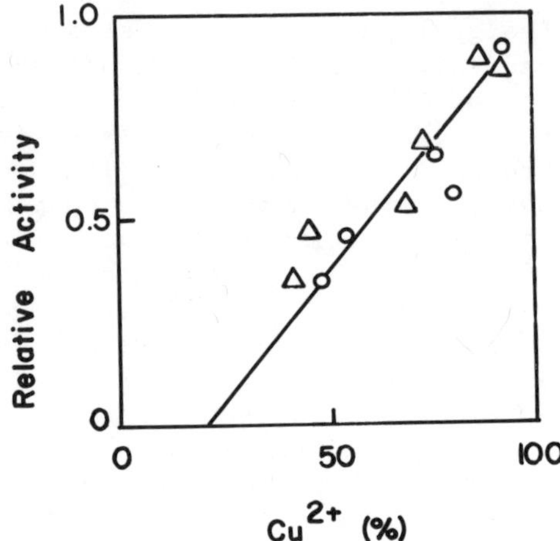


Fig. 8. Plot of relative activity (relative value of the rate of $O_2$ uptake/mg protein) versus $Cu^{2+}$ content (%) of various laccase samples obtained from latexes of different Chinese origin. (○,△), Plot of the relative activity against $Cu^{2+}$%. (○, $Cu^{2+}$ estimated by EPR; △, by cuproine method). ——, Theoretical line drawn by the use of equation 3(relative activity) = 4(%$Cu^{2+}$/100) - 1.

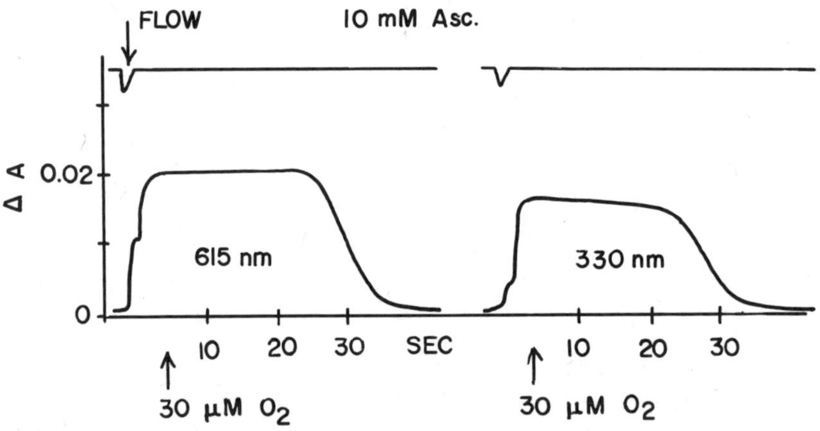


Fig. 9. Stopped flow experiment of laccase at pH 7.0, 25°. 6 μM laccase (Chu Shi, 73% $Cu^{2+}$: i.e. 3.8 μM active fraction according to the equation shown in the legend for Fig. 8) in the reduced form was mixed with 30 μM $O_2$ in the presence of 10 mM ascorbate, and the rapid appearance of the absorption at 615 or 330 nm were recorded.

during the steady state, most of the Cu(615-330) and Cu(330) in the active molecule is in cupric state (Fig. 9, see also Fig. 1 in ref. (10)).

(f) It is clear (Table III) that rapidity in reduction velocity of Cu by substrate is not a sufficient condition for a copper protein to be an oxidase. Rather, the oxidase activity is provided by the reactivity of the copper protein with $O_2$.

(g) Involvement of the 330 nm chromophore Cu of Rhus laccase in enzymatic activity was also shown by Malmström, Reinhammar and Vänngård (18). Andrèasson et al. (19) have observed a biphasic change in the spectrophotometric time course of anaerobic reduction of the copper ion by substrate in a fungal (polyporus versicolor) laccase. Their observation that the relative amount of rapidly reducing copper increased after mild denaturation of the enzyme seems to fit our proposal that the rapidly reducing copper may be due to the presence of inactive enzymes in the preparation. Malmström and coworkers have claimed that the copper oxidases ceruloplasmin, fungal- and Rhus laccases all possess so-called type-II Cu (non-blue, showing wide and normal hyperfine splitting in EPR spectrum), in addition to type-I Cu (blue and narrow hyperfine splitting in EPR spectrum, in agreement with our data on Rhus laccase). The former type of EPR spectrum has not been detected in our preparations of Rhus laccase. It was reported by this group that type-I Cu could be converted to type-II Cu by moderate denaturation (19,20) and it was reported that type-II Cu is hardly reducible in their preparations by substrates (18,20). It may be also mentioned that so-called type-II Cu is due to denatured enzyme.

(h) The detailed mechanism of reduction of $O_2$ to $H_2O$ by the reduced laccase is still unknown. Makino and Ogura (9) have observed a linear increase of absorbance both at 615 and 330 nm in oxidative titration of reduced laccase with $O_2$, and this result supports the view that laccase is reactive to $O_2$ only when all 4 copper ions are in cuprous state. This is in agreement with the fact that no $H_2O_2$ is formed under aerobic conditions during the laccase reaction. It may be presumed that a conformational change of laccase protein occurs when 4 Cu ions in the molecule are fully reduced and this structural change causes the molecule to become reactive to molecular oxygen.

## ACKNOWLEDGEMENT

Thanks are due to the Saito Company Osaka for their kindness in supplying crude latex used in this investigation.

SUMMARY

1) It was demonstrated by colorimetric as well as EPR measurements that the native (aerobic, resting state) Rhus vernicifera laccase contains both $Cu^{2+}$ and $Cu^{+}$ (total Cu content was 4.0 gram atoms/mole). The ratio of $Cu^{2+}$ to total Cu in laccase varied (42-90%) in samples of latex collected from various districts. The absorption maximum at 615 nm was proportional to the content of total Cu in the enzyme sample. Laccase activity was found to almost parallel the content of the $Cu^{2+}$ form. The oxidized minus reduced difference absorbance of the enzyme at 330 nm shoulder was proportional to the amount of $Cu^{2+}$.

2) Steady state level of oxidation of laccase copper during the laccase copper catalytic action, the rates of reduction by substrates and the oxidation by $O_2$ were determined by following absorbance changes at 615 and 330 nm by the stopped flow method.

3) All the results from titrimetric and kinetic experiments were consistent with the laccase model previously proposed by Makino and Ogura in which a laccase molecule contains 1 Cu(615) and 3 Cu(330). Our expanded model states that a laccase sample originally contains active as well as inactive enzymes. In the active enzyme, Cu ions are reactive to $O_2$ but in the inactive enzyme, Cu can be oxidized only by oxidizing agents such as $H_2O_2$ or ferricyanide, or by a slow intermolecular electron transfer from Cu(615) to the active enzyme. In both species of enzyme rapid reduction of $Cu^{2+}$ ions by substrate takes place. In comparative studies of the reactivities of Cu ions in various copper proteins, we would like to suggest that oxidatic activity of a copper protein is due to the $Cu^{+}$ form of the enzyme ions with $O_2$.

REFERENCES

(1) Miyama, K., Tokyo Kagakukaishi, 27, 1191 (1906).

(2) Majima, R., Ber., 55, 172 (1922).

(3) Sunthankar, S. and Dawson, C.R., J. Am. Chem. Soc., 76, 5070 (1954).

(4) Yoshida, H., J. Chem. Soc., 43, 473 (1883).

(5) Bertrand, G., Bull. Soc. Chim., 3, 11, 17 (1894).

(6) Keilin, D. and Mann, T., Nature, 143, 23 (1939).

(7) Yakushiji, E., Acta Phytochim., 12, 227 (1941).

(8) Nakamura, T., Biochim. Biophys. Acta, 30, 44 (1958).

(9) Makino, N. and Ogura, Y., J. Biochem., 69, 91 (1971).

(10) Nakamura, T. and Ogura, Y., J. Biochem., 59, 449 (1966).

(11) Nakamura, T., Ikai, A. and Ogura, Y., J. Biochem., 57, 808 (1965).

(12) Nakamura, T., Biochim. Biophys. Acta, 30, 640 (1958).

(13) Nakamura, T., Biochim. Biophys. Acta, 30, 538 (1958).

(14) Reinhammar, B.R.M. and Vänngård, T.I., Eur. J. Biochem., 18, 463 (1971).

(15) Nakamura, T. and Ogura, Y., J. Biochem., 64, 267 (1968).

(16) Poillon, W.N. and Dawson, C.R., Biochim. Biophys. Acta, 77, 37 (1963).

(17) Nakamura, T., Biochem. Biophys. Res. Commun., 2, 111 (1960).

(18) Malmström, B.G., Reinhammar, B. and Vänngård, T., Biochim. Biophys. Acta, 205, 48 (1970).

(19) Andrèasson, L.E., Malmström, B.G., Strömberg, C.S. and Vänngård, T., Eur. J. Biochem., 34, 434 (1973).

(20) Malmström, B.G., Reinhammar, B. and Vänngård, T., Biochim. Biophys. Acta, 156, 67 (1968).

# RECENT STUDIES ON COPPER CONTAINING OXIDASES

B. Mondoví, L. Morpurgo, G. Rotilio and A. Finazzi-Agró

Institutes of Applied Biochemistry and of Biological Chemistry, University of Rome and Center of Molecular Biology, National Research Council, Rome, Italy

The role of copper in the oxidizing machinery of life has been recognized long ago. However a detailed knowledge of its function inside proteins is still far from complete and depends on parallel understanding of structure. Malmström et al. (1) first proposed three different types of protein-bound copper, usually referred to as Type 1, Type 2 and Type 3 copper, with special reference to multicopper enzymes such as laccases or ceruloplasmin (2) which contain all the three types. For sake of simplicity we will use this terminology throughout the present paper though, as we will see below, it appears to be rather naive.

Type 1 copper is currently called "blue" copper, with reference to its fairly high extinction coefficient (> $10^3$) in the 600 nm region. It also shows an unusually narrow hyperfine structure in the $g_{\parallel}$ region of the EPR spectrum. These peculiar features of the optical and EPR spectrum have been given different interpretations (3-7). However there are differences among the various blue centers in blue proteins (see Table I), concerning energy and extinction coefficients of optical transitions, EPR spectra (see for instance azurin and stellacyanin) and, most striking, redox potentials. There are different blue centers even inside the same protein, as in ceruloplasmin (8). Blue copper reacts very differently with NO in similar proteins, i.e. laccase (9) and ascorbate oxidase (10). Ascorbate oxidase, which contains 4 EPR detectable coppers (11), shows a variable distribution between blue and non blue coppers (10,12).

TABLE I

Optical, Magnetic and Oxidation-Reduction Properties of Type 1 Copper in Some Blue Copper Proteins (from Ref. 2)

| Protein | Wavelength (nm) | $\varepsilon(M^{-1}\ cm^{-1})$ | $g_{\parallel}$ | $g_{\perp}$ | $A_{\parallel}$ ($cm^{-1}$) | $E'_o$ (pH) (volts) |
|---|---|---|---|---|---|---|
| Azurin (Pseudomonas fluorescens) | 459<br>625<br>781 | 285<br>3500<br>320 | 2.261 | 2.052 | 0.0058 | 0.328 (6.4) |
| Stellacyanin (Rhus vernicifera) | 448<br>604<br>845 | 554<br>3820<br>700 | 2.288 | 2.08(a)<br>2.025(b) | 0.0031 | <0.3 (6.5) |
| Laccase (Rhus vernicifera) | ∿450<br>614<br>788 | ?<br>5200<br>900 | 2.298 | 2.047 | 0.0037 | 0.415 (6.8) |
| Ceruloplasmin (Human plasma) | 459<br>610<br>794 | 1200<br>11300<br>2200 | 2.21 | 2.05 | 0.0083 | 0.5-0.6(6.25) |

(a) = value for $g_y$; (b) = value for $g_x$.

In a strict sense, the expression, "Type 2 copper" should label the non blue EPR detectable copper in multicopper oxidases. It is a common practice to extend such terminology to other proteins in spite of evidence that this class is not at all homogeneous. In fact the only common features are an extinction coefficient in the 500-800 nm region an order of magnitude lower than that of Type 1 copper ($\simeq 10^2$) and "normal" hyperfine structure in the $g_{\parallel}$ region.

Spectroscopic parameters of some copper proteins usually considered as containing Type 2 copper are reported in Table II. It is evident that these proteins are consistently different. In particular the high rhombic distortion of the EPR spectrum, differentiates the superoxide dismutase copper from other Type 2 copper sites. It is also evident that galactose oxidase has a high extinction coefficient in spite of its "normal" EPR spectrum. The so-called artificial copper proteins, i.e. proteins which contain another metal when isolated from tissues but may bind copper in the place of the original metals in some stoichiometric and specific way, also belong to this group. Among them the most useful for the purpose of understanding the copper sites of proteins are those whose X-ray structure has already been worked out (13) as carbonic anhydrase. The spectroscopic data for two artificial copper proteins, namely carbonic anhydrase and carboxypeptidase are reported in Table II. No X-ray data are available for other copper proteins except bovine superoxide dismutase (14).

Even less can be said about Type 3 copper which does not show any clear spectroscopic feature. It is defined as a center made of a couple of copper ions (probably cupric) which is diamagnetic up to room temperature (EPR non detectable copper (2)). Some proteins which contain EPR non detectable copper are reported in Table III. Strictly, Type 3 is the EPR non detectable copper of the blue oxidases, even though there is evidence that the situation is very similar in oxytyrosinase (15) and oxyhemocyanin (16).

It can be concluded that a fairly clear cut classification of copper proteins in terms of spectroscopic features is not possible. Nonetheless to compare the properties of copper proteins between each other in search of possible homologies is a reasonable approach to the understanding of their structure and working mechanisms. In the following part of the paper we will show in particular how the properties of anionic derivatives of Cu(II) carbonic anhydrase, a non blue protein, can be used to build up a model for the site of blue copper.

Carbonic anhydrase is a metal protein which contains Zn(II) in the native state. X-ray data (13) have shown that the metal coordination sphere is made up of three histidyl residues, each

TABLE II

Optical and Magnetic Properties of Type 2 Copper in Some Copper Proteins

| Protein | Wavelength | ε ($M^{-1}$ $cm^{-1}$)(a) | $g_{\parallel}$ | $g_{\perp}$ | $\lvert A_{\parallel} \rvert$ ($cm^{-1}$) |
|---|---|---|---|---|---|
| Superoxide (b) dismutase (bovine plasma) | 680 | 150 | 2.265 | 2.108(c)<br>2.023(d) | 0.016 |
| Diamine oxidase (e) (pig kidney) | 470-500 | 1250 | 2.29 | 2.06 | 0.016 |
| Galactose oxidase (f) | 445<br>630<br>775 | 1155<br>1015<br>905 | 2.28 | 2.04 | 0.0185 |
| Laccase (g) (Rhus vernicifera) | | | 2.237 | 2.053 | 0.0178 |
| Ceruloplasmin (g) (human plasma) | | | 2.28 | 2.04 | 0.0136 |
| $Cu^{2+}$-Carboxi-peptidase A (h) | 795 | 124 | 2.327 | 2.057 | 0.0135 |
| $Cu^{2+}$-Carbonic anhydrase (i) | 719 | 110 | 2.310 | 2.077 | 0.0133 |

(a): referred to copper; (b): G. Rotilio et al. (1972), Biochemistry, 11, 2187; (c) = $g_y$; (d) = $g_x$; (e): B. Mondovi et al. (1967), J. Biol. Chem., 242, 1160; (f): The optical data are from M.J. Ettinger (1974) Biochemistry, 13, 1242 and EPR data from W.E. Blumberg et al. (1965), Biochim. Biophys. Acta, 96, 336; (g): from ref. 2; (h): R.C. Rosemberg et al. (1975), Amer. Chem. Soc., 97, 2092; (i) The optical data are from S. Lindskog and P.O. Nyman (1964), Biochim. Biophys. Acta, 85, 462 and EPR data from ref. 2.

TABLE III

Optical Properties of Type 3 Copper in Some Copper Proteins

| Protein | $\lambda_{max}$ | $\varepsilon(M^{-1}\ cm^{-1})$ |
|---|---|---|
| Oxyhemocyanin (Mollusc)(a) | 347 | 8900 |
| Oxytyrosinase (isozyme α from Agaricus bisporicus)(b) | 345 | 7800 |
| Laccase (Rhus vernicifera)(c) | 330 | 2600 |
| Ceruloplasmin (Human plasma)(c) | 332 | 4100 |

(a): K.E. Van Holde (1967), Biochemistry, 6, 93; (b): R.L. Jolley et al. (1974), J. Biol. Chem., 249, 335; (c): from ref. 2.

TABLE IV

Optical Absorption Bands of Cu(II)-Bovine Carbonic Anhydrase Anion Complexes and of Stellacyanin

| | L → M nm (ε) | L → L' nm (ε) | d - d nm (ε) |
|---|---|---|---|
| none (a) | - | - | 780 (130) |
| $I^-$ (a) | 355 (3200) | 445 (2800) | 790 (400) |
| $Br^-$ (a) | - | ∿400 (Sh) | 790 (300) |
| $SH^-$ | 375 (2700) | 475 ( 900) | 700 (250) |
| $CNS^-$ (a) | 360 ( 900) | - | 735 (150) |
| $N_3^-$ (a) | 400 (2700) | - | 735 (330) |
| $CN^-$ (a) | - | - | 690 (190) |
| $(SEtOH)^-$ | 410 (2000) | 515 ( 900) | 690 (330) |
| Stellacyanin (b) | 450 (1000) | 604 (4080) | 850 (790) |

(a): From ref. 19; (b) From ref. 22.

bound through one of their imidazole nitrogens and of a water molecule, and that Co(II) and Cu(II) are bound in the same site within 0.1 A. A tetrahedral geometry, trigonally distorted, is assigned to the Co(II) derivative (17), while the Cu(II) derivative is axial (18). All carbonic anhydrases have a very high affinity for monovalent anions. In the case of the Cu(II) derivatives this affinity is much greater (19) than that of superoxide dismutase, a copper protein with relatively similar spectroscopic properties (Table II) and whose metal coordination sphere is made up of four histidyl residues and of a water molecule (14) in a rhombically distorted arranagment (20). In Table IV are reported the relevant data for the visible and near U.V. absorption bands characteristic of some Cu(II) carbonic anhydrase anionic derivatives. The anions can roughly be divided into two groups: the first one includes $CN^-$, $N_3^-$, $CNO^-$, $CNS^-$; that is anions with good $\sigma$-donor and $\pi$-acceptor properties, which cause a shift of the d-d bands to shorter wavelengths and therefore are thought to bind in the place of the water molecule, coplanar with the three histidyl nitrogens. Since water and the sulfonamides (powerful inhibitors of the native enzyme) are known by X-ray structures (13) to bind to the metal and to be also hydrogen bonded to the -OH of threonine 197, by analogy it is suggested that the derivatives with these anions are stabilized by such a hydrogen bond.

To the second group belong the halides which have weak $\sigma$- and strong $\pi$-donor properties. The hydrogen bonding ability of these anions decreases from $F^-$ to $I^-$. On the contrary the stability of their complexes with Cu(II) carbonic anhydrase increases in this order (19), so that they are not likely to be stabilized by hydrogen bonding. The presence of intense charge transfer bands in the iodide derivative (Fig. 1a) and the linear dependence of the logarithm of the stability constant on the halide redox potential (Fig. 2) suggest that the complexes with these anions are stabilized by "donor-acceptor" or "charge-transfer" interactions. NMR and spectroscopic data also suggest (19) that the halides do not coordinate "in the place" of the water molecule but enter into the metal coordination sphere as a fifth ligand, giving rise to pentacoordinate metal species.

The assignment of the sulfide and 2-mercaptoethanol complexes to either group is not possible since they partake the characteristics of both groups. In fact they cause a shift of the copper d-d band to shorter wavelength (Fig. 3), as do the anions of the first group, and give rise to two intense charge-transfer bands (Fig. 1b) like the anions of the second group do. This is not unreasonable as the sulfides are good $\sigma$-donor and good $\pi$-donor anions. So we would suggest that they give rise to tetracoordinate metal species, stabilized by charge-transfer interactions.

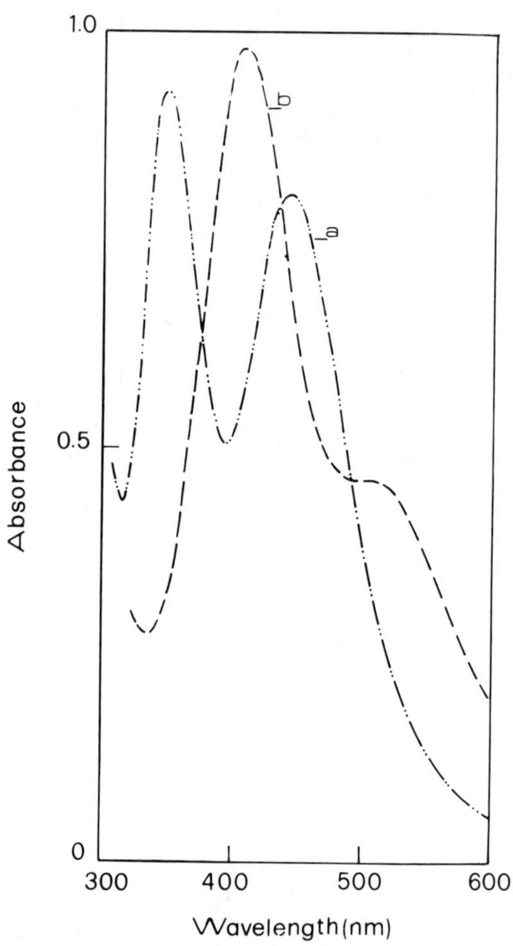


Figure 1. Charge transfer bands of iodide and 2-mercaptoethanol complexes of Cu(II) bovine carbonic anhydrase. a) 0.34 mM enzyme, 20 mM $I^-$; b) 0.6 mM enzyme, 1.0 mM 2-mercaptoethanol. Unbuffered aqueous solutions, pH about 6.0.

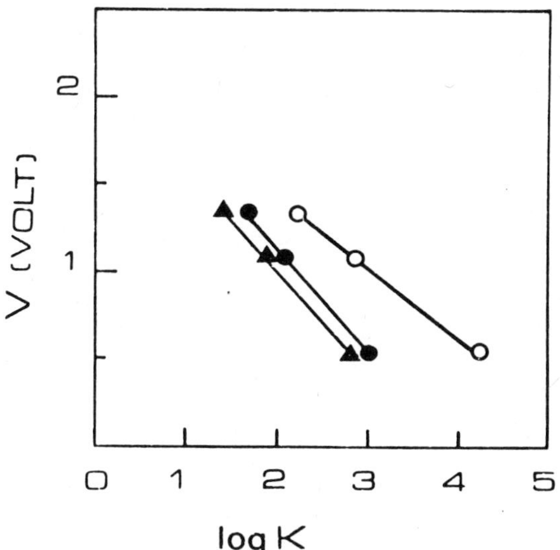


Figure 2. Relationship between the stability constant of Zn(II) ( ▲ - ▲ ), Co(II) (● - ●), and Cu(II) BCA (o - o) complexes with halides and the redox potential of $X_2/X^-$ couples.

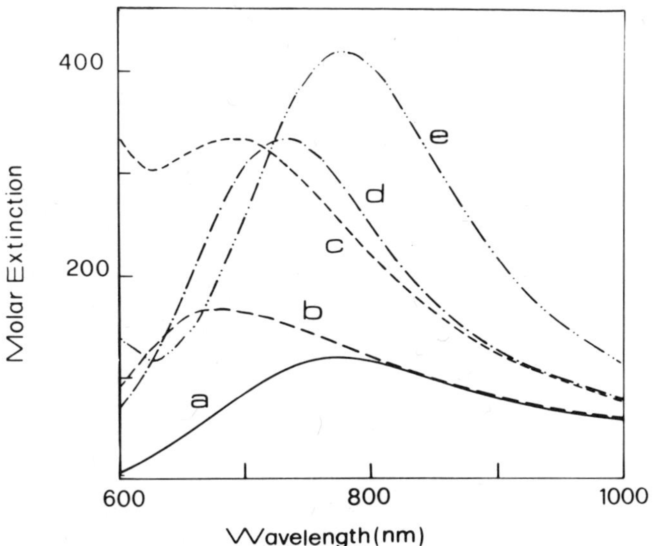


Figure 3. d-d absorption bands of Cu(II) bovine carbonic anhydrase and of its anion complexes. a) 1.3 mM enzyme; b) +3.0 mM $CN^-$; c) +1.3 mM 2, mercaptoethanol; d) +2.0 mM $N_3^-$ ; e) +20 mM $I^-$. Unbuffered aqueous solutions, pH about 6.0.

In Table IV are also reported the tentative assignments of the anion complexes-charge transfer bands, whose position correlates reasonably well with the donor properties of the anions.

The band of higher energy occurs in a region where anion → Cu(II) transitions are generally found (20) and is accordingly assigned. The band at lower energy, which only occurs in the complexes of the "charge-transfer" type is assigned to charge-transfer from the anion to a protein group with acceptor properties, possibly an imidazolium (21) or a Cu(II) bound imidazole. A more obvious assignment for the latter band would be a charge-transfer from the anion π-orbital to the metal. We are, however, more in favor of the first view since this band is missing in the much less stable iodide complexes of superoxide dismutase and since the halide complexes of Zn(II) carbonic anhydrase appear to be stabilized by interaction of the same type as in the Cu(II) enzyme (Fig. 2), despite the poor acceptor properties of Zn(II).

In Table V are reported the position and intensities of the ultraviolet absorption bands of some anion complexes of Co(II) carbonic anhydrase, obtained from difference spectra against the Co(II) protein itself. The position of the first band is uncertain and at the limit of instrumental resolution. By analogy with the Cu(II) carbonic anhydrase anionic derivatives, the two bands are assigned to anion → Co(II) and anion → imidazole-Co(II) charge transfer.

TABLE V

Optical Absorption Bands of Co(II)-Bovine Carbonic Anhydrase Anion Complexes and of Co(II)-Stellacyanin

| | L → M<br>nm (ε)(a) | L → M<br>nm (ε)(a) |
|---|---|---|
| $I^-$ | 310 ( 500) | 340 (350) |
| $SH^-$ | 308 ( 700) | 340 (350) |
| $N_3^-$ | 312 ( 750) | 325 (680) |
| $(SEtOH)^-$ | 300 (1750) | 390 (250) |
| Co(II)-Stellacyanin (b) | 310 (2600) | 360 (800) |

(a) Values obtained from differential spectra; (b) from ref. 23.

We will now compare the spectra of stellacyanin (22) and of Co(II) substituted stellacyanin (23) with the spectral data shown in Table IV and Table V respectively. The striking similarity of the d-d spectra of Co(II) stellacyanin and the sulfide derivative of Co(II) carbonic anhydrase has already been noted by Mc Millin et al. (23). From Table V it also appears that the ultraviolet portions of the spectra are similar. The metal coordination is believed to be tetrahedral in both cases. The similarity of native Cu(II) containing stellacyanin and of the sulfide complexes of Cu(II) carbonic anhydrase is not so striking, although some analogies can be found towards longer wavelength (see Table IV). The stellacyanin spectrum appears to be shifted by about a hundred nm with respect to the derivative. A reason for the shift can be due to differences in the geometries of the two metal sites which are approximately square planar in the case of the carbonic anhydrase derivative and which is tetrahedral in stellacyanin. A further assumption was that stellacyanin has the same geometry as the Co(II) enzyme, since the two metals were shown to bind at the same site. Tetrahedral complexes are in fact expected to absorb at longer wavelength (5). However, the data of Table VI where the differences in energy are reported between the L → L' bands of homologous Co(II) - Cu(II) couples, show that this difference is much larger for stellacyanin than for any other couple, especially for those complexes which have a more similar geometry such as probably the iodide complex.

TABLE VI

Energy Differences Between L → L' Bands of Homologous Co(II)-Cu(II) Couples of Protein Derivatives

| | |
|---|---|
| Stellacyanin: | |
| Co(II) - Cu(II) | = 27.4 - 16.5 = 10.9 kK |
| Bovine carbonic anhydrase: | |
| Co(II)SH - Cu(II)SH | = 29.4 - 21.0 = 8.4 kK |
| Co(II)I - Cu(II)I | = 29.4 - 22.4 = 6.9 kK |
| Co(II)(SEtOH) - Cu(II)(SEtOH) | = 25.6 - 19.6 = 6.0 kK |

Furthermore, it appears that the energy of the L → L' bands is very similar for Co(II) carbonic anhydrase-iodide and the -sulfide derivatives (or Co(II) stellacyanin) complexes in spite of the greater donor properties of the sulfide. The energy difference is much larger between Cu(II) carbonic anhydrase-iodide complex and native stellacyanin. All these data can be rationalized by assuming that the metal site of stellacyanin is pentacoordinate. In this case an L → L' charge transfer could certainly occur more easily than in a tetrahedral complex and hence the lower energy of this transition. Such a hypothesis is supported by the position and intensity of the stellacyanin band assigned as a d-d transition in Table IV. This band is actually an envelop of two transitions as shown by circular dichroism spectra (8) and by the optical spectrum of stellacyanin measured at high pH (Fig. 4). The two bands in the 800-900 nm region of the spectrum are in fact known to occur in low molecular weight pentacoordinate Cu(II) complexes, and their intensities can be quite high (24).

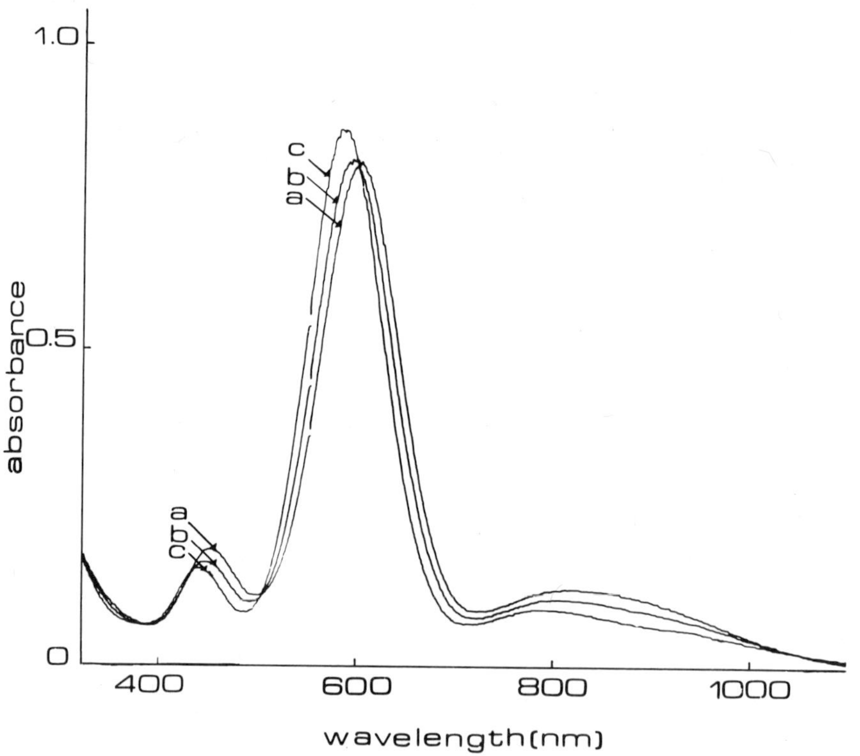


Figure 4. Optical absorption spectra of 0.2 mM Stellacyanin in 0.67 M carbonate buffer pH 9.0 (a), pH 10.0 (b), pH 11.0 (c).

An important implication of this pentacoordinate model is that reduction of Cu(II) must involve release of at least one ligand from the metal coordination sphere since Cu(I) is not likely to form pentacoordinate complexes. It is at present hard to speculate how this can, in fact, occur. However, we can take advantage of the Co(II)-derivative of the enzyme (23) which is known to have a tetrahedral active site and suggest that this complex is a model of the reduced copper site. The release of one ligand could occur with different ease in different blue proteins and would be dependent on the environment of the metal site. The environmental effect could account for the different redox potentials (25) of otherwise similar copper sites. Small differences in the nature or geometry of ligands could give rise to the observed variable spectroscopic properties of the blue copper centers as well. One might presuppose that in some proteins, such as for instance ascorbic oxidase which for physiological reasons may undergo seasonal variations, could show contents of blue and non-blue copper enzymes as a result of subtle conformational changes.

A further consideration arising from this work is the finding that the crystal field requirements of the metal in a particular oxidation state may play an important role in determining the structure of a metal binding site in metallo-proteins. Therefore metal substitution can cause considerable perturbation of the metal binding site but nonetheless can provide valuable information about this site.

## REFERENCES


1) Broman, L., Malmström, B.G., Aasa, R. and Vänngard, T. (1962) J. Mol. Biol. 5, 301.

2) Malkin, R. and Malmström, B.G. (1970) Advances in Enzymology 33, 177-244.

3) Blumberg, W.E. (1966) "Biochemistry of Copper" Blumberg, W.E., Peisach, J. and Aisen, P., Eds., Academic Press Inc., N.Y., pp. 49-66.

4) Brill, A.S. and Brice, G.F. (1968) J. Chem. Phys. 48, 4398-4404.

5) Morpurgo, G. and Williams, R.J.P. (1968) "Physiology and Biochemistry of Haemocyanins" Ghiretti, F., Ed., Academic Press, N.Y., pp. 113-130.

6) Gray, H.B. (1971) Adv. Chem. Ser. N. 100, 365.

7) Miskowsky, V., Tang, S.P.W., Spiro, T.G., Shapiro, E. and Moss, T.H. (1975) Biochemistry 14, 1244-1250.

8) Falk, K.E. and Reinhammar, B. (1972) Biochim. Biophys. Acta 285, 84-90.

9) Rotilio, G., Morpurgo, L., Graziani, M.T. and Brunori, M. (1975) FEBS Letters 54, 163-166.

10) Van Leeuwen, F.R., Wever, R., Van Gelder, B.F., Avigliano, L. and Mondoví, B. (1975) Biochim. Biophys. Acta 403, 285-291.

11) Avigliano, L., Gerosa, P., Rotilio, G., Finazzi-Agró, A., Calabrese, L. and Mondoví, B. (1972) It. J. Biochem. 21, 248-255.

12) Mondoví, B., Avigliano, L., Rotilio, G., Finazzi-Agró, A., Gerosa, P. and Giovagnoli, C. (1975) Mol. Cell. Biochem. 7, 131-135.

13) Lindskog, S., Henderson, L.E., Kannan, K.K., Liljas, A., Nyman, P.O. and Strandberg, B. (1971) "The Enzymes" Vol. 5, pp. 587-665.

14) Richardson, T.S., Thomas, K.A., Rubin, B.H. and Richardson, D.C. (1975) Proc. Nat. Acad. Sci. U.S.A., 72, 1349-1353.

15) Jolley, R.L., Evans, L.H., Makino, N. and Mason, H.S. (1974) J. Biol. Chem. 249, 335-345.

16) Van Holde, K.E. (1967) Biochemistry 6, 93-99.

17) Coleman, J.E. and Coleman, R.V. (1972) J. Biol. Chem. 247, 4718-4728.

18) Taylor, J.S. and Coleman, J.E. (1971) J. Biol. Chem. 246, 7058-7067.

19) Morpurgo, L., Rotilio, G., Finazzi-Agró, A. and Mondoví, B. (1975) Arch. Biochem. Biophys. 170, 360-366.

20) Morpurgo, L., Giovagnoli, C. and Rotilio, G. (1973) Biochim. Biophys. Acta 322, 204-210.

21) Shinitzky, M., Katchalski, E., Grisaro, V. and Sharon, N. (1966) Arch. Biochem. Biophys. 116, 332-343.

22) Malmström, B.G., Reinhammar, B. and Vänngard, T. (1970) Biochim. Biophys. Acta 205, 48-57.

23) Mc Millin, D.R., Holwerda, R.A. and Gray, H.B. (1974) Proc. Nat. Acad. Sci. U.S.A. 71, 1339-1341.

24) Elliott, H., Hataway, B.J. and Slade, R.C. (1966) J. Chem. Soc. (A) 1443-1445.

25) Reinhammar, B. (1972) Biochim. Biophys. Acta 275, 245-259.

# COLLAGEN CROSS-LINKING: THE SUBSTRATE SPECIFICITY OF LYSYL OXIDASE

Robert C. Siegel, Joseph C.C. Fu and Yu-Hua Chang

University of California, San Francisco

Third & Parnassus, San Francisco, California 94143

Lysyl oxidase is a specific amine oxidase that contains copper and is irreversibly inhibited by compounds called lathyrogens (1,2,3). β-Aminopropionitrile (BAPN) is the most widely studied and best known of these compounds. When growing animals are either fed lathyrogens or made copper deficient they develop a wide variety of connective tissue and skeletal abnormalities such as dissecting aneurysms of the thoracic aorta, kyphoscoliosis and abdominal hernias. The pathogenesis of these abnormalities is related to decreased collagen cross-linking secondary to decreased lysyl oxidase activity which results from either lack of the copper cofactor(4) or irreversible inhibition by BAPN. Since clinical abnormalities in these disorders have been noted as a result of deranged biosynthesis in both elastin and collagen with lysyl oxidase deficiency, it seemed appropriate to define the substrate specificity of highly purified lysyl oxidase in more detail and to determine whether the same enzyme is active on collagen and elastin substrates (5,6,7,8).

Under the influence of lysyl oxidase, the ε amino groups of certain lysyl residues in collagen and elastin and of certain hydroxylysyl groups in collagen are oxidatively deaminated to form the corresponding δ-semialdehydes, allysine and hydroxyallysine(9,10) (Figure 1, Equations 1 and 2). In elastin, three allysyl and one lysyl residue condense to form the desmosines. In collagen, allysine and hydroxyallysine condense by Schiff base formation with the ε-$NH_2$ groups of certain lysyl residues in other molecules to form the reducible intermolecular cross-links, dehydrohydroxylysinohydroxynorleucine (de HLHNL) and dehydrohydroxylysinonorleucine (de HLNL) as outlined in Figure 1. Two allysyl residues may also undergo an aldol condensation and form an intramolecular cross-link. In the collagen fibril, these reducible cross-links are further metabolized

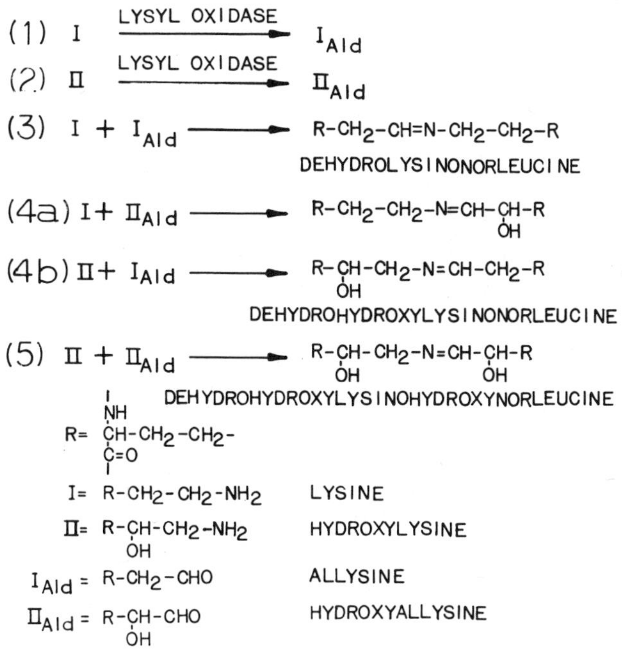

$$(1)\quad \mathrm{I} \xrightarrow{\text{LYSYL OXIDASE}} \mathrm{I_{Ald}}$$

$$(2)\quad \mathrm{II} \xrightarrow{\text{LYSYL OXIDASE}} \mathrm{II_{Ald}}$$

$$(3)\quad \mathrm{I} + \mathrm{I_{Ald}} \longrightarrow \text{R-CH}_2\text{-CH=N-CH}_2\text{-CH}_2\text{-R}$$

DEHYDROLYSINONORLEUCINE

$$(4a)\quad \mathrm{I} + \mathrm{II_{Ald}} \longrightarrow \text{R-CH}_2\text{-CH}_2\text{-N=CH-CH(OH)-R}$$

$$(4b)\quad \mathrm{II} + \mathrm{I_{Ald}} \longrightarrow \text{R-CH(OH)-CH}_2\text{-N=CH-CH}_2\text{-R}$$

DEHYDROHYDROXYLYSINONORLEUCINE

$$(5)\quad \mathrm{II} + \mathrm{II_{Ald}} \longrightarrow \text{R-CH(OH)-CH}_2\text{-N=CH-CH(OH)-R}$$

DEHYDROHYDROXYLYSINOHYDROXYNORLEUCINE

R= –NH–CH(–C=O–)–CH$_2$–CH$_2$–

I= R-CH$_2$-CH$_2$-NH$_2$ LYSINE

II= R-CH(OH)-CH$_2$-NH$_2$ HYDROXYLYSINE

$\mathrm{I_{Ald}}$ = R-CH$_2$-CHO ALLYSINE

$\mathrm{II_{Ald}}$ = R-CH(OH)-CHO HYDROXYALLYSINE

Figure 1. Summary of the biosynthesis of bifunctional collagen cross-links.

to form polyfunctional cross-links such as dehydrohydroxymerodesmosine which is formed from an aldol condensation product and a hydroxylysyl residue and dehydrohistidinohydroxymerodesmosine (de HHMD) formed by Michael addition of histidine to hydroxymerodesmosine. Although the mechanism is unknown, the concentration of the Schiff base cross-links decreases in the collagen fibril as it ages. Presumably, this is due to synthesis of polyfunctional cross-links from the Schiff bases (11,12,13,14,15).

From this brief review of collagen cross-linking, it should be apparent that a wide variety of collagen cross-links are synthesized at least in part from the aldehyde cross-link intermediates generated by lysyl oxidase. The principal questions to be examined in this study are whether cross-link biosynthesis is regulated by lysyl oxidase through synthesis of the aldehyde intermediates or whether other factors might be required for collagen cross-linking.

Lysyl oxidase was initially detected in phosphate saline extracts of embryonic chick cartilage by a tritium release assay with insoluble elastin biosynthetically labelled with 6-$^3$H lysine (1). As aldehydes formed at the 6 positions, tritium was released and could be isolated by distillation and measured. It was not possible to use similarly labelled collagen substrates for a sensitive assay until we noted that the enzyme has much higher activity with collagen precipitated as reconstituted, native fibrils than for collagen monomers or even insoluble elastin. In the work discussed below, chick calvaria collagen precipitated as fibrils was used as the collagen substrate; chick aorta elastin as the elastin substrate and 17 day old chick embryo cartilage was used as the enzyme source(6).

Purification of Lysyl Oxidase. The chick cartilage was initially extracted with 0.15 M NaCl 0.10 M $NaH_2PO_4$ pH 7.8 (PBS) twice and then twice with 6 M urea, 0.05 Tris, pH 7.6. For this study, the phosphate saline extracts were discarded and the two urea extracts were pooled. These were 7 fold more pure than the phosphate-saline extracts with respect to lysyl oxidase activity and contained approximately 70% of the total recoverable activity. They were dialyzed against PBS and then assayed with collagen and elastin substrates. Two peaks of activity were found and these were similar with both substrates. The second peak which contained most of the activity was further purified by absorption to an "affinity" resin that consisted of lathyritic rat skin collagen covalently linked to sepharose 4B. The enzyme was eluted from the resin in 6 M urea, 0.05 Tris pH 7.6 and then chromatographed on a second DEAE column (Figure 2). A single peak of activity was found that had a similar activity profile with either the collagen or elastin substrate. This appeared to be a single protein species with molecular weight 64,000 when run on 5% SDS-acrylamide gels with mercaptoethanol reduction (Figure 3). This material had specific activity approximately 1000 fold greater than the initial urea extract and 7000 fold greater than the first lysyl oxidase preparation reported. Further purification by recycling through another "affinity" and DEAE step did not increase the specific activity further. Approximately 1.0 microgram of the purified enzyme was used in each experiments described below.

As illustrated in Table I, highly purified lysyl oxidase is active with both collagen and elastin substrates. If collagen fibril formation is inhibited by addition of arginine, little activity is observed. If fibril formation is accelerated by addition of aspartic acid to the bone substrate before preincubation, there is a slight increase in activity. If embryonic chick skin collagen is used as substrate, there is low activity with highly purified lysyl oxidase. However, this is due to lack of collagen precipitation as fibrils rather than high tissue specificity of the enzyme. With preincubation to form fibrils, only 3% of the radioactivity precipitated from the partially purified skin collagen solution as compared to 78% with pure bone collagen. There was no precipitation with arginine. However, with aspartic acid, fibril formation increased to 35% and apparent enzyme activity increased similarly. Amino acid analysis indicated that this preparation was approximately

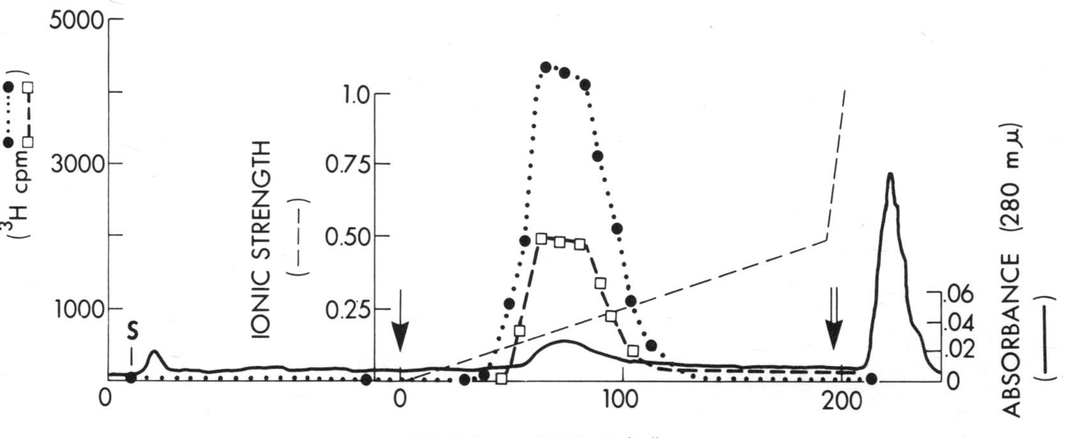


Figure 2. DEAE-cellulose chromatography of partially purified lysyl oxidase. The enzyme was loaded onto a 1.5 x 10 cm DEAE-52 column that had been previously equilibrated with 6 M urea, 0.05 M Tris, pH 7.6 at 25°. Elution was achieved with a 0-0.5 M NaCl gradient in 6 M urea, 0.05 M Tris pH 7.6 in a total volume of 200 ml. Assays were done with 6-$^3$H lysine collagen fibrils (●——●) or 4,5$^3$H lysine elastin (□——□) substrates with 500,000 cpm per assay tube.

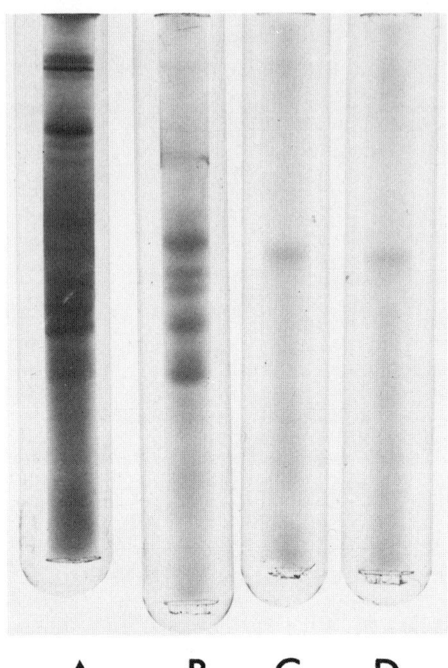


Figure 3. SDS-acrylamide gel electrophoresis of enzyme fractions during purification of lysyl oxidase. All samples were applied after reduction with mercaptoethanol and were run in 5% acrylamide gels at 8 ma/tube for 4 hours.

A. Initial urea extract.
B. Major peak of lysyl oxidase activity after first DEAE chromatogram.
C. Fraction with lysyl oxidase activity after elution with 6 M urea, 0.05 M Tris, pH 7.6 from an "affinity" resin consisting of sepharose 4B with rat skin collagen that was covalently linked after CNBr activation.
D. Fraction with maximum lysyl oxidase activity after chromatography of fractions similar to gel C on DEAE again.

one-third collagen. These experiments provide evidence that the conformation of the substrate rather than the tissue of origin of the collagen is the principal determinant of enzyme activity.

In addition, the relative activity of the purified enzyme with intact, native fibrils, denatured collagen and renatured $\alpha$ chains

TABLE I. LYSYL OXIDASE ACTIVITY WITH VARIOUS COLLAGEN SUBSTRATES

| | Net $^3H$ Release[a] Per 5 x $10^5$ cpm Substrate | % $^3H$ Release Compared to Reconstituted Bone Collagen Fibril |
|---|---|---|
| Insoluble Elastin | 1145 | 46.0 |
| Bone Collagen Fibrils | 2490 (75)[f] | 100 |
| +0.05 M Arginine[b] | 124.5 (5.0)[f] | 5.0 |
| +0.05 M Aspartic Acid[b] | 2590 (81)[f] | 104 |
| Skin Collagen Fibrils | 450 (3.3)[f] | 18.1 |
| +0.05 M Arginine[b] | 75 (0)[f] | 3.0 |
| +0.05 M Aspartic Acid[b] | 1025 (35.3)[f] | 41.2 |
| Denatured Bone Fibrils[c] | 142 | 5.7 |
| Intact, Native Bone Fibrils[d] | 54.1 | 2.2 |
| Renatured Collagen, labelled α1 Chains[e] | 317.9 | 12.8 |
| Renatured Collagen, labelled α2 Chains[e] | 997.3 | 40.0 |

[a]Assays for two hours at 37°. All substrates preincubated for one hour.

[b]Added to substrate prior to preincubation.

[c]Heated to 60° to 15 minutes.

[d]Unhomogenized chick calvaria.

[e]Prepared by renaturing $6^3H$ lysine α1 or α2 chains with unlabelled α chains of the opposite type from chick bone collagen.

[f]Percent radioactivity precipitating from solution after one hour preincubation is given in parentheses.

from collagen was also studied. There was little activity with intact, native fibrils prepared from labelled calvaria. There was low activity with denatured collagen but there was significantly higher activity with renatured, labelled $\alpha2$ chains than with labelled $\alpha1$ chains renatured with unlabelled $\alpha2$ chains. In either case, the activity was considerably less than that with collagen fibrils. The greater activity with renatured labelled $\alpha2$ chains than $\alpha1$ chains suggest that the initial site of enzyme action in the intact collagen molecule may be on the $\alpha2$-chain. At present, the reason for this is unknown although it might be related to the fact that the $\alpha2$ chain has twice as much hydroxylysine as the $\alpha1$ chain.

Finally, we studied the synthesis of aldehyde cross-link intermediates and collagen cross-links with highly purified lysyl oxidase. For these experiments $^{14}C$ lysine labelled chick calvarial collagen was used. This substrate was pure collagen as judged by amino acid analysis. Formation of allysine and hydroxyallysine intermediates was examined by amino acid analysis after oxidizing the collagen with performic acid. Cross-links derived from $^{14}C$ lysyl residues were measured by amino acid analysis after borohydride reduction. As illustrated in Table II, both hydroxyallysine and allysine intermediates were synthesized in the presence of lysyl oxidase. Synthesis was inhibited by BAPN. The two major reducible cross-links, dehydrohydroxylysinohydroxynorleucine and dehydrohydroxylysinonorleucine, were also synthesized. In addition, levels of the tetrafunctional cross-link, dehydrohistidinohydroxymerodesmosine, were detectable after 24 hours incubation. By forty-eight hours, the levels of the reducible Schiff base cross-links de HLHNL and de HLNL were lower than at 4 hours. This decrease was significant and is similar to the decrease which occurs with maturation of collagen fibrils in vivo.

In summary, these studies provide evidence that lysyl oxidase, purified to homogeneity on SDS-acrylamide gel electrophoresis and to constant specific activity is active with an elastin substrate as well as with collagen substrates from at least two different tissues. In collagen the activity is greatest with loosely packed, reconstituted fibrils rather than intact native fibrils isolated from bone. The activity is greater with renatured labelled $\alpha2$ chains than $\alpha1$ chains. This may indicate that the initial site of enzyme activity in the collagen molecule is on the $\alpha2$ chain. With reconstituted bone fibrils, collagen cross-link synthesis catalyzed by the highly purified enzyme closely resembles biosynthesis in vivo. There is synthesis of both hydroxyallysine and allysine cross-link intermediates as well as synthesis of the two principal intermolecular cross-links, dehydrohydroxylysinohydroxynorleucine and dehydrohydroxylysinonorleucine. With prolonged incubation, the concentration of the reducible Schiff base cross-links decreases and the concentration of the tetrafunctional cross-link, dehydrohistidinohydroxymerodesmosine increases. Synthesis of all of these is

TABLE II. LYSYL OXIDASE DEPENDENT SYNTHESIS OF COLLAGEN CROSS-LINKS AND ALDEHYDE INTERMEDIATES

| A. Aldehyde Intermediates | $^{14}C$ cpm/$10^5$ (Lys + Hylys) cpm | | |
|---|---|---|---|
| Incubation Time | 5° | 24° | 24° + BAPN[c] |
| Allysine[a] | 510 | 580 | 270 |
| Hydroxyallysine[b] | 760 | 1150 | 500 |

| B. Collagen Cross-Links | $^{14}C$ cpm/$10^5$ (Lys + Hylys) cpm | | | |
|---|---|---|---|---|
| Incubation Time | 4° | 24° | 48° | 48° + BAPN[c] |
| Dehydrohydroxylysinonorleucine[d] | 1347 | 1205 | 1122 | 278 |
| Dehydrohydroxylysinohydroxynorleucine[e] | 1179 | 752 | 584 | 104 |
| Dehydrohistidinohydroxymerodesmosine[f] | 131 | 194 | 370 | 135 |

[a]Measured after performic acid oxidation as α-aminoadipic acid.

[b]Measured after performic acid oxidation as glutamic acid.

[c]50 mcg/ml.

[d]Measured after borohydride reduction as hydroxylysinonorleucine.

[e]Measured after borohydride reduction as hydroxylysinohydroxynorleucine.

[f]Measured after borohydride reduction as histidinohydroxymerodesmosine.

inhibited by BAPN or by compounds that inhibit fibril formation. These results indicate that lysyl oxidase and collagen fibrils are all that is required for the biosynthesis of collagen cross-links in vivo.

## ACKNOWLEDGEMENTS

The original research described in this paper was supported by National Institute of Health Grants AM16424 and AM18237 and a grant from the National Foundation. R.C.S. is recipient of NIH Research Career Development Award K01 AM00114.

## REFERENCES

1. Pinnel, S.R. and Martin, G.R. (1968) Proc. Nat. Acad. Sci. U.S. 61, 708.
2. Narayanan, A.S., Siegel, R.C., and Martin, G.R. (1974) Arch. Biochem. Biophys. 162, 231.
3. Chvapil, M., McCarthy, D.W., Misiorowski, R.L., Madden, J.W., and Peacock, E.E. (1974) Proc. Soc. Exp. Biol. Med. 146, 688.
4. Harris, E.D., Gonnerman, W.A., Savage, J.E., and O'Dell, B.L. (1974) Biochim. Biophys. Acta 341, 332.
5. Shieh, J.J., Tamaye, R., and Yasunobu, K.T. (1975) Biochim. Biophys. Acta 377, 229.
6. Siegel, R.C. (1974) Proc. Nat. Acad. Sci. U.S. 71, 4826.
7. Martin, G.R., Pinnell, S.R., Siegel, R.C. and Goldstein, E.R. (1970) Chemistry and Molecular Biology of the Intercellular Matrix, Balazs, E.A., Ed., Academic Press, New York.
8. Vidal, G., Shieh, J.J., and Yasunobu, K.T. (1975) Biochim. Biophys. Res. Commun., 64, 989.
9. Fowler, L.J., Peach, C.M., and Bailey, A.J. (1970) Biochem. Biophys. Res. Commun. 41, 251.
10. Siegel, R. C. and Martin, G. R. (1970) J. Biol. Chem. 245, 1653.
11. Tanzer, M.L. (1973) Science 180, 561.
12. Davis, N.R., Risen, O.M., and Pringle, G.A. (1975) Biochemistry 14, 2031.
13. Gallop, P.M. and Paz, M. (1975) Phys. Rev. 55, 418.
14. Robins, S.P., Shimokomaki, M., and Bailey, A.J. (1973) Biochem. J. 131, 771.
15. Traub, W. and Piez, K.A. (1971) Adv. Prot. Chem. 25, 243.

# PURIFICATION AND PROPERTIES OF LUNG LYSYL OXIDASE, A COPPER-ENZYME

James J. Shieh and Kerry T. Yasunobu
Dept. of Biochem-Biophysics, University of Hawaii
School of Medicine, Honolulu, Hawaii 96822

## INTRODUCTION

Lysyl oxidase is an enzyme which converts peptidyl lysyl or hydroxyllysyl to peptidyl α-amino adipic or δ-hydroxyl-α-amino-adipic ε-semialdehyde. These so called allysine residues are the precursors of desmosine and isodesmosine which form cross links in elastin. The enzyme was detected for the first time in chick bone by Pinnell and Martin (1). At the present time Harris et al (2), Siegel et al (3), Stassen (4) and Vidal et al (5) have reported the isolation of highly purified preparations of the enzyme from chick aorta, chick cartilage, embroyonic chick cartilage and bovine aorta, respectively. In the present investigation, highly purified preparation of bovine lung lysyl oxidase was isolated and some of its properties are described in this report.

## MATERIALS AND METHODS

### Substrate preparation

T-4,5-lysine labelled tropoelastin was prepared from chick embryo aorta slices, by the method of Pinnell and Martin (1). Intrinsic lysyl oxidase in the chick aorta was inactivated by dialyzing the soluble extract against 0.1 M HCl for 24 hours. Contamination of the substrate with radioactive lysine was minimized by dialyzing against 0.01 M potassium phosphate buffer containing 0.9% NaCl.

### Assay for lysyl oxidase

The reaction mixtures contained enzyme, 300,000 cpm of tritium labelled tropoelastin and 5 μmole of potassium phosphate buffer containing 0.9% NaCl in a total volume of 1.3 ml. Samples were incubated for 5 hrs at 37°C in a water bath with moderate shaking. After incubation the reaction was stopped by adding 50 μg β-aminopropionitrile. After distillation, the water was counted in a Packard Tri-carb Liquid Scintilation Spectrometer. After appropriate corrections, the enzyme specific activity was expressed as counts per minute per mg of protein as described by Pinnell and Martin (1). The counting efficiency of tritiated water was found to be about 31%.

Protein concentration

It was determined spectrophotometrically and an $E^{1\%}_{1cm}$ value of 15.0 at 280 nm was used.

Collagen-Sepharose 4B affinity chromatography

It was prepared by coupling 200 mg collagen (Sigma Co.) with 200 ml of Sepharose 4B (Pharmacia Co.) in borate buffer or bicarbonate buffer at pH 9.0 as described by Cuatrecasas and Anfinsen (6).

Gel electrophoresis

Electrophoresis was carried out in a solution of glycine-diethylbarbiturate buffer, pH 7.0, and 4.5% polyacrylamide gel was used (7). About 50 μg of enzyme was applied to the gel. SDS-gel electrophoresis was performed as described by Weber and Osborn (8).

## RESULTS

### Preparation of highly purified enzyme

Preparation of the homogenate

About 2000 g of lung was sliced and homogenized for 5 minutes with 8 ℓ of extraction medium (0.03 M potassium phosphate buffer, pH 7.6 - 4 M urea solution). The suspension was then centrifuged for 15 min at 16,000 x g. The supernatant fraction was found to contain 8,820,000 units of the enzyme with a specific activity of 59.

Fractionation with ammonium sulfate

The supernatant solution was fractionated with ammonium sulfate and the fraction precipitating between 30-70% saturation was collected. The precipitate was dialyzed against 0.01 M potassium phosphate buffer, pH 7.6, for two days and found to contain 2,889,600 units of enzyme with a specific activity of 127.

Chromatography on DEAE-cellulose

The enzyme was placed on a DEAE-cellulose column (4.5 x 150 cm). The column was washed with about 4 liter of 0.01 M potassium phosphate buffer and then was followed by about 4 ℓ of 0.01 M potassium buffer-4 M urea and about 4 ℓ of 0.07 M potassium phosphate buffer-4 M urea solution to elute lysyl oxidase A and B as shown in Fig. 1.

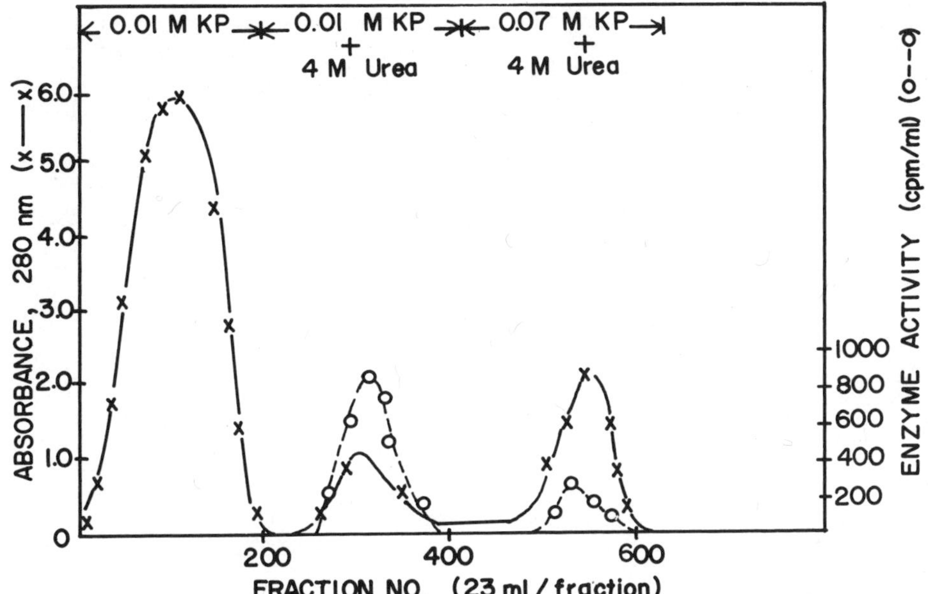


Fig. 1. DEAE-cellulose chromatography of lung lysyl oxidase. The enzyme, 22.64 g containing 2,889,600 units was applied to a column (4.5 x 150 cm) after it had been purified by ammonium sulfate fractionation. Fractions eluted by 0.01 M and 0.07 M potassium phosphate - 4 M urea, pH 7.6 were designed as lysyl oxidase A and B, respectively.

The enzyme in the urea-buffer solution, was precipitated with ammonium sulfate and the resulting precipitate was dialyzed against 0.01 M potassium phosphate buffer, pH 7.6. About 95,260 units of lysyl oxidase A was present at this step and the specific activity was 897. Total lysyl oxidase B units at this step was 96,000 and the specific activity of the enzyme was 320.

<u>Chromatography on Collagen-Sepharose 4B</u>

The resulting enzyme from the previous step was applied to a 4.0 x 50 cm affinity column. Following the removal of unbound inert protein by means of 0.1 M buffer, pH 7.6 containing 0.5 M NaCl solution, the column was eluted with the same buffer containing 6 M urea as shown in Figs. 2 and 3. The eluated enzyme fractions which contained activity were pooled and were then precipitated with 75% ammonium sulfate. The precipitate was suspended in 0.1 M potassium phosphate buffer, pH 7.6, and was dialyzed against the same buffer. About 61,200 units of enzyme A with a specific activity of 16,721 and 94,500 units of enzyme B with a specific activity of 7365 were isolated.

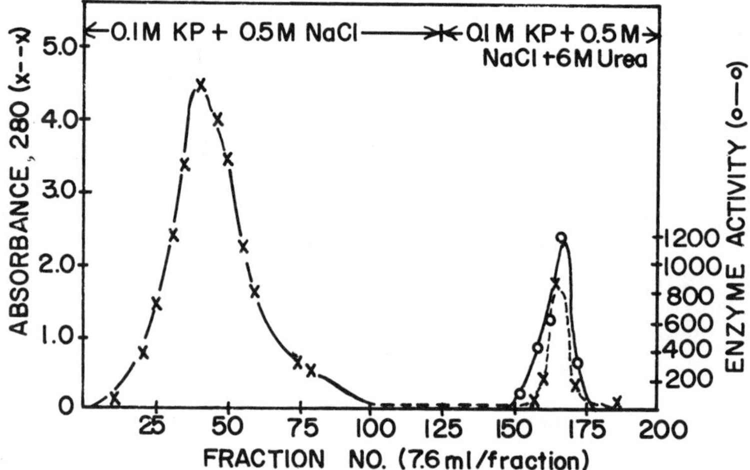


Fig. 2. Collagen Sepharose 4B affinity chromatography of purified lysyl oxidase A. The enzyme A, 106.2 mg containing 95,260 units was applied to a column (4.0 x 50 cm). After unbound proteins had passed through the column, the enzyme was eluted by the same buffer containing 6 M urea. The fractions which contained enzyme were pooled.

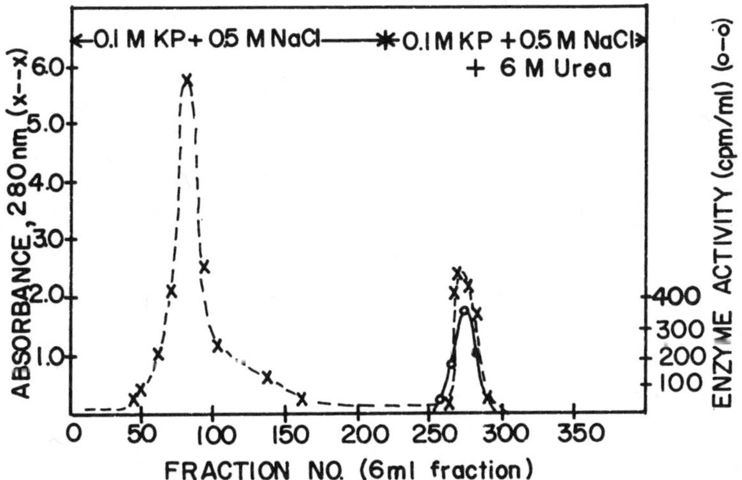


Fig. 3. Collagen Sepharose 4B affinity chromatography of purified lysyl oxidase B. The enzyme (300 mg, 96,000 units) was applied to a column (4.0 x 50 cm) equilibrated with 0.1 M potassium phosphate buffer-0.5 M NaCl, pH 7.6. After unbound proteins was removed, buffer containing 6 M-urea was used to elute the enzyme. The fractions which contained enzyme were pooled.

A summary of the purification of lysyl oxidase is shown in Table I.

TABLE I

Purification of Lysyl Oxidase from Bovine Lung

| Procedure | Total Volume (ml) | Total Activity (cpm) | Total Protein (mg) | Specific Activity (cpm/mg) |
|---|---|---|---|---|
| Crude Extract | 10,000 | 8,820,000 | 149,000 | 59 |
| 1. Ammonium sulfate | 2,000 | 2,889,600 | 22,640 | 127 |
| 2. DEAE-cellulose a. 0.01 M eluate | 20 | 95,260 | 106.2 | 897 |
| DEAE-cellulose b. 0.07 M eluate | 40 | 96,000 | 300 | 320 |
| 3. Collagen-Sepharose a. 0.01 M eluate | 4 | 61,200 | 3.66 | 16,721 |
| Collagen-Sepharose b. 0.07 M eluate | 5 | 94,500 | 12.83 | 7,365 |

Purity checks

The purified enzyme exhibited only one protein band when examined by the disc electrophoresis. One protein band was also observed in SDS-disc electrophoresis experiments in which 2-mercaptoethanol was omitted.

Survey of the lysyl oxidase in different parts of the lung.

The lung was dissected into trachea, combined bronchi, pleura and parenchymal tissues. About 50 g of each tissue was homogenized in 0.1 M potassium phosphate buffer, pH 7.6. Lysyl oxidase activity was mainly located in the parenchymal tissue (21.6 units/mg of protein.) Slight activity was found in the bronchial tissues (4.0) but no significant amounts of activity were detected in the trachea and pleura tissues as shown in Table II.

TABLE II

Distribution of Lysyl Oxidase in Various Portions of Bovine Lung

| Tissue | Volume (ml) | Activity (cpm/ml) | Protein Concentration (mg/ml) | Specific Activity |
|---|---|---|---|---|
| Trachea | 25 | --- | 8.67 | --- |
| Bronchi | 26 | 61 | 15.71 | 4.0 |
| Pleura | 27 | --- | 7.13 | --- |
| Parenchyma | 30 | 368 | 17.00 | 21.6 |

Lysyl oxidase content of the bovine lung as a function of age.
The results obtained in this investigation on the enzyme content of the lung with age are summarized in Fig. 4.

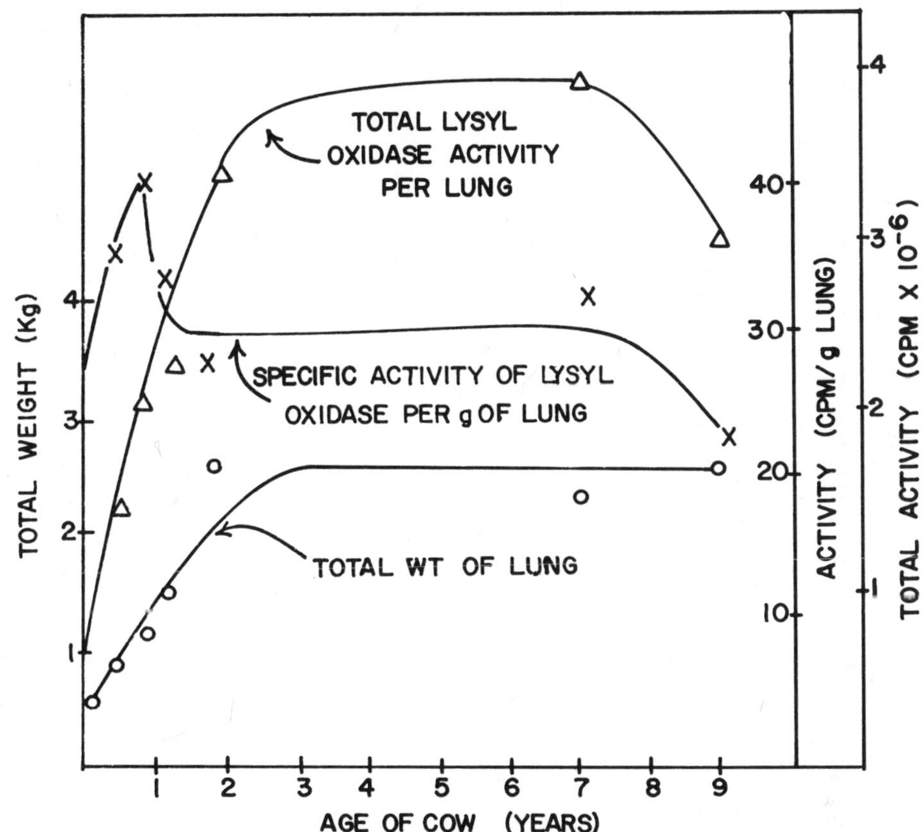


Fig. 4. The lysyl oxidase content of whole lung at different stages of development.

The weight of the lung reached a fairly constant value after 22 months. The total lysyl oxidase content of the lung increased up to 2 years, remaining constant till about 7 years and appeared to be decreasing after this age. The soluble protein present in the homogenate followed a similar pattern. The specific activity of the enzyme in the homogenate started to drop after 10 months and reached a constant level at about 70% of the initial value.

Proportionality of lysyl oxidase activity with incubation time.
Lysyl oxidase activity was linear with respect to time when the reaction mixture was incubated for periods for two or more hours as shown in Fig. 5. There was a lag period for 1-2 hours but the reason for it is unknown. The incubation time chosen for the subsequent experiments was five hours.

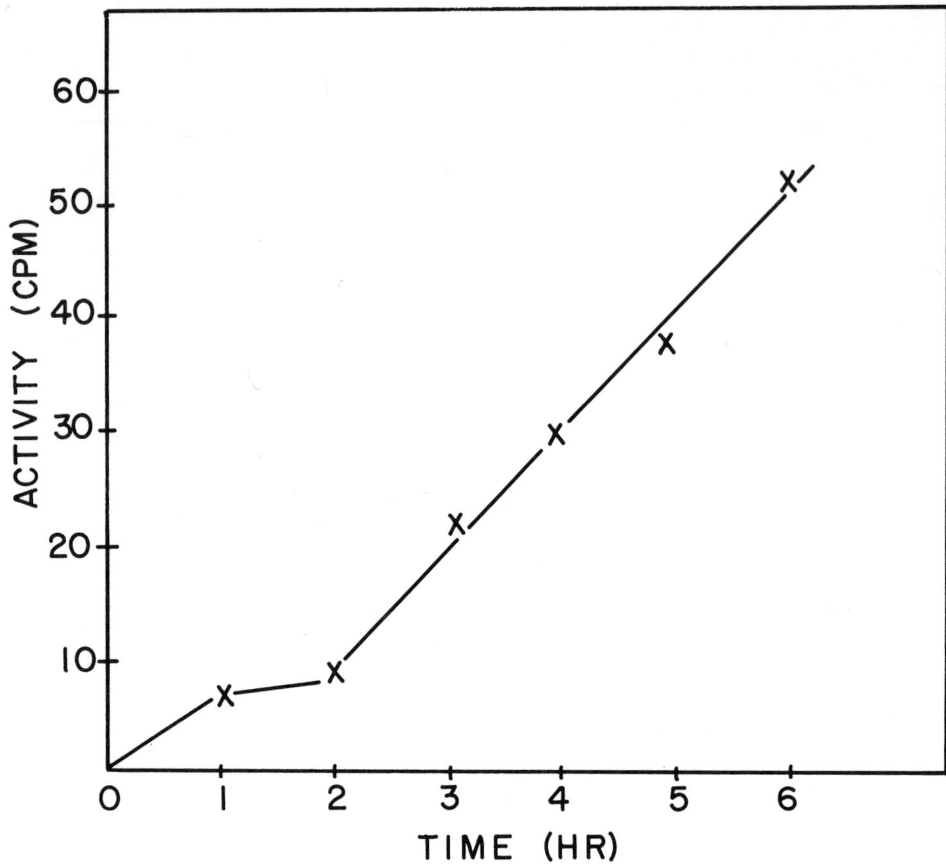


Fig. 5. Reaction rate of lysyl oxidase in relation to incubation time. Standard enzyme assay procedures were used and the enzyme assay mixture contained 20 μg protein.

Proportionality of lysyl oxidase activity and protein concentration. A linear relationship between protein concentration and enzyme activity was observed up to about 1,300 units of enzyme activity as shown in Fig. 6a and 6b.

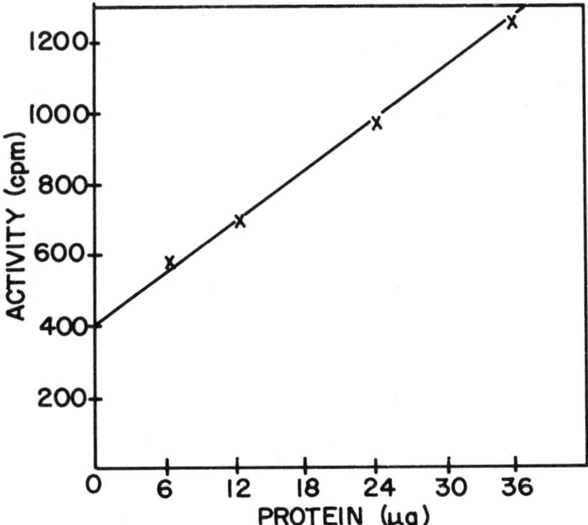


Fig. 6a. Reaction rate of lysyl oxidase A with respect to protein concentration. The standard enzyme assay procedures were used.

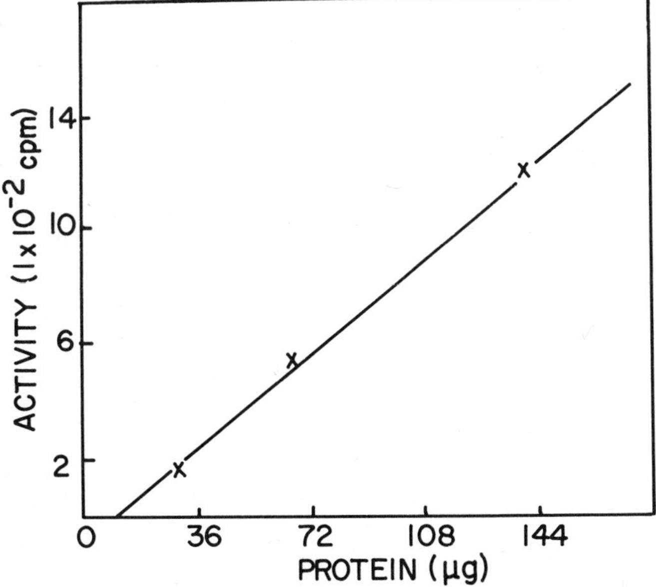


Fig. 6b. Reaction rate of lysyl oxidase B with respect to protein concentration. The standard enzyme assay procedures were used.

pH optimum of lysyl oxidase.

Lysyl oxidase was measured in acetate, phosphate and Tris-HCl buffers within a pH range from 6.5 to 9.0 at intervals of 1.0 pH unit. Highest activity was observed between pH 7 to 8 as shown in Fig. 7.

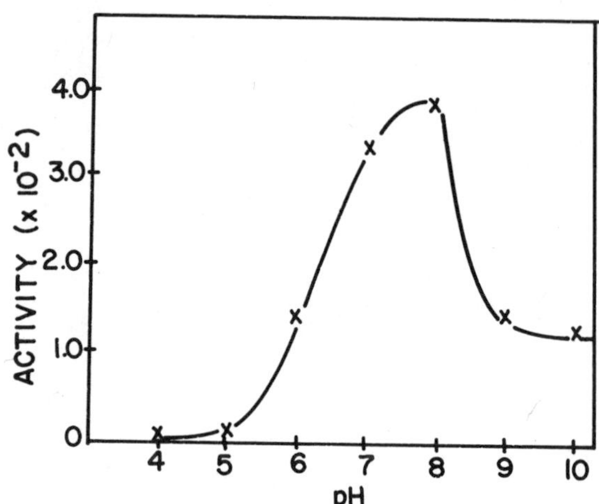


Fig. 7. Reaction rate of lysyl oxidase A in relation to pH. Acetate, phosphate and Tris-HCl buffer (0.1 M) were used in the reaction mixture and the standard assay procedures were run. The assay mixture contained 6 μg protein.

Stability of lysyl oxidase.

Only a trace of enzyme activity remained when the enzyme was heated at 60°C for 15 minutes. The purified enzyme could be stored at 0°C for months without losing significant enzyme activity. Urea was found to inhibit the enzyme 50% when the urea concentration was about 0.7 M.

Effect of substrate concentration.

Tritiated chick embryo soluble protein (1) was used as the substrate for the experiment. When the assay was run in the presence of various concentrations of substrate which contained from 0.75 x $10^5$ to 6.5 x $10^5$ cpm, the results shown in Fig. 8 were obtained. Substrate saturation was noted when the substrate contained about 320,000 cpm.

Effect of β-aminopropionitrile on lysyl oxidase activity.

The compound β-aminopropionitrile is known to be a lathyrogen and an inhibitor of lysyl oxidase (9). About 50% inhibition of the enzyme was observed when the concentration of β-aminopropionitrile was 0.4 mM was shown in Fig. 9. This enzyme required much more inhibitor then reported by Harris *et al.* (2).

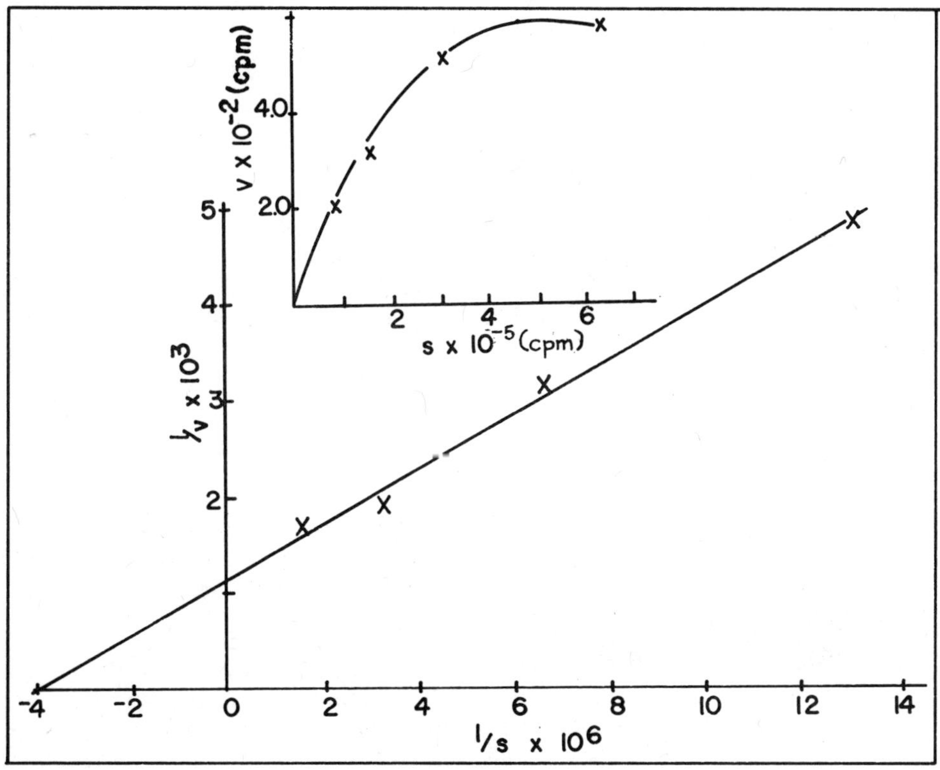


Fig. 8. Rate of lysyl oxidase reaction and substrate concentration. The reaction mixtures including 6 μg protein and various concentrations of the substrate were incubated using the standard assay procedures.

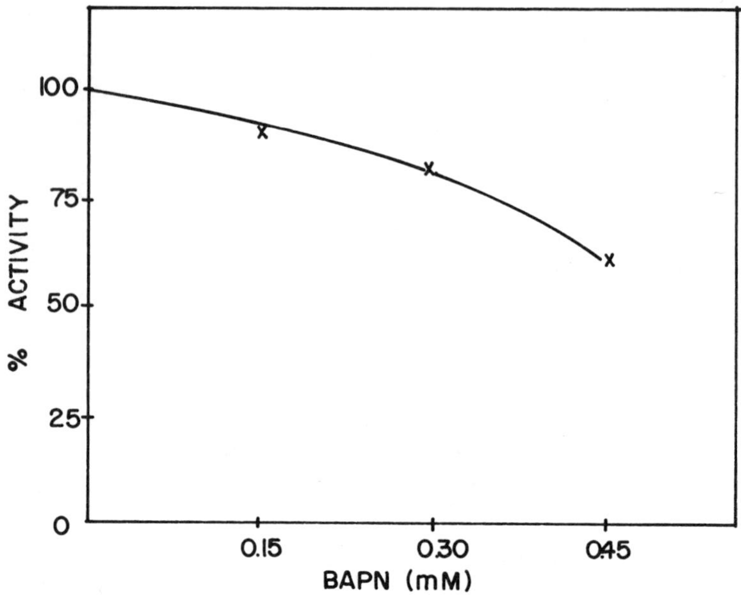


Fig. 9. Effect of β-aminopropionitrile on lysyl oxidase activity. β-Aminopropionitrile was preincubated with enzyme(6 μg) for 30 min. Then the standard assay procedures were used.

Reactivation of enzyme treated with diethyldithiocarbamate.

The enzyme was first dialyzed against 0.05 M phosphate buffer pH 7.6 which contained 0.1 M diethyldithiocarbamate until the enzyme activity was lost. The resulting enzyme sample was then dialyzed against 0.01 M Tris-HCl buffer, pH 7.6 to remove the excess DDC. About 87% and 75% of the original activity was restored by the addition of $5 \times 10^{-5}$ M copper sulfate and ferrous sulfate, respectively. However, no reactivation was observed upon the addition of $Mn^{2+}$, $Zn^{2+}$, and $Co^{2+}$ at the same final concentrations.

Some physicochemical properties of lysyl oxidase.

The molecular weight of the lysyl oxidase A and B were estimated by exclusion diffusion chromatography on samples of lysyl oxidase which were pure by ordinary and SDS-disc eletrophoresis in which β-mercaptoethanol was omitted. The molecular weights of lysyl oxidase A and B were determined to be 80,000 and 160,000, respectively. However other preparations have yielded different molecular weights and this may be a consequence of the polymerization of the enzyme protomer to higher molecular weight aggregates.

When the subunit structure of the enzyme was examined by SDS-disc electrophoresis as described by Weber and Osborne (8), both forms of lysyl oxidase showed a major band with a molecular weight of about 53,000 and a minor band of molecular weight of 28,000. The ratio of the heavier to lighter subunit was about 4:1.

The copper content of the enzyme was determined by using the atomic absorption spectrometer. The copper content for the A and B-forms of the enzyme were 0.97 and 0.88 g atoms per 70,000 g of enzyme.

The purified enzyme showed a typical absorption maximum at 280 nm and general absorption in the visible wavelength region. A slight maximum was observed at about 400-410 nm, which may be due to slight contamination by a heme protein. The $E^{1\%}_{1cm}$ values at 280 nm for the A- and B-forms of the enzyme were estimated to be about 15.3 amd 14.1, respectively. The ESR spectrum of the enzyme was determined and is shown in Fig. 10 and 11. The $g_{max}$ values of 2.12 and 2.11 were determined for the A- and B-forms of the enzyme.

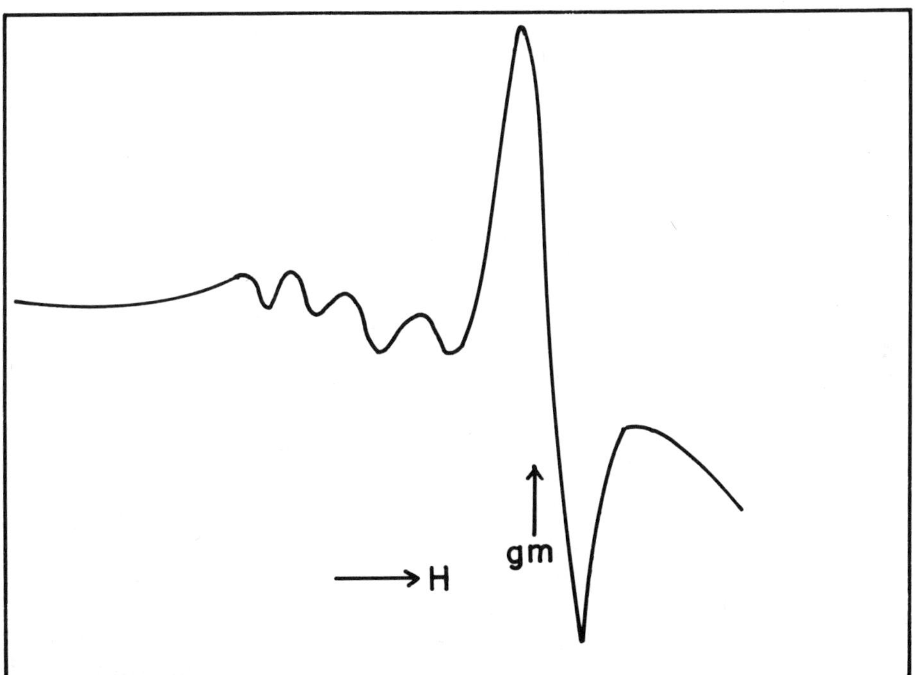


Fig. 10. ESR spectrum of lysyl oxidase B. About 11.5 mg protein in 0.1 M potassium phosphate buffer, pH 7.6 was run at a temperature of $-165^{o}$, modulation amplitude of 10.0 gauss, microwave power of 5 mW, scanning rate of 217 g/min and receiver gain of $6 \times 10^{3}$.

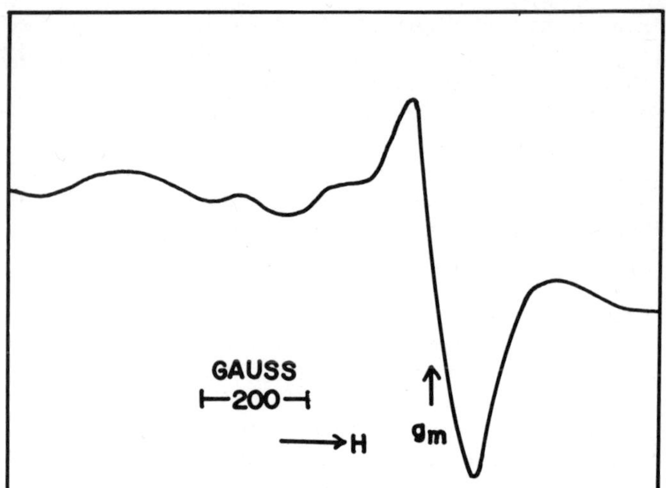


Fig. 11. ESR spectrum of lysyl oxidase B. About 13.8 mg protein in 0.1 M potassium phosphate buffer, pH 7.6 was run at a temperature of $-165^{\circ}$, modulation amplitude of 10.0 gauss, microwave power of 9.0 mW, scanning rate of 500 g/min and receiver gain of $4 \times 10^3$.

## DISCUSSION

Lysyl oxidase has been shown to catalyze the first step in the conversion of lysine and hydroxyllysine residues in the tropocollagen and tropoelastin to collagen and elastin. The enzyme possibly catalyses the following reaction:

$$\text{—— Lysine ——} + O_2 + H_2O \longrightarrow \text{Allysine} + NH_3 + H_2O_2$$

It is not definitely established that water is a substrate and that $NH_3$ and $H_2O_2$ are products of the reaction. In subsequent reactions, the allysine (or hydroxyllysine) are converted to various cross-linking compounds which cross link the subunits of the structural proteins. In certain diseases, as well as in experimental animals fed the lathyrogen, β-aminopropionitrile, the cross-links fail to form because of the lack or inhibition of lysyl oxidase (9). The enzyme from chick cartilage and the aorta have been highly purified but the lung enzyme has not been studied. Since there are a number of diseases of the lung in which there may be an alteration of the structural proteins, it is important first of all to demonstrate the presence of the enzyme in the lung, to study the localization of the enzyme in the various parts of the lung, to study the change of the enzyme content with age and to isolate the highly purified enzyme so that its molecular properties could be studied.

The enzyme was found to be mainly localized in the parenchymal tissue. In the cow lung, the enzyme content initially increased in parallel with the weight of the lung, remained constant from about 2-7 years and then decreased. Since the cow life expectancy is about 18-20 years, the level of the lysyl oxidase in the lung appears to decrease somewhat before the mid-life of the cow. It was found that the lung enzyme could be purified by the procedures developed by the earlier investigators (1-4). However, in the affinity chromatography step, it was found that a number of different proteins could be coupled to Sepharose 4B. Functional proteins included the water soluble protein obtained from bovine aorta tissue culture slices (2), soluble collagen purchased from Sigma Chemical Co., and even gelatin could be used although the gelatin-Sepharose was very inefficient in purifying the enzyme. Moreover, the assay procedure, which utilized $^{3}$H-lysine labelled affinity column tropoelastin was very inaccurate and very odd kinetics were observed at times. A simpler, more accurate assay is badly needed.

However, the purification procedure devised has led to the isolation of the so-called lung lysyl oxidase A and B which are pure by ordinary disc electrophoresis as well as by SDS-electrophoresis in which mercaptoethanol is not added. At the present time, it cannot be stated what the differences in the chemical properties of lysyl oxidase A and B are. The term is mainly an operational one used to indicate that two forms of the enzyme can be isolated. The specific activity of lysyl oxidase A is much higher than the specific activity of lysyl oxidase B. In addition, the net charge of the two forms of the enzyme differs since the mobilities of lysyl oxidase A and B are different when they are examined by ordinary disc electrophoresis. The A-form of the enzyme was more positively charged then the B-form. The electrophoretic mobilities of the two forms of the enzyme are in agreement with their order of elution from the DEAE-cellulose columns. In addition, the molecular weight of the A- and B-forms of the enzyme when estimated by exclusion-diffusion chromatography on Biogel A-1.5 m also disclosed that the molecular weights of the isolated enzymes were different, namely 80,000 and 160,000, respectively. However, more highly aggregated forms of the enzyme were observed in other preparations. When the purified A- and B-forms of the enzyme were examined by SDS-disc electrophoresis as described by Weber and Osborne (8), both forms of the enzyme showed two components with molecular weights of about 52-53,000 and 28,000. However, the estimated ratio of the two components were not 1:1 but more like 1 part of the 28,000 component to 4 parts of the 52,000 component. Thus at the present time, it is not clear whether or not the two components are subunits of the enzyme or whether the lysyl oxidase contains an impurity which is linked to the lysyl oxidase by means of a disulfide bond. As far as the reported molecular weights of lysyl oxidase from other sources are concerned, values of 170,000 (3), 59-61,000 (2) and 28,000 (4) have reported. Therefore, additional studies are definitely required to clearly establish the true molecular weight and subunit structure of the enzyme. However, it is clear that the enzyme exists in polymerized forms as shown by the exclusion-diffusion chromatography experiments.

The majority of the enzymatic properties of lysyl oxidase determined by previous investigators do agree with the properties of lung lysyl oxidase determined in the present study. However, much higher quantities of the lathyrogen, β-aminopropionitrile, were required to inhibit the enzyme than has been reported (1,3,4).

In agreement with the report of Harris et al. (2), lysyl oxidase is a copper protein. Copper was shown to be present by chemical and ESR experiments. The ESR spectrum of the lysyl oxidase resembles the spectrum of certain other enzymes such as plasma amine oxidase (9) but the $g_{max}$ values appear to be usually high. Further experiments are in progress to determine why such high $g_{max}$ values are found. Siegel et al. (3) have reported that the copper in the enzyme could be replaced by ferrous and cobaltous ions. A similar result was obtained in the present investigation with ferrous ion, but cobaltous ion was incapable of reactivating the apoenzyme. These experiments will have to be repeated since the metal content of the apoenzyme and the metal content of the enzyme after reconstitution were not determined. It is possible that the apoenzyme was not formed and instead an enzyme-diethyldithiocarbamate complex was present. The addition of metals would simply restore enzyme activity by removing the chelating agent from the enzyme.

Finally, no evidence has been obtained thus far for the presence of any organic cofactors. The visible absorption spectrum of the enzyme does not resemble the reported spectra of any known heme-, flavo- or pyridoxal-phosphate enzymes.

## ACKNOWLEDGEMENTS

This research project was supported by Contract Grant NO1 HR 42944 from the National Institutes of Health.

## REFERENCES

1. Pinnell, S.R. and G.R. Martin. (1968) Proc. Nat. Acad. Sci. USA 61,708.
2. Harris, E.D., W.A. Gonnerman, J.E. Savage and B.L. O'Dell (1974) Biochim. Biophys. Acta 341, 332.
3. Siegel, R.C., S.R. Pinnell and G.R. Martin (1970) Biochem. 9, 4486.

4. Stassen, F.L.H. (1975). Fed. Proc. 34, 2717 (abstract).
5. Vidal, G.P., J.J. Shieh and K.T. Yasunobu (1975) Biochem. Biophys. Res. Commun. 64, 989.
6. Cuatrecasas, P. and C.B. Anfinsen (1971) Methods in Enzymology Vol. XXII, p. 234, Academic Press.
7. David, B.J. (1964) Ann. N.Y. Acad. Sci. 121, 404.
8. Weber, K. and M. Osborn (1969). J. Biol. Chem. 244, 4406.
9. Yamada, H., Yasunobu, K.T., Yamano, T., and Mason, H.S. (1963) Nature, 198, 1092.

# BINUCLEAR COPPER CLUSTERS AS ACTIVE SITES FOR OXIDASES

Howard S. Mason

Department of Biochemistry, School of Medicine
University of Oregon Health Sciences Center
Portland, Oregon

I would like to review briefly the evidence supporting the concept that binuclear clusters of copper ions form an important category of active site for the enzymes catalyzing the reactions of dioxygen. In addition, I will summarize some results of my coworkers relating to the properties of the binuclear copper clusters in *Cancer magister* methemocyanin and *Agaricus bispora* tyrosinase. In these copper proteins the properties of the binuclear clusters are not complicated by the presence of copper ions of other types.

The subject of binuclear clusters of copper ions was last reviewed by Malkin and Malmstrom in 1970 (Adv. Enzymol. 33, 177), primarily in connection with blue oxidases containing "Type 3" copper. The operational criteria for recognition of binuclear cupric ion clusters was then: (1) that these electron-accepting sites have a redox potential greater than 0.5V(cf. Fee *et al.*, J. Biol. Chem. 244, 4200 (1969)); (2) that they be diamagnetic in the oxidized protein at room temperature (cf. Ehrenberg *et al.*, J. Mol. Biol. 5, 450 (1962)); (3) that they exhibit no EPR signal in the oxidized or reduced forms of the enzyme at 77°K; and (4) that they be associated with a near-ultraviolet absorption band of moderate intensity (e.g., 330 nm, $\varepsilon$ = 2,800 $M^{-1}cm^{-1}$ (*Polyporus laccase*); 332 nm, $\varepsilon$ = 4,100 $M^{-1}cm^{-1}$ (ceruloplasmin)). These criteria have been modified and extended in the six years since the appearance of the Malkin and Malmstrom review, and I will propose new ones.

CERULOPLASMIN (HUMAN). This blue copper protein, mw 140,000 (McKie and Frieden, Biochemistry 10, 3880 (1971)) but somewhat controversial, consists of Light (mw 16,000) and Heavy (mw 53,000) subunits as $L_2H_2$ (Freeman and Daniel, Biochemistry 12, 4806, (1973)) comprising 1 or 2 Type 1 cupric ions, 2 or 3 Type 2 cupric ions, and two binuclear copper clusters, Type 3 (Ehrenberg et al., (1962); Osaki et al., J. Biol. Chem. 246, 3018 (1971), and others). Its absorption spectrum contains four bands: 610 nm, $\epsilon$ = 11,300 $M^{-1}cm^{-1}$ (Type 1 cupric ion), 332 nm, $\epsilon$ = 4,100 $M^{-1}cm^{-1}$ (Type 3 cluster), 459 nm, $\epsilon$ = 1,200 $M^{-1}cm^{-1}$; and 794 nm, $\epsilon$ = 2,200 $M^{-1}cm^{-1}$. Type 1 cupric ion reacts with NO by charge transfer (Van Leeuwen et al., BBA 302, 236 (1973)), but Type 2 does not. Reduced Type 3 cupric ion clusters react with NO, giving EPR signals at g = 2 and g = 4, the latter having seven hyperfine lines characteristic of magnetic dipolar interaction among two cupric ions at a distance somewhat closer than 6A. This constitutes direct physical evidence of cupric-cupric interaction in the nitrosyl derivative of reduced ceruloplasmin, $NO^-$-Cu(II)...Cu(II)-$NO^-$. ( Byers et al., BBA 310, 38 (1973)) suggest on the basis of spectral changes accompanying anion binding that resting ceruloplasmin (330 nm band) contains a bicuprous disulfide structure, capable of reduction and oxidation. Velsema and Van Gelder (BBA 293, 322 (1973)) showed that ceruloplasmin is capable of accepting only 4 electrons at pH 7.0; Ryden and Eaker (FEBS Lett. 53, 279 (1975)) showed that reduced enzyme and resting enzyme contain the same number of sulfhydryl groups. Thus the model of Byers et al. does not apply. A third possibility, based upon the suggestion of Hamilton et al. (BBRC 55, 333 (1973)) that galactose oxidase functions through trivalent copper, seems unlikely since the $Cu^{1+}$...$Cu^{3+}$ couple required by this model should have great redox stability.

HEMOCYANIN (ARTHROPODAL). This $O_2$-transporting copper protein, from the Pacific crab, Cancer magister, has mw 960,000 and contains 24 $Cu^{1+}$ ions, or 2 $Cu^{+}$ per subunit mw 80,000 or 78,000 (Loehr and Mason, BBRC 51, 741 (1973)). The stoichiometry of $O_2$ binding is 2Cu/$O_2$ (Van Holde and Van Bruggen, Biological Macromolecules Series 5, 1 (1973); Lontie et al., Metal Ions in Biological Systems 3,183 (1974)),and for CO, 2Cu/CO (Vanneste and Mason, Biochem. Copper. p. 465 (1966), Academic Press )). The oxy-form has absorption bands at 340 nm; $\epsilon$ = 10,000 $M^{-1}cm^{-1}$; 580 nm, $\epsilon$ = 600 $M^{-1}cm^{-1}$(Moss et al., Biochemistry 12, 2444 (1973)), and is diamagnetic. Oxyhemocyanin is EPR-silent (Nakamura and Mason, BBRC 3, 297 (1960)). The NO derivative gives EPR signal with g = 2 (very broad) and g = 4 (seven hyperfine lines), definitive evidence for magnetic dipole coupled $Cu^{2+}$ ions in the nitrosyl derivative (Schoot Uiterkamp, FEBS Lett. 20, 93 (1972); Schoot Uiterkamp and Mason, PNAS 70, 993 (1973)). Simulation of this EPR signal indicates the Cu-Cu distance to be 6A (Schoot Uiterkamp et al., BBA 372, 407 (1973). Thus, the dinitrosyl derivative of hemocyanin contains contiguous Cu ions.

Hemocyanin is oxidized by $H_2O_2$: $(Cu^{1+})_2 + H_2O_2 = (Cu^{2+})_2 + 2OH^-$ (Felsenfeld et al., JACS 81, 6259 (1959)); the bi-cupric product has no EPR spectrum (Simo, Thesis, Univ. Oregon Medical School, 1966; Makino et al., unpublished results, 1976). Methemocyanin is diamagnetic (Gould et al., unpublished results 1976) between 1.4°K - 200°K. Methemocyanin has an absorption band, 680 nm, $\varepsilon = 60\ M^{-1}cm^{-1}$. These properties are consistent with a pair of cupric ions either interacting directly (2A apart), or by superexchange through intervening ligand(s). Loehr et al., found that the Raman spectrum is best interpreted if oxygen acts as peroxide bound to a bi-cupric center (BBRC 56, 510 (1974)).

TYROSINASE (MUSHROOM). This copper-containing mixed function oxidase, mw 120,000, consists of subunits, Light, mw 13,500, and Heavy, mw 43,000, in LH and $(LH)_2$ aggregates (Strothkamp et al., unpublished, 1976). It contains 4 $Cu^{2+}$ / 120,000, $E_o' = +0.36V$, n = 2 (Makino et al., JBC 249, 6062 (1974)); the protein is diamagnetic. The bicupric form reacts with $H_2O_2$ to form reduced tyrosinase: $(Cu^{2+})_2 + H_2O_2 = (Cu^{1+})_2 + O_2 + 2H^+$. The NO derivative prepared from reduced tyrosinase has g = 2 (broadened) and g = 4 (seven hyperfine lines) diagnostic of magnetic dipolar coupling. Simulation of the EPR spectrum indicated a Cu-Cu distance of 5.7A (Schoot Uiterkamp et al., (1974)). In mettyrosinase the coupling is much stronger and the protein is diamagnetic and gives no EPR signal from 8°K to 77°K (Makino et al., (1974)). As in the case of methemocyanin, this can mean that the $Cu^{2+}$ ions are about 2A apart, or that there is strong superexchange through intervening ligands not present in the nitrosyl dimer.

LACCASE (FUNGAL). These blue copper proteins, mw 62,000, consist of more than one polypeptide chain (Malkin and Malmstrom, Adv. Enzymol. 33, 177 (1970)). They contain 4 cupric ions per molecule, as follows: 1 Type 1 "Blue" copper, 610 nm band, $\varepsilon = 4{,}600\ M^{-1}cm^{-1}$, $E_o' = 395mV$, n = 1; 1 Type 2 non-blue copper, $E_o' = 365$ mV, n = 1; 2 Type 3 clustered $Cu^{2+}$, 330 nm band, $\varepsilon = 2{,}800\ M^{-1}cm^{-1}$, $E_o' = 435$ mV, n = 2 (Reinhammar et al., BBA 275, 245 (1972)). The Type 3 copper is reducible by 2 electron equivalents but shows no EPR spectrum and is diamagnetic (Ehrenberg et al., J. Mol. Biol. 5, 450 (1962); Moss and Vanngard, Life Science, 1974; Malmstrom et al., BBA 156, 1968; Malkin et al., Eur. J. Biochem. 10, 324 (1969); Fee et al., J. Biol. Chem. 244, 4200 (1969)). Byers et al., (BBA 310, 38 (1973)) proposed three types of structure for such a center: (1) a binuclear spin-coupled cupric ion complex. (2) a binuclear cuprous ion complex of disulfide, or (3) a binuclear system containing $Cu^{1+}$ and $Cu^{3+}$ (cf. ceruloplasmin). All copper ions in laccase become EPR detectable upon anaerobic denaturation, and appear to be in an oxidation state equivalent to $4Cu^{2+}$. Only one

-SH group and 2 disulfide bonds are detectable after anaerobic denaturation of fungal laccase whether or not reduced with 4 equivalents of ascorbate. Since the copper is recovered as $Cu^{2+}$ (EPR) after denaturation, no oxidation of -SH to -S-S- occurred and the dithiolate model of Type 3 copper appears to be eliminated (Briving and Deinum, FEBS Lett. 51, 43 (1975)). Thus magnetic susceptibility measurements, redox titrations, $Cu^{2+}$, cysteine -SH, and cystine -S-S- determinations upon native and reduced laccases anaerobically denatured show it to contain antiferromagnetically stabilized binuclear cupric ion pairs having an absorption band at 330 nm and a redox potential of +435 mV, n = 2.

ASCORBATE OXIDASE (ZUCCHINI SQUASH). This blue copper protein, mw 140,000, consists of Light subunits, mw 28,000, and Heavy subunits, mw 38,000; a functional LH has mw 65,000 and the native form is $(LH)_2$, mw 132,000 (Strothkamp *et al.*, Biochemistry 13, 434 (1974)). The protein contains 8 - 10 copper ions per molecule (i. e., on average, 2 copper ions per subunit polypeptide). They consist of Type 1 $Cu^{2+}$, Blue, 3 copper ions; Type 2 $Cu^{2+}$, non-blue, 1 copper ion; and Type 3 clusters (330 nm, $\varepsilon$ = 5,800 $M^{-1}cm^{-1}$ which disappears upon reduction) 2 pairs or 4 copper ions. 50% of the total copper (4 atoms) is EPR detectable (Types 1 and 2 $Cu^{2+}$). The Type 3 $Cu^{2+}$ is EPR-indetectable (Deinum *et al.*, FEBS Lett. 42, 241 (1973); Lee *et al.*, JBC 248, 6594, 6603 (1973); Avigliani *et al.*, Ital. J. Biochem. 21, 248 (1972)).

Reduced ascorbate oxidase reacts with NO but does not produce an EPR signal characteristic of magnetic dipolar interactions. Hence, the cuprous ions may be closer together than in ceruloplasmin, since the oxidized enzyme fails to show an EPR signal for its two Type 3 copper centers. In this respect the enzyme resembles fungal laccase (Van Leeuwen *et al.*, BBA 403, 285 (1975)). Since ascorbate oxidase has the 330 nm absorption band ascribed to diamagnetic (Cu(II)-Cu(II) pairs in laccase and ceruloplasmin, since this band is associated with 4 cupric ions showing no EPR spectrum, and since it is lost upon reduction with substrate, it appears that two binuclear cupric ion clusters per molecule are present in ascorbate oxidase.

DOPAMINE-BETA HYDROXYLASE (MAMMALIAN). This non-blue copper protein, mw 250,000-290,000, consists of 8 subunits, mw 37,500, aggregated into dimers, mw 75,000; two dimers form a tetramer, mw 150,000, held together by disulfide bonds (Wallace, Krantz, and Lovenberg, PNAS 70, 2253 (1973); Craine, Daniels, and Kaufman JBC 248, 7838 (1973); Aunis, Miras-Portugal, and Mandel,BBA 365, 259 (1974)). Dopamine-β-hydroxylase contains 4-8 Cu ions per molecule of enzyme, mw 290,000 (Aunis *et al.*, (1974); the precise value is uncertain and appears to depend upon the method of preparation of enzyme. High Cu/protein ratios have been obtained with concanavalin A-affinity chromatography (Aunis et al., (1974); Rush et al., BBRC 57, 1301 (1974); Wallace, et al., PNAS 71, 3217 (1974); Blackburn and Mason, unpublished data). On stoichiometric and kinetic grounds, it has been proposed that two atoms of Cu act cooperatively to bind and activate dioxygen for this mixed function oxidation (Friedman and Kaufman, JBC 240, 4763 (1965); Goldstein and Joh and Garvey, Biochemistry 7, 2724, 1968)). In an uncharacterized preparation of dopamine-β-hydroxylase, treated with Chelex (a reagent which removes extraneous $Cu^{2+}$) the EPR-detectable $Cu^{2+}$ (axial or Type 2 Cu), was 100% of the total copper (Blumberg et al., BBA 99, 187 (1965)). No magnetic interaction among cupric ions was detectable. However, it remains possible that EPR-indetectable Cu present, or cuprous copper in fully reduced enzyme, may be contiguous pairs capable of binding dioxygen in the mode proposed by Friedman and Kaufman (1965).

CYTOCHROME OXIDASE (MAMMALIAN). This copper-heme a protein, mw 200,000, consists of seven polypeptide subunits, and contains two cupric ions, one EPR-detectable, and a second EPR-invisible, an EPR-visible low spin cytochrome a, and an EPR-invisible cytochrome $a_3$ (Hartzell, Beinert, BBA 336, 318 (1974)). Gibson and Greenwood studied the relative rates of oxidation-reduction of cytochromes and visible copper (820 nm band); the copper was found to be kinetically distinct from cytochromes a and $a_3$ (JBC 240, 2694 (1965). Lindsay and Wilson found that the EPR-indetectable Cu participates in CO binding by cytochrome $a_3$, inasmuch as the binding is associated with a two-equivalent reduction (FEBS Lett. 48, 45 (1974)). Lindsay, Owen and Wilson determined the redox potentials of the components of cytochrome oxidase and found $E_o'$ (pH 7.2), cytochrome a = 220 mV, n = 1; visible Cu, 245mV, n = 1; invisible Cu, 336 mV, n = 1; cytochrome $a_3$, 375 mV, n = 1 (Arch, Biochem. Biophys. 169, 492 (1975)). While the route and mechanism of electron transport from cytochrome c through cytochrome oxidase to dioxygen is uncertain, it seems probable on thermodynamic grounds that at some stage of the catalytic cycle electrons flow from visible copper to invisible copper, and in this sense the copper ions may be considered clustered.

SUMMARY. Binuclear cupric ion clusters have been established in: human ceruloplasmin, hemocyanin, and mushroom tyrosinase. Substantial evidence makes it very probable that fungal laccase and zucchini ascorbate oxidase contain this cluster. Some evidence makes it possible that copper clusters function in the catalytic cycles of cytochrome oxidase (mammalian) and dopamine-β-hydroxylase.

These studies throw light on the criteria which must be employed to establish the existence of functional binuclear copper clusters in enzymes: (1) Stoichiometric Criteria: binding of $O_2$ and CO with Cu/ligand = 2; redox titrations with n = 2;

(2) Physical and Chemical Criteria: magnetic evidence of diminished paramagnetism of cupric centers, EPR evidence of broadened or absent absorptions, EPR evidence of magnetic dipolar interactions among cupric ions; absorption bands characteristic of Cu(II)-Cu(II) complexes; laser resonance raman scattering characteristic of peroxidic dioxygen in the oxyforms.

This work was supported in part by grants AM 0718 from the National Institutes of Health, and BC-10 from the American Cancer Society.

# An Interesting Reaction of Cupric Ions With Ferricyanide and Ferrocyanide

P. McMahill, N. Blackburn, and H. S. Mason

Department of Biochemistry, School of Medicine

University of Oregon Health Sciences Center

Portland, Oregon 97201

The possibility that Cu(III), or cupryl ion [Cu(I)O°], participates in enzymic reactions of copper with dioxygen and other oxidizing substrates, as Dr. Hamilton suggests in the preceeding paper, is an extremely interesting one which opens new vistas of mechanism. However, absolutely unambiguous criteria for the 3+ oxidation level of copper in these processes are required. That is the principal point of this note. Here we show that in the presence of excess ferricyanide ion, the EPR spectrum of $Cu^{2+}$ ions disappears, and reappears when excess ferrocyanide ion is added. As Dr. Hamilton suggests for galactose oxidase, this may be explicable by the reversible oxidation-reduction of $Cu^{2+}$ to low spin $Cu^{3+}$. On the other hand, $Cu^{2+}$ may form ion pairs with ferricyanide, the magnetic interaction in the pairs broadening the EPR spectrum to indetectability. The addition of (low spin) ferrocyanide would then so dilute the ferricyanide that EPR-detectable $Cu^{2+}$-ferrocyanide ion pairs are observed. Alternatively, $Cu^{2+}[Fe(CN)_6]^{3-}$ complexes, within which charge transfer between $Cu^{2+}$ and $Fe(CN)_6^{3-}$ takes place, would also have a broadened EPR spectrum; if $Fe(CN)_6^{4-}$ competed for the $Cu^{2+}$ to form complexes with similar dissociation constants, the EPR spectrum of $Cu^{2+}$ would reappear upon addition of $Fe(CN)_6^{4-}$ (Figures 1 and 2). The structures of the ion pairs and complexes are not specified, since with high ferricyanide/$Cu^{2+}$ ratios, the compounds remain soluble. However, the EPR signal shapes and g-values of the regenerated paramagnetic species are different; therefore structural differences among the initial and final species must exist.

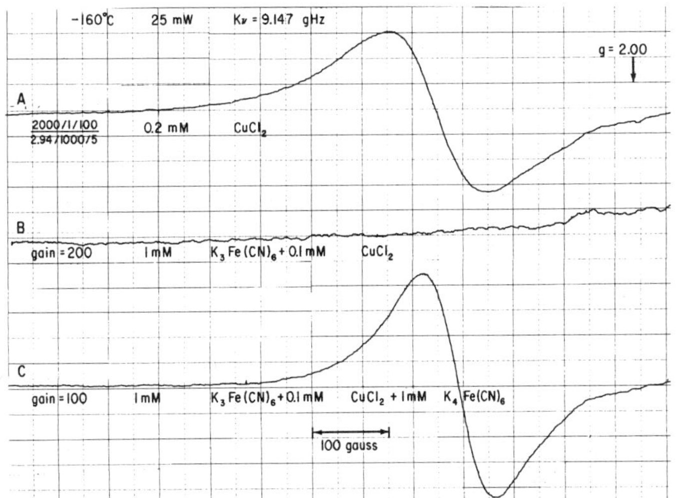


Fig. 1. EPR spectra of complexes of ferricyanide and ferrocyanide with cupric chloride. Curve A, 99 nM $CuCl_2$. Curve B illustrates the disappearance of the g = 2 region signal of cupric chloride in the presence of ferricyanide. Curve C represents the recovery of the signal when $K_4Fe(CN)_6$ solid was added to the solution B in an equimolar ratio with the $K_3Fe(CN)_6$. Modulation amplitude, 8.0 gauss; microwave power, 25 mW; scanning rate, 200 g/min; temperature, -160° C; magnetic field increased from left to right.

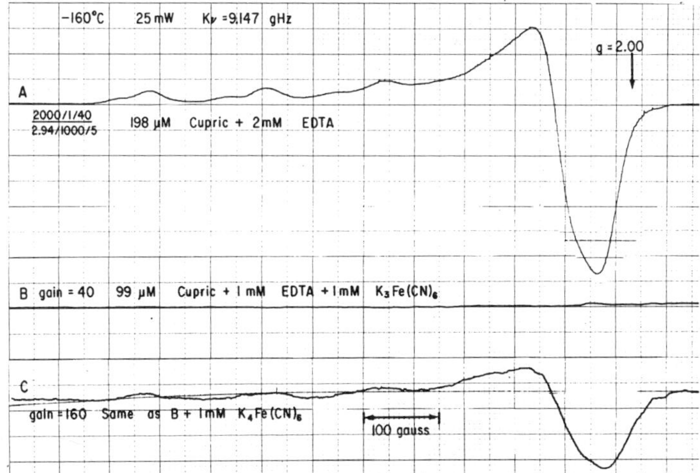


Fig. 2. EPR spectra of cupric complexes of ferricyanide and ferrocyanide in the presence of EDTA. Curve A, 198 mM $Cu^{2+}$, 2mM EDTA. Curve B shows the loss of g = 2 region cupric EPR signal on treatment with ferricyanide. Curve C represents a 29% recovery of the EPR signal when solid $K_4Fe(CN)_6$ was added to solution B. Same instrumental conditions as in Figure 1.

# FORMAL CATALYTIC MECHANISM OF ASCORBATE OXIDASE

Stephen R. Burstein, Brenda Gerwin,
Hugh Taylor, and John Westley

Department of Biochemistry
The University of Chicago
Chicago, Illinois 60637

## INTRODUCTION

Ascorbate oxidase (EC 1.10.3.3), a copper enzyme widely distributed in the plant kingdom, catalyzes the oxidation of ascorbate by oxygen to the final products dehydroascorbate and water [1]. Yamazaki and Piette [2,3] showed that the catalyzed reaction proceeds first to an ascorbyl radical, which dismutes spontaneously to ascorbate and dehydroascorbate. They also found that the enzyme has the same maximal velocity with reductate or ascorbate as donor substrate [2]. Nakamura *et al.* [4] have shown that double reciprocal plots for ascorbate oxidase with ascorbate as the varied substrate at different oxygen concentrations are a family of parallel straight lines. This pattern implies that the points of entry of the reductant and oxidant substrates into the catalytic cycle are separated by a reaction that is irreversible under initial velocity conditions. The results of stopped flow experiments by Nakamura and Ogura [5,6] have been cited as evidence that oxidation of the reduced enzyme by molecular oxygen is intrinsically a faster process than reduction of the oxidized enzyme by donor substrates [7]. Gerwin *et al.* [8] showed that the active form of the reductant substrate is the monoanion. These workers also showed that the maximal velocity, but not the Michaelis constant for ascorbate as reductant substrate, varies with pH in accord with a pK' near 8.

Different parts of this body of data appear to suggest two different, mutually exclusive double displacement formal mechanisms. The present study was undertaken to extend the kinetic observations to the point that an explicit formal mechanism could be proposed consistent with all of the foregoing evidence.

## EXPERIMENTAL

Ascorbate oxidase was purified to apparent homogeneity as reported previously [8]. D-Araboascorbic acid (erythorbic acid) and polygalacturonic acid (practical) were purchased from Eastman Kodak. Reductic acid was synthesized by isolation from the pyrolysis products of polygalacturonic acid, according to the procedure of Reichstein and Oppenauer [9]. A light tan product with a melting point of 212° was obtained in over-all yield of 1.8%. One gram of this product was further purified by sublimation at a pressure of less than 10 μ of mercury in the temperature range 130° to 150°. Sublimation was discontinued when yellow material began to condense onto the white crystals in the condensation chamber. The sublimed product melted at 212.3° to 213.0° d (Lit. 211°-213.5° d [9,10]).

Each of the reductant substrates was dried to constant weight and samples were dissolved in $10^{-3}$ M EDTA for titration with an $I_2$-KI solution standardized against a National Bureau of Standards primary standard sample of arsenious trioxide. By this method the ascorbic acid was determined to be 99.69±0.05% pure; the araboascorbic acid, 99.78±0.08% pure; the reductic acid 98.90±0.06% pure. The melting point of the ascorbic acid was 193°-194° d (Lit. 190° [11]); that of the araboascorbic acid 171°-172° (Lit. 168° [12]).

The molar absorbancies of the reductant substrates were determined by recording the ultraviolet absorption spectrum of an EDTA-phosphate buffer solution (pH 7.0) of each against an appropriate blank solution with a Cary 15 recording spectrophotometer. The spectral characteristics (corrected for purity) found under these conditions are given in Table I. Light absorption is due entirely to the monoanions at this pH.

TABLE I

Spectral characteristics of reductant substrates

| Reductant substrate | $\lambda_{max}$ | $\varepsilon_{max}$ |
|---|---|---|
| | nm | $M^{-1}$ $cm^{-1}$ |
| L-ascorbate | 267 | $1.38 \times 10^4$ |
| D-araboascorbate | 266 | $1.10 \times 10^4$ |
| Reductate | 282 | $1.73 \times 10^4$ |

Enzyme assay and kinetic study conditions were 0.05 M phosphate buffer, pH 7.0, containing 5.0 x $10^{-4}$ M EDTA, at 25°. Unless specified otherwise, the ascorbate concentration was 7.2 x $10^{-5}$ M and the oxygen was 0.2 atmospheres (that is, the reaction mixture was saturated with air at ambient atmospheric pressure). No citrate or chloride species were present and ascorbate solutions were prepared directly in the phosphate-EDTA buffer [8]. The time course of absorbance change was followed with a Cary 15 recording spectrophotometer or a Zeiss PMQ II spectrophotometer equipped for kinetic recording. Initial rates in this assay system were linear with enzyme concentration over the range employed in this study.

For kinetic studies, initial rates of oxidation with ascorbate as substrate were determined spectrophotometrically at various initial ascorbate concentrations. Initial oxygen concentration was varied by mixing pure oxygen or air with pure nitrogen in a precisely known ratio by the use of a pair of calibrated Gilmont No. 1 flowmeters. Variation of the pH of the kinetic assay system in the pH range 7.5 to 9.5 at a constant ionic strength of 0.2 M was accomplished by use of phosphate-borate buffers containing $10^{-3}$ M EDTA. Borate was found not to inhibit the enzyme-catalyzed oxidation of ascorbate.

At each oxygen concentration, initial rate data for all reductants as variable substrates were fitted to a hyperbolic function as suggested by Wilkinson [13], assuming the error function to yield a homogeneous envelope in a hyperbolic fit, by the use of a BASIC program written in this laboratory for this purpose.

## RESULTS

Double reciprocal plots with ascorbate as varied substrate at oxygen partial pressures from 0.095 to 1.0 atmosphere had slopes identical within 2.5%, confirming the parallel nature of such patterns documented by Nakamura *et al*. [4]. From the data, the following kinetic constants were calculated for ascorbate oxidase activity: $K_m^{ascorbate}$ at ambient oxygen partial pressure ($K_{amb.\ O_2}^{AH^-}$) = (2.1 ± 0.1) x $10^{-4}$ M; $K_m^{ascorbate}$ at saturating oxygen concentration ($K_m^{AH^-}$) = (5.3 ± 0.4) x $10^{-4}$ M; $K_m^{O_2}$ with ascorbate as reductant substrate ($K_{AH^-}^{O_2}$) = (2.40 ± 0.06) x $10^{-4}$ M. The average maximal specific velocity at ambient oxygen partial pressure for the enzyme preparation used for the kinetic studies was (1.49 ± 0.02) x $10^3$ M $sec^{-1}$ per molar protein as ascorbate oxidase. The enzyme could be activated by reduction, but this procedure alters no kinetic parameters other than maximal velocity (see reference [8] and further data given below). Data obtained by similar procedures with reductate as variable substrate in the range of substrate concentrations accessible by the spectrophotometric assay method did not permit calculation of the kinetic coefficients for reductate as substrate. However, these data were fully compatible with the report of Yamazaki and Piette [2] that the maximal velocities for reductate and ascorbate are identical, although this finding was previously questioned by us [7] on the basis of our preliminary work with this system. Similar double reciprocal plots obtained with araboascorbate as variable substrate had slopes identical within 5% and permitted computation of the following kinetic constants: $K^{ArH^-}_{amb.\ O_2} = (1.7 \pm 0.1) \times 10^{-3}$ $\underline{M}$; $K^{ArH^-}_{m} = 3.2 \times 10^{-3}$ $\underline{M}$; $K^{O_2}_{ArH^-} = 1.9 \times 10^{-4}$ $\underline{M}$. The maximal specific velocity with this substrate at ambient oxygen partial pressure was determined to be $(1.38 \pm 0.04) \times 10^3$ $\underline{M}$ $sec^{-1}$ per molar protein as ascorbate oxidase. This value is equal within experimental error to the same constant for the identical enzyme preparation with ascorbate as reductant substrate, $(1.33 \pm 0.06) \times 10^3$ $\underline{M}$ $sec^{-1}$ per molar protein as ascorbate oxidase.

Measurement of steady state velocities with oxygen as varied substrate at pH values between 7 and 9.5 showed that the slopes of double reciprocal plots for oxygen do not change with pH as do the apparent maximal velocities. Further, the activation of ascorbate oxidase by reduction with excess ascorbate at pH 5 in the presence of citrate and chloride and the absence of oxygen was found not to affect $K^{O_2}_{AH^-}$, just as it previously was found not to affect $K^{AH^-}_{amb.\ O_2}$ [8]. It should be noted that these observations verify the earlier interpretation [8] of the activation phenomenon as involving generation of additional catalytically competent sites for ascorbate oxidation. In any formal mechanism for the enzyme, an activation that operated by changing a rate constant rather than by creating or unmasking entire new sites would have to affect the value of the Michaelis constant for one or the other substrate.

## DISCUSSION

The steady state kinetic data that were obtainable by the methods used in the present study are all compatible with the invariance of the maximal velocity and $K^{O_2}_{m}$ for three quite dissimilar reductant substrates with very different $K^{reductant}_{m}$ values. These findings substantiate and extend the invariance of maximal velocity reported by Yamazaki and Piette [2]. Since the kinetic constants for reactions involving these three different reductant substrates with different $E_o'$ values and other chemical properties are unlikely to be the same coincidentally, these results imply certain constraints on the formal mechanism for ascorbate oxidase-catalyzed reactions. Specifically it is to be expected that the unimolecular (or pseudo-unimolecular) reaction that limits the maximal velocity must involve as reactant an enzymic form that does not contain the reductant substrate or a product structurally related to that substrate. These results eliminate from consideration mechanisms such as the double displacement form (Fig. 1A) previously suggested from this laboratory [7] in which ascorbate is in equilibrium with the ascorbate-enzyme complex and electron transfer from ascorbate to the enzymic copper controls the maximal velocity. In fact, all mechanisms having kinetically significant complexes of enzyme with the reductant substrates, including single displacement mechanisms, are eliminated. Since there is a finite maximal velocity, however, there must be a rate-limiting unimolecular step, and this must then involve either an enzyme-oxygen complex or a free enzymic form.

On the other hand, the data forming the basis for the suggestion of the mechanism in Fig. 1A (pH-variation kinetic and activation data [8] and stopped-flow measurements [5,6]) remain valid. Furthermore, this body of evidence is not consistent with the alternative simple double displacement form (Fig. 1B) in which the rate-limiting step at substrate saturation is the breakdown of a cuprous enzyme-oxygen complex, even though this form would be in accord with the identity of maximal velocities for the alternate reductant substrates. Specifically, the mechanism in Fig. 1B requires that variations in maximal velocity (equal to $k_{+3}E_o$) also occur in $K_m^{reductant}$ (equal to $k_{+3}/k_{+1}$). It was previously shown, however, that V diminishes toward zero as the pH is increased above a pK' near pH 8, while $K_{amb.\ O_2}^{AH^-}$ is little affected [8].

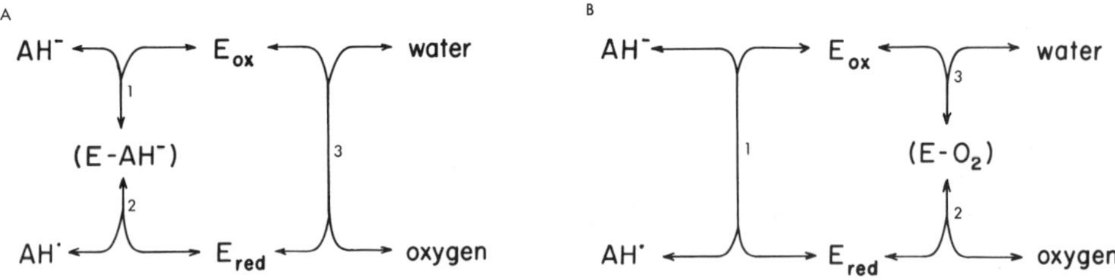


Fig. 1. Double displacement mechanisms for ascorbate oxidase. A: Form suggested by the results of previous stopped-flow and pH-variation studies. B: Form suggested by data for maximal velocities of alternate reductant substrates.

Since all of the simple single and double displacement forms thus appear to be at odds with the over-all body of evidence, a new class of formal mechanisms is indicated.

## A New Class of Mechanisms

Any formal catalytic mechanism proposed for ascorbate oxidase must now take into account all of the experimental phenomena detailed above. In addition, it should involve the participation of four reductant monoanions for each oxygen molecule, as substrates, and four reductant free radicals and two molecules of water as products. Provision must be made for parallel, linear double reciprocal plots for reductant substrates at different oxygen concentrations. All secondary plots of slopes and intercepts are also linear experimentally. Further, there must be a formal basis for the experiments of Nakamura and Ogura [5,6], in which the rate of disappearance of the blue absorption band of the enzyme as oxygen concentration fell after ascorbate addition was far less than the rate of reappearance of the color when oxygen was added to the reduced enzyme. It would also be useful if a formal basis were provided for the reaction inactivation of the enzyme, which has been shown to be caused by the production of $H_2O_2$ as a side-product of catalysis irrespective of enzyme purity [14].

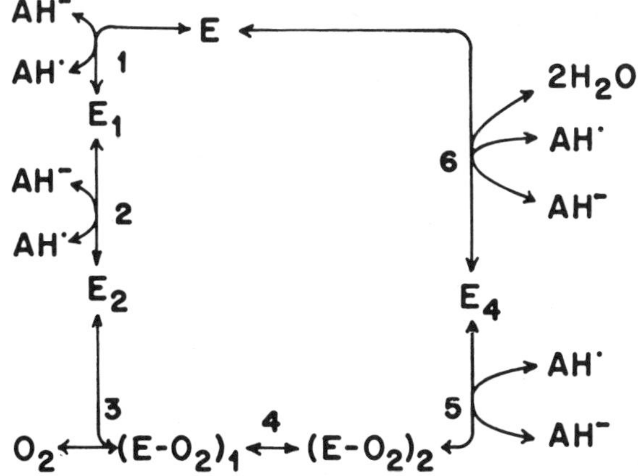


Fig. 2. A basic multiple displacement mechanism capable of satisfying the requirements for ascorbate oxidase action.

A formal mechanism capable of fulfilling these requirements is presented in Fig. 2. All permutations of this mechanism yield a rate equation of this form:

$$E_o/v_o = \left(\frac{1}{k_{+1}} + \frac{1}{k_{+2}} + \frac{1}{k_{+5}K_4} + \frac{1}{k_{+5}} + \frac{1}{k_{+6}}\right)(AH^-)^{-1} + \frac{k_{+4} + k_{-3}}{k_{+3}k_{+4}}(O_2)^{-1}$$
$$+ \frac{k_{-3}k_{-4}}{k_{+3}k_{+4}k_{+5}}(AH^-)^{-1}(O_2)^{-1} + (k_{+4})^{-1}.$$

Several features of this form require discussion. First, it is necessary to show that this formulation is in fact consistent with the results reported for stopped-flow experiments based on the appearance and disappearance of the blue color of the enzyme. Second, it is clear from the rate equation that these mechanisms will give parallel double-reciprocal patterns only if $\left(\frac{k_{-3}k_{-4}}{k_{+3}k_{+4}k_{+5}}\right)$ is near zero. That is, mechanisms of this form are applicable only when either reaction 3 or reaction 4 is practically irreversible. It will be argued that reaction 3, which includes an electron transfer to oxygen from enzymic cuprous copper, could be expected on thermodynamic grounds to be essentially irreversible. Further, mechanisms of the form suggested in Fig. 2 give rise to some additional predictions concerning the covariance of kinetic coefficients with pH that can be verified experimentally, providing a further basis for distinguishing between this form and the conventional double displacement forms given in Fig. 1. Finally, it is noted that the particular permutation of multiple displacement forms given in Fig. 2 involves accumulation of a peroxide-level intermediate that could account for the appearance of $H_2O_2$ as a by-product of the catalytic cycle.

## The Stopped-Flow Experiments of Nakamura and Ogura

The class of mechanisms presented here is clearly at variance with the conclusions of Nakamura and Ogura [5,6] that the oxidation of the enzyme by oxygen is much faster than its reduction by ascorbate. However, careful examination of the evidence on which this conclusion was based suggests that there is in fact no discrepancy. The visible adsorption of ascorbate oxidase solutions was followed by these investigators using stopped-flow techniques. In one set of experiments, the disappearance of absorption at 607 nm was followed as oxygen concentration decreased after addition of ascorbate. The following experimental results were reported:

Where $k_{red}$ is a pseudofirst-order reaction constant for the disappearance of absorption at 607 nm,

$$\frac{1}{k_{red}} = (\text{constant})(AH^-)^{-1} + \frac{1}{k_{red}^{max}}.$$

That is, the rate of disappearance of the blue absorption band of ascorbate oxidase as oxygen was exhausted displayed saturation behavior with respect to ascorbate. The value for $k_{red}^{max}$ obtained at 1.7° was $4.30 \times 10^2$ $sec^{-1}$ per molar enzymic copper, which may be converted to enzyme-molecular units of $1.72 \times 10^3$ M ascorbate per second per molar ascorbate oxidase. This value agrees satisfactorily with the molecular maximal velocity of $7.5 \times 10^3$ $\underline{M}$ $sec^{-1}$ per molar ascorbate oxidase calculated [8] from the steady state data obtained by Dawson for pure, fully active ascorbate oxidase at ambient oxygen partial pressure, considering the higher temperature of 25°.

This close correlation of $k_{red}^{max}$ and the molecular maximal velocity identifies the rate-limiting step in these stopped-flow experiments as the reaction which is rate-limiting for the total catalytic cycle. It must be remembered that the actual observations record only the time course of conversion of some blue form of the enzyme to a colorless form as oxygen is being exhausted. The enzyme in the absence of reductant substrate (E of Fig. 2) is blue; the form in the presence of excess ascorbate ($E_2$ in Fig. 2) is colorless; the color of any of the other forms is uncertain. Under these circumstances the rate-limiting step could still be that demanded by the total catalytic cycle. Only in the complete absence of oxygen would the rate necessarily correspond to the rate of reduction of the enzyme.

For the mechanism of Fig. 2 explicitly, if v is the rate of $E_2$ formation, and if $(AH^\cdot) \simeq 0$ [Reference 2],

$$v = k_{+2}(E_1)(AH^-) + k_{-3}(E\text{-}O_2)_1 - k_{+3}(O_2)(E_2), \quad \text{and}$$

$$k_{-3}(E\text{-}O_2)_1 = k_{+3}(E_2)(O_2), \quad \text{since } k_{+4} \text{ is small.}$$

Therefore,

$$v = k_{+2}(E_1)(AH^-).$$

By assuming a steady-state for all enzymic species except $E_2$ and $(E\text{-}O_2)_1$, it may be shown that as $(AH^-)$ gets large,

$$v_{max} = k_{+4}(E)_o, \quad \text{and}$$

$$k_{red}^{max} = k_{+4}.$$

However, if $(AH^-)$ is not very large, then steps involving $AH^-$ become rate-limiting rather than step 4. In this case it may be shown that

$$v_{(AH^-)\to 0} = k_{+5}(AH^-)(E\text{-}O_2)_2 = k_{+5}K_4(AH^-)(E\text{-}O_2)_1 \simeq k_{+5}K_4(AH^-)(E)_o.$$

Then,

$$\frac{1}{v} = \frac{1}{v_{(AH^-)\to 0}} + \frac{1}{v_{max}}$$

$$\frac{1}{v} = \frac{1}{(E_o)}\left(\frac{1}{k_{+5}K_4(AH^-)} + \frac{1}{k_{+4}}\right), \text{ or}$$

$$\frac{1}{k_{red}} = \frac{1}{k_{+5}K_4}(AH^-)^{-1} + \frac{1}{k_{red}^{max}},$$

which was the experimentally determined relationship. All the permutations of the mechanism in Fig. 2 yield results of this same form.

The foregoing considerations show that a proper interpretation of Nakamura and Ogura's first set of experiments is that reduction of the enzyme is not rate-limiting, rather a reaction like step 4 in our class of mechanisms is rate-limiting.

In a second set of experiments Nakamura and Ogura watched the reaction when colorless, reduced ascorbate oxidase free of any excess ascorbate was mixed with an oxygen solution, and the rate of appearance of the blue color was monitored. Where $k_{ox}$ is a pseudo-second-order rate constant for the appearance of absorption at 607 nm in these experiments,

$$v_{ox} = k_{ox}(O_2).$$

The value of $k_{ox}$ determined at 1.7° was $4.8 \times 10^6$ $sec^{-1}$ per molar $O_2$ per molar enzymic Cu. Conversion to enzyme-molecular units yields $1.92 \times 10^7$ $\underline{M}$ $sec^{-1}$ per molar oxygen per molar ascorbate oxidase. In terms of the mechanism of Fig. 2, in the absence of any excess ascorbate the blue color cannot be attributed to the blue "resting" enzymic form. Moreover, since there was no saturation behavior seen with respect to oxygen, form $(E\text{-}O_2)_1$ in the mechanism of Fig. 2 must be blue, as must its product, $(E\text{-}O_2)_2$, to yield the relationship in the above equation.

Analytically, if $(E\text{-}O_2)_1$ and $(E\text{-}O_2)_2$ were colorless, a blue color would never reappear. If $(E\text{-}O_2)_1$ were colorless and $(E\text{-}O_2)_2$ were blue, then

$$v'_{ox} = \frac{d(E\text{-}O_2)_2}{dt} = k_{+4}(E\text{-}O_2)_1,$$

$$(E\text{-}O_2)_1 = (E_2)(O_2)K_3, \quad \text{and}$$

$$v'_{ox} = k_{+4}K_3(E_2)(O_2).$$

Now, $k_{ox} = 1.92 \times 10^7$ and $(O_2) = 1.2 \times 10^{-4}$ $\underline{M}$ in these experiments. Therefore $v_{ox} = 2.3 \times 10^3\ sec^{-1}$. Initially $(E_2)$ equals $(E_2)_o$ and $v_{ox_o} = \frac{v'_{ox}}{(E_2)_o} = k_{+4}K_3(O_2)$. Substituting the value of $k_{+4} = 1.7 \times 10^3\ sec^{-1}$ from the data of Nakamura and Ogura, then $K_3 = 1.4$ $\underline{M}$. $K_3$ must be greater than or equal to $(K_m^{O_2})^{-1}$ in this mechanism, but $(K_{AH^-}^{O_2})^{-1}$ is in fact equal to $4.2 \times 10^3$ $\underline{M}$, a value significantly greater than 1.4 $\underline{M}$. Hence the presumption that $(E\text{-}O_2)_1$ is colorless must be incorrect. On the other hand, if the first species is blue, then $v'_{ox} = k_{+3}(E_2)(O_2)$; $(E_2) = (E_2)_o$; and $k_{+3} = 1.9 \times 10^7\ sec^{-1}$, which is a reasonable value for such a rate constant. Therefore, the enzymic forms on each side of step 4 must be blue, and the value $1.9 \times 10^7\ sec^{-1}$ at 1.7° must be assigned to to $k_{+3}$.

In summary, the data obtained in these stopped-flow experiments, when analyzed in conjunction with other data presented here, indicate that reduction of the enzyme is not rate-limiting at maximal velocity, that a (pseudo)-unimolecular step (step 4 in our class of mechanisms) is rate-limiting, and that both kinetically significant oxygenated enzymic forms immediately following the entry of oxygen are blue.

## The Question of Reversibility of Reaction 3 in the Mechanism of Fig. 2

It is interesting to consider that the blue color of Cu(II) complexes is generally considered to be a result of an unshared pair of electrons in square-planar Cu(II) complexes. Since $(E\text{-}O_2)_1$ is blue, this implies that electron transfer to oxygen occurs very rapidly in step 3 and is not rate-limiting in step 4. This also implies that $K_3$ is a function of the redox potential for that step. Since step 3 includes electron transfer from cuprous copper to oxygen, we can estimate its value from model systems.

For the reaction (compare to step 3 in mechanism of Fig. 2)

$$2\ Cu(I) + O_2 + 2\ H^+ \rightleftharpoons 2\ Cu(II) + H_2O_2$$

one may calculate from data in standard sources an $E'_o$ of -6.91 kcal at pH 7.0, corresponding to an apparent equilibrium constant of $1.0 \times 10^5$. In general Cu(I)-amine complexes have stability constants one to two orders of magnitude less than Cu(II)-amine complexes [15]; in particular it is known that the reduced copper-enzyme bonds in ascorbate oxidase are much less stable than those in the oxidized enzyme [16]. Therefore $1.0 \times 10^5$ is a lower limit for $K_3$, with the actual value probably somewhere between $10^6$ and $10^7$. This value appears to justify the assumption that step 3 is practically irreversible, an assumption necessary to explain how the suggested class of mechanisms can yield parallel double reciprocal patterns.

## Reaction Inactivation and the Significance of Enzymic Protonation with a pK' of 7.9

It has been demonstrated that, during its catalytic cycle, ascorbate oxidase has an ionizable group with a pK' of 7.9, deprotonation of which causes a decrease of maximal velocity to zero without affecting $K_m^{reductant}$ [8]. This protonated group is required for the unimolecular rate-limiting step in the mechanisms to which we are constrained by previous considerations. This implies that steps 3 and 4 can be resolved into three conceptually distinct events: The entry of oxygen and oxidation of the enzyme, a subsequent protonation, and the unimolecular step, which is rate-limiting at maximal velocity.

That such a protonation would result in the reduction of maximal velocity to zero as the pH crosses the transition is intuitively apparent from the form of the mechanism in Fig. 2. Only that fraction of intermediate $(E\text{-}O_2)_1$ which is protonated can proceed to enzymic form $(E\text{-}O_2)_2$ and complete the catalytic cycle. A protonation which is kinetically significant immediately before the unimolecular rate-limiting step imposes constraints upon the values of other kinetic constants because the $K_m^{AH^-}$ was not affected.

According to a development due to Cha [17] the rate equations for mechanisms with the assumption of rapid steps at equilibrium

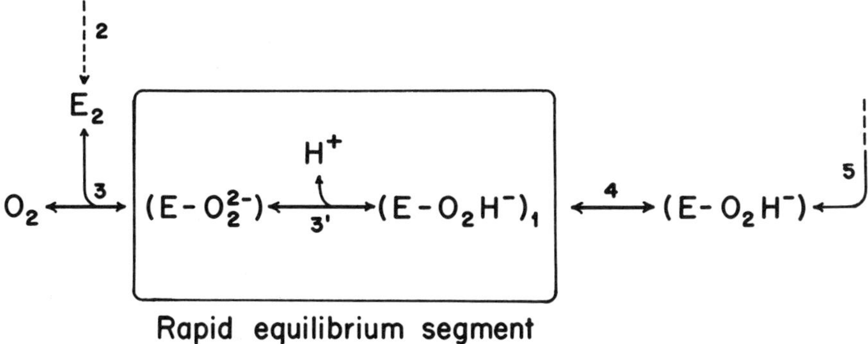


Fig. 3. Cha diagram for inclusion of a protonation step in the catalytic cycle of Fig. 2.

may be represented for this case in terms of Fig. 3 as follows, where $f_x$ is that fraction of enzyme in the rapid equilibrium segment which is in the form reactive in step x:

$$\frac{E_o}{v_o} = \left(\frac{1}{k_{+1}} + \frac{1}{k_{+2}} + \frac{1}{f_{+4}\ K_4\ k_{+5}} + \frac{1}{k_{+5}} + \frac{1}{k_{+6}}\right) \frac{1}{(AH^-)} +$$

$$\left(\frac{f_{+4}\ k_{+4} + [1 - f_{+4}]\ k_{-3}}{f_{+4}\ k_{+3}\ k_{+4}}\right) \frac{1}{(O_2)} +$$

$$\left(\frac{[1 - f_{+4}]\ k_{-3}\ k_{-4}}{f_{+4}\ k_{+3}\ k_{+4}\ k_{+5}}\right) \frac{1}{(AH^-)(O_2)} + \frac{1}{f_{+4}\ k_{+4}} .$$

Multiplication through by $V/E_o$ yields a cross-term of the form

$$\frac{(1 - f_{+4})\ k_{-3}\ k_{-4}}{k_{+3}\ k_{+5}} .$$

Clearly this term is approximately zero if the original cross-term is approximately zero. That is, the slopes are no more affected here than in the original case.

The effect of the pH transition on the maximal velocity is as predicted above:

$$\frac{V}{E_o} = f_{+4}\ k_{+4} .$$

Far below the pK', $f_{+4}$ is practically equal to one and the relationship reduces to

$$\frac{V}{E_o} = k_{+4}.$$

Far above the pK', $f_{+4}$ is practically equal to zero and the maximal velocity reduces to zero.

However,

$$K_m^{O_2} = \frac{f_{+4}\ k_{+4} + (1 - f_{+4})\ k_{-3}}{k_{+3}} .$$

As $f_{+4}$ goes to one, $(1 - f_{+4})$ goes to zero and

$$K_m^{O_2} = \frac{f_{+4}\ k_{+4}}{k_{+3}} .$$

Therefore we would predict that $K_m^{O_2}$ should be a function of pH. The general expression for this function, assuming only that $K_3$ is very large, is

$$K_m^{O_2} = \left(\frac{(H^+)}{K_a' + (H^+)}\right) \frac{k_{+4}}{k_{+3}} .$$

The slope of the oxygen double reciprocal plots, on the other hand, should not show the pH variation of the maximal velocity. The prediction has been tested experimentally and verified, as noted in RESULTS. In this connection, it should be noted that the double displacement mechanism shown in Fig. 1B leads to the opposite prediction. According to this mechanism, the slopes of the oxygen double reciprocal plots should vary as does maximal velocity. Again, the multiple displacement, but not the simple double displacement, formal mechanism is in accord with the observations.

The effect of these considerations on the $K_m$ for reductant yields the following relationship:

$$K_m^{AH^-} = f_{+4}\ k_{+4}\ \left(\frac{1}{k_{+1}} + \frac{1}{k_{+2}} + \frac{1}{f_{+4}\ K_4\ k_{+5}} + \frac{1}{k_{+5}} + \frac{1}{k_{+6}}\right).$$

Experimentally, however, $K_m^{AH^-}$ is not a function of pH, implying that only $1/f_{+4}K_4k_{+5}$ is significant in the sum term. In that case

$K_m^{AH^-}$ reduces to $\frac{k_{-4}}{k_{+5}}$ and to this ratio may be assigned the experimental value of $5 \times 10^{-4}$ M.

It is interesting to consider why protonation of the enzyme should be significant after the entry of oxygen and before the unimolecular rate-limiting step. Two experimental facts are germane to this consideration: The protonation with a pK' near 8 does not have an electrostatic effect on $K_m^{AH^-}$ [8], and we can find no change in either the position or the magnitude of the absorption peak of the enzyme at 610 nm across this pH range. Since the ionizable group that affects the rate-limiting step in kinetic studies presumably must be in the active site, however, the implication is that the ionizable group is not present until oxygen is bound. Dawson et al. have observed that hydrogen peroxide is a side-product of ascorbate oxidase catalysis, as has been mentioned previously [1,14], and its production results in "reaction inactivation," apparently by subsequent denaturation of the enzyme. For this reason, it is tempting to postulate an enzymic form which might generate peroxide. Such an opportunity presents itself in the mechanism of Fig. 2. The outline of such a mechanism can be seen in Fig. 4. Here $(E\text{-}O_2^{2-})$, the enzymic form generated in step 3, must be protonated in a step which is pseudo-unimolecular at constant pH, to yield $(E\text{-}O_2H^-)_1$, the enzymic form which accumulates before the rate-limiting step. Further protonation through step 7 at a very slow rate results in the regeneration of free enzyme and the release of hydrogen peroxide.

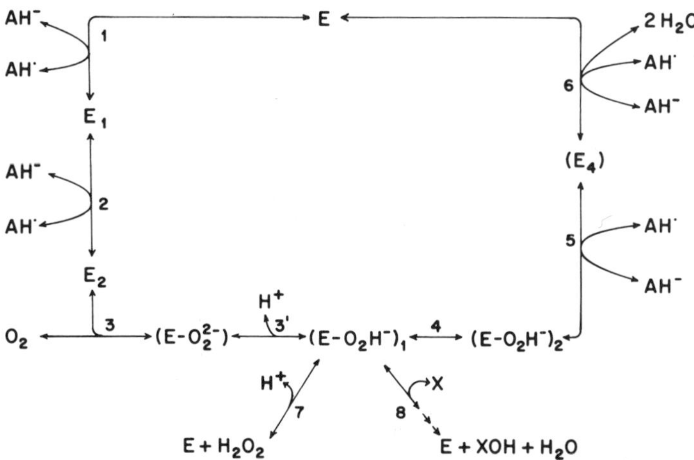


Fig. 4. A formal mechanism proposed for ascorbate oxidase catalysis. Reaction 7 leads to the non-stoichiometric production of hydrogen peroxide. Pathway 8 is postulated for catalysis of an oxygenase reaction of unknown substrate X.

It would be of interest to compare the pK' of 7.9 to that of a peroxide anion bound to a strongly electron-withdrawing Cu(II) in a Cu(II) chelate. Although no values for such species could be found in the literature, the values for some peroxides and peroxyacids which are available [18,19] range from near 5 for a phosphoric peroxide to near 12 for $H_2O_2$ itself and suggest that a pK' of about 8 is not unreasonable for the proposed cupric peroxide species.

## Values of Some Rate Constants in the Mechanism of Fig. 2

The highest value for maximal velocity at saturation in oxygen is that calculated from the maximal specific activity published by Lee and Dawson [20], $1.9 \times 10^4$ molar ascorbate/sec/molar ascorbate oxidase. Since $\frac{V}{E_o} = k_{+4}$, the value of $k_{+4}$ is $1.9 \times 10^4$ $sec^{-1}$.

$$K_m^{O_2} = \frac{k_{+4} + k_{-3}}{k_{+3}} = \frac{k_{+4}}{k_{+3}}$$, and this Michaelis constant is about $2 \times 10^{-4}$ $\underline{M}$. Therefore $k_{+3} = \frac{k_{+4}}{K_m^{O_2}} = 9.5 \times 10^7\ M^{-1}\ sec^{-1}$. The values of any other rate constants are functions of the value of $k_{-4}$, for which we have no estimate.

## The Physiological Significance of Ascorbate Oxidase

In connection with the mechanism proposed in Fig. 4 it is interesting to speculate about the physiological function of ascorbate oxidase. The oxidation of ascorbate does not appear to be teleonomically useful. Dehydroascorbate can only degrade to 2,3-diketogulonate irreversibly [21]. It is certainly wasteful of biological energy if it serves simply as a hydrogen shuttle to oxygen, with rapid rehydrogenation of dehydroascorbate as the other half of the cycle, as first proposed by Szent-Györgyi [22]. Indeed, since the only biological system known which reduces dehydroascorbate is an enzyme which uses NAD(P)H either directly or ultimately as a substrate [21], we would be in the untenable position of arguing for a profitless drain on NAD(P)H synthesis. The ascorbate oxidase function is unreasonable even as a mechanism for the catabolism of excess ascorbate; there is no correlation between the ascorbic acid content and the ascorbate oxidase content of many plants [23], and further, ascorbic acid is widely acknowledged to be nontoxic even in biological systems which lack ascorbate oxidase.

In this regard it is useful to review the physiological functions which have been ascribed to ascorbate. These include reduction of oxidized glutathione and NAD(P) as a poorly rationalized electron transport system in both respiration and photosynthesis, participation as a substrate in steroid biosynthesis (perhaps as a substrate for hydroxylation), and hydroxylation of compounds as a reductant substrate [24]. Mono-oxygenases known to use ascorbate as a substrate in well-defined systems include dopamine-β-hydroxylase [25] and proline hydroxylase [26].

Much of the mechanism in Fig. 4 is known in model catalytic systems. Many ferric and cupric chelates, including the aquo- and EDTA complexes, have been shown to catalyze the oxidation of ascorbate monoanion by oxygen to ascorbyl free radical and hydrogen peroxide [27-29]. An iron (III)-EDTA complex catalyzes the oxidation of ascorbate by a free radical mechanism with ascorbate monoanion and hydrogen peroxide as the reactants [30]; the same complex catalyzes the hydroxylation of salicylate to 2,3- and 2,5-dihydroxybenzoic acid and of benzoic acid to a mixture of o-, m-, and p-hydroxybenzoic acids via a free-radical mechanism involving ascorbate monoanion, oxygen and the aryl acid as reactants [31]. By analogy perhaps ascorbate oxidase is in fact an oxygenase, employing the kind of mechanism suggested by pathway 8 in Fig. 4.

## ACKNOWLEDGMENTS

The work reported here was supported by Research Grants GB-8578 and GB-29097 from The National Science Foundation and by United States Public Health Service Training Grants GM-424 and HD-1.

## REFERENCES

1. H.G. Steinman and C.R. Dawson, J. Amer. Chem. Soc., 64, 1212 (1942).
2. I. Yamazaki and L.H. Piette, Biochim. Biophys. Acta, 50, 62 (1961).
3. I. Yamazaki, J. Biol. Chem., 237, 224 (1962).
4. T. Nakamura, N. Makino, and Y. Ogura, J. Biochem., 64, 189 (1968).
5. T. Nakamura and Y. Ogura, J. Biochem., 59, 449 (1966).
6. T. Nakamura and Y. Ogura, J. Biochem., 64, 267 (1968).
7. J. Westley, Enzymic Catalysis, p. 124, Harper and Row, New York (1969).
8. B. Gerwin, S.R. Burstein, and J. Westley, J. Biol. Chem., 249, 2005 (1974).

9. T. Reichstein and R. Oppenauer, Helv. Chim. Acta, 16, 988 (1933).
10. G. Hesse, G. Krehbill, and F. Rümisch, Ann. Chem., 592, 137 (1955).
11. R.G. Ault, D.K. Baird, H.C. Carrington, W.N. Haworth, R. Herbert. E.L. Hirst, E.G.V. Percival, F. Smith, and M. Stacey, J. Chem. Soc., 1419 (1933).
12. D.K. Baird, W.N. Haworth, R.W. Herbert, E.L. Hirst, F. Smith, and M. Stacey, J. Chem. Soc., 62 (1934).
13. G.N. Wilkinson, Biochem. J., 80, 324 (1961).
14. K. Tokuyama and C.R. Dawson, Biochim. Biophys. Acta, 56, 427 (1962).
15. J.F. Fisher and J.L. Hall, Anal. Chem., 39, 1550 (1967).
16. R.J. Magee and C.R. Dawson, Arch. Biochem. Biophys., 99, 338 (1962).
17. S. Cha, J. Biol. Chem., 243, 820 (1968).
18. A.J. Everett and G.J. Minhoff, Trans. Faraday Soc., 49, 410 (1953).
19. D.H. Fortnum, C.J. Battaglia, S.R. Cohen, and J. Edwards, J. Amer. Chem. Soc., 82, 778 (1960).
20. M.H. Lee and C.R. Dawson, J. Biol. Chem., 248, 6596 (1973).
21. J.J. Burns, in Metabolic Pathways (D.M. Greenberg, ed), 3rd ed., Vol. I, pp. 394-411, Academic Press, New York (1967).
22. A. Szent-Györgyi, Science, 72, 125 (1930).
23. E. Silverblatt and C.G. King, Enzymologia, 2, 222 (1938).
24. J.J. Burns and G. Ashwell, in The Enzymes (P.D. Boyer, H. Lardy, and K. Myrbäck, eds), 2nd ed., vol. 3, part B, pp. 388-392, Academic Press, New York (1960).
25. E.Y. Levin, B. Levenberg, and S. Kaufman, J. Biol. Chem., 235, 2080 (1960).
26. K.I. Kivirikko and D.J. Prockop, J. Biol. Chem., 242, 4007 (1967).
27. A. Weissberger, J.E. LuVallee, and D.S. Thomas, Jr., J. Amer. Chem. Soc., 65, 1934 (1943).
28. M.M.T. Khan and A.E. Martell, J. Amer. Chem. Soc., 89, 4176 (1967).
29. M.M.T. Khan and A.E. Martell, J. Amer. Chem. Soc., 89, 7104 (1967).
30. R.R. Grinsted, J. Amer. Chem. Soc., 82, 3464 (1960).
31. R.R. Grinsted, J. Amer. Chem. Soc., 82, 3472 (1960).

# THE INVOLVEMENT OF SUPEROXIDE AND TRIVALENT COPPER IN THE GALACTOSE OXIDASE REACTION

Gordon A. Hamilton, Gary R. Dyrkacz, and R. Daniel Libby

Department of Chemistry, The Pennsylvania State University

University Park, Pennsylvania 16802

## INTRODUCTION

Until the present work, Cu(I) and Cu(II) were considered to be the only valence states of copper which are involved in the various cuproenzymic reactions. In the investigations to be summarized here we have obtained the first direct evidence that Cu(III), or some mono-copper species which behaves like it, is a catalytically active intermediate in an enzymic reaction, in this case that catalyzed by galactose oxidase. In addition, the role of superoxide in the galactose oxidase reaction has been clarified to a greater extent than has been the case for most other enzymic reactions in which superoxide has been implicated. When we began this work, we certainly did not anticipate all these results, especially the involvement of Cu(III). Thus, before beginning a detailed summary of the work, it seems appropriate to briefly discuss some of our reasons for studying this enzyme, and to point out the mechanistic dilemma which faced us when we began.

In non-enzymic systems, the direct oxidation of an alcohol by $O_2$ (eq. 1) does not occur readily even though it is a considerably

$$R_2CHOH + O_2 \longrightarrow R_2C{=}O + H_2O_2 \qquad (1)$$

exothermic reaction. The main reason for this is because $O_2$ is a triplet molecule (has 2 unpaired electrons) whereas the other reactant and both products are singlets (all electrons paired). The direct reaction of a triplet molecule with a singlet to give only singlet products is a spin forbidden process (spin conservation rule) and will not occur readily (1). It is spin allowed for the reaction of eq. 1 to occur in two or more steps with free radicals being intermediates; one simple possibility is that shown in eq. 2. However, the reaction does not proceed rapidly by this pathway because the first step is now considerably endothermic (by up to

$$R_2CHOH + O_2 \longrightarrow R_2\dot{C}OH + HO\dot{O} \longrightarrow R_2C{=}O + H_2O_2 \quad (2)$$

50 kcal/mole) (1). The same is true for all other radical pathways in the absence of catalysis. Thus, alcohols and $O_2$ do not react in non-catalytic systems because there is no low energy allowed pathway.

There are at least two classes of enzymes which catalyze reactions of the type shown in eq. 1. A flavin coenzyme is a required constituent of one of these classes, and there is now considerable evidence that such reactions proceed as outlined in eq. 3. In one sequence of steps, apparently not involving free radical

$$E.F \xrightarrow{R_2CHOH \;\to\; R_2C{=}O} E.FH_2 \xrightarrow{O_2 \;\to\; H_2O_2} E.F \quad (3)$$

intermediates (2), the oxidized enzyme-bound flavin (E.F) reacts with the alcohol to give the carbonyl compound and fully reduced flavin ($E.FH_2$). Subsequently, $O_2$ reacts with the reduced flavin to give the oxidized form and $H_2O_2$. In non-enzymic systems it is quite clear (3) that this latter reaction proceeds by a free radical mechanism (with superoxide as an intermediate) as expected from the foregoing considerations. Such a pathway is reasonable because, unlike $R_2\dot{C}OH$ (eq. 2), the intermediate flavin semiquinone radical is highly resonance stabilized, and consequently the step involving the formation of the radicals does not have a high energy barrier. Although there is no direct evidence that the flavooxidase-catalyzed reactions involve superoxide and the flavin radical as intermediates, the implication is strong that the reaction proceeds in that way, with the further reaction of the radicals to give $H_2O_2$ and oxidized flavin occurring before superoxide has an opportunity to separate from the enzyme. Thus, the flavoenzyme-catalyzed reaction can at least be rationalized in terms of the known reactivities of alcohols and $O_2$.

The other class of enzymes which catalyze the reaction of eq. 1 is exemplified by galactose oxidase (D-galactose: $O_2$ oxidoreductase; 1.1.3.9). In its name reaction this enzyme (4-16) catalyzes the oxidation of the primary alcohol group at the 6-position of galactose to an aldehyde with the concomitant reduction of $O_2$ to $H_2O_2$. However, it is a relatively unspecific enzyme with $\alpha$- and $\beta$-galactosides being good substrates, and simpler molecules such as dihydroxyacetone, glycerol, and benzyl alcohols among others being oxidized at respectable rates. Galactose oxidase has no known cofactor other than one atom of copper per molecule (m.w. approximately 68,000 (15)). Removal of the copper gives an inactive apoenzyme, and activity is restored by incubation with copper ions but not by other metal ions. Early EPR results (6) indicated that the enzyme as isolated has the copper mainly as Cu(II), and no change in the intensity of the EPR signal was observed on adding galactose in the presence or absence of $O_2$. These results suggested either, that some other group on the enzyme is undergoing reversible oxidation and reduction, or that galactose oxidase catalyzes its reaction by a mechanism fundamentally different from that used by the flavoenzymes. Regardless of which of these or other alternatives is correct, it was felt that an understanding of this basically simple overall reaction might help to clarify the mechanisms of considerably more complex enzymic reactions involving $O_2$. Thus, a few years ago we began a detailed investigation of the mechanism of the galactose oxidase reaction, and some of our results are summarized here. In brief, the results indicate that the overall mechanism bears a marked similarity to that of the flavoenzymes (eq. 3).

## EXPERIMENTAL METHODS

For all experiments, essentially homogeneous enzyme, prepared by slight modifications (17) of methods previously described (11) was used. Enzyme concentrations were determined from the absorption at 280 nm assuming a molar extinction coefficient of 105,000 (15). Kinetic results were obtained by following oxygen uptake at 25° with a Gilson Oxygraph equipped with a Clark electrode. EPR spectra were taken on a Varian E-9 spectrometer under the following conditions: frequency, 9 GHz; microwave power, 30 mW; modulation amplitude, 12.5 gauss; time constant, 0.3 sec; scanning rate, 125 gauss/min.; temperature, 100°K.

## SOME KINETIC CHARACTERISTICS

In early experiments (11,12) designed to characterize some of the basic kinetics of the galactose oxidase reaction, it was found that alterations in reaction conditions (sometimes changed by only a small amount), and the presence or absence of various additives frequently affect the rate of uptake of $O_2$ by large amounts. Also, in many cases a linear uptake of $O_2$ with time was not observed even though the reactions were being studied under what should have been zero order conditions. In some cases, especially those which eventually led to a rapid uptake of $O_2$, an induction period was observed, whereas in other cases, especially those which ultimately gave a relatively slow uptake of $O_2$, a burst was noted; the reaction started off rapidly but eventually slowed down to some linear rate. Among the additives which increase the rate of $O_2$ uptake are the following: peroxidase, catalase (at high concentrations), EDTA, ferricyanide, and superoxide. Some species which inhibit the reaction are: $H_2O_2$, any mono-negatively charged anion such as chloride or acetate, and superoxide dismutase (SOD) (18). In addition, the above additives affect the initial induction period or burst in various ways, some eliminating these effects and others prolonging them. The effects of $H_2O_2$ are particularly noteworthy. Although a product of the enzymic reaction, it is also an activator (eliminates any induction period) and an inactivator; when its concentration rises much above millimolar the enzyme is irreversibly inactivated. This inactivation by $H_2O_2$ only occurs in systems where the enzyme is actively catalyzing its reaction. In the absence of turnover the enzyme is stable for periods of hours in the presence of $H_2O_2$. All of the above results suggested that the enzyme can exist in 2 or more states, some of which are catalytically active and some inactive, and the various additives alter the rate of interconversion or the equilibrium concentrations of the various states. The peroxide inactivation results indicate that a state is present during active catalysis which is not present in the resting enzyme.

The quantitative effects of some of the additives under a given set of conditions are shown by the typical data (12) summarized in Table I. In the absence of all additives the rate varies from day to day but in the presence of EDTA a reproducible rate is obtained. Trace metal ion impurities, whose concentrations could vary, thus apparently cause some inhibition. The absence of any effect of EDTA on the inhibition by SOD indicates that the trace metal ions are probably inhibiting the reaction by the same mechanism that SOD does, i.e., by removing $O_2^{\bar{\cdot}}$ from solution. It is known that trace metal ions catalyze the disproportionation of $O_2^{\bar{\cdot}}$ while EDTA complexes of these ions do not (19). In those experiments (Table I) showing inhibition by SOD, the rate starts off more rapidly but slows to the reported rate after about a minute reaction time.

The activation by ferricyanide is concentration dependent and at $10^{-3}$M ferricyanide the rate is near the maximum achievable. Since both ferricyanide and $O_2^{\bar{\cdot}}$ activate the enzyme by comparable amounts, ferricyanide apparently performs the same function that $O_2^{\bar{\cdot}}$ usually does. The lack of any effect of EDTA and SOD on the ferricyanide activation is consistent with this conclusion. The following two observations indicate respectively that neither the $O_2^{\bar{\cdot}}$ nor the ferricyanide is reacting stoichiometrically when they increase the rate of $O_2$ uptake: (1) the control for the final experiment listed in Table I contained no galactose oxidase and less than 5% as much $O_2$ reacts compared to when the enzyme is present (the same amount of $O_2^{\bar{\cdot}}$ should be formed from the flavin system in both experiments), (2) in the experiment where all of EDTA, SOD, and ferricyanide are present, more than 800 molecules of $O_2$ react per molecule of ferricyanide converted to ferrocyanide (limit of detection).

TABLE I Effects of Some Additives on the Rate of the Galactose Oxidase Catalyzed Reaction[a]

| Additives | | | | Rate of $O_2$ uptake |
|---|---|---|---|---|
| EDTA | SOD | $Fe(CN)_6^{-3}$ | $O_2^{\bar{\cdot}}$[b] | (nmoles/min/ml) |
| - | - | - | - | ~4 |
| + | - | - | - | 12 |
| - | + | - | - | ~1 |
| + | + | - | - | ~1 |
| - | - | + | - | 23 |
| + | - | + | - | 23 |
| + | + | + | - | 24 |
| +[b] | - | - | + | 20 |

[a]Reaction Conditions: 25°C; 0.01 to 0.02M phosphate, pH 7.0 to 7.1; $[O_2]=2.5\times10^{-4}$M; [galactose]=0.09 M; [catalase]=2μg/ml; [galactose oxidase]= $2\times10^{-9}$M; and when present, [EDTA]=$5\times10^{-4}$M; $[Fe(CN)_6^{-3}]=10^{-3}$M; [SOD]= $10^{-9}$ M. [b]$O_2^{\bar{\cdot}}$ generated photochemically in situ by a flavin-EDTA system (3) consisting of 10 mM EDTA and 3.3μM flavin (3-carbethoxymethyl-10-methylisoalloxizine)]. [c]The reported rate is twice the observed rate because catalase is present.

In summary, the above results indicate that removal of $O_2^{\bar{\cdot}}$ from the actively catalyzing system gives an inactive form of the enzyme, while the addition of $O_2^{\bar{\cdot}}$ or ferricyanide increases the amount of an active form. The resting enzyme, and the actively catalyzing system in the absence of additives, apparently have a mixture of active and inactive forms. The observed stoichiometry of the activation, and the lag in the inhibition by SOD, indicate that at least 800 turnovers occur with the active form (or forms) of the enzyme before a molecule of $O_2^{\bar{\cdot}}$ leaks out to give the inactive form.

## SOME EPR EXPERIMENTS

Despite the earlier EPR results (6), the above kinetic experiments strongly implied that the copper of galactose oxidase is capable of existing in different valence states, at least one of which is catalytically active and one which is not. With conditions defined for obtaining enzyme which is apparently fully active (in the presence of ferricyanide) and almost completely inhibited (with SOD present), it thus became obvious to reexamine the EPR of the enzyme under these conditions (12). In controls performed at the same time the earlier EPR results (6) were completely confirmed; the resting enzyme gives a strong signal indicating that its copper exists largely in the Cu(II) state, and the addition of galactose does not appreciably change the intensity of the signal. In the presence of SOD and under conditions (galactose and $O_2$ present) where thousands of turnovers have occurred, the surprising result obtained (12) is that the intensity of the galactose oxidase Cu(II) EPR signal increases 20 to 30%. Since these are conditions which give almost completely inhibited enzyme (Table I), the result indicates that the Cu(II) form of the enzyme is inactive catalytically. In the presence of millimolar ferricyanide (conditions which give fully active enzyme) the EPR signal given by the galactose oxidase essentially completely disappears (12). This result implies that a state of the enzyme more highly oxidized than the Cu(II) state is a catalytically active species.

## REDOX POTENTIAL OF THE OXIDIZED ENZYME

In the above studies ferricyanide alone was used. Subsequently, it was found that ferrocyanide inhibits the enzymic reaction and the ferricyanide-ferrocyanide effects are reversible. Thus, if the enzymic reaction is initiated with 1 mM ferricyanide present, the addition of 1 mM ferrocyanide about a minute later causes a decrease in the rate of $O_2$ consumption. Correspondingly, if the reaction is initiated with 1 mM ferrocyanide present a very slow uptake of $O_2$ is observed but the addition of 1 mM ferricyanide increases the rate to the same final value as above. As some results obtained at pH 7 and shown in Fig. 1(a) indicate, the amount of activation, and the intensity of the EPR signal due to enzymic Cu(II), depend only on the ratio of ferricyanide to ferrocyanide concentrations. For any given ratio the same results within experimental error (10%) are obtained regardless of the total concentrations of ferricyanide and ferrocyanide (in the range 0.5 to 2 mM). The results shown in Fig. 1(a) can be used to calculate an oxidation-reduction potential for the enzymic group (Fig. 1(b)). Assuming the ferricyanide-ferrocyanide couple has a potential of 424 mV (20) one obtains a potential at pH 7 for the enzymic group of 440 mV from the EPR data and 410 mV from the kinetic data. The fact that the lines in Fig. 1(b) have a slope of 1 indicates that the catalytically active, but EPR inactive oxidized form of the enzyme and the catalytically inactive, but EPR active reduced form differ by only one electron.

The results shown in Fig. 1 were obtained at pH 7. Kinetic experiments performed at other pH's (Fig. 2) indicate that the redox potential increases at lower pH's but appears to remain constant at pH's above 7.5. This indicates that at high pH's the redox couple is that shown in eq. 4 whereas at low pH's it is the reaction of eq. 5. The data can be used to calculate a standard redox potential

$$\text{Enz(oxidized)}^{+n} + e^- \longrightarrow \text{Enz(reduced)}^{+n-1} \quad (4)$$

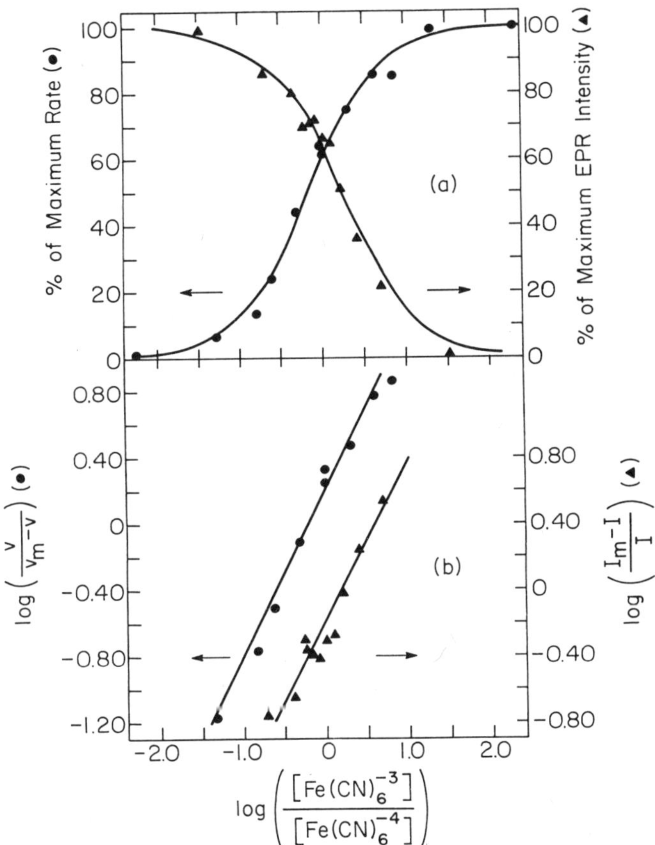


Figure 1: Effects of different ratios of ferricyanide to ferrocyanide concentrations on the rate (●) of the galactose oxidase reaction and on the intensity (▲) of the EPR signal given by the enzymic Cu(II). All solutions had 0.10 M phosphate, pH 7.0, air atmosphere and had differing ferricyanide and ferrocyanide concentrations varying from 0 to 1 mM (total iron concentration 1 to 2 mM). The kinetic results were obtained using 0.10 M galactose and 2.7 nM galactose oxidase. For the EPR experiments the galactose oxidase concentration was 30μM. In the bottom figure (b) the ordinates are essentially $\log([E_{ox}]/[E_{red}])$; v is the initial rate of $O_2$ uptake for a particular ratio of ferricyanide to ferrocyanide, and $v_m$ the rate obtained with 1 mM ferricyanide alone. Similarly, I is the EPR signal intensity with a particular ratio of ferricyanide to ferrocyanide and $I_m$ the intensity obtained with 1 mM ferrocyanide. The curves in the upper part (a) and the lines in the lower part (b) of the figure are theoretical calculated for a one-electron change and using the oxidation-reduction potentials given in the text.

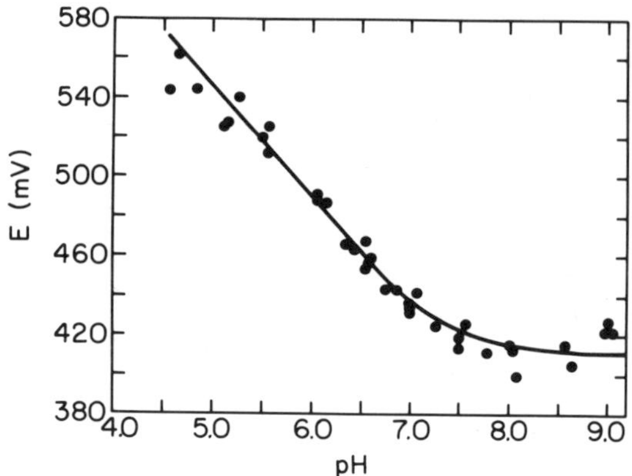


Figure 2: Effect of pH on the redox potential of the enzymic group as determined by kinetic methods. The experimental values of E were calculated from data obtained at various pH's using the following expression:

$$E = 424 + 59 \log \frac{[Fe(CN)_6^{-3}](v_m - v)}{[Fe(CN)_6^{-4}]\ v}$$

where v is the initial rate of $O_2$ uptake for the particular ratio of ferricyanide to ferrocyanide, and $v_m$ the rate obtained with 1 mM ferricyanide alone (424 is the redox potential of the ferricyanide-ferrocyanide couple (20) and is independent of pH over this range). All reaction solutions contained: 0.01 M buffer (phosphate or EDTA; no effect of buffer was noted), 0.10 M galactose, 2.04 nM galactose oxidase, 0.25 mM $O_2$, and 1 mM total concentrations of the iron hexacyanides. Most of the values of v were obtained using a 1:1 ratio of ferricyanide to ferrocyanide but at some pH's data was obtained for several ratios and the point shown is an average value. The illustrated line is a theoretical one (see text).

($E_o$), the redox potential for the couple of eq. 4, and an ionization constant ($K_a$) for the ionization of EnzH(reduced)$^{+n-1}$ by using the

$$Enz(oxidized)^{+n} + H^+ + e^- \longrightarrow EnzH(reduced)^{+n} \qquad (5)$$

relation shown in eq. 6. The best fit of the data of Fig. 2 gives

$$E = E_o + 59 \log(1 + \frac{[H^+]}{K_a}) \qquad (6)$$

$E_o = 410 \pm 5$ mV and $pK_a = 7.25 \pm 0.15$. The line shown in Fig. 2 was calculated using these values.

## WHAT IS THE NATURE OF THE OXIDIZED ENZYME?

All the foregoing results obtained using ferricyanide-ferrocyanide redox buffers indicate that a catalytically active form of the enzyme is a species which is 1 electron higher in oxidation state than the Cu(II) enzyme and that the Cu(II) enzyme is inactive catalytically. Before discussing possible structures for the oxidized species, some alternative explanations which were considered and can be discarded will be briefly mentioned. The possibility existed that the observed effects might somehow be due to dimerization or polymerization of the enzyme. Although the kinetic results did not seem consistent with such an interpretation, this possibility was directly disproven by observing that the sedimentation coefficient for the enzyme in the presence of excess ferricyanide is the same within experimental error as that (approximately 4.5 S) for the native enzyme (we thank Prof. A. T. Phillips for performing this experiment).

It has been suggested (16) that the loss of the EPR signal from the enzymic copper on adding ferricyanide is due to the enzyme and ferricyanide forming a complex with the ferricyanide bound in the immediate vicinity of the copper. The present results eliminate such a possibility because they show that the intensity of the resultant EPR signal is not a function of the ferricyanide concentration (which the above explanation requires) but only of the ratio of ferricyanide to ferrocyanide. Furthermore, there is nothing unique about ferricyanide (which again the above explanation would require). Ferricyanide and ferrocyanide were used for most of the experiments because they are stable and their redox potential is close to that of the enzyme. However, similar kinetic effects are observed using sodium chloroiridate as the oxidant. In fact 2μM $Na_2IrCl_6$ gives the same amount of activation as 1 mM ferricyanide, and this is consistent with the higher redox potential for chloroiridate (899 mV (21)).

A Cu(II) superoxide complex is another possibility for the structure of the oxidized enzyme which was considered but is unlikely. The only reasonable way by which this could be formed upon addition of ferricyanide to the Cu(II) enzyme would be if $H_2O_2$ was also present. However, similar kinetic effects of ferricyanide are observed in the presence or absence of catalase. Furthermore, after dialyzing a galactose oxidase solution versus a catalase solution for 36 hours, it was found that the subsequent addition of ferricyanide to the enzyme led as usual to the complete disappearance of the enzymic Cu(II) EPR signal. Thus, if $H_2O_2$ is necessary it must be very tenaciously bound to the Cu(II) enzyme.

All our results point to a Cu(III) state of the enzyme being the best description for the catalytically active species. Structures such as Cu(II).$\dot{X}$, where X is some enzymic group complexed to the copper, may contribute to the stabilization of the Cu(III), but it should be emphasized that these are just two limiting forms of a system which would be a resonance hybrid, and the only question is which limiting structure contributes more to the hybrid. Little is known about the ligands involved in the binding of copper to galactose oxidase. Spectral measurements on the Cu(II) enzyme have been interpreted (14,16) to indicate ligation by several nitrogens but the identity of these is unknown. There are indications but no definitive proof that an SH group of cysteine may be involved in copper binding; no sulfhydryl groups are titratable with the native enzyme but the apoenzyme has one such group (5). If the metal is complexed to a sulfur it is expected that structures such as Cu(II).ṠE would contribute considerably to stabilizing the Cu(III) state.

Until recently, Cu(III) was considered to be a relatively esoteric valence state for copper, encountered in only a few special cases in inorganic chemistry. However, it is now apparent that Cu(III) is a readily accessible state, and several such complexes involving organic ligands have now been well characterized (22). Of particular note are some tetrapeptide complexes studied by Margerum and coworkers (22e-g) which have redox potentials in the range of 600 mV. This is not much different from the potential obtained for the enzymic group. On this basis then the possibility of the oxidized enzyme being a Cu(III) species seems eminently reasonable.

Depending on the ligands involved, the non-enzymic Cu(III) complexes show absorbance maxima (molar extinction coefficients of several thousands) at various wavelengths in the ultraviolet and visible region (22,23). An optical difference spectrum obtained using a solution of galactose oxidase, which had been dialyzed against 1 mM ferricyanide, in one cuvette versus native enzyme and the dialysate in separate cuvettes showed peaks at 318 and 443 nm with molar extinction coefficients of approximately 5000 to 7000 at each wavelength. However, there is relatively strong absorption throughout the entire 300 to 500 nm region with the minimum at about 350 nm having an extinction coefficient of over 3000. This spectrum for the oxidized enzyme species is not identical to that of any of the known non-enzymic Cu(III) compounds but peaks in this region are common.

## MECHANISM

Illustrated in eq. 7 is an overall mechanism for the galactose oxidase reaction which is consistent with all the data now available. No specific sequence of addition of reactants and loss of products is meant to be implied in this mechanism; the illustrated steps merely outline the redox states of the enzymic species which oxidize or reduce the two reactants. In the usual catalytic cycle it is suggested that the Cu(III) form of the enzyme reacts with the alcohol to give the aldehyde and a Cu(I) form. The Cu(I) enzyme is then reoxidized by $O_2$ to the Cu(III) form and $H_2O_2$ with an enzymic Cu(II) superoxide species as a fleeting intermediate.

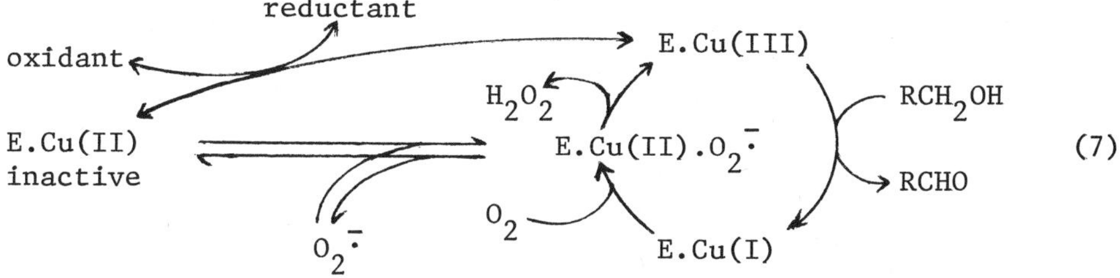


(7)

In outline this mechanism is very similar to that believed to hold for related reactions catalyzed by flavoenzymes (see introduction). As in the flavoenzyme case, the alcohol oxidation undoubtedly occurs without an organic free radical or an enzymic Cu(II) species being intermediates. As has been discussed previously (2), it is thermodynamically unreasonable to suggest that the alcohol oxidation could occur in 1-electron steps. In order for the alcohol oxidation step to proceed by a direct 2-electron reaction, E.Cu(III) and E.Cu(I) must have the same number of unpaired electrons (spin conservation rule, see introduction). Since E.Cu(I) is a $d^{10}$ system and has no unpaired electrons, the E.Cu(III) must be a spin paired $d^8$ system. This is certainly consistent with the observation that the oxidized enzyme shows no EPR signal.

The proposed oxidation of an alcohol to an aldehyde by the E.Cu(III) species is very similar to the conversion of an amine to an imine, a reaction known to occur in several non-enzymic Cu(III) systems (22b,24). A reasonable mechanism for the alcohol oxidation is that shown in eq. 8; loss of a proton from the α-carbon of the

$$\text{E.Cu(III)}-\text{O}-\overset{\text{H}}{\text{CHR}} \longrightarrow \text{E.Cu(I)}\ldots\text{O=CHR} \qquad (8)$$

complexed alkoxide, and electron migration as shown, would give the products directly. Such a step is very similar to one which occurs in the oxidation of alcohols by chromate (25). There is evidence (14) that galactose and other externally added ligands bind directly to the copper of the E.Cu(II) species. Since anions inhibit the enzymic reaction (11), an alkoxide complex of the E.Cu(III) thus is a reasonable intermediate. It seems possible that the $pK_a$ associated with the redox potential is due to ionization of an E.Cu(II) alcohol substrate complex; experiments are continuing to clarify this point.

Although there is considerable evidence that the Cu(III) form is in the usual catalytic cycle, there is no direct evidence that

E.Cu(I) is. However, in addition to one's expectation (from the above considerations) that it is involved, the inactivation by $H_2O_2$ is indirect evidence for its existence. In non-enzymic systems Cu(I) reacts very rapidly with $H_2O_2$ to generate the hydroxyl radical (HO·). Such a species formed near the active site of the enzyme would be expected to react unspecifically with many different amino acid residues, and this would ultimately lead to inactivation of the enzyme.

It is not known why the enzyme leaks a molecule of superoxide once in every few thousand turnovers (best estimate from the available data is once in 2 to 5 thousand turnovers at pH 7 and 25°) rather than once every turnover or not at all. However, the observation (inhibition by SOD) that it does in fact leak superoxide with the concomitant formation of the catalytically inactive E.Cu(II) indicates that an E.Cu(II) superoxide species is probably a catalytic cycle intermediate. Such an intermediate is also expected on the basis of the spin conservation rule; if E.Cu(I) and E.Cu(III) are both spin-paired systems their interconversion involving triplet $O_2$ and singlet $H_2O_2$ must occur in a two-step mechanism involving radical intermediates. There is no evidence indicating whether the superoxide in the catalytic intermediate is bound directly to the copper or to other groups in the vicinity of the copper. Either inner sphere or outer sphere electron transfers may be involved.

Since the discovery of SOD, several investigators have obtained evidence for the involvement of superoxide in various enzymic reactions (18). However, in few cases has its role been elucidated in the detail of the present investigation. Because of the triplet nature of $O_2$, it seems very likely that most enzymic reactions of $O_2$ involve superoxide as an intermediate. In many of these cases no evidence for superoxide has been obtained presumably because the enzyme-bound intermediate reacts to give spin-paired products before superoxide has a chance to dissociate. If the reaction to give spin-paired products is not fast enough to trap all the superoxide then one will have a 'leaky' enzyme, as is apparently the case with galactose oxidase. It is suggested that, in other enzymic examples where evidence for superoxide has been obtained, the main conclusion to be derived is that the enzyme is leaky. One should not conclude that such enzymes catalyze the $O_2$ reaction by a mechanism basically different from that of enzymes not giving evidence for superoxide. The reaction involving $O_2$ is probably very similar in both systems. Why some enzymes are leaky and others (for example, the flavin-containing oxidases) not is another question; perhaps it is related to the specific chemistry involved or to some control mechanism.

In the presence of ferricyanide-ferrocyanide redox buffers, the results indicate that thermodynamic equilibrium is rapidly established and maintained between the E.Cu(II) and E.Cu(III) species even during active catalysis. The E.Cu(III) form is presumably the predominant catalytic cycle intermediate during turnover because there is a large kinetic deuterium isotope effect (deuterium substituted for hydrogen at the carbon undergoing oxidation) associated with alcohol oxidation (10), i.e., the rate determining step in the catalytic cycle is the alcohol oxidation step. In the absence of redox buffers there is apparently a much slower interconversion of E.Cu(II) and catalytic cycle intermediates but it apparently still occurs to some extent. This is evidenced by the fact that under any given set of conditions (not leading to irreversible enzyme inactivation) a pseudo equilibrium between catalytically inactive and catalytic cycle intermediates is eventually established, i.e., although the reaction may show an induction period or a burst, it eventually gives a linear uptake of $O_2$. The induction period and burst phenomena are almost certainly due to the position of the pseudo equilibrium eventually established during catalysis being different from the ratio of the catalytically inactive and catalytic cycle intermediates in the resting enzyme (the enzyme as isolated and stored in 0.1 M phosphate buffer, pH 7). A burst would be expected if the resting enzyme has a higher ratio of catalytic cycle intermediates than the ultimately established pseudo equilibrium, and an induction period is expected if the opposite were the case. All our evidence indicates that the resting enzyme is a mixture of the E.Cu(II) and E.Cu(III) forms in a ratio of about 3 or 4 to 1, but this ratio appears to vary somewhat with method of isolation and storage. The above number is obtained from the observed increase in the intensity of the EPR signal due to the enzymic copper on adding excess ferrocyanide to the resting enzyme, or the intensity increase observed on SOD inhibition. Even in the earliest EPR experiments, the data of Blumberg, Peisach _et al._ (6) indicated only 70% of the enzymic copper could be accounted for as Cu(II), but they assumed the accuracy of their measurement was only ±30%, and thus that all the copper was present as Cu(II). Their experiments were probably much more accurate than they claimed. It is not immediately clear why the resting enzyme is a mixture of the two forms with the above approximate ratio, but it seems possible that it may be controlled by the oxygen-hydrogen peroxide couple. Very low levels of $H_2O_2$ and an air atmosphere would give about the right redox potential.

## IS Cu(III) INVOLVED IN OTHER ENZYMIC REACTIONS?

Since the Cu(III) state of galactose oxidase and various nonenzymic complexes can be attained so readily (low redox potential), it seems certain that the Cu(III) state will be shown to occur in some other cuproenzymic reactions as well. Probably the main importance of the present work is that in the future investigators will be sensitized to consider the possibility of this valence state with each individual enzyme, whereas in the past only the Cu(I) and Cu(II) states were considered. One enzyme for which the Cu(III) state is strongly indicated is tyrosinase. For the tyrosinases with several coppers it has been especially difficult to assign valence states because of the possibility of copper dimers and clusters being involved. However, one tyrosinase has been characterized which has only one copper per molecule (26), and it can exist in two redox states (differing by 2 electrons) neither of which is EPR active. Very probably these are the Cu(I) and Cu(III) states as in case of galactose oxidase. The oxidation of catechol by the Cu(III) state presumably occurs by a mechanism very similar to that shown in eq. 8, i.e., electron migration to the copper of a Cu(III) phenolate complex with concomitant loss of a proton from the other phenolic group.

## ACKNOWLEDGMENTS

This research was supported by a research grant (AM 13448) from the National Institute of Arthritis, Metabolism, and Digestive Diseases, Public Health Service. The EPR spectrometer used in this research was purchased using funds partially supplied by an equipment grant from the National Science Foundation to the Chemistry Department.

## REFERENCES

1. G. A. Hamilton in "Molecular Mechanisms of Oxygen Activation," O. Hayaishi (ed), Academic Press, New York, 1974, p. 405.

2. G. A. Hamilton, Progress in Bioorganic Chemistry, 1, 83 (1971).

3. V. Massey, G. Palmer and D. Ballou, in "Oxidases and Related Redox Systems," T. E. King, H. S. Mason and M. Morrison (eds), University Park Press, Baltimore, Maryland, 1973, p. 25; V. Massey, S. Strickland, S. G. Mayhew, L. G. Howell, P. C. Engel, R. G. Matthews, M. Shuman and P. A. Sullivan, Biochem. Biophys. Res. Commun., 36, 891 (1969).

4. J. A. D. Cooper, W. Smith, M. Bacilla and H. Medina, J. Biol. Chem., 234, 445 (1959).

5. D. Amaral, F. Kelly-Falcoz, and B. L. Horecker, Meth. Enzymol., 9,87 (1966); G. Avigad, D. Amaral, C. Asensio, and B. L. Horecker J. Biol. Chem., 237, 2736 (1962); D. Amaral, L. Bernstein, D. Morse, and B. L. Horecker, J. Biol. Chem., 238 2281 (1963); F. Kelly-Falcoz, H. Greenberg, and B. L. Horecker, J. Biol. Chem.

240, 2966 (1965).

6. W. E. Blumberg, B. L. Horecker, F. Kelly-Falcoz, and J. Peisach, Biochim. Biophys. Acta, 96, 336 (1965).

7. R. A. Schlegel, C. M. Gerbeck, and R. Montgomery, Carbohyd. Res., 7, 193 (1968).

8. G. T. Zancan and D. Amaral, Biochim. Biophys. Acta, 198, 146 (1970).

9. S. Bauer, G. Blauer, and G. Avigad, Israel J. Chem., 5, 126p (1967).

10. A. Maradufu, G. M. Cree and A. S. Perlin, Canadian J. Chem., 49, 3429 (1971).

11. G. A. Hamilton, J. de Jersey and P. K. Adolf in "Oxidases and Related Redox Systems," T. E. King, H. S. Mason and M. Morrison (eds), University Park Press, Baltimore, Maryland, 1973, p. 103.

12. G. A. Hamilton, R. D. Libby and C. R. Hartzell, Biochem. Biophys. Res. Commun., 55, 333 (1973).

13. G. R. Dyrkacz, R. D. Libby and G. A. Hamilton, J. Amer. Chem. Soc., 98, 0000 (1976).

14. L. D. Kwiatkowski and D. J. Kosman, Biochem. Biophys. Res. Commun., 53, 715 (1973); D. J. Kosman, R. D. Bereman, M. J. Ettinger and R. S. Giordano, Biochem. Biophys. Res.Commun., 54, 856 (1973); M. J. Ettinger, Biochemistry, 13, 1242 (1974); M. J. Ettinger and D. J. Kosman, Biochemistry, 13, 1247 (1974); R. S. Giordano and R. D. Bereman, J. Amer. Chem. Soc., 96, 1019 (1974); R. S. Giordano, R. D. Bereman, D. J. Kosman and M. J. Ettinger, J. Amer. Chem. Soc., 96, 1023 (1974).

15. D. J. Kosman, M. J. Ettinger, R. E. Weiner and E. J. Massaro, Arch. Biochem. Biophys., 165, 456 (1974).

16. L. Cleveland and L. Davis, Biochim. Biophys. Acta, 341, 517 (1974); L. Cleveland, R. E. Coffman, P. Coon and L. Davis, Biochemistry, 14, 1108 (1975).

17. R. D. Libby, Ph.D. Thesis, The Pennsylvania State University, 1974.

18. J. M. McCord and I. Fridovich, J. Biol. Chem., 244, 6049 (1969); I Fridovich in ref. 1, p. 453.

19. D. Klug, J. Rabani and I. Fridovich, J. Biol. Chem., 247, 4839 (1972).

20. J. E. O'Reilly, Biochim. Biophys. Acta, 292, 509 (1973).

21. E. Jackson and D. A. Pantony, J. Applied Electrochem., 1, 113 (1971).

22. (a) J. J. Bour, P. J. M. W. L. Birker, and J. J. Steggerda, Inorg. Chem., 10, 1202 (1971); (b) D. C. Olson and J. Vasilevskis, ibid. 10, 463 (1971); (c) L. F. Warren, and M. A. Bennett, J. Amer. Chem. Soc., 96, 3340 (1974); (d) F. J. Hollander, M. L. Caffery, D. Coucouvanis, ibid., 96, 4682 (1974); (e) D. W. Margerum, K. L. Chellappa, F. P. Bossu and G. L. Burce, J. Amer. Chem. Soc., 97, 6894 (1975); (f) G. L. Burce, E. B. Paniago, and D. W. Margerum, Chem. Comm., 261 (1975); (g) Chemical and Engineering News, 53, no. 49, Dec. 8, 1975, p. 26.

23. D. Meyerstein, Inorg. Chem., 10, 2244 (1971).

24. (a) M. Anbar, R. A. Munoz and P. Rona, J. Phys. Chem., 67, 2708 (1963); (b) A. Levitzki, M. Anbar, and A. Berger, Biochemistry, 6, 3757 (1967).

25. (a) J. Rocek and A. E. Radkowsky, J. Amer. Chem. Soc., 95, 7123 (1973); M. Rahman and J. Rocek, ibid, 95, 5455, 5462 (1971); (b) K. B. Wiberg and S. K. Mukherjee, ibid, 96, 1884 (1974).

26. S. Gutteridge and D. Robb, Eur. J. Biochem., 54, 107 (1975).

# THE BIOLOGICAL ROLE OF CERULOPLASMIN AND ITS OXIDASE ACTIVITY

Earl Frieden and H. Steve Hsieh

Dept. of Chemistry, Florida State Univ.

Tallahassee, Fla. 32306

## INTRODUCTION AND EVOLUTIONARY IMPLICATIONS

A blue protein from the $\alpha_2$-globulin fraction of human serum which possessed oxidase activity was first reported by Holmberg (1) in 1944. It was named ceruloplasmin (Cp), meaning a blue (substance) from plasma, by Holmberg and Laurell (2, 3) in later papers describing the purification and other basic observations on its chemical properties. Since Cp accounts for over 95% of the circulating copper in a normal mammal and fluctuates greatly in numerous disease and hormonal states, its study has excited the imagination of scientists who have generated several thousand papers on its chemistry and biology. Cp is an attractive protein for study also because, like serum albumin, it appears to be multifunctional, as the principal copper transport protein, as a molecule directly involved in iron mobilization from the iron storage sites to the plasma via its ferroxidase activity and, possibly, as a regulator of circulating biogenic amine levels through its oxidase activity. Several recent reviews summarize the chemical and catalytic properties of Cp (See our review (4) for further references.).

Ceruloplasmin may represent a current endpoint in the parallel development of copper and iron biochemistry in the natural selection of aerobic cells during the past 3 billion years. That copper and iron as metals have played a dominant part in the recent unfolding of human civilization as well as in the lengthy evolution of essential metalloproteins and metalloenzymes is a remarkable coincidence. In a recent review, tracing the evolution of the essential metal ions, Frieden (5) has pointed out the many close associations of copper and iron that have evolved in aerobic cells. A primary role was assigned to the development of enzymes to protect cells from unavoidable toxic oxygen byproducts - superoxide ion, singlet oxygen and hydrogen peroxide. This resulted in the ubiquitous occurrence of the Cu-Zn enzyme (superoxide dismutase) and the heme enzymes (catalase and peroxidase). The success of the aerobes was accompanied by the development of more sophisticated iron and copper enzymes, notably the cytochromes, cytochrome oxidase, and the numerous electron transferases in plants. With the increasing complexity of organisms, the cellular machinery utilizing iron and copper expanded greatly for the production of the oxygen carrying proteins, hemoglobins, hemerythrins and hemocyanins. This adaptation required the elaboration of storage and transport proteins exclusively for copper and iron - ceruloplasmin, ferritin, and transferrin. Later stages of evolution were accompanied by the appearance of crucial biosynthetic enzymes associated with connective tissue and other more specific processes. An example of the continuing close connection between iron and copper in the vertebrates is the ability of the copper protein of plasma, ceruloplasmin, to mobilize iron into transferrin for iron transport and distribution (6). Finally in man there appears to be a unique reciprocal relationship between serum copper and iron in pathology and stress.

## THE CATALYTIC ACTIVITY OF CERULOPLASMIN

### Background

In 1948, Holmberg and Laurell (2) explored the oxidase activity of Cp on numerous reducing substances. Their preparations of Cp enhanced the oxidation of aryldiamines, diphenols and other reducing substances including ascorbate, hydroxylamide, and thioglycolate. Later Curzon (7) found that the oxidation of aryldiamines in the presence of Cp could be activated or inhibited by certain transition metal ions. Finally Curzon and O'Reilly (8) reported that Fe(II) could reduce Cp and suggested a coupled iron-Cp oxidation system (9).

Our interest in this enzyme was originally stimulated by the possibility that Cp was a mammalian ascorbate oxidase, an enzyme that had been clearly identified in plants but had eluded detection in animal tissues. First we showed that the ascorbate oxidase activity of Cp was not due to traces of free Cu(II) (10). However it was found that at low concentrations of ascorbate, oxidation was greatly stimulated by traces of iron ions, present in most Cp preparations unless special precautions were taken to eliminate the iron impurity. At that time (11) we proposed three major groups of substrates to describe the oxidase action of ceruloplasmin:

1. Fe(II), the substrate with the highest Vm, and the lowest Km.

2. An extensive group of bifunctional aromatic amines and phenols, which do not depend on traces of iron ions for their activity. This group includes the two classes of biogenic amines, the epinephrine and 5-hydroxyindole series, and the phenothiazine series.

3. A third group of pseudosubstrates comprising numerous reducing agents which can repidly reduce Fe(III) or partially oxidized (free radical) intermediates of class 2. We consider these compounds to be secondary substrates by way of an iron cycle or an aromatic diamine acting as a shuttle.

These three classes of substrates are shown in Fig. I. In principle, any reductant can be a substrate if it can transfer an electron to oxidized ceruloplasmin without poisoning or blocking the autooxidizability of reduced ceruloplasmin. For example, in our laboratory, D.J. McKee has shown that $VOSO_4$ rapidly bleaches blue in Cp in its conversion to $VO_2^+$. Thus $VO^{2+}$ represents an additional Cp substrate that fits the first group along with Fe(II).

There has been some uncertainty as to whether certain organic compounds were true substrates (group 2) or pseudosubstrates (group 3). The issue is complicated by the fact that the cyclical iron catalyzed reactions are faster, in general, than direct electron transfer to Cp. After considerable early controversy and uncertainty, Curzon and Young (12) maintain that ascorbate is a true ceruloplasmin substrate with a rather high Km, 5.2 mM, and a typical Vm, 4.0 3/Cu/min. In their experiments, the role of iron impurities was assumed to be eliminated by using 100μM EDTA (13). Young and Curzon (13) also found that catechol was a true Cp substrate with a very large Km, 282mM. Similarly, Lovstad (14) reported that D- or L- Dopa could also be catalytically oxidized, though very weakly, in the presence of the iron chelator, Desferal. A summary of Km and Vm data on several classes of organic substrates is presented in Table I.

The reactions illustrated by group 3 pseudosubstrates have been used extensively to study the kinetics of Cp. Ascorbate at concentrations well below its effective substrate range (100μM) was used by Huber and Frieden (15) to study Fe(II) oxidation as in reaction sequence in Fig. 1. Young and Curzon (13) also used ascorbate (50μM) as the reducing agent in reaction sequence 3b to study the oxidation of N, N-dimethyl-p-phenylenediamine. Walaas and Walaas (16) introduced the use of NADH and NADPH to provide the electrons necessary to reduce partially oxidized of free radical intermediates resulting from the action of Cp on aromatic diamines, phenols or other oxidizable substrates. Since NADH does not react directly with Cp it has been widely used as an electron donor to study group 2 substrates.

TABLE I

SUBSTRATE GROUPS OF CERULOPLASMIN

| | Km, μM | Vm, e/Cu |
|---|---|---|
| Ferrous ion | 0.6,50 | 22 /min |
| p-Phenylenediamine | 21-3,000 | 1-7 |
| Aminophenols | > 180 | 4-15 |
| Catecholamines | > 2550 | 2-11 |
| 5-Hydroxyindoles | > 900 | 1.5-6 |
| Phenothiazines | > 900 | 1-10 |

While the role of Cp in iron mobilization is now widely documented, its catalytic activity towards any other class of substrates has not been related, as directly, to its biological function. We have suggested, therefore, that the name ferroxidase be used when describing the activity of Cp as an enzyme (17). It was further proposed that the enzyme be designed as a ferro-$O_2$ oxido-reductase and be assigned the I.U.B. number of E.C.1.12.3.1. It was realized, however, when referring to the copper transport protein of the plasma, that the name ceruloplasmin would be retained because of its historical significance and widespread familiarity.

The presence of a p-phenylenediamine oxidase activity has been reported in a wide variety of vertebrate sera (18), although the necessity of adequate precautions regarding the presence of iron and other contaminants was not appreciated at that time. Despite the fact that Seal (18) found no evidence for an oxidase activity in bullfrog sera, Inaba and Frieden (19) were able to isolate from frog sera a blue oxidase resembling human Cp very closely in several oxidase parameters. The ferroxidase activities of human, pig and rat ceruloplasmin were recently compared by Williams *et al.* (20) who estimated ratios of ferroxidase activity to plasma Cu and p-phenylenediamine oxidase activity (Table II). The pig and human enzymes compared more closely than did rat Cp. Despite the low ferroxidase activity, it was still possible to show that Cp was essential for the flow of iron from R.E. cells to transferrin in rat plasma (20).

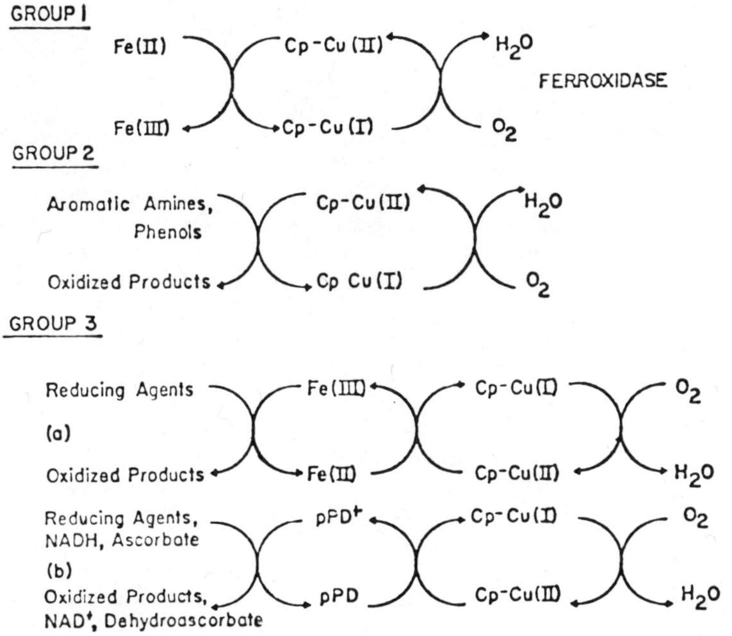

FIG. 1 The various substrate groups of ceruloplasmin and how they react. Groups 1 and 2 are true substrates, since they react directly with the oxidized form of Cp. Groups 3a and 3b may be considered pseudosubstrates since their reactions are mediated by a group 1 or 2 substrate.

TABLE II

Plasma ferroxidase activity in pig, man, and rat, phosphate buffer, pH 6.7 (300 μM ascorbate), Williams et al (20)

| Determination | Pig | Man | Rat |
|---|---|---|---|
| Plasma copper, μg/100 ml | 170±13.5 | 117±3.2 | 129±4.8 |
| pPD oxidase, A 530/mμm | .600±.0653 | .317±.0073 | .431±.0123 |
| Ceruloplasmin ferroxidase, μmol/ml per h | 118±8.6 | 43±1.3 | 6±1.2 |
| pPD-ox:Cu ratio | 0.35 | 0.27 | 0.33 |
| Cp ferroxidase:Cu ratio | 69 | 37 | 5 |
| Cp ferroxidase:pPD-ox ratio | 197 | 136 | 14 |

## The Ferroxidase Activity of Ceruloplasmin

The ferroxidase activity of human Cp was first reported by Curzon and O'Reilly (8, 9), but the appreciation of its significance in iron metabloism and its substrate characteristics have been explored almost exclusively at Florida State University. In our first paper, the effect of oxygen and Fe(II) concentration on the Cp and non-enzymic reaction was compared under normal serum conditions (17). The non-enzymic rate of Fe(II) oxidation was first order with respect to both Fe(II) and oxygen concentrations. In contrast, Fe(II) oxidation catalyzed by Cp showed typical saturation kinetics, reaching zero order at > 10μM $O_2$ and > 50μM Fe(II). From estimates of normal serum oxygen and Fe(II) levels, it was estimated that the Cp catalyzed oxidation of Fe(II) was 10-100 times faster than the non-enzymic oxidation. This estimate does not include any correction for the presence of reducing metabolites, such as ascorbate at 40μM, which did not affect the Cp-catalyzed oxidation of Fe(II), but significantly reduced the net rate of the non-enzymic oxidation of Fe(II).

Further studies of the kinetics of ferroxidase revealed biphasic curves in v vs v/Fe(II) plots (21) with two Km values, 0.6 M and 50 M, which differed by almost two orders of magnitude. While these data were originally interpreted in terms of two binding sites, Huber and Frieden (15) reported an excellent fit between experimental points and calculated values based on a rate-determining substrate-activation mechanism. Curves calculated on this basis fit the experimental data within the allowed error at five temperatures. Consistent with the activation mechanism are the loss of blue color (at 610nm) with low concentrations of Fe(II), and the activation observed by other divalent metal ions which are not substrates of ferroxidase (15, 22).

The mechanism of iron oxidation has been studied using rapid stopped flow methods to determine kinetic parameters (23). A minimum reaction sequence with rate constants indicated was proposed as follows:

$$\text{Cp-Cu(II)+Fe(II)} \underset{}{\overset{k_1}{\rightleftharpoons}} \text{Cp-Cu(II)-Fe(II)} \quad k_1=1.2\text{x}10\ M^{-6}\text{sec.}^{-1} \quad (1)$$

$$\text{Cp-Cu(II)-Fe(II)} \overset{k_2}{\rightleftharpoons} \text{Cp-Cu(I)-Fe(III)} \quad (2)$$

$$\text{Cp-Cu(I)-Fe(III)} \overset{k_3}{\rightleftharpoons} \text{Cp-Cu(I)+Fe(III)} \quad (3)$$

$$\text{Cp-Cu(I)} \overset{k_4}{\rightleftharpoons} [\text{Cp-Cu(I)}]' \quad k_4=1.1\text{ sec.} \quad (4)$$

$$[\text{Cp-Cu(I)}]'+O_2+4H \xrightarrow{k_5} \text{Cp-Cu(II)}+2H_2) \quad k_5=5.7\text{x}10^5M^{-1}\text{sec.}^{-1} \quad (5)$$

These data suggest that Cp reacts with Fe(II) much faster than with other substrates and indicated the presence of a substrate independent rate-determining step, e.g. reaction (4), with the smallest rate constant, $k_4$. That the slowest step involved a conformational change as depicted in reaction (4), was suggested by the large entropy change, $\Delta S^{+}$=-23 cal./mol./deg. estimated from the effect of temperature on the ferroxidase reaction (23). As emphasized in the discussion of other Cp substrates, this rate-limiting appears to be common to all substrates and appears to be relatively independent of the chemical nature of the substrate. Carrico, et al. (24), also, have shown that the Fe(II) mediated anaerobic reduction of Cp by sub-equivalent ascorbate leads to a rapid reduction of the blue chromophore (reaction 2). Later Gunnarsson, et al. (25) proposed an even more simplified mechanism for the oxidation of aromatic substrates, which will be considered later.

The inhibitory effects of trivalent and other metal ions on Cp activity were investigated by Huber and Frieden (26). All trivalent cations tested inhibited ferroxidase activity but the strongest trivalent inhibitors have an ionic radius of 0.81 A or less. The inhibition by Al(III) was mixed competitive and uncompetitive with respect to one of the substrates, Fe(II). The uncompetitive portion of the inhibition was not the result of competition by Al(III) with the other substrate, oxygen. A mechanism for the inhibition by Al(III) was proposed consistent with these results. A comparison of the strong cationic inhibitors provided the following series in order of decreasing effectiveness of inhibition: In (III) > ZrO(II) > Al(III) > Sc(III) > Ga(III). Insofar as they relate, the results are consistent with those from earlier studies on the inhibition of aryldiamine oxidase activity of human Cp reported by Curzon (7), McDermott et al. (11) and McKee and Frieden (22).

## Other Oxidase Activities of Ceruloplasmin

The relationship between ferroxidase activity and the iron mobilization properties of Cp have focused much recent attention on Fe(II) as a substrate. However, stimulated by the fact that Group 2 substrates include two important types of biogenic amines, i.e. the epinephrine and 5-hydroxindole series, and the phenothiazine series (tranquilizers), numerous investigators, particularily Barass, Coult, Curzon, Lovstad and Pettersson and their coworkers, have vigorously pursued the study of the Cp catalyzed oxidation of arylamines and phenols. Much of this work has been summarized by Young and Curzon (13), Fee (27), and Gunnarsson (28).

## Oxidation of Aromatic Amines and Phenols

A comprehensive study of group 2 substrates was presented by Young and Curzon (13) and is extended and summarized in Table I. The basic parameters, Km and Vm, were estimated from reciprocal plots under standard conditions at 25°, pH 5.5 in the presence of 100mM EDTA and 50mM ascorbate. EDTA was used to eliminate trace Fe effects and the ascorbate used to assure linear kinetics by preventing the accumulation of free radical intermediates. The data is remarkably consistent for an extensive series of p-phenylenediamines amino-phenols, catechols and 5-hydroxyindoles. All the Vm values except that for Fe(II), fall in the range of 1-10 $e^-$/Cu/min, with few exceptions. This seems to emphasize a common substrate-independent rate-determining step, mentioned earlier. Young and Curzon (13) pointed out a negative correlation between log Km and log Vm. Replacement of a benzene ring by an indole group had little effect. The variation within a group of 20 p-amino compounds was only about twofold. The presence of a side chain reduced the Vm for both the catechol and hydroxyindole series. No significant relation was noted to pK or the size, position, or electronic character of substituents on any of the ring systems studied.

In contrast to the limited range of the Vm's the Km's varied over a $10^4$-fold range. With the exception of the special role of Fe(II), a group of substituted p-phenylenediamines (pPD) appear to have the smallest Km values and presumably, a more favorable interaction with Cp. The three preferred substrates, with Km's less than $10^{-4}$M all have an additional benzene ring n(p-methoxyphenyl)-pPD, N-phenyl-pPD, and N-ethyl-N-2(S-methylsulfonamido-ethyl)-pPD. Km values increased, also, in the order para, ortho, meta suggesting that electronic rather than steric factors are dominant. In fact, rigid steric restrictions for Cp substrates are not indicated. For example, the tetra N-methyl pPD has a Km comparable to that of pPD or its partially substituted derivatives. However, ring substituted pPD derivatives with strongly electron withdrawing 2-nitro and 2-sulfonic acid groups had much greater Km values than these of other 2-substituted pPD derivatives. There is some evidence that negatively charged groups can increase Km values, possibly because of the repulsion of a negative charge on the enzyme near the active site.

Gunnarsson, _et al_. (29) reported that within a limited comparison group, eight Cp substrates had especially high energies of their highest occupied molecular orbital. The higher orbital energies were correlated with the lower Km values. While most Cp substrates have at least two electron-donating groups, Gunnarsson _et al_. (30), from their orbital energy calculations, explored a special group of N-alkylated anilines as Cp substrates. N-methyl aniline, N,N-dimethylaniline, N,N-diethylaniline, N,N-dipropylaniline, and N,N-dibutylaniline were all oxidized first to yellow and then to blue pigments. The reaction sequence including the Cp steps is shown in Fig. 2. However, no quantitative data on these substrates were reported. A detailed study of the Cp catalyzed oxidation of a related oxidation product, o-dianisidine (4,4'-diamino -3,3'-dimethoxybiphenyl), has been published (31). These data emphasize that electronic, rather than steric characteristics, are of prime importance for the activity of a Cp substrate.

## Newer Substrate Groups

Lovstad (32) has discovered a new class of Cp substrates, the phenothiazine derivatives. A highly suggestive clue was provided by Barrass and Coult (33) when they reported that the phenothiazines activated the Cp-catalyzed oxidation of catecholamines. The data included in Table I on the phenothiazines was adapted from a recent paper by Lovstad (32). As with other organic substrates, the Vm's do not vary greatly. In this series, the Km also varies only over a ten fold range. At low concentrations, compounds with a piperazinylpropyl side chain are more rapidly oxidized than those with an aliphatic side chain suggesting greater enzyme affinity for the former compounds. Phenothiazines with three-carbon side chains (promazine, alimemazine) are more rapidly oxidized by Cp than those with only two carbon atom side chains in the 10-position (promethazine, diethazine). Lovstad also confirmed that these phenothiazine derivatives activated the Cp-catalyzed oxidation of catecholamines. The substrates most rapidly oxidized by Cp also activated the oxidation of dopamine most effectively.

Barrass, Coult, et al. (34) have made a comprehensive study of the phenylalkylamines and the indoles and their isosteres as substrates for Cp. However, much of this work is qualitative with no Vm's and a limited number of Km's reported. Further, the enzyme and other components used in the reaction mixture were not always carefully screened for traces of iron ions or other trace elements. Therefore it is possible that some of these substrates might be of the group 3 type rather than the group 2 type. Barrass, et al. found that all of the 3,4 dihydroxyphenylalkylamines are substrates. Numerous compounds of the substituted amphetamines series also were oxidized. Only 3,4-dihydroxyphenylalanine and 3-aminotyrosine were oxidized among eight substituted phenylalanines tested. The best substrates had a 3,4-dioxygen pattern with at least one free OH group. A primary amino group could replace one of the OH groups. The alkylamine side chain was essential for maximum substrate activity, but the length of the side chain was not critical. Higher homologs of dopamine were effective substrates with the smallest Km (40μM) observed for the propylamine side chain. Monosubstitution of the

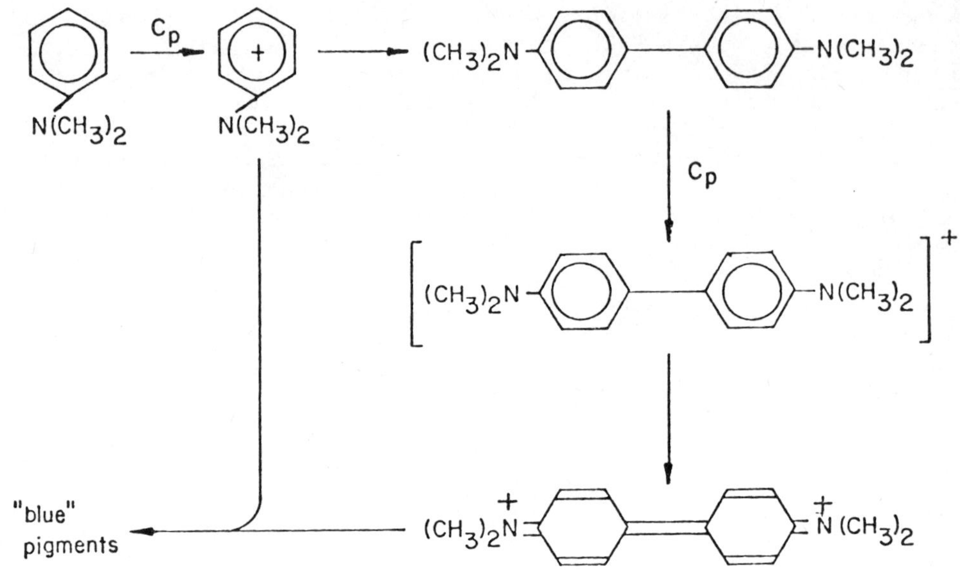

FIG. 2. Sequence of reactions in the oxidation of N,N-dimethylaniline to a blue benzidine derivative. At least two of the reactions involve Cp; the other reactions are believed to be spontaneous. This is a modification of Fig. 1 of Gunnarson, *et al.*(30).

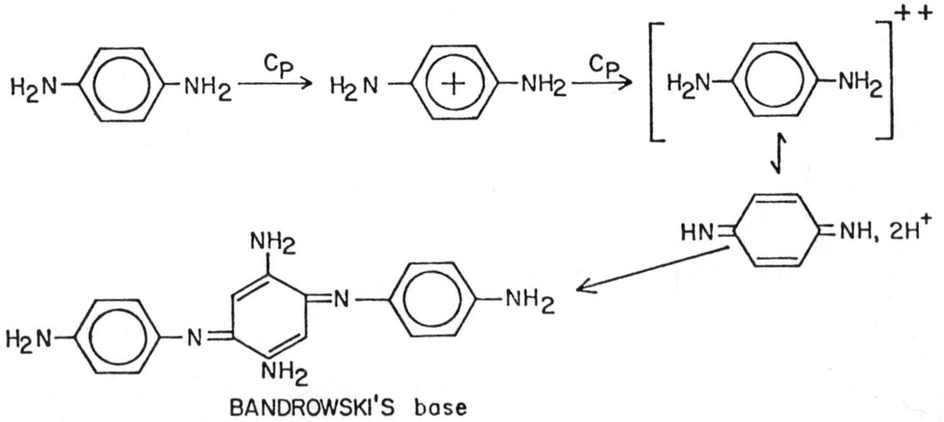

FIG. 3. Sequence of oxidative reactions by which Cp and $H_2O_2$ convert p-phenylenediamine (pPD) to Bandrowski's base, identified by Rice (38) as the ultimate oxidation product. Part of the $pPD^{2+}$ is formed by a rapid disproportionation reaction: $2pPD^{+} \rightarrow pPD + pPD^{2+}$.

α-carbon atom of the side chain had little effect, but disubstitution at this point greatly reduced substrate activity.

In a survey of the indole deries, Barrass, et al. (35) confirmed the essentiality of a hydroxy group on the aromatic ring in the 4,5, or 6 position. An aminoalkyl chain free, or substituted, at position 3 is necessary for high substrate activity, but again, the distance separating the basic group of the side chain from the indole ring is not critical. Among indole isosteres, analogs of 5-hydroxytryptamine, only an imino group at C.-1 of the bicyclic system is compatible with substrate activity.

## Mechanism of Oxidation of Aromatic Diamines

In a study of the pH-dependence of the Cp catalyzed oxidation of N,N-dimethyl-pPD (DPD), Gunnarsson et al. (36) were able to resolve some of the kinetic parameters of the four major substrate forms of DPD. They reported Km's for these substrates as DPD°= 0.2μM, DPD=45μM, $DPD^+$=70μM, and DPD, $H^+$-1100μM. The extremely low Km for DPD compares favorably with that of Fe(II). The argument of Gunnarsson et al. that this challenges the designation of Cp as a ferroxidase is still not valid since DPD is not a native substrate and no other potentially physiological substrate has been shown to have a Vm 1/4 as great or a Km one-thousandth as small as that of Fe(II).

A simplified mechanism to describe the Cp catalyzed oxidation of organic substrates has been proposed by Gunnarsson et al. (36) as follows:

$$E + S \xrightarrow{k_1} E' + P$$

$$E' \xrightarrow{k_2} E$$

Here $k_1$ is identical with the second order rate constant for reduction of the 610nm chromophore and $k_2$ is a pseudo first order rate constant corresponding to the rate limiting reaction step. Thus Vm=$k_2$ and Km= $\frac{k_2}{k_1}$ and Vm/Km=$k_1$. This mechanism assumes the substrate-independent rate limiting step, the kinetic insignificance of the rate of formation of enzyme-substrate complexes in these reactions and the saturation of the system by oxygen [a Km value of 3.9 μM for $O_2$ was reported earlier by Frieden, et al. (37)]. Support for this was claimed by the correspondence of $k_1$ values determined directly by reduction of the 610nm chromophore and $k_1$ values estimated from Vm/Km ratios obtained from steady state data.

Little information is available on the identity of the oxidation products produced when aromatic amines and phenols are exposed to Cp. Phenols are oxidized to the corresponding quinones which frequently react further to form cyclization products such as adrenochrome. The oxidation of dimethylaniline produces byphenyl derivatives that are oxidized further, probably spontaneously, to a blue dye (Fig. 2). For p-phenylenediamine, Rice (38) showed that the principal product for both Cp and $NH_3$-$H_2O_2$ oxidations was Bandrowski's base (Fig. 3), known since 1889.[2] It was proposed that the molar absorption of this compound at 540nm (E = 1910) could serve as a basis for defining the molecular activity of Cp.

## Inhibition of Oxidase Activity

The inhibition of the catalytic activity of Cp toward aromatic diamines has been reviewed extensively by Curzon, et al. (7-9), Gunnarsson (28), and Fee (27). Curzon and Cumings (39) identified seven categories of inhibitors, including inorganic anions, carboxylate anions, -SH compounds, chelating agents, hydrazines, 5-hydroxyindoles and a miscellaneous group, which included metal ions. Probably the most useful group is that of the inorganic anions which includes two of the strongest inhibitors, cyanide and azide. Holmberg and Laurell (3) had observed that the oxidase activity of Cp was affected by virtually any anion. In Curzon's extensive studies (39) at pH 5.5, 10mM acetate buffer and 4μM EDTA, the order of inhibition of human Cp was:

$$CN^- \geqq N_3^- > F^- > I^- > NO_3^- > Cl^- > Br^- > OCN^- > SCN^- > HPO_4^- > SO_4^=.$$

A similarily ordered series was observed for rat Cp by Lovstad and Frieden (40). The metal binding feature of these ions is the strongest but not the only factor establishing the inhibitory impact. The two most powerful inhibitors are cyanide and azide with inhibitory constants ($K_i$) of about $2x10^{-6}$M.

Azide is obviously the most convenient of the inhibitors and has been used frequently in attempts to distinguish between Cp catalysis of Fe(II) oxidation and other ferroxidase activity in biological media such as plasma. To assess the importance of other ferroxidases in human serum, Sexton (41) compared the effect of 1.0mM azide on the ferroxidase activity of fresh human serum with crystalline Cp, with and without bovine serum albumin (7%). The per cent non-azide-inhibited activity was 1.5 ± 0.2% in all samples, suggesting that 98.5% of the ferroxidase activity was due to Cp.

The mechanism of azide inhibition has been the most thoroughly studied of any of the inhibitors of Cp. Curzon and Cumings (39) reported that azide was a reversible but virtually stoichometric inhibitor of Cp-an azide concentration not much greater than that of Cp itself was required for inhibition at 25°. They also noted that only one azide reacted with each Cp molecule, regardless of the number of coppers at the active site. Reciprocal plots (1/v vs 1/s) at various azide concentrations were linear and parallel, suggesting that azide binds to an intermediate form of the enzyme during catalysis. A variety of methods by different authors (27) supports the view that azide inhibits primarily by impeding the breakdown of the reduced form of the enzyme in the oxidative reaction sequence.

## THE BIOLOGICAL FUNCTION OF CERULOPLASMIN:

### Ferroxidase and Iron Mobilization

Ceruloplasmin has been shown to be a direct molecular link between copper and iron metabolism. Copper deficiency results in low plasma Cp and iron, reduced iron mobilization, and eventually anemia, even with high iron storage in the liver. Cp controls the rate of iron uptake by transferrin. Transferrin plays a key role in the availability of iron for the biosynthesis of hemoglobin in the reticulocytes. The ferroxidase activity of Cp may result in the reduction of free iron ion generating a concentration gradient from the iron stores to the capillary system, thus promoting a rapid iron efflux in the reticuloendothelial system. It has been confirmed both _in vivo_ and in the perfused liver that $10^{-7}$ M Cp specifically induces a rapid rise in plasma iron. This evidence and its implications has been extensively reviewed (42).

How does Cp mobilize iron? Is iron mobilization due to its ferroxidase activity or is some specific receptor site(s) on the membrane of the iron storage cells activated by the Cu-protein? Cp has two properties which may be related to this question: its ferroxidase activity and its relatively unique ability to complex strongly with Fe(II) in preference to Fe(III). As pointed out earlier, the formation of the Fe(II)-Cp complex is the first reaction in the ferroxidase sequence and is extremely fast, $k = 10^7$ $M^{-1}$, $sec.^{-1}$. This reaction may be a limiting factor in iron mobilization from the iron storage cells in the liver. The binding of Fe(II) could provide the impetus for the removal of Fe(II) from the liver iron stores as the iron is reductively released from ferritin (Fig. 4). The role of Cp in the movement of iron from R.E. cells to transferrin has also been described as follows (20): Fe(II) occupies specific iron binding sites on the membranes of R.E. cells. Cp interacts with these iron binding sites and then forms a Fe(II)-Cp intermediate which transfers iron to apotransferrin by a specific ligand exchange reaction.

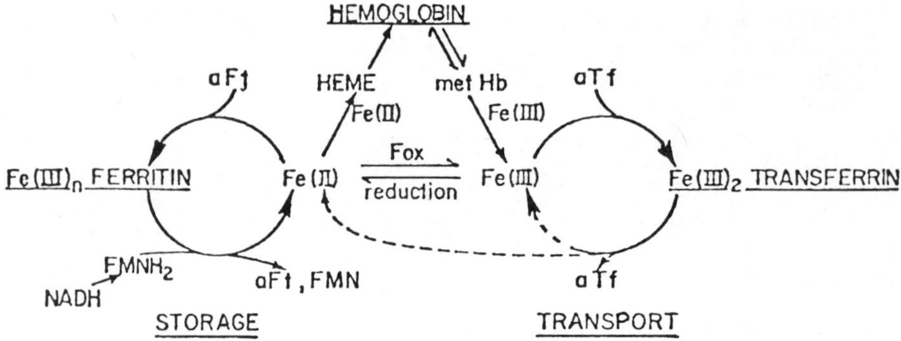


FIG. 4 The central role ceruloplasmin (ferroxidase, Fox) plays in regulating the ferrous to ferric cycles which in turn affect the storage, transport, biosynthesis and catabolism of iron compounds.

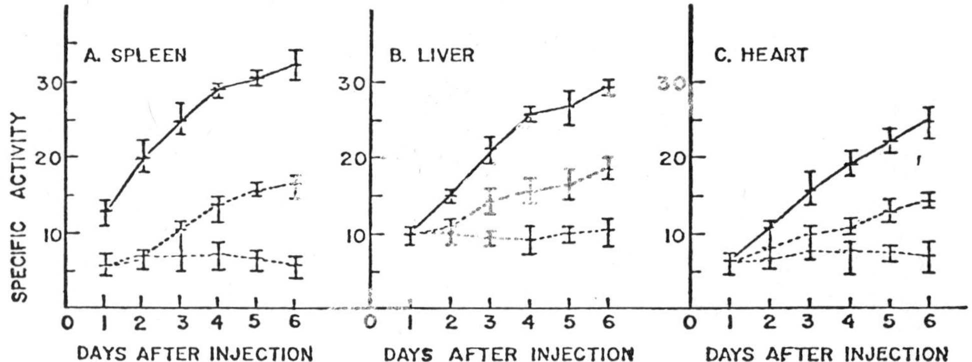


FIG. 5. Effect of ceruloplasmin (——), Cu-his (—-—) and saline (---) on cytochrome C oxidase activity in spleen (A), liver (B), and heart (C). Rats were injected intravenously with Saline or copper containing compounds corresponding to 12.3 μg Cu/100g body weight and three rats from each group were sacrificed at indicated intervals after injections. Specific activity is defined as n mole cytochrome C oxidized/min/mg protein.

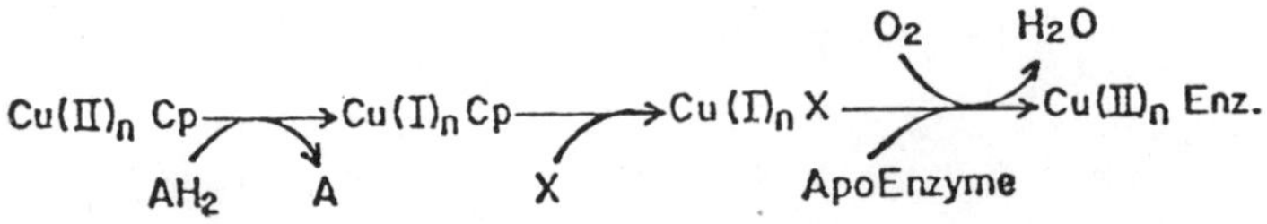

$$Cu(II)_n\,Cp \xrightarrow[AH_2 \,\to\, A]{} Cu(I)_n\,Cp \xrightarrow[X]{} Cu(I)_n\,X \xrightarrow[ApoEnzyme]{O_2 \,\to\, H_2O} Cu(II)_n\,Enz.$$

FIG. 6 Proposed mechanism for the transfer of copper from ceruloplamsin to an intracellular Cu-enzyme. $AH_2$ is a reducing substrate and X is a hypothetical intracellular Cu(I) acceptor and/or ligand exchanger.

## Ceruloplasmin as the Transport Form of Copper

Equally important is the role of Cp in the transport function of copper from the liver to the blood, where it provides copper in a stable form for distribution to the tissues to fulfill the need for this vital metal in Cu-enzymes and Cu-proteins. To understand the transport function of Cp, it is necessary first to review some essential facts about the circulating forms of copper and the biosynthesis and turnover of Cp.

Plasma copper is only slightly more concentrated than the Cu of the erythrocyte (111±30 vs 90±20μg%). However the red cell copper, of which about 60% is associated with superoxide dismutase (formerly, erythrocuprein), is relatively inert metabolically and is not known to be involved in transport. The plasma copper is composed of two major fractions when classified on the basis of the strength of copper binding. First, there is Cp in which the copper ion is an integral part of the molecule and, in the normal animal, accounts for over 95% of the plasma copper. The other category is the dialyzable, or more dissociable, form of copper which has been shown to include Cu(II)-albumin, albumin-Cu(II)-histidine complexes, Cu(II)-histidine, and, perhaps, ternary complexes of Cu with other amino acids. The latter fraction is believed to be a highly labile transport form, primarily involved in the pre-Cp movement of copper from the gut to the liver after absorption. However a possible role for these labile complexes in subsequent Cu transport has not been excluded.

## Biosynthesis and Turnover of Ceruloplasmin

It has been shown that ingested copper disappears rapidly from plasma with a concomitant increase in hepatic copper which is incorporated into Cp and then released into the blood (43). The appearance of Cp copper in the plasma reaches a maximum within 24 hours after copper intake (44). It is known that the plasma Cp levels are under a wide range of humoral and hormonal controls (45). In the rat dramatic increases in serum Cp are observed within three weeks after birth at the expense of copper stores, after estrogen administration, and after stress and inflammation (45, 46). But virtually nothing is known about the intracellular mechanism involved in Cp biosynthesis including both the apo-Cp moiety and how and when the copper is inserted into the protein. We know that the vertebrate hepatocyte has an impressive capacity to synthesize Cp given sufficient available copper. The metabolic defect in Wilson's disease, wherin copper accumulates, particularly in the liver and the brain, with low plasma Cu and Cp, could involve either site. The lower availability of copper might arise from stronger binding by intracellular carrier proteins as reported by Evans *et al*. (48), or there could be a block in the biosynthetic utilization of copper in the last step of Cp production (47). The probable sequence of the biosynthesis of Cp involves the synthesis of several peptide chains, their assembly, glycosylation and finally, copper addition (48). While the amount of copper storage proteins may be affected, the synthesis of apo-Cp appears to be independent of copper status (49). Apo-Cp was found to be released into the plasma of copper deficient rats at the same rate as that of Cp in rats kept on a diet adequate in copper (49). The question of whether the injection of copper will induce Cp synthesis is still open since the response seems to vary with the dose of copper (45). Obviously, excessive copper is capable of producing a toxic response via hemolysis or some other inflammatory process. However, copper has been reported to induce the biosynthesis of Cp in human and monkey liver slices (50).

Once it reaches the blood, Cp has a survival $t_{1/2}$ of 54 hours in the rabbit (51), shorter than that of other staple proteins of the plasma. A $t_{1/2}$ of about 12 hours in the rat has been estimated (51). These turnover times are adequate to account for the rate of utilization of copper by the tissues. In a series of ingenious papers Ashwell and Morell (51) have shown that the survival of Cp in the plasma depends upon an intact carbohydrate moiety, particularly sialic acid. Desialation of Cp with neuraminidase reduces the $t_{1/2}$ of asialo-Cp to less than 1/2 hour (51).

The life cycle of the copper in Cp is a one time journey to the tissues or a return to the liver for resynthesis. The ability to add copper to the protein moiety seems to be an exclusive property of certain cells, particularly in the liver. Holtzman and Gaumnitz (49) showed that once apo-Cp reaches the circulation, it is incapable of adding copper to form an active Cp molecule. However Owen (52) recently found that *in vitro* Cp could exchange its copper for ionic copper at pH 7.4 in the presence of a reducing substrate, p-phenylenediamine. This observation is in accord with earlier findings that the copper ion in Cu-proteins exchanges much more readily in the cuprous state. In fact, reducing agents usually precede chelating agents in the preparation of apoproteins from Cu-proteins (53).

## Transport of Copper

The role of Cp as a copper transport protein has been proposed for more than a decade. Broman (54) particularly advocated such a role because of the abundance of Cp in the blood. Shokeir and Shreffler (55) found that leucycytes from Wilson's disease patients showed a great reduction in cytochrome oxidase activity but only a moderate decrease in heterozygous carriers compared to normal controls. A similar trend was observed for plasma Cp concentration. Based on these limited data, Shokeir and Shreffler proposed that Cp functioned as a Cu-donor to copper containing proteins. Along this line, Owen (56) observed that after the intravenous injection of radioactive copper, rats did not accumulate radioactivity in extrahepatic organs until after the emergence of [$^{64}$Cu]-ceruloplasmin. Following intravenous injection of plasma containing [$^{67}$Cu]-ceruloplasmin, which was prepared from donor rats pretreated with [$^{67}CuCl_2$], Marceau and Aspin (57) found that [$^{67}$Cu] activity in rat plasma decreased with the increase of radioactivity in various organs. A similar result was obtained by Owen (58). Following the same method of treatment, Marceau and Aspin (59) recently reported that radioactive copper was tightly bound to cytochrome C oxidase in the liver and brain after injection of [$^{67}$Cu] labeled plasma. In addition, the [$^{67}$Cu] activity was also found to be incorporated into liver cytocuprein (superoxide dismutase) (60). Their results, however, do not exclude the possibility that the radioactive copper found in cytochrome C oxidase or superoxide dismutase may come from sources other than Cp. Although approximately 95% of [$^{67}$Cu] in the plasma preparation they used was Cp bound, there is still 5% radioactive copper which is bound to albumin or amino acids. Thus the radioactive copper found in cytochrome C oxidase or superoxide dismutase after injection of isolated plasma might be derived from other sources since no quantitative data to the contrary were presented. Marceau and Aspin (59) did observe that copper derived from [$^{67}$Cu]-plasma was tightly bound to cytochrome C oxidase but copper from [$^{64}$Cu]-albumin was only loosely bound to the oxidase. This, however, was not an appropriate comparison since rats injected with [$^{67}$Cu]-plasma were killed after 6 days while rats injected with [$^{64}$Cu]-albumin were sacrificed after 2 hours.

## New Evidence for Ceruloplasmin as a Copper Transport Protein

A critical test of the transport role of Cp was provided by a series of experiments in the authors' laboratory (61). Rats were fed a copper deficient diet for about 2 months to reduce the cytochrome C oxidase activity. The reduction was especially profound in the heart, but no significant change was ovserved in kidney (Table III). Injection of either human or rat Cp caused an 8-fold increase in cytochrome C oxidase activity in spleen, a 4-fold increase in liver and heart, and more than a 2-fold increase in lung while there was only a small increase in cytochrome C oxidase activity in pancreas or kidney, compared to rats fed a deficient diet and injected with saline only. In spleen, lung and kidney, injection of Cp raised the oxidase levels above the copper control. On the other hand, rats receiving $CuCl_2$, Cu-histidine, and Cu-albumin produced a smaller increase in cytochrome C oxidase compared to Cp treated animals. Since the addition of the injected copper

TABLE III

PREFERENTIAL RESTORATION OF CYTOCHROME C OXIDASE ACTIVITY IN TISSUES OF COPPER-DEFICIENT RATS[a]

| Group | 1 | 2 | 3 | 4 | 5 | 6 | 7 |
|---|---|---|---|---|---|---|---|
| Diet | +Cu | -Cu | -Cu | -Cu | -Cu | -Cu | -Cu |
| Injection | saline | saline | $CuCl_2$ | Cu-alb.[b] | Cu-his.[c] | human Cp | rat Cp |
| Tissue | Cytochrome c Oxidase Activity[d] | | | | | | |
| Spleen | 25.4 | 3.57 | 13.6 | 14.0 | 16.5 | 32.3 | 30.7 |
| Liver | 29.8 | 8.13 | 17.0 | 18.0 | 21.3 | 30.5 | 34.5 |
| Heart | 102.5 | 6.13 | 11.3 | 14.2 | 15.9 | 26.0 | 24.8 |
| Lung | 16.5 | 8.8 | 12.9 | 17.0 | 16.9 | 21.4 | 22.7 |
| Pancreas | 38.3 | 12.1 | 15.1 | 10.7 | 13.4 | 17.0 | 16.0 |
| Kidney[e] | 22.9 | 20.4 | 23.0 | 22.3 | 27.1 | 29.2 | 30.1 |

[a]Rats were injected intravenously with the indicated compounds corresponding to 4.1 μg Cu/100g body weight on days 1, 3 and 5. On day 6 all rats were sacrificed and 10% homogenates were prepared from each tissue for measurement of enzyme activity.

[b]Cu-albumin

[c]Cu-histidine

[d]Specific activity is defined as nmoles cytochrome c oxidized/min/mg protein. Each value represents the average of data from three rats.

[e]It was also observed that copper deficiency produced no significant reduction in brain cytochrome c oxidase.

compounds into the supernatant of various tissues did not alter cytochrome C oxidase activity, the observed increase in cytochrome C oxidase activity represents an increased in vivo synthesis of this enzyme. The copper required for additional cytochrome C oxidase should come from the injected copper compounds as the animals were fed a copper deficient diet. Therefore, the results of Table III suggest that in all tissues Cp was a much better copper source for cytochrome C oxidase than $CuCl_2$, Cu-histidine and Cu-albumin.

Cytochrome C oxidase in spleen showed a faster and greater response to Cp injection than that in the liver or heart (Fig. 5); the enzyme activity in spleen increased one day after Cp injection while that in the liver or heart remained unchanged until two days later. The effect of Cu-histidine was apparently smaller and slower than that of Cp; no increase in cytochrome C oxidase activity was observed until three days after injections. Furthermore, in rats treated with Cu-histidine the increase in cytochrome C oxidase activity occurred after Cp activity in the blood reached a maximal level. This indicates that Cu-histidine and probably Cu from $CuCl_2$ or Cu-albumin are first converted into Cu-Cp which in turn is transferred to cytochrome C oxidase, in accord with the implications of radioactive copper studies. These data offer the strongest evidence to date supporting a major role of Cp as the copper transport protein from which copper is transferred to cytochrome C oxidase and probably other copper containing proteins. Other forms of circulating copper, albumin or amino acid (histidine) bound copper, are believed to be mainly involved in the transport of the newly absorbed copper from the gut to the liver.

The transport of copper by Cp probably requires specific receptor mechanisms in the various target tissues. The greater lability of Cu(I) in proteins, mentioned earlier, strongly suggests a reductive step in copper release. Ample reductive mechanisms are available once Cp is within the reactive sphere of the cell. First there are the numerous endogenous substrates described earlier. Secondly, Cp has been shown to be able to tap the electron transport machinery of the cell. Brown and White (62) have reported that in the presence of cytochrome C and typical oxidative substrates, e.g., succinate and NADH, heart muscle particles can reduce Cp. The reaction occurs under anaerobic conditions and is reversed by oxygen and was sensitive to cyanide, carbon monoxide and Antimycin A. Under aerobic conditions, Cp inhibited the electron transport system, possibly by a reaction with essential -SH groups.

Based on these ideas we propose a simplified mechanism for the incorporation of the copper from Cp into an intracellular Cu-enzyme or Cu-protein (Fig. 6). The first step is the reduction of the Cu(II) of Cp by any of the substrates or reaction sequences described earlier. If this occurs at the cell membrane, the Cu(I) is likely to be transferred to an intracellular Cu(I) acceptor, X. If Cp penetrates to the inside of the cell, X might not be as necessary for intracellular transport. In its Cu(I) form, the copper is added to an apoenzyme, where it is fixed into the holo-enzyme in the Cu(II) state with the aid of oxygen. This mechanism takes into account the primary role of Cp as the copper transport and donor molecule, the high exchangability of Cu(I)-ligands, and the greater stability of copper ion in Cu(II)-proteins.

## Regulation of Plasma or Tissue Levels of Biogenic Amines

Barrass and Coult (63) have summarized the effects of drugs used in the treatment of mental illness, e.g., tranquilizers, anti-depressants, on the Cp catalyzed oxidation of the biogenic amines, noradrenaline and 5-hydroxytryptamine. The suggestion here, is that Cp, or an enzyme with similar properties, may be of importance in affecting the relative concentrations of noradrenaline and 5-hydroxytryptamine in the serum and eventually, in those areas of the brain where these compounds act as neuro-transmitters. Thus a Cp-like enzyme, by its effect on the life-time of biogenic amines, could play an important role in the regulation of brain chemistry necessary for mental function; and interference with this enzyme may lead to the appearance of abnormal mental states.

A varied spectrum of drug effects are observed (Table IV). Hallucinogens such as LSD accelerated the Cp catalyzed oxidation of noradrenaline but inhibited the enzymic oxidation of 5-hydroxytryptamine. Tranquilizers of the phenothiazine type accelerated the enzymic oxidation of both substrates. As discussed earlier, Lovstad (32) found that many phenothiazines were effective substrates but only at relatively high concentrations ($10^{-2}$M). However, some phenylethylamines and anticholinergics with CNS activity showed no effect on the enzymic oxidation of the two biogenic amines.

The mode of action of LSD was of particular interest since this drug elevates brain 5-hydroxytryptamine levels and depresses brain catecholamines. When tested with Cp, LSD inhibited the oxidation of 5-hydroxytryptamine by 50% at a concentration of 1/10 of the substrate and enhanced the oxidation of noradrenaline 4 fold when treated at 1/10 of the substrate. The Km values for noradrenaline and 5-hydroxytryptamine are similar, 3 mM and 1 mM respectively. Cp could exert close control over the relative concentrations of these two compounds in key parts of the brain. Thus LSD might produce its central effects via Cp by perturbing the balance between these two groups of biogenic amines.

TABLE IV

EFFECT OF DRUGS ON BIOGENIC AMINE OXIDATION

| Drug | Effect on Cp Activity | |
|---|---|---|
| | nor-adrenaline | serotonin |
| Trifluperazine $2x10^{-4}M$ | 200% | 200% |
| LSD $2x10^{-4}M$ | 400% | -50% |
| Imipramine $10^{-2}M$ | -50% | -50% |

Barrass, _et al._ (64) also proposed a possible involvement of Cp in Parkinson's disease. The compound, 3-hydroxy-4-methoxyphenethylamine, is one of the endogenous toxins which accumulates in Parkinsonians, producing tremors and hypokinesia. _In vitro_, it was shown that this compound enhances dopamine oxidation catalyzed by Cp. Since Parkinson's disease may be associated with decreased catecholamine and increased 5-hydroxytryptamine levels and an elevated serum Cp, the latter has been suggested as a basis for the etiology of this disease.

Most recently Shokeir (65) has reported reduced Cp levels in the sera of patients whth Huntington's disease. This has long been suspected, since many of the early symptons of Huntington's chorea resemble those of another genetic disorder, Wilson's disease. This hepatolenticular disease is accompanied by copper accumulation and toxicity in the brain and liver and decreased Cp biosynthesis with almost invariably low serum Cp levels.

## SUMMARY

Ceruloplasmin (ferroxidase) the blue Cu-protein of vertebrate plasma, possesses significant oxidase activity towards Fe(II) and numerous aromatic amines and phenols. Its ferroxidase activity has led to the discovery that it is a molecular link between copper and iron metabolism. Ceruloplasmin mobilizes iron into the plasma from iron storage cells in the liver. An additional role of Cp may be as a contributor to the regulation of the balance of biogenic amines through its oxidase action on the epinephrine and the hydroxyindole series. Ceruloplasmin also serves as a major copper transport vehicle, comparable to transferrin for iron. Evidence is presented that the copper atoms of Cp are a prerequisite for copper utilization in the biosynthesis of cytochrome oxidase. The ability of Cp to release copper at specific cellular sites is believed to be related to its broad substrate spectrum of biological reducing agents. Thus Cp is a serum protein with several important functions, all of which are directly related to its oxidase activity.

## REFERENCES

1. Holmberg, C.G., Acta Physiol. Scand 8, 227 (1944).

2. Holmberg, C.G., and Laurell, C.B. Acta Chem. Scand. 2, 550 (1948).

3. Holmberg, C.G., and Laurell, C.B. Acta Chem. Scand. 5, 476 (1951).

4. Frieden, E. and Hsieh, H.S. Adv. Enzym. 44, in press (1976).

5. Frieden, E. in Protein-Metal Interactions, M. Friedman, Ed., Plenum Press, N.Y. 1974 pp. 1-31.

6. Frieden, E. in Adv. in Chem. Series: Bioinorganic Chem. 100, 292-321 (1971).

7. Curzon, G., Biochem. J. 77, 66 (1960).

8. Curzon, G., O'Reilly, S., Biochem. Biophys. Res. Commun. 2, 284 (1960).

9. Curzon, G., Biochem. J. 79, 656 (1961).

10. Osaki, S., McDermott, J.A., and Frieden, E. J. Biol. Chem. 239, 3570 (1964).

11. McDermott, J.A., Huber, C.T., Osaki, S., and Frieden, E., Biochem. Biophys. Acta 151, 541 (1968).

12. Curzon, G., and Young, S.M., Biochim. Biophys. Acta 268, 41 (1972).

13. Young, G., and Curzon, G., Biochem. J. 129, 273 (1972).

14. Lovstad, R.S., Acta Chem. Scan. 26, 2832 (1972).

15. Huber, C.T., Frieden, E., J. Biol. Chem. 245. 3973 (1970).

16. Walaas, O. and Walaas, E., Arch. Biochem. Biophys. 95, 151 (1961).

17. Osaki, S., Johnson, D.A., Frieden, E., J. Biol. Chem. 241, 2746 (1966).

18. Seal, U.S., Comp. Biochem. Biophys. 13, 143 (1964).

19. Inaba, T., Frieden, E., J. Biol. Chem. 242, 4789 (1967).

20. Williams, D.M., Lee, G.R., and Cartwright, G.E., Am. J. Physiol. 227. 1094 (1974).

21. Osaki, S., J. Biol. Chem. 241, 5053 (1966).

22. McKee, D., and Frieden, E., Biochem. 10, 3880 (1971).

23. Osaki, S., and Walaas, O., J. Biol. Chem. 242, 2653 (1967).

24. Carrico, R.J., Malmstrom, B.G., and Vanngard, T. Eur. J. Biochem. 22, 127 (1971).

25. Gunnarsson, P.O., Nylen, V., Pettersson, G., Eur. J. Biochem. 37, 41 (1973).

26. Huber, C.T., Frieden, E., J. Biol. Chem. 245, 3979 (1970).

27. Fee, J.A., Structure and Bonding, 23, 1 60 (1975).

28. Gunnarsson, P.O., Ph.D. Thesis, Univ. Lund, 1974, Lund, Sweden.

29. Gunnarsson, P.O., Pettersson, G., and Pettersson, I., Eur. J. Biochem. 17, 586 (1970).

30. Gunnarsson, P.O., Lindstrom, A., and Pettersson, G. Acta Chem. Scand. 25, 770 (1971).

31. Schosinsky, K.H., Lehman, H.P., and Beeler, M.G., Clin. Chem. 20, 1556 (1974).

32. Lovstad, R.A., Biochem. Pharm. 24, 475 (1975).

33. Barrass, B.C., and Coult, D.B., Biochem. Pharm. 21, 677 (1972).

34. Barrass, B.C., Coult, D.B., Rich, P., and Tutt, K.J. Biochem. Pharm. 23, 47 (1974).

35. Barrass, B.C., Coult, D.B., Pinder, R.M., and Skeels, M., Biochem. Pharm. 22, 2891 (1973).

36. Gunnarsson, P.O., Nylen, V., and Pettersson, G., Eur. J. Biochem. 27, 572 (1972).

37. Frieden, E., Osaki, S., Kobayashi, H., J. Gen. Physiol. 49, 213 (1965).

38. Rice, E.W., Analyt. Biochem. 3, 542 (1962).

39. Curzon, G., and Cumings, J.N., in The Biochem. of Copper, J. Peisach, P. Aisen, W.E. Blumberg, Eds., Academic Press, N.Y., 1966 p. 545-558.

40. Lovstad, R.A., and Frieden, E., Acta Chem. Scan. 27, 121 (1973).

41. Sexton, R.C. Ph.D. Dissertation, Florida State Univ., 1974.

42. Frieden, E., and Osaki, S. in Protein-Metal Interactions M. Friedman, Ed., Plenum Press, 1974, pp. 235.

43. Scheinberg, I.H., and Sternlieb, I., Pharmacol. Rev., 12, 355 (1960).

44. Holtzman, N.A., and Gaumnitz, B.M., J. Biol. Chem., 245, 2350 (1970).

45. Evans, G., Physiol. Rev. 53, 535-70 (1973).

46. Linder, M.C. and Munro, H.N., Enzyme 15, 111, (1973).

47. Scheinberg, I.H., and Morell, A.G., in Inorganic Biochemistry, G.I. Eichorn, Ed. Elsevien, N.Y. (1973) pp. 306-319.

48. Evans, B.W., Dubois, R.S., and Hambidge, K.M., Science, 181, 1175 (1973).

49. Holtzman, N.A., and Gaumnitz, B.M., J. Biol. Chem., 245, 2354 (1970).

50. Neifakh, S.A., Nomkhov, N.K., Shaposhnikov, A.M., and Zubzhitski, Y.N., Experientia, 25, 337 (1969).

51. Ashwell, G. and Morell, A.G., Adv. Enzymol., 41, 99 (1974).

52. Owen, C.A.J., Proc. Soc. Expt. Med. 144, 681 (1975).

53. Erickson, J., Gray, R., and Frieden, E., Proc. Soc. Exptl. Biol. Med., 134, 117 (1970).

54. Broman, L., in Molecular Basis of Some Aspects of Mental Activity, O. Walaas, Ed. Academic Press, 1967, Vol. 2, p.131.

55. Shokeir, M.H.K., and Shreffler, D.C., Proc. Natl. Acad. Sci., 62, 867 (1969).

56. Owen, C.A.J., Am. J. Physiol., 209, 900 (1965).

57. Marceau, N., and Aspin, N., Am. J. Physiol., 222, 106 (1972).

58. Owen, C.A.J., Am. J. Physiol., 221, 1722 (1971).

59. Marceau, N., and Aspin, N., Biochim. Biophys. Acta 328, 338 (1973).

60. Marceau, N., and Aspin, N., Biochim. Biophys. Acta, 328, 351 (1973).

61. Hsieh, H.S., and Frieden, E. Biochem. Biophys. Res. Com., 1976 in press.

62. Brown, F.C., and White, Jr., J. Biol. Chem. 236, 911 (1961).

63. Barrass, B.C. and Coult, O.B., in Progress in Brain Research, Vol. 36, P.B. Bradley and R.W. Brimble, Eds., Elsevier Pub. Co., Amsterdam, 1972, pp. 97-104.

64. Barrass, B.C., Coult, D.B., and Pinder, R.M., J. Pharm. Pharmac. 24, 499-501 (1972).

65. Shokeir, M.H.K., Clin. Gen. 7, 354 (1975).

## ACKNOWLEDGEMENTS

This work was supported by NSF Grant GB 42379 and NIH Grant HL 08344. We are especially grateful to Dr. James A. Fee, Biophysics Div., U. of Michigan, who made a galley proof of his comprehensive review available to us (27) in advance of publication.

Paper no. 51 in a series on the Biochemistry of The Copper and Iron Metalloproteins.

# SUPEROXIDE DISMUTASES: STUDIES OF STRUCTURE AND MECHANISM

Irwin Fridovich

Department of Biochemistry, Duke University Medical Center, Durham, North Carolina 27710

## INTRODUCTION

Superoxide dismutases are enzymes of defense which serve to protect respiring cells against a product of their own respiration. This is achieved by a catalytic scavenging of the superoxide radical which, surprisingly, is a rather commonplace intermediate of the reduction of oxygen. The reaction catalyzed is:

$$O_2^- + O_2^- + 2H^+ \rightarrow H_2O_2 + O_2$$

Enhancing the rate of this particular reaction may appear to be a peculiar undertaking for an enzyme, since this reaction is reasonably rapid, even in the absence of catalysis and since several transition metal cations serve as effective catalysts. Nevertheless superoxide dismutases are indispensable because they operate with ultimate efficiency and are present in abundance, whereas free transition metal cations are less effective and are not plentiful inside cells. It is a fact that all of the superoxide dismutase activity, which can be measured in crude extracts of such diverse materials as erythrocytes, mammalian liver or *Escherichia coli*, can be accounted for in terms of the superoxide dismutases which they contain. There is thus no significant superoxide dismutase activity in these extracts save that due to these enzymes. The evidence which establishes the biological importance of superoxide dismutases has been reviewed (1-9). We shall therefore now eschew discussion of such matters and shall concentrate on structure and mechanism.

Living forms contain several distinct superoxide dismutases. These fall naturally into two families. Within each family the enzymes are very similar, whereas comparisons between families show no likeness save that of a common catalytic effect. These two families of superoxide dismutases probably evolved independently in response to the common selection pressure imposed by the oxygenation of the biosphere. One of these groups comprises enzymes containing copper and zinc whereas the members of the other family contain iron or manganese.

## COPPER AND ZINC-CONTAINING SUPEROXIDE DISMUTASES

The cytosols of eukaryotic cells contain a superoxide dismutase whose molecular weight is 32,000. It is a homodimer and contains one $Cu^{++}$ and one $Zn^{++}$ per subunit. The properties of these enzymes has been remarkably resistant to evolutionary change. Thus the enzymes isolated from yeast, *Neurospora crassa*, spinach, chicken or cow (10-15) are hardly distinguishable and show only minor differences in amino acid composition, in the super-hyperfine details of their esr spectra (16) and in their optical spectra in the ultraviolet (17). One indication of the similarities among these enzymes is their ability to be isolated from diverse sources by application of essentially the same purification procedure (10, 12,14,15). These enzymes were isolated and were characterized, as proteins, well before their catalytic activity was discovered. Thus the first to be described, in 1938, was a "hemocuprein" from ox blood (18). It and similar "cupreins" were isolated on the basis of copper content and were thought to function in copper storage or transport. The relatively recent realization (15) that these cupreins were enzymes of vital importance to respiring cells caused a resurgence of interest, which has culminated in a thorough knowledge of the structure of one of them. Thus the complete amino acid sequence of the superoxide dismutase from bovine erythrocytes has been reported (19-21) and structural analysis by X-ray diffraction has progressed to 3Å resolution (22,23). The Richardsons are at this moment refining this structural analysis to 2Å resolution (24).

The most prominent structural feature of the subunit of this enzyme is a cylinder whose wall is made up of eight strands of the peptide chain arranged in an antiparallel β structure. The segments of the sequence which constitute the slats of this β barrel are 2-11, 13-23, 26-35, 38-47, 80-88, 91-100, 112-118 and 142-149. Two non-helical coils, involving residues 48-79 and 119-141, protrude from one side of the β barrel and together enclose and constitute the active site. Each subunit is stabilized by an intrachain disulfide bond between cys-55 and cys-144. There is a free thiol residue on each subunit, that of cys-6, which is remarkably unreactive in the native molecule (19). The enzyme is unusually stable and retains activity in the presence of 9.0 M urea or 4% sodium dodecyl sulfate (36). Removal of the metal prosthetic groups or reduction of the intrachain disulfide bond diminishes the resistance of the enzyme towards denaturing stresses (19,37). The subunits are joined by non-covalent interactions and the large area of contact helps explain the strength of this association.

The $Cu^{++}$ and $Zn^{++}$, at the active sites, are in close proximity, as was predicted from physico-chemical studies (25-28). Indeed the $Cu^{++}$ and $Zn^{++}$ are joined by a common ligand, which is the imidazolate ring of histidine 61. The $Cu^{++}$ is relatively exposed to the solvent whereas the zinc is more buried within the structure. The $Cu^{++}$ appears to be in contact with a freely exchangeable molecule of water and this contact point is probably the site of direct interaction with the substrate, $O_2^-$ (29). In addition to the bridging imidazolate of histidine 61, the other groups liganded to the $Cu^{++}$ are the imidazole rings of histidines 44, 46, and 118; while the other ligands of the $Zn^{++}$ are the imidazole rings of histidines 78 and 69 and the carboxyl group of aspartate 81 (22). Analysis of the epr spectrum of the enzyme and studies with group specific reagents had indicated that nitrogen atoms of imidazole rings formed the ligand field of the $Cu^{++}$, before the X-ray structure was in hand (25,30-33).

The $Cu^{++}$ and the $Zn^{++}$ can be reversibly removed with concomitant loss and regain of catalytic activity. Indeed activity can be largely restored to the apoenzyme by $Cu^{++}$ alone (15,16,34). Attempts to substitute other metals for $Cu^{++}$, with retention of activity, have been uniformly unsuccessful; whereas the $Zn^{++}$ may be replaced by $Co^{++}$, $Hg^{++}$ or $Cd^{++}$, without gross modification of activity. It is clear that catalytic activity is centered at the $Cu^{++}$ whereas the $Zn^{++}$ plays a secondary role. It is certainly the case that the $Zn^{++}$ contributes to the stability of the enzyme. Replacement of $Zn^{++}$ by $Hg^{++}$ gave an active enzyme whose stability actually exceeded that of the native enzyme (36).

The catalytic mechanism of the bovine erythrocyte enzyme has been probed by pulse radiolysis. This technique allows the rapid introduction of $O_2^-$ into buffered solutions of the enzyme, under conditions which allow monitoring either the rate of decay of the $O_2^-$ or the degree of bleaching of the enzyme. The enzyme has been found to react with $O_2^-$ at a rate of approximately $2.2 \times 10^9\ M^{-1}\ sec^{-1}$. This rate was unresponsive to changes of pH in the range 5.5-9.5. Since the $pK_a$ for $HO_2\cdot$ is 4.8 we must conclude that $O_2^-$, rather than $HO_2\cdot$, is the substrate. If the enzyme was fully in the cupric form when first exposed to $O_2^-$ then a partial bleaching of the enzyme occurred during catalysis. If the enzyme was fully reduced prior to exposure to $O_2^-$, then a partial oxidation of the enzyme, with augmentation of its 680 nm absorption occurred during the catalytic cycle. This behavior indicates that the copper was alternately reduced and reoxidized during the catalytic cycle (38-43) as follows:

$$E\text{-}Cu^{++} + O_2^- \rightarrow E\text{-}Cu^+ + O_2 \quad (I)$$

$$E\text{-}Cu^+ + O_2^- + 2H^+ \rightarrow E\text{-}Cu^{++} + H_2O_2 \quad (II)$$

One group reported the rate constants for reactions I and II to be $1.2 \times 10^9\ M^{-1}\ sec^{-1}$ and $2.2 \times 10^9\ M^{-1}\ sec^{-1}$, respectively (40), while another group (43), in substantial agreement, found $2.4 \times 10^9\ M^{-1}\ sec^{-1}$ for both reactions. They furthermore concluded that the reaction was truly diffusion limited, since the energy of activation was only 4.6 Kcal/mole and since the rate constant decreased with increasing viscosity of the solvent, without changes in the energy of activation. Given the enormous turnover rate of this enzyme, it is not surprising that the reaction showed no signs of saturation by substrate up to $2.4 \times 10^{-4}\ M\ O_2^-$; which was the maximum concentration of $O_2^-$ achievable. If the rate constants for reactions I and II are in fact equal then the steady-state bleaching of the enzyme during its catalytic cycle should be 50%. The bleaching actually observed was closer to 25% and on this basis it has been proposed (43) that the enzyme exhibits 'half of the sites' reactivity. Since the active sites are on opposite sides of the molecule, approximately 36 Å apart, the interactions needed to explain 'half of the sites' behavior are difficult to envision in this case and one is tempted to seek other explanations for the less than expected bleaching during catalysis.

Enzymes facilitate the forward and reverse aspects of any given reaction to the same degree. The superoxide dismutase should therefore catalyze the oxidation of $H_2O_2$ by oxygen. It was possible to demonstrate this effect by using tetranitromethane as a very efficient scavenger of $O_2^-$. Superoxide dismutase was able to catalyze an oxygen-dependent reduction of tetranitromethane by $H_2O_2$ (44). The reactions which explain this are:

$$E\text{-}Cu^{++} + H_2O_2 \leftrightarrow E\text{-}Cu^+ + 2H^+ + O_2^- \quad (III)$$

$$E\text{-}Cu^+ + O_2 \leftrightarrow E\text{-}Cu^{++} + O_2^- \quad (IV)$$

$$C(NO_2)_4 + O_2^- \rightarrow C(NO_2)_3^- + O_2 + NO_2 \quad (V)$$

Reactions III and IV are simply the reverse of reactions I and II. In view of reaction III it is not surprising that $H_2O_2$ has been seen to reduce the $Cu^{++}$ and thus to bleach the visible color of the enzyme (30,40,45-48). Less expected was the irreversible inactivation of the enzyme by $H_2O_2$ (45-49). It has been reported that one histidine residue per subunit is destroyed during this activation (47).

The enzyme acts as a peroxidase towards a variety of compounds which, in being peroxidized, protect the enzyme against inactivation by $H_2O_2$ (50). Inactivation of the enzyme was accompanied by chemiluminescence. The peroxidation of imidazole or of xanthine, by the superoxide dismutase, was also accompanied by luminescence. In addition, it was observed that oxygen protected the enzyme against inactivation by $H_2O_2$. The explanation offered for these observations (50) envisions the generation of a potent oxidant as a consequence of the interaction of the cuprous enzyme with $H_2O_2$. This oxidant, which remains bound to the copper, could then attack an adjacent imidazole and thus inactivate the enzyme; or it could oxidize various small molecules, in which case we would observe peroxidase action rather than inactivation. The luminescence comes from electronically-excited products during their return to their ground states. The scheme proposed (50) was:

$$\text{Enz-Cu}^{++} + H_2O_2 \rightleftharpoons \text{Enz-Cu}^{+} + O_2^{-} + 2H^{+} \quad \text{(III)}$$

$$\text{Enz-Cu}^{+} + H_2O_2 \rightleftharpoons \text{Enz-Cu}^{++}\text{-OH} + OH^{-} \quad \text{(V)}$$

$$\text{Enz-Cu}^{++}\text{-OH} + \text{ImH} \rightarrow \text{Enz-Cu}^{++} + \text{Im}\cdot + H_2O \quad \text{(VI)}$$

Reaction III is the reduction of the enzyme by $H_2O_2$ already discussed. Reaction V is analagous to the reaction of $Fe^{++}$ with $H_2O_2$, as in Fenton's reagent, which generates $OH\cdot$. In this case the $OH\cdot$ is considered to remain bound to the copper. If the bound oxidant attacked an adjacent imidazole ring (ImH), which is part of the enzyme, then the active site would be destroyed as in reaction VI. If, on the other hand, the bound oxidant attacked exogenous imidazole or xanthine or any one of a number of other compounds, we would observe peroxidase action rather than inactivation. The ability of oxygen to protect the enzyme against destruction by $H_2O_2$ can also be explained. Thus if the cuprous enzyme reacts with oxygen, as in reaction IV, then the oxygen would be in competition with $H_2O_2$. That is, reaction IV competes with reaction V for the cuprous enzyme and thus diminishes the rate of generation of bound oxidant and hence diminishes the rate of inactivation of the enzyme.

The inactivation of the enzyme by one of its reaction products, $H_2O_2$, is very interesting in that it aids studies of the intimate mechanism. Is this inactivation of any consequence _in vivo_? Since the concentration of $H_2O_2$ needed in order to see reasonable rates of inactivation is approximately 10 µM and since the highest concentrations of $H_2O_2$ reached _in vivo_ are less than this by several orders of magnitude (51) we would answer this question in the negative. Furthermore the inactivation by $H_2O_2$ is most rapid at elevated pH and is really rather sluggish close to neutrality (50). Finally, the compounds which protect the enzyme against this inactivation, such as glutathione, are abundant within cells. The level of superoxide dismutase activity within human erythrocytes does not diminish during the life of these cells, whereas the levels of numerous less stable enzymes, do decrease with time (52). It is apparent then that inactivation of superoxide dismutase by $H_2O_2$ is not a biological problem.

The dismutation of superoxide radicals to $H_2O_2$ plus $O_2$ requires two protons. It is surprising therefore that the enzymatic dismutation is independent of pH over the range 5.5-9.5 (38-43) and furthermore shows no deuterium isotope effect at pH 10 (50). One way to explain this is to suppose that only one proton is transferred during the catalyzed dismutation, such that the product is $HO_2^-$ rather than $H_2O_2$ and further to provide for some means of facilitating this proton transfer. The imidazolate of histidine 61, which bridges the $Cu^{++}$ and $Zn^{++}$ in the resting enzyme, could function to conduct protons as follows (50):

$$[-Zn^{++}-Im^{-}-Cu^{++}] + O_2^- \rightarrow O_2 + [-Zn^{++}-Im^{-} \quad Cu^{+}-] \qquad (VII)$$

$$[-Zn^{++}-Im^{-} \quad Cu^{+}] + H^{+} \rightarrow [-Zn^{++}-ImH \quad Cu^{+}] \qquad (VIII)$$

$$[-Zn^{++}-ImH \quad Cu^{++}] + O_2^- \rightarrow HO_2^- + [-Zn^{++}-Im^{-}\text{——}Cu^{++}] \qquad (IX)$$

$$HO_2^- + H^+ \rightarrow H_2O_2 \qquad (X)$$

In reaction VII reduction of the $Cu^{++}$ by $O_2^-$ is accompanied by release of the imidazolate bridge from its attachment to the copper. The imidazolate, which remains attached to $Zn^{++}$, is a strong base and it protonates as in reaction VIII. When $O_2^-$ then interacts with the reduced enzyme it simultaneously acquires an electron from the $Cu^+$ and a proton from the imidazole of histidine 61. This $O_2^-$ is thus converted to $HO_2^-$ while the bridging of $Cu^{++}$ and $Zn^{++}$ is reestablished, as in reaction IX. Finally $HO_2^-$ becomes protonated in free solution as in reaction X.

Cyanide and azide both ligate to the active site of this enzyme (25) yet have disparate effects. Thus cyanide is an effective inhibitor, whereas azide is not (30,38,50,53). It has been suggested (25) that azide binds to the $Zn^{++}$ while cyanide binds to the $Cu^{++}$. It is possible that both cyanide and azide bind to the $Cu^{++}$ but that azide merely displaces one of the non-bridging ligands whereas cyanide binds to that site on the copper which is ordinarily available for interaction with $O_2^-$. Binding of $N_3^-$ in this way would merely replace one nitrogenous ligand by another and should have relatively minor effects on the epr spectrum and on the activity. Cyanide, in contrast, binds by its carbon end (54) and should thus grossly change the epr spectrum, as it does (25).

Specific inhibitors of superoxide dismutase could be very useful both in advancing our knowledge of its catalytic mechanism and in more firmly establishing its biological functions. Unfortunately the list of effective inhibitors is short and does not include compounds expected to exhibit specificity. Thus, as already stated, cyanide inhibits. Since it is only the copper-zinc enzymes which are cyanide sensitive, whereas the mangani and the ferri superoxide dismutases are not, cyanide has been useful in distinguishing these families of enzymes (14,55). Several copper-chelating agents, such as diethyldithiocarbamate, xanthogenate and diphenylthiocarbazone have recently been reported to inhibit the copper-zinc superoxide dismutase (56).

The generality that copper-zinc superoxide dismutases are characteristic of the cytosol of eukaryotes has been challenged. Thus a superoxide dismutase bearing one atom of copper and two of zinc per molecule has been reported in a symbiotic luminescent bacterium (57). The ultraviolet absorption spectrum of this enzyme differed strikingly from that seen with the copper-zinc enzymes from eukaryotes. Its place in the evolution of superoxide dismutases will remain uncertain until some knowledge of its amino acid sequence is available.

The manganese and the iron-containing superoxide dismutases and their evolutionary relationships will be discussed by Dr. Joe McCord.

## REFERENCES

1. Fridovich, I. (1972) Accounts Chem. Res. 5, 321-326.

2. McCord, J. M., Beauchamp, C. O., Goscin, S., Misra, H. P., and Fridovich, I. (1973) in Proc. 2nd Int. Symp. Oxidases and Related Redox Systems, Memphis, 1971 (King, T. E., Mason, H. S., and Morrison, M., eds) pp. 51-76, University Park Press, Baltimore.

3. Fridovich, I. (1974) Advan. Enzymol. 41, 35-97.

4. Fridovich, I. (1974) in Molecular Mechanisms of Oxygen Activation (Hayaishi, O., ed.) pp. 453-477, Academic Press, New York.

5. Sawada, Y., and Yamazaki, I. (1974) Tampakushktsu Kakusan Koso 19, 527-536.

6. Halliwell, B. (1974) New Phytology 73, 1075-1086.

7. Bors, W., Saran, M., Lengfelder, E., Stöttl, R., and Michel, C. (1974) Curr. Top. Radiat. Res. Q. 9, 247-309.

8. Fridovich, I. (1975) Ann. Rev. Biochem. 44, 147-159.

9. Bouanchaud, D. (1975) Recherche 6, 370-373.

10. Misra, H. P., and Fridovich, I. (1972) J. Biol. Chem. 247, 3410-3414.

11. Rapp, U., Adams, W. C., and Miller, R. W. (1973) Canad. J. Biochem. 51, 158-171.

12. Goscin, S. A., and Fridovich, I. (1972) Biochim. Biophys. Acta 289, 276-283.

13. Asada, K., Urano, M., and Takehashi, M. (1973) Europ. J. Biochem. 36, 257-266.

14. Weisiger, R. A., and Fridovich, I. (1973) J. Biol. Chem. 248, 3582-3592.

15. McCord, J. M., and Fridovich, I. (1969) J. Biol. Chem. 244, 6049-6055.

16. Beem, K., Rich, W. E., and Rajagopalan, K. V. (1974) J. Biol. Chem. 249, 7298-7305.

17. Calabrese, L., Frederici, G., Bannister, W. H., Bannister, J. V., Rotilio, G., and Finazzi-Agro, A. (1975) Europ. J. Biochem. 56, 305-309.

18. Mann, T., and Keilin, D. (1938) Proc. Roy. Soc. (London) B126, 303.

19. Abernethy, J. L., Steinman, H. M., and Hill, R. L. (1974) J. Biol. Chem. 249, 7339-7347.

20. Evans, H. J., Steinman, H. M., and Hill, R. L. (1974) J. Biol. Chem. 249, 7315-7325.

21. Steinman, H. M., Naik, V. R., Abernethy, J. L., and Hill, R. L. (1974) J. Biol. Chem. 249, 7326-7338.

22. Richardson, J. S., Thomas, K. A., Rubin, B. H., and Richardson, D. C. (1975) Proc. Nat. Acad. Sci. U.S.A. 72, 1349-1353.

23. Richardson, J. S., Thomas, K. A., and Richardson, D. C. (1975) Biochem. Biophys. Res. Commun. 63, 986-992.

24. Richardson, J. S., and Richardson, D. C., personal communication.

25. Fee, J. A., and Gaber, B. P. (1972) J. Biol. Chem. 247, 60-65.

26. Fee, J. A. (1973) J. Biol. Chem. 248, 4229-4234.

27. Fee, J. A. (1973) Biochim. Biophys. Acta 295, 107-116.

28. Rotilio, G., Calabrese, L., Mondovi, B., and Blumberg, W. E. (1974) J. Biol. Chem. 249, 3157-3160.

29. Gaber, B. P., Brown, R. D., Koenig, S. H., and Fee, J. A. (1972) Biochim. Biophys. Acta 271, 1-5.

30. Rotilio, G., Morpurgo, L., Giovagnoli, C., Calabrese, L., and Mondovi, B. (1972) Biochemistry 11, 2187-2192.

31. Forman, H. J., Evans, H. J., Hill, R. L., and Fridovich, I. (1973) Biochemistry 12, 823-827.

32. Stokes, A. M., Hill, H. A., Bannister, W. H., and Bannister, J. V. (1973) FEBS Lett. 32, 119-123.

33. Haffner, P. H., and Coleman, J. E. (1973) J. Biol. Chem. 248, 6626-6629.

34. Fee, J. A., and Briggs, R. G. (1975) Biochim. Biophys. Acta 400, 439-450.

35. Rotilio, G., Calabrese, L., and Coleman, J. E. (1973) J. Biol. Chem. 248, 3855-3859.

36. Forman, H. J., and Fridovich, I. (1973) J. Biol. Chem. 248, 2645-2649.

37. Beauchamp, C. O., and Fridovich, I. (1973) Biochim. Biophys. Acta 317, 50-64.

38. Rotilio, G., Bray, R. C., and Fielden, E. M. (1972) Biochim. Biophys. Acta 268, 605-609.

39. Klug, D., Rabani, J., and Fridovich, I. (1972) J. Biol. Chem. 247, 4839-4842.

40. Klug-Roth, D., Fridovich, I., and Rabani, J. (1973) J. Am. Chem. Soc. 95, 2786-2790.

41. Fielden, E. M., Roberts, P. B., Bray, R. C., and Rotilio, G. (1973) Biochem. Soc. Transactions 1, 52-53.

42. Bannister, J. V., Bannister, W. H., Bray, R. C., Fielden, E. M., Roberts, P. B., and Rotilio, G. (1973) FEBS Lett. 32, 303-306.

43. Fielden, E. M., Roberts, P. B., Bray, R. C., Lowe, D. J., Mautner, G. N., Rotilio, G., and Calabrese, L. (1974) Biochem. J. 139, 49-60.

44. Hodgson, E. K., and Fridovich, I. (1973) Biochem. Biophys.

Res. Commun. 54, 270-274.

45. Symonyan, M. A., and Nalbandyan, R. M. (1972) FEBS Lett. 28, 22-24.

46. Rotilio, G., Morpurgo, L., Calabrese, L., and Mondovi, B. (1973) Biochim. Biophys. Acta 302, 229-235.

47. Bray, R. C., Cockle, S. H., Fielden, E. M., Roberts, P. B., Rotilio, G., and Calabrese, L. (1974) Biochem. J. 139, 43-48.

48. Rotilio, G., Calabrese, L., Bossa, F., Barra, D., Finazzi-Agro, A., and Mondovi, B. (1972) Biochemistry 11, 2182-2187.

49. Beauchamp, C. O., and Fridovich, I. (1973) Biochim. Biophys. Acta 317, 50-64.

50. Hodgson, E. K., and Fridovich, I. (1975) Biochemistry, in press.

51. Oshino, N., Jamieson, D., Sugano, T., and Chance, B. (1975) Biochem. J. 146, 67-77.

52. McCord, J. M., personal communication.

53. Tyler, D. D. (1975) Biochem. J. 147, 493-504.

54. Haffner, P. H., and Coleman, J. E. (1973) J. Biol. Chem. 248, 6626-6629.

55. Beauchamp, C. O. (1973) Ph.D. dissertation, Duke University.

56. Hirata, F., and Hayaishi, O. (1975) J. Biol. Chem. 250, 5960-5966.

57. Puget, K., and Michelson, A. M. (1974) Biochem. Biophys. Res. Commun. 58, 830-838.

# IRON- AND MANGANESE-CONTAINING SUPEROXIDE DISMUTASES: STRUCTURE, DISTRIBUTION, AND EVOLUTIONARY RELATIONSHIPS

Joe M. McCord

Duke University Medical Center

Durham, North Carolina 27710

## INTRODUCTION

In 1969, a well-studied family of mammalian copper-containing proteins was discovered to possess superoxide dismutase activity (1). This rather bizarre activity was postulated to play a protective role for the oxygen-metabolizing organism. Its ubiquity among the various mammalian tissues led us to attempt isolating the activity from an evolutionarily distant species, *Escherichia coli* (2). Although cell-free extracts of *E. coli* contained roughly the same amount of superoxide dismutase activity as mammalian tissue extracts, the enzyme's behaviour during purification bore no resemblance to that of the bovine enzyme. When the enzyme was purified to homogeniety and concentrated, we did not see the familiar blue-green color of the mammalian copper-containing superoxide dismutases. The enzyme was pink. Undaunted, and perhaps comforted by the fact that some copper proteins are pink (3), we set out to determine the copper content of the new superoxide dismutase. No copper could be detected, either by a colorimetric method or by electron paramagnetic resonance (EPR). In fact, no EPR signal at all could be detected in the native protein. Upon denaturation by boiling in 0.1 N HCl, however, the characteristic six-pronged spectrum of manganese(II) appeared. This, then, became the first member to be isolated from a widespread family of non-copper superoxide dismutases.

A similar surprise and yet another color were experienced by Yost and Fridovich upon the purification of a second superoxide dismutase from *E. coli* (4). This time the enzyme was yellow and contained no copper, no zinc, and no manganese. EPR spectroscopy and atomic absorption spectroscopy revealed one mole of iron per mole of enzyme. The amino acid composition of this enzyme was very similar to that of the manganese enzyme, but both were quite unlike that of the cuprozinc enzyme.

## STRUCTURE AND METAL CONTENT

In the last five years, at least nine more iron- or manganese-containing superoxide dismutases have been isolated from a wide variety of sources, both prokaryotic and eukaryotic. Some of the structural properties of these eleven enzymes are presented in Table I. The seven enzymes from prokaryotes have all been isolated as dimers, with a subunit molecular weight of close to 20,000. Sequence data suggests that four are composed of identical subunits. Three are composed of subunits of equal size which are probably identical, although this has not been shown. Four contain iron; three contain manganese. Five were found to contain one metal atom per subunit; the metal content of one was not quantitated; one (the periplasmic enzyme from *E. coli*) was found to contain one (1.02) atom of iron per dimer. In the case of the latter, it is considered possible that the native enzyme may contain two atoms of iron per enzyme molecule. This would seem to be most likely, in view of the other members of the family, and given that the two subunits are identical (see sequence data below). A similar situation has been seen in the case of the iron-containing enzyme from *Plectonema boryanum*. This enzyme from a blue-green alga has been isolated by two groups (7,13). Misra and Keele found it to contain 0.94 g atoms Fe/mole of enzyme (13), whereas Asada *et al.* found 2.0 g atoms Fe/mole enzyme (7).

The four enzymes in Table I from eukaryotic sources are all manganese-containing superoxide dismutases, and all are tetrameric with molecular weights approximately twice those of the prokaryotic enzymes. The rather outstanding member of this family of four is the enzyme from the luminous fungus *Pleurotus olearius* (9). The tetramer is claimed to consist of two each of two chromatographically distinct kinds of subunits. The molecule contains only 2 g atoms of manganese, so presumably only one kind of subunit is catalytically active. This kind of structure usually indicates that the enzyme is under some sort of metabolic regulation, but a reason for any such regulation of a protective enzyme is difficult to propose. The enzyme was found to contain insignificant amounts of Cu, Zn, and Fe, so the possibility of a Mn-Fe hybrid appears to be ruled out.

The remaining three members of the eukaryotic group are tetramers of identical subunits, and two of them clearly contain 1 g atom of Mn per subunit. The enzyme isolated from chicken liver mitochondria contained 2.3 Mn/mole, which probably reflects the loss of some Mn upon purification (11).

TABLE I

STRUCTURAL PROPERTIES OF SOME MANGANESE- OR IRON-CONTAINING SUPEROXIDE DISMUTASES

| *SOURCE* | *MOLECULAR WEIGHT* | *SUBUNITS* | *METAL CONTENT* | *REFERENCE* |
|---|---|---|---|---|
| Prokaryotes | | | | |
| *Escherichia coli B* | 39,500 | 2 identical | 2 Mn | (2) |
| *Escherichia coli B* (periplasmic space) | 38,700 | 2 identical | 1 Fe *(2?)* | (4) |
| *Streptococcus mutans* | 40,250 | 2 of equal size | 2 Mn | (5) |
| *Bacillus stearothermophilus* | ∿40,000 | 2 identical | Mn *(2?)* | (6) |
| *Plectonema boryanum* | 41,700 | 2 identical | 2 Fe | (7) |
| *Photobacterium sepia* | 40,660 | 2 of equal size | 2 Fe | (8) |
| *Photobacterium leiognathi* | 40,660 | 2 of equal size | 2 Fe | (8) |
| Eukaryotes | | | | |
| *Pleurotus olearius* | 76,000 | 4 ($\alpha_2\beta_2$) | 2 Mn | (9) |
| *Saccharomyces cerevisiae* | 96,000 | 4 identical | 4 Mn | (10) |
| Chicken liver mitochondria | 79,400 | 4 identical | 2.3 Mn *(4?)* | (11) |
| Human liver | 84,100 | 4 identical | 4 Mn | (12) |

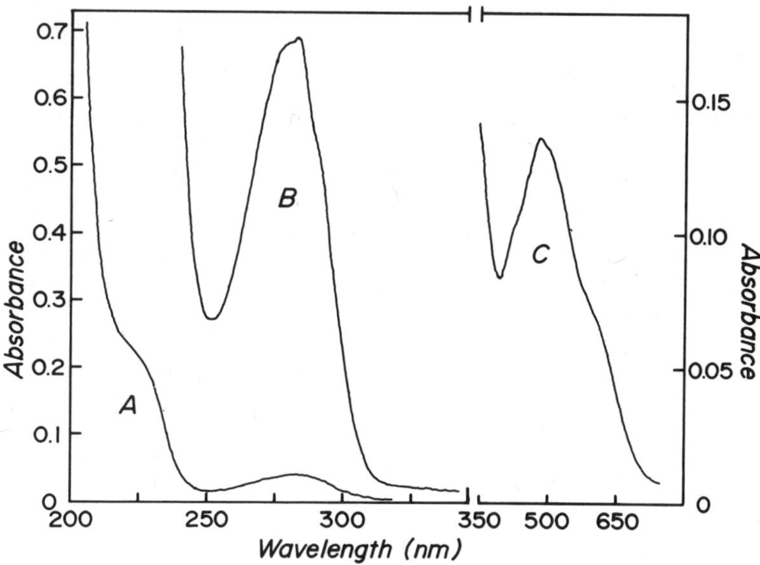


Figure 1. Absorption spectrum of human liver manganese superoxide dismutase. Enzyme concentrations were as follows: A, 14.6 μg/ml; B, 0.293 mg/ml; C, 5.85 mg/ml. Absorption maxima occur at 283.5 nm and at 480 nm.

## SPECTRAL PROPERTIES

Figure 1 shows the optical absorption spectrum of the manganese-containing superoxide dismutase isolated from human liver (12), which is representative of the family of manganese enzymes. The shoulder at 225 nm, probably due to the manganese chromophore, results in an anomalously low ratio of absorbances at 215 and 225 nm of 1.31. The 215/225 ratio for other proteins is very close to 2.0. Obviously, if this ratio is used to determine the concentration of unknown solutions of manganese superoxide dismutase by the method of Murphy and Kies (14), erroneous results will be obtained unless a correction is introduced. It is not presently known whether the iron superoxide dismutases also exhibit this anomalous shoulder at 225 nm. In the region from 250 nm to 350 nm, both the manganese- and iron-containing enzymes are undistinguished, showing normal-looking peaks near 280 nm. The human enzyme shown in Figure 1 has a molar absorptivity of $1.97 \times 10^4$ at 283.5 nm. In the visible range, there is a fairly weak maximum at 480 nm with a shoulder at about 600 nm. For the tetrameric human enzyme which was isolated with a full complement of four manganese per molecule, the molar absorptivity at 480 nm is 2049, or about 500 per manganese-containing subunit. This is qualitatively and quantitatively similar to results observed with the other manganese-containing superoxide dismutases (2,5,11).

The absence of an EPR signal from the native "resting" manganese superoxide dismutase could be consistent with either Mn(II) or Mn(III). Although Mn(II) is a paramagnetic species, it need not present an observable signal if its rotational motion is highly restricted by the protein (15). The red-purple color of the visible chromophore, however, is much more a characteristic of Mn(III) complexes than of Mn(II) complexes (16).

The iron-containing superoxide dismutases, yellow-brown in color, show a nearly featureless visible absorption consisting of a shoulder in the range of 325-350 nm, falling off progressively from 350 to 500 nm. The molar absorptivities at 350 nm have been in the range of 1200 to 1600 per iron-containing subunit (2,7,8). The EPR signal observed with the resting iron-containing enzymes indicates that the iron is present as high-spin Fe(III) (7).

## SEQUENCE HOMOLOGIES

In 1973 Steinman and Hill (17) reported the amino acid sequences for the 29 N-terminal residues of three members of the iron-or-manganese superoxide dismutase family: the two enzymes from *E. coli* and the enzyme from chicken liver mitochondria. A high degree of homology was evident among these enzymes, but no homology could be seen between these and the bovine cuprozinc superoxide dismutase, for which the total sequence is now known (18). Steinman has preliminary N-terminal sequence data for three more Fe or Mn enzymes (19), and Bridgen *et al.* (6) have reported the sequence of 60 residues from the N-terminus of the manganese enzyme from *Bacillus stearothermophilus*. These seven sequences (residues 1 through 29) are presented in Table II. An even higher degree of homology is now apparent, with homologous residues appearing at every position, as well as sub-groups of homologous residues tending to fall among either the prokaryotic or eukaryotic enzymes. Residues in at least five positions are totally conserved through all seven sequences. At seven positions, sub-homologies are seen exclusively among the three enzymes from eukaryotic sources. At eleven positions, sub-homologies are seen exclusively among the four enzymes from prokaryotes. Although identities exist between the prokaryotic family and the eukaryotic family at 23 of 29 loci, sub-homologies bridge eukaryotic and prokaryotic enzymes at only two loci, 17 and 21.

The chicken enzyme shares identical residues with one or both the other eukaryotic enzymes at 26 of the 27 known loci. The *E. coli* manganese enzyme shares identical residues with one or all the other prokaryotic enzymes at 25 of the 27 known loci.

TABLE II. SEQUENCE HOMOLOGIES AMONG MANGANESE- OR IRON-CONTAINING SUPEROXIDE DISMUTASES

| | | 1 | | | | 5 | | | | | 10 | | | | | 15 |
|---|---|---|---|---|---|---|---|---|---|---|---|---|---|---|---|---|
| † Human Mn SOD | H | N-*LYS* | *HIS* | s* | LEU | PRO | *ASP* | LEU | PRO | TYR | *ASP* | TYR | *GLY* | ALA | LEU | GLU |
| § Chicken Mn SOD | H | N-*LYS* | *HIS* | THR | LEU | PRO | *ASP* | LEU | PRO | TYR | *ASP* | TYR | *GLY* | ALA | LEU | GLU |
| † Yeast Mn SOD | H | N-*LYS* | v* | THR | LEU | PRO | *ASP* | LEU | k | w | *ASP* | f* | *GLY* | ALA | LEU | GLU |
| † *P. boryanum* Fe SOD | H | N- a* | TYR | THR | q* | PRO | p* | LEU | PRO | f* | *pro* | *lys* | *asp* | ALA | LEU | GLU |
| ¶ *B. stearo.* Mn SOD | H | N- p* | *phe* | *glu* | LEU | PRO | *ala* | LEU | PRO | TYR | *pro* | TYR | *asp* | ALA | LEU | GLU |
| § *E. coli* Mn SOD | H | N-*ser* | TYR | THR | LEU | PRO | s* | LEU | PRO | TYR | ALA | TYR | *asp* | ALA | LEU | GLU |
| § *E. coli* Fe SOD | H | N-*ser* | *phe* | *glu* | LEU | PRO | *ala* | LEU | PRO | TYR | ALA | *lys* | *asp* | ALA | LEU | a* |

| | 16 | | | | 20 | | | | | 25 | | | | 29 |
|---|---|---|---|---|---|---|---|---|---|---|---|---|---|---|
| Human Mn SOD | PRO | HIS | ILE | n* | ALA | *GLN* | *ILE* | MET | *GLN* | LEU | HIS | ? | s* | LYS |
| Chicken Mn SOD | PRO | HIS | ILE | SER | ALA | GLU | *ILE* | MET | *GLN* | LEU | HIS | ? | ? | LYS |
| Yeast Mn SOD | PRO | *tyr* | ILE | SER | g* | *GLN* | *ILE* | n | GLU | LEU | HIS | TYR | THR | ? |
| *P. boryanum* Fe SOD | PRO | *tyr* | ? | m | ? | a* | e | ? | f | ? | ? | ? | ? | ? |
| *B. stearo.* Mn SOD | PRO | HIS | ILE | *asp* | *lys* | GLU | *thr* | MET | n | i* | HIS | *his* | THR | LYS |
| *E. coli* Mn SOD | PRO | HIS | f* | *asp* | *lys* | *GLN* | *thr* | MET | GLU | LEU | ? | *his* | ? | LYS |
| *E. coli* Fe SOD | PRO | HIS | ILE | SER | ALA | GLU | ? | i* | GLU | y | HIS | TYR | g | LYS |

† H. M. Steinman, personal communication.
§ H. M. Steinman and R. L. Hill, Proc. Nat. Acad. Sci. USA 70, 3725-3729 (1973).
¶ J. Bridgen, J. I. Harris, and F. Northrop, FEBS Lett. 49, 392-395 (1975).

Table II. Homologous residues are designated by three-letter abbreviations. Subgroupings of homologous residues are shown in italic and are boxed. Non-homologous residues are indicated by their one-letter abbreviations. Non-homologous residues which may be accounted for by changing a single base in the codon are designated with an asterisk.

Of the 188 residues known among the seven sequences, 26 residues are non-homologous (14%). Seventeen of these non-homologous residues (65%) can be accounted for by a single base change in the appropriate RNA codon.

It is of some interest to note that the two iron superoxide dismutases show no particular homologies to each other. In fact, the only exclusive sub-homology linking the two is a single locus, the lysine at position 11. The iron enzymes show greater homologies with the manganese enzymes than with each other. This suggests that the differences in sequence between iron and manganese superoxide dismutases may be fairly subtle. Any definitive statement on this matter must, of course, await more complete sequencing data.

## PHYLOGENETIC DISTRIBUTION

It appears that the manganese-containing superoxide dismutases enjoy the widest distribution among the diversity of organisms, being found in both prokaryotes and eukaryotes. The cuprozinc superoxide dismutases, which have been reviewed by Doctor Fridovich, seem to be found exclusively in eukaryotic organisms, although one possible exception has been raised (20). The iron-containing superoxide dismutases, although only four have been isolated, have all been found in prokaryotes. Given the high degree of sequence homology, however, between the iron and manganese enzymes, it would not be too surprising to find an iron-containing superoxide dismutase in a higher organism.

## SUBCELLULAR DISTRIBUTION

The first examples of both manganese and iron superoxide dismutases were isolated from *E. coli*. Gregory *et al.* (21) determined that the two enzymes were present in roughly equal amount, but were localized in different parts of the *E. coli* cell. The iron enzyme could be preferentially released by osmotically shocking the cells, indicating a location between the cell membrane and the cell wall known as the periplasmic space. The manganese enzyme, on the other hand, was released only upon disruption of the cells, indicating a localization in the matrix of the cell. *Streptococcus mutans* was found to contain two distinct superoxide dismutases, both of which were manganese enzymes (5). *P. boryanum* was found to contain both an iron and a manganese enzyme, with the iron enzyme accounting for 90% of the total (7). *Photobacterium sepia*, a bioluminescent marine organism, was found to have only an iron-containing superoxide dismutase (8). Thus, among the prokaryotes, it may be impossible to make any generalizations about which enzyme is likely to be found in a particular subcellular location, or, for that matter, in a particular organism.

Subcellular distribution of the manganese superoxide dismutases in eukaryotic organisms is a matter which seems to vary from species to species. Three recent independent studies on the subcellular distribution of superoxide dismutases in rat liver are in excellent agreement, all finding that the manganese enzyme accounts for 7 to 8% of the total activity present, and that it is apparently located exclusively in the matrix of the mitochondria (23-25). The first example of an eukaryotic manganese superoxide dismutase was isolated from chicken liver mitochondria (11). The enzyme was unequivocally shown to be present in the matrix space of the mitochondria (21), but the question was not addressed as to whether this were the *exclusive* subcellular location (11).

Our recent isolation and characterization of the manganese enzyme from human liver produced a rather astonishing observation: human liver contains 20 times the amount of manganese superoxide dismutase as does an equal weight of rat liver, but less than half as much cuprozinc enzyme. A comparison of the relative amounts of manganese and cuprozinc superoxide dismutases in liver homogenates from four species may be seen in Figure 2. All gels received equivalent amounts of liver homogenate and were stained for

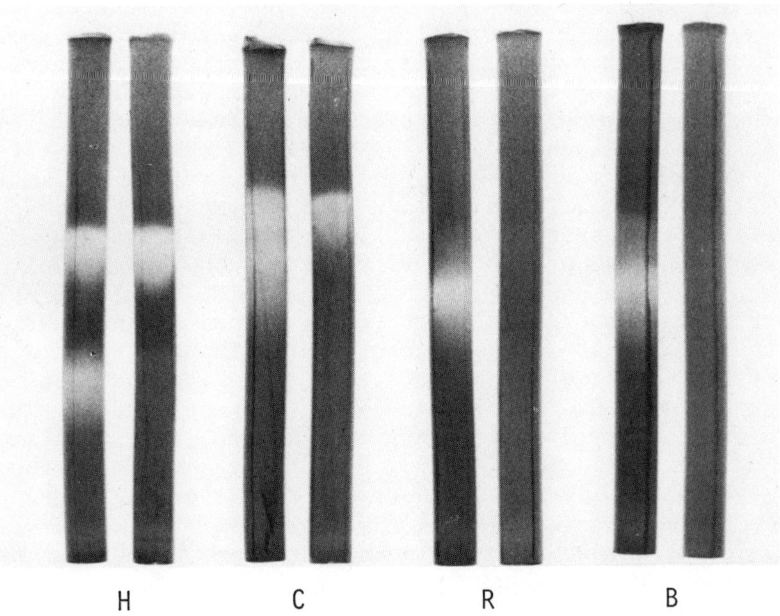


Figure 2. Polyacrylamide gels stained for superoxide dismutases in liver homogenates. All gels received equivalent amounts of supernate from homogenized human (H), chicken (C), rat (R), or bovine (B) liver. In each pair, the gel on the right was stained in the presence of 2 mM cyanide to inhibit the cuprozinc dismutases.

superoxide dismutase activity by a modification (26) of the method of Beauchamp and Fridovich (27), in the presence and absence of cyanide. Cyanide inhibits the cuprozinc superoxide dismutases, but does not affect the manganese enzymes, thereby permitting an identification of the bands on the gels. The human liver homogenate shows two distinct bands of activity, the slower moving band being the cyanide-insensitive manganese enzyme, and accounting for well over half the total activity. The two types of enzyme are less well separated in the chicken liver homogenate, but again the cyanide-insensitive manganese superoxide dismutase accounts for a sizable fraction of the total activity, about 50%. In marked contrast, however, the homogenates of rat and bovine liver show only a cyanide-sensitive cuprozinc enzyme. Under optimal conditions, a very faint cyanide-insensitive band may be seen.

By spectrophotometric assay, we found the manganese superoxide dismutase accounts for 60-70% of the total activity in the human liver, with some variation among individuals. In the rat, the manganese enzyme accounts for only 7-8% of the total, and is located exclusively in the mitochondria (23-25). Subcellular fractionation of chicken liver indicates that the majority of the manganese enzyme is present in the cytosol-- three times as much as in the mitochondria. Because of the difficulty in obtaining human liver fresh enough for a good subcellular fractionation, we have examined the subcellular distribution of manganese superoxide dismutase in another primate, the baboon *Papio ursinus*. Here we found an even greater fraction of the manganese enzyme in the cytosol fraction-- about six times as much as in the mitochondria. Tyler calculated that in rat liver the concentration of manganese superoxide dismutase in the matrix space of the mitochondria is about 11 μM, or 0.9 mg/ml (23). In primates the matrix concentration must be 2 or 3 times this value, despite the fact that most of the activity resides in the cytosol.

By atomic absorption spectrometry, ashed human liver was found to contain $3.1 \times 10^{-5}$ moles Mn/kg liver. The manganese was completely extractable into the soluble fraction of a liver homogenate. Human liver was also found to contain 614 mg manganese superoxide dismutase per kg liver, or 7.3 μmoles enzyme/kg liver. This quantity of enzyme contains $2.9 \times 10^{-5}$ moles of Mn, accounting for 94% of the total manganese in the liver! By contrast, rat liver was found to contain 3 times the manganese of human liver, but only 5% as much manganese superoxide dismutase. Thus, in the rat, the enzyme accounts for only 1.5% of the total manganese present (12).

By examining a petite mutant of *Saccharomyces cerevisiae* which lacked mitochondrial DNA but did possess manganese superoxide dismutase, Weisiger and Fridovich deduced that the enzyme is encoded in the nuclear DNA (22). The protein is presumably synthesized in the cytosol, then transported into the mitochondrion. It is perhaps easier to understand the finding of such a protein in both the mitochondria and the cytosol than one which is encoded in the mitochondrial DNA.

## ACKNOWLEDGMENT

This work was supported by Research Grant AM 17091 from the National Institute of Arthritis, Metabolism, and Digestive Diseases.

## REFERENCES

1. McCord, J.M. and Fridovich, I. (1969) J. Biol. Chem. 244:6049-6055.

2. Keele, B.B., Jr., McCord, J.M. and Fridovich, I. (1970) J. Biol. Chem. 245:6176-6181.

3. Reed, D.W., Passon, P.G. and Hultquist, D.E. (1970) J. Biol. Chem. 245:2954-2961.

4. Yost, F.J. and Fridovich, I. (1973) J. Biol. Chem. 248:4905-4908.

5. Vance, P.G., Keele, B.B., Jr. and Rajagopalan, K.V. (1972) J. Biol. Chem. 247:4782-4786.

6. Bridgen, J., Harris, J.I. and Northrop, F. (1975) FEBS Lett. 49:392 395.

7. Asada, K., Yoshikawa, K., Takahashi, M., Maeda, Y. and Enmanji, K. (1975) J. Biol. Chem. 250:2801-2807.

8. Puget, K. and Michelson, A.M. (1974) Biochimie 56:1255-1267.

9. Lavelle, F., Durosay, P. and Michelson, A.M. (1974) Biochimie 56:451-458.

10. Ravindranath, S.D. and Fridovich, I. (1975) J. Biol. Chem. 250:6107-6112.

11. Weisiger, R.A. and Fridovich, I. (1973) J. Biol. Chem. 248:3582-3592.

12. McCord, J.M., Rizzolo, L., Boyle, J.A., Day, E.D., Jr. and Salin, M.L., to be published.

13. Misra, H.P. and Keele, B.B., Jr. (1975) Biochim. Biophys. Acta 379:418-425.

14. Murphy, J.B. and Kies, M.W. (1960) Biochim. Biophys. Acta 45:382-384.

15. Reed, G.H. and Cohn, M. (1970) J. Biol. Chem. 245:662-664.

16. Sastry, G.S., Hamm, R.E. and Pool, K.H. (1969) Anal. Chem. 41: 857-858.

17. Steinman, H.M. and Hill, R.L. (1973) Proc. Nat. Acad. Sci. USA 70:3725-3729.

18. Steinman, H.M., Naik, V.R., Abernethy, J.L. and Hill, R.L. (1974) J. Biol. Chem. 249:7326-7338.

19. Steinman, H.M., preliminary unpublished results.

20. Puget, K. and Michelson, A.M. (1974) Biochem. Biophys. Res. Comm. 58:830-838.

21. Gregory, E.M., Yost, F.J., Jr., and Fridovich, I. (1973) J. Bacteriol. 115:987-991.

22. Weisiger, R.A. and Fridovich, I. (1973) J. Biol. Chem. 248: 4793-4796.

23. Tyler, D.D. (1975) Biochem. J. 147:493-504.

24. Peeters-Joris, C., Vandevoorde, A. and Baudhuin, P. (1975) Biochem. J. 150:31-39.

25. Panchenko, L.F., Brusov, O.S., Gerasimov, A.M. and Loktaeva, T.D. (1975) FEBS Lett. 55:84-87.

26. Salin, M.L. and McCord, J.M. (1974) J. Clin. Invest. 54:1005-1009.

27. Beauchamp, C.O. and Fridovich, I. (1971) Anal. Biochem. 44: 276-287.

# SUPEROXIDE DISMUTASES IN PHOTOSYNTHETIC ORGANISMS

Kozi Asada, Sumio Kanematsu, Masa-aki Takahashi and Yasuhisa Kona

The Research Institute for Food Science, Kyoto University
Uji, Kyoto 611, Japan

Although molecular oxygen is an effective electron acceptor, for most organisms oxygen is harmful unless they have a protective mechanism against oxygen toxicity. This is especially the case when the organisms are irradiated by visible light at high intensity mainly due to photodynamic action of endogeneous pigments. For example, almost all of non-photosynthetic microorganisms are not able to survive under sun light.

On the contrary, photosynthetic organisms have developed a system to transform light into chemical energy. As a first step for such a development, these organisms should acquire a protection mechanism against light-oxygen hazards. It is generally accepted that most of the molecular oxygen in biosphere has accumulated by photosynthetic activity of blue-green algae followed by the eukaryotic algae and then land plants. Therefore, it seems likely that the most primitive oxygen-evolving organisms, blue-green algae, were the first to develop the protection system against toxicity due to oxygen evolved by themselves.

In this communication, first we briefly summarized the results on the formation of superoxide radical in chloroplasts on illumination and reactivity of superoxide with the chloroplast components. Then, the characterization of superoxide dismutases in photosynthetic organisms at different levels of evolution are described. Superoxide dismutases in photosynthetic bacteria and prokaryotic and eukaryotic algae are the cyanide-insensitive Fe- or Mn-enzyme and these organisms lack the cyanide-sensitive Cu, Zn-enzyme. Cu, Zn-superoxide dismutase appears only in land plants including fern and moss. The relation between oxygen concentration in the earth's atmosphere and appearance of various forms of superoxide dismutase is discussed.

I) FORMATION OF SUPEROXIDE IN CHLOROPLASTS ON ILLUMINATION

In 1951 Mehler established the reduction of molecular oxygen into hydrogen peroxide by illuminated chloroplasts (1). During the decades it had been believed that illuminated chloroplasts donate two electrons to molecular oxygen, however, the following observations indicate the absence of the divalent reduction of molecular oxygen in chloroplasts.

Figure 1 shows the photoreduction of cytochrome c by spinach chloroplasts and its inhibition by superoxide dismutase, confirming the production of superoxide in illuminated chloroplasts (2). The same observation were also presented by Nelson et al. (3). Formation of superoxide in chloroplasts has been confirmed and also deliniated from the other superoxide-induced reactions by chloroplasts on illumination and by its inhibition by superoxide dismutase. Figure 2 shows the oxidation of sulfite induced by illuminated spinach chloroplasts. This reaction is a oxidative chain reaction induced by superoxide and completely inhibited by superoxide dismutase, a scavenger of hydroxyl radical which participates in the chain reaction (4). Photooxidation of ascorbate (5,6), epinephrine (7), $Mn^{2+}$ (6,8), hydroxylamine (9) and tiron (10) are also induced by chloroplasts and all of them are inhibited by superoxide dismutase. Recently, EPR-evidence for the production of superoxide in chloroplasts is presented using the spin trapping method (11).

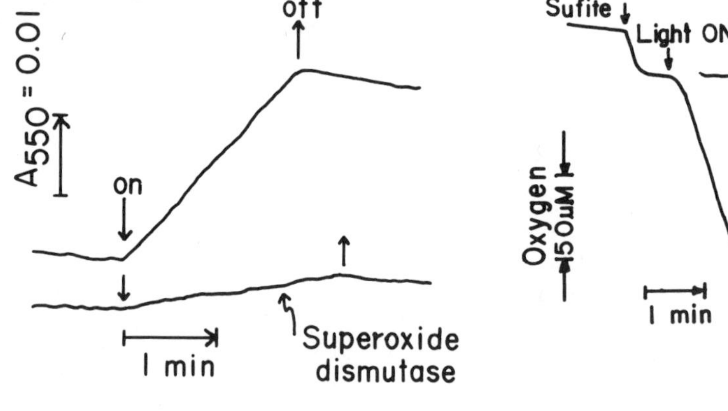


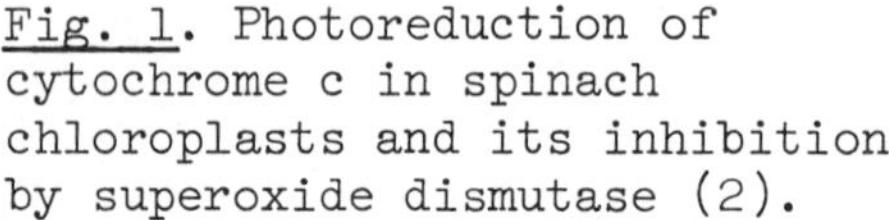

Fig. 1. Photoreduction of cytochrome c in spinach chloroplasts and its inhibition by superoxide dismutase (2).

Fig. 2. Photooxidation of sulfite in spinach chloroplasts (4). $O_2$-uptake was measured by a Clarke oxygen electrode.

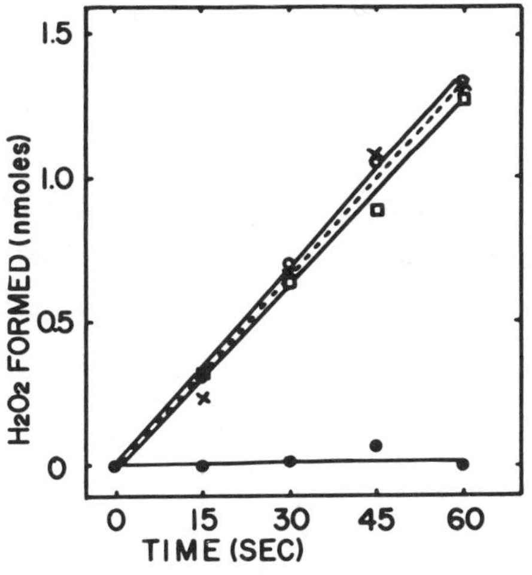


Fig. 3. Effects of cytochrome c and superoxide dismutase on the production of hydrogen peroxide by illuminated spinach chloroplasts. The reaction mixture contained, in 2 ml, 50 mM phosphate, pH 7.8, 10 mM NaCl and chloroplasts containing 10 µg of chlorophyll (O——O) (2).
●——●: + 20 µM Cyt. c
X——X: + 20 µM Cyt. c + 0.53 µM superoxide dismutase
□——□: + 0.53 µM superoxide dismutase

Fluorometric determination of hydrogen peroxide formed by illuminated chloroplasts (2) indicates the absence of the divalent reduction of molecular oxygen (Fig. 3). Hydrogen peroxide was formed without any additive, however; in the presence of cytochrome c hydrogen peroxide was not detected. The addition of superoxide dismutase to cytochrome c restored the production of hydrogen peroxide at the same rate as with chloroplasts alone. Photoreduction rate of cytochrome c is equivalent to the formation of hydrogen peroxide, indicating that cytochrome c is reduced by superoxide and in illuminated chloroplasts hydrogen peroxide is formed only through the disproportionation of superoxide. Thus, illuminated chloroplasts donate one electron to molecular oxygen to form superoxide. $H_2O_2$ is formed by the Mehler reaction, which is composed of the following two reactions:

$$O_2 + [e^-] \longrightarrow O_2^-$$

$$2\,O_2^- + 2\,H^+ \longrightarrow H_2O_2 + O_2$$

In chloroplasts electron donor for molecular oxygen is the primary electron acceptor of photosystem I (2) which has been proposed to be a bound ferredoxin, P-430 (12).

Formation of hydroxyl radical in chloroplasts is expected through the Harber-Weiss reaction from superoxide and hydrogen peroxide and recently this has been confirmed by Elstner and Konze (13). Further, formation of singlet oxygen in illuminated chloroplasts has been shown by Takahama and Nishimura (14) and singlet oxygen is supposed to be formed through the oxidation of superoxide by electron transport system in chloroplasts.

Recent $^{18}O$-experiments show the occurrence of photoreduction of molecular oxygen (15), possibly by the above mechanism under physiological conditions, indicating the formation of superoxide *in vivo*. Superoxide thus formed would disproportionate in a reaction catalyzed by superoxide dismutase contained in chloroplast stroma (16) or react with the components of chloroplasts as described below. Under physiological conditions, the deteriorative effects of superoxide and the other "activated" oxygen derived from it would be prevented by superoxide dismutase and the other protection mechanisms in chloroplasts. Toxicity of superoxide anion to plant cells can be observed by application of dipyridyl compounds. Several dipyridyl compounds (paraquat or diquat) cause the death of plant leaf cells only under light and have been used as a herbicide. This herbidical action has been deduced to be due to the formation of superoxide by the reaction of molecular oxygen and reduced dipyridyl radicals, which are produced by the photoreduction in chloroplasts (2,17). Photooxidative death of algae has been observed under high concentration of oxygen, especially in the absence of carbon dioxide which is a physiological electron acceptor of photosynthesis (18).

## II) SCAVENGING OF SUPEROXIDE IN CHLOROPLASTS

For photosynthetic organisms, in addition to photodynamic action, light brings about the damage due to superoxide as described above. The participation of superoxide dismutase is supposed to be due to the induction of superoxide dismutase by light. Table I indicates the highest superoxide dismutase activity in *Euglena* cultured under photoautotrophic conditions. In non-photosynthetic organisms the induction of superoxide dismutase by oxygen during the culture has been observed (19).

Table I. Effect of growth conditions on superoxide dismutase in *Euglena gracilis*.

| Growth conditions | Protein (mg/culture bottle) | Superoxide dismutase (units/mg protein) |
|---|---|---|
| Photoautotrophic | 197 | 20.5 |
| Photoheterotrophic[a)] | 160 | 11.9 |
| Heterotrophic (Dark)[a)] | 60 | 7.0 |

a) Glucose was added.

In addition to enzymatic scavenging, superoxide could be removed by the reaction with the compounds in chloroplasts. Figures

4 and 5 indicate the oxidation of $Mn^{2+}$ (8) and the reduction of plastocyanin by superoxide. Free manganese is found in chloroplasts at about 0.4 mM and plastocyanin is an electron carrier between photosystems I and II. In addition, ascorbate, which is also contained in chloroplasts, is photooxidized through superoxide (5,6).

Blue protein from rice bran (20) was also reduced by superoxide. The blue protein and plastocyanin are both "blue" Cu-protein and both of them did not catalyze the dismutation of superoxide. It is interesting to note that galactose oxidase which has a low dismutase activity (21) and superoxide dismutase are both "non-blue" Cu-protein.

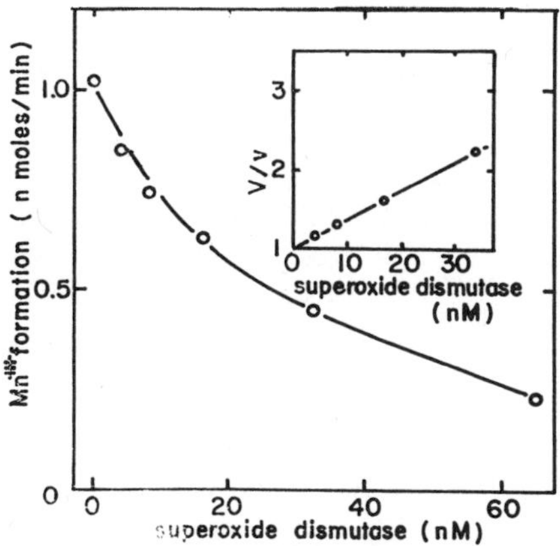


Fig. 4. Effect of superoxide dismutase on $Mn^{3+}$ formation with the superoxide generated by xanthine-xanthine oxidase system. The reaction mixture, in 1 ml, contained 0.1 mM xanthine, 10 μM $MnCl_2$, 50 mM pyrophosphate, pH 7.8 xanthine oxidase and superoxide dismutase.

n moles Plastocyanin reduced·min⁻¹
0.5
0.4
0.3
0.2
0.1
0
0 1 2 3 4 5
Superoxide Dismutase (nM)
V/v
3
2
1
0 2 4
SOD (nM)

Fig. 5. Effect of superoxide dismutase on the oxidation of plastocyanin by superoxide. The reaction mixture, in 1 ml, contained 0.1 mM xanthine, 3.8 μM plastocyanin, 50 mM phosphate, pH 7.8, xanthine oxidase and superoxide dismutase.

The second order reaction rate constants are shown in Table II. All of them were determined from the competition of superoxide between superoxide-induced reaction and superoxide dismutase (22), assuming that the rate constant between superoxide and superoxide dismutase is $2.3 \times 10^9$ $M^{-1}x^{-1}$ (23). Figures 4 and 5 indicate the inhibition data of superoxide-induced reactions by superoxide dismutase which were used for the determination of the rate constant. V and v are the reaction rates in the absence and the presence of superoxide dismutase, respectively.

Table II. The second order reaction rate constant between superoxide and several compounds found in chloroplasts.

| | k ($M^{-1}s^{-1}$) |
|---|---|
| $Mn^{2+}$ | 6.0 X $10^6$ |
| Plastocyanin | 1.1 X $10^6$ |
| Ascorbate[a)] | 2.7 X $10^5$ |
| a) Data of Nishikimi (24) | |

## III) SUPEROXIDE DISMUTASES IN PHOTOSYNTHETIC ORGANISMS AT DIFFERENT LEVELS OF EVOLUTION

For photosynthetic organisms superoxide dismutase plays a role of scavenging superoxide produced by light in chloroplasts and also superoxide formed by oxidases and reductants of oxygen in both nonphotosynthetic and photosynthetic organisms. Superoxide dismutase has been purified from various organisms and three forms of the enzyme have been found with respects to the metal in the enzyme, i.e., Cu, Zn-, Mn- and Fe-containing enzymes. The Cu, Zn-enzyme is sensitive to cyanide and lacks or contains only a low content of tyrosine and tryptophan. In contrast, Fe- and Mn-superoxide dismutases are both insensitive to cyanide and have the similar amino acid composition which is different from that of the Cu, Zn-enzyme (25).

The Cu, Zn-enzyme has been found in eukaryotes, however, recently this form of the enzyme was also detected in a prokaryote, *Photobacterium leiognathi* (26). The Mn-enzyme has been isolated from prokaryotes and also from mitochondria of yeast and chicken liver. A high homology of amino acid sequence between the Mn-enzymes from both sources was found supporting the symbiotic theory of mitochondria (25,27). The Fe-enzyme has been found only in prokaryotes. However, recently the occurrence of soluble cyanide-insensitive Mn- or Fe-enzyme in eukaryotic organisms, flagellates (28) and a mushroom (29) has been reported.

Under these circumstances we had an interest to survey what form of superoxide dismutase is contained in photosynthetic organisms at different levels of evolution. For characterization of the enzyme we used two criteria: 1) sensitivity of crude extract to cyanide to distinquish Cu, Zn-enzyme and cyanide-insensitive Mn- or Fe-enzyme. 2) effect of antibodies against spinach Cu, Zn-enzyme (16) and *Plectonema* Fe-enzyme (30). The results indicate the absence of Cu, Zn-enzyme in photosynthetic bacteria and prokaryotic and eukaryotic algae.

HIGHER PLANTS (Angiospermae)

In higher plants major soluble superoxide dismutase is Cu,Zn-enzyme and we purified it to a crystalline state from spinach (16). Spinach Cu,Zn-enzyme is similar to those from the other sources except for minor differences in amino acid composition. However, immunologically, plant and mammalian enzymes are distinguishable. Although antibody against the spinach enzyme inhibits the enzymatic activity in the extracts of all of land plants, erythrocuprein was not affected. For comparison of the structure with mammalian enzymes (31), a crystallographic study has been carried out by Morita (32).

Fractionation of spinach leaf by sucrose density centrifugation revealed the localization of the enzyme in chloroplasts. Cu,Zn-enzyme was released by osmotic shock indicating the localization in stroma (16). On the contrary lamellae-bound enzyme is cyanide-insensitive suggesting the occurrence of Fe- or Mn-enzyme in spinach lamellae. In kidney bean leaves, three isozymes (A,B and C) of superoxide dismutase was obtained in soluble form. Isozymes A and B were cyanide-sensitive, however, isozyme C was insensitive. Figure 6 indicates only cyanide-insensitive isozyme C is not inhibited by the spinach enzyme's antibody. Accordingly, in higher plants, Cu,Zn-enzyme occurs in stroma and bound-enzymes in chloroplasts lamellae and possibly in mitochondria are cyanide-insensitive.

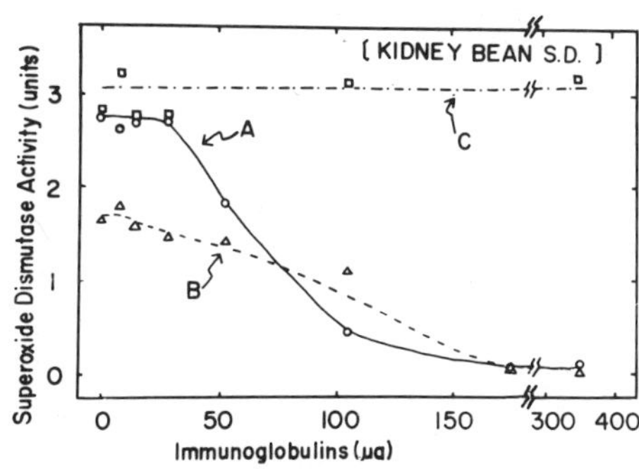


Fig. 6. Effect of spinach Cu,Zn-enzyme's antibody on superoxide dismutase in kidney bean. A,B and C refer to different superoxide dismutase.

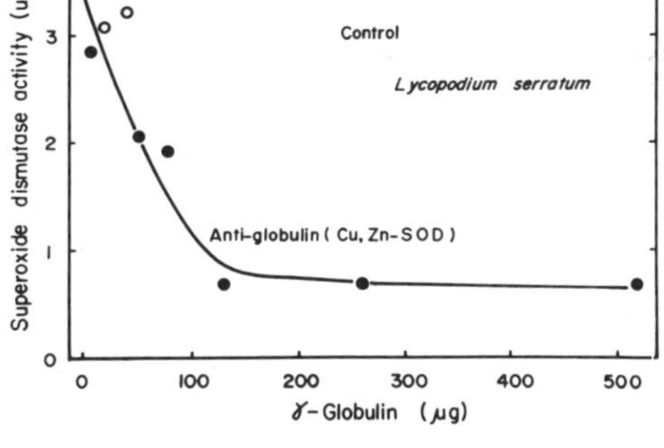


Fig. 7. Effect of spinach Cu,Zn-enzyme's antibody on superoxide dismutase in extract of fern, Lycopodium serratum.

## FERN AND MOSS (Pteridophyta and Bryophyta)

Superoxide dismutase in extract of moss (3 species) was inhibited partially by both cyanide and the spinach enzyme's antibody. This is also the case of fern (7 species) as shown in Fig. 7, indicating the occurrence of common antigenic determinants in superoxide dismutases of spinach, fern and moss.

## EUKARYOTIC ALGAE (Charophyta, Euglenophyta, Rhodophyta, Phaeophyta and Chlorophyta)

Typical results on the effects of cyanide and the spinach enzyme's antibody on superoxide dismutase in the extracts of green algae (Figs. 8 and 9 ) indicate the ineffectiveness of both reagents. The same results were obtained for the other eukaryotic algae including Charophyta (2 sp.), Euglenophyta (1 sp.), brown (4 sp.), red (4 sp.) and green algae (5 sp.). Disc gel electrophoresis of the extracts of several algae revealed the occurrence of multiple forms of superoxide dismutase, however, all of the isozymes were not affected by cyanide. These observations indicate, in contrast to land plants, the absence of the Cu, Zn-enzyme in eukaryotic algae (30). By the same test it was found that aquatic seed plants, *Lemna perpusilla* and *Halophila ovalis*, contained the Cu,Zn-enzyme indicating that a form of superoxide dismutase is not affected by growing habitat but reflects an evolutionary level. Lumsden and Hall also have confirmed the absence of the Cu,Zn-enzyme in eukaryotic algae (33).

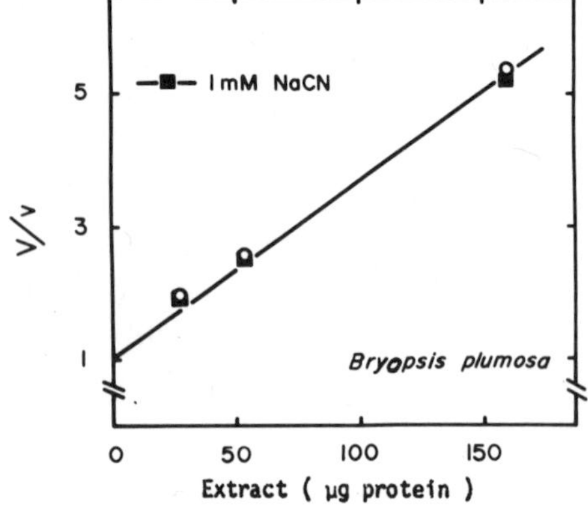


Fig. 8. Effect of 1 mM cyanide on superoxide dismutase in extract of *Bryopsis plumosa*. V and v are cytochrome c reduction rate by $O_2^-$ generated with xanthine-xanthine oxidase system in the presence and absence of the extract, respectively.

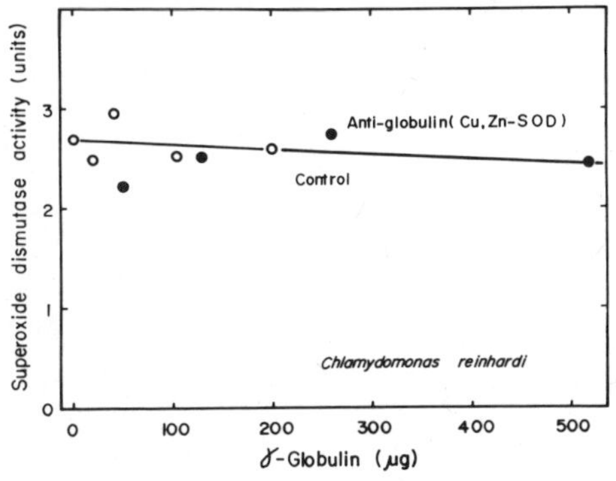


Fig. 9. Effect of spinach Cu, Zn-enzyme's antibody on superoxide dismutase in extract of Chlamydomonas reinhardi.

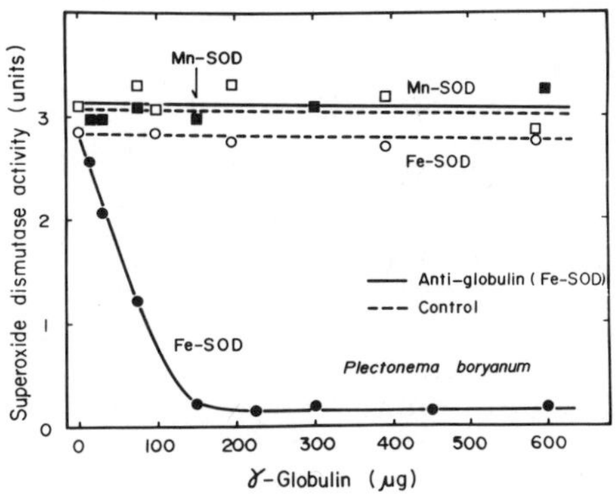


Fig. 10. Effect of Plectonema Fe-enzyme's antibody on Plectonema Fe- and Mn-superoxide dismutases.

PROKARYOTIC ALGAE (Blue-green algae, Cyanophyta)

Four species of blue-green algae were tested and the absence of the Cu, Zn-enzyme was confirmed in the prokaryotic algae and eukaryotic algae. In Plectonema boryanum the Fe-enzyme is major superoxide dismutase (30,34) and we have isolated it also from Spirulina (35). A minor superoxide dismutase in Plectonema is the Mn-enzyme (30). Antibody against Plectonema Fe-enzyme has a very narrow reactivity compared with the antibody to the spinach Cu, Zn-enzyme. As shown in Fig. 10, the Fe-enzyme's antibody inhibited the Fe-enzyme, but did not show any effect on the Mn-enzyme from the same algae. This antibody also did not inhibit superoxide dismutase in the extract of Nostoc verrucosum. These observations suggest the absence of common antigenic determinants in Fe- and Mn-enzymes from Plectonema and superoxide dismutase from the same division.

Using these properties of the antibody for localization of Fe- and Mn-superoxide, the dismutases in Plectonema cells were studied. The protoplasts of the algal cells were prepared by incubation with lysozyme and the cells were disrupted in water. Cell fragments were subjected to sucrose density centrifugation and lamellae and supernatant fractions were obtained. The Fe-enzyme's antibody inhibited superoxide dismutase activity in the supernatant fraction but not the enzyme in the lamellae fraction indicating the binding of Mn-superoxide dismutase to the lamellae membrane of blue-green algae.

All attempts to see the effect of the spinach Cu, Zn- and Plectonema Fe-enzyme's antibodies on the Hill reactions of spinach chloroplasts and the algal protoplasts failed, suggesting that Cu, Zn- and Fe-superoxide dismutases do not participate in the water oxidation (oxygen evolution) system in photosystem II. Localization of Mn-superoxide dismutase in algal lamellae is interesting because of the essentiality of manganese for photosystem II (36).

PHOTOSYNTHETIC BACTERIA (Bacteriophyta)

Superoxide dismutase in the extracts of both purple nonsulfur and sulfur bacteria, *Rhodospirillum rubrum*, *Rhodopseudomonas capsulatus* and *Chromatium vinosum*, were insensitive to cyanide and the spinach enzyme antibody. The cyanide-insensitive superoxide dismutase has recently reported also by Hewitt and Morris (37) in green sulfur bacterium, *Chlorobium thiosulfatophilum*. Thus, so far in all prokaryotes of photosynthetic organisms tested the Cu, Zn-superoxide dismutase is absent, as well as in most prokaryote except for *Photobacterium* (26). Since the nonsulfur bacteria can grow under aerobic and dark conditions consuming organic substrates, the occurrence of superoxide dismutase is inferred by its enzymatic activity. On the other hand, although the sulfur bacteria grow using sulfide as an electron donor for photosynthesis under anaerobic condition, the level of superoxide dismutase is the same as that in the nonsulfur bacteria on a protein basis. Because of the interest in surveying the physiological function of superoxide dismutases in anaerobic sulfur bacteria and for following the evolution of superoxide dismutase, we have purified the *Chromatium* enzyme.

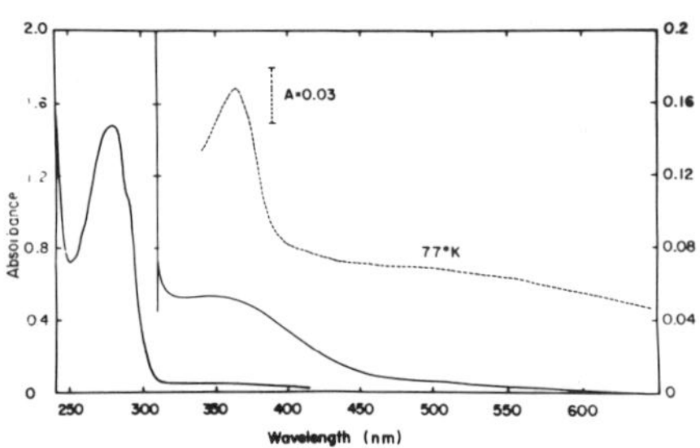


Fig. 11. Absorption spectra of *Chromatium* superoxide dismutase at room temperature and 77 K.

Absorption spectrum of the purified enzyme (Fig. 11) and the metal analysis clearly indicates that *Chromatium* superoxide dismutase is Fe-enzyme. The molecular weight is about 41,000 and the enzyme is composed of two subunits of equal size. In these respects, *Chromatium* superoxide dismutase is similar to the *E. coli* and blue-green algae's Fe-enzymes (30,38). Physiological function of this enzyme in anaerobic sulfur bacteria is totally unknown. However, recently superoxide dismutase has been found also in anaerobic sulfate reducing and fermentative bacteria (37). Clarification of the function of superoxide dismutase in these organisms is an interesting future project.

IV) CONCLUDING REMARKS

The distribution of superoxide dismutase in photosynthetic organisms at different levels of evolution is summarized as follows.

Photosynthetic bacteria and prokaryotic and eukaryotic algae contain only the cyanide-insensitive enzyme and lack the Cu,Zn-enzyme. Superoxide dismutases of Chromatium and blue-green algae are Fe-enzymes. In addition to the Fe-enzyme, blue-green algae also contain the Mn-enzyme in lamellae-bound form. The Cu, Zn-superoxide dismutase is found only in land plants and its appearance starts from the most primitive land plants, Bryophyta and Peteridophyta. Plants of both divisions contain the Cu,Zn-enzyme and cyanide-insensitive enzyme. Seed plants also contain both forms of the enzyme, however, in some plants like spinach, the cyanide-insensitive enzyme binds to chloroplant lamellae.

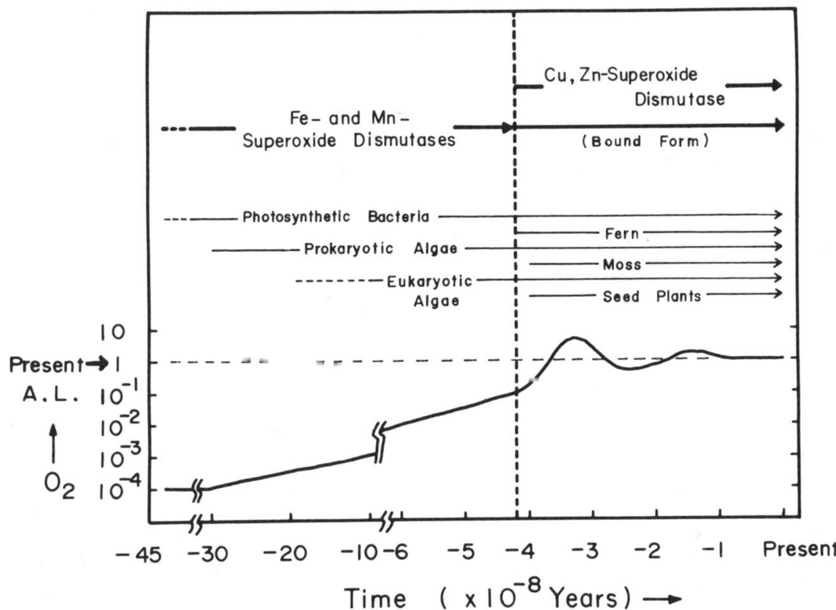


Fig. 12. Evolution of photosynthetic organisms, forms of superoxide dismutase, and oxygenic concentration in the earth's atmosphere. Oxygen content values in the atmosphere were obtained from the paper of Berkner and Marshall (39).

In Fig. 12 the forms of superoxide dismutase, evolution of photosynthetic organisms and increase of oxygen in the atmosphere are schematically illustrated. It is not known when the photosynthetic bacteria appeared on the earth. However, the appearance of the primitive blue-green algae has been deduced to be about $30 \times 10^8$ years ago. Geological record indicates the appearance of the first land plants at the late Silurian age, $4.2 \times 10^8$ years ago. According to Berkner and Marshall (39) at this age, the oxygen level in the atmosphere was 10% of the present atmosphere. This allowed the plants to emerge on land. Filtration of lethal ultraviolet light by the ozone layer could there take place at the known oxygen content of the atmosphere at that time.

The present observations indicate that photosynthetic organisms which developed up to the late Silurian age did not contain the Cu,Zn-enzyme and this form of the enzyme was acquired by the plants which developed after that. We propose that the Cu,Zn-superoxide dismutase was acquired only from organisms developing after the late Silurian Age. Further survey of superoxide dismutase in the other organisms, especially, in Eumycota, would be important for the establishment of this working hypothesis, because though yeast and *Neurospora* contain the Cu,Zn-enzyme (25), *Pleurotus* (29) contains only the cyanide-insensitive enzyme. From the reactivity of cytochrome c with cytochrome oxidases Yamanaka has supposed that yeast and *Neurospora* appeared after algae (40).

The absence of copper proteins in anaerobes may have been due to the proposed inavailability of copper to the organisms. It has been supposed that until oxygen had accumulated in the atmosphere by photosynthetic activity copper existed as extremely insoluble Cu(I) sulfides because of the low environmental redox potential (41,42). For photosynthetic bacteria this was probably the case. However, this theory does not apply to prokaryotic and eukaryotic algae since the two algae contain plastocyanin despite the absence of the Cu,Zn-superoxide dismutase. Little differences exist in the enzymatic properties of the different Cu,Zn-superoxide dismutase. However, the Fe- and Mn-enzymes have lower reaction rate constants with superoxide than the Cu,Zn-enzyme (43). At present we cannot answer why at the late Silurian age plants added the new form of superoxide dismutase which require different metals and which have different amino acid sequences. As described in part I and II of this paper, the Cu, Zn-superoxide dismutase in seed plants and the cyanide-insensitive enzymes in algae may have a role of scavenging the superoxide which is produced by light in the chloroplasts or the lamellae. Further characterization of the three forms of the superoxide dismutase and further investigations of the functions of the Mn-enzyme in the lamellae of blue-green algae and the Fe-enzyme in anaerobic photosynthetic bacteria may provide clues about the evolution of the different types of superoxide dismutases present in the organisms investigated.

accumulation of the protein upon subsequent injection of copper. It was thus apparent that the Cu-dependent formation of the protein was a true induction process and required the formation of a specific messenger RNA. Treatment with the antibiotics did not inhibit the uptake of copper by the liver.

The copper-containing protein, called Cu-chelatin, was purified by a method involving gel filtration and acetone fractionation (Table III). The purified protein contained 4.2 percent copper and negligible amounts of other metals. Acrylamide gel electrophoresis revealed the presence of a single protein band (Fig. 3) which corresponded to the zone of highest copper concentration on the gel. This experiment showed the marked difference in the electrophoretic properties of Cu-chelatin and metallothionein.

Chromatography of Cu-chelatin on DEAE-cellulose revealed a further difference in behavior from that of metallothionein. The chelatin was strongly adsorbed to the gel and was not eluted in the same manner as thionein. In order to determine whether this difference reflected the presence of copper versus cadmium on the same protein, production of a copper-complex of rat liver thionein <u>in vivo</u> was attempted. Rats initially given injections of cadmium (0.5 mg Cd per Kg body weight) were administered actinomycin D 30 minutes later to inhibit RNA synthesis. One hour later 2.5 mg of copper (as $CuCl_2$) per Kg was injected subcutaneously. Six hours later the animals were killed, and the livers used for purification of thionein. Chromatography of the sample on DEAE-cellulose revealed the presence of two copper-containing fractions corresponding exactly to the two forms of thionein (Fig. 4). The molar Cu/Cd ratio of the fraction exceeded 3, demonstrating the predominance of binding of copper to the thionein variants. Cu-chelatin could not

Table 3. Purification of Cu-Chelatin from Rat Liver

| Step | | Total Cu µg | Protein[a] mg | Cu/protein ratio |
|---|---|---|---|---|
| 60° heat step | | 221 | -- | -- |
| Sephadex G-75 | α | 70 | 9.7 | 0.72 |
| | β | 92 | 6.6 | 1.39 |
| 60-80% acetone fraction | α' | 48 | 2.2 | 2.18 |
| | β' | 55 | 2.2 | 2.50 (4.2)[b] |

[a] Value based on protein determination by the Lowry procedure
[b] Based on quantitative amino acid analysis

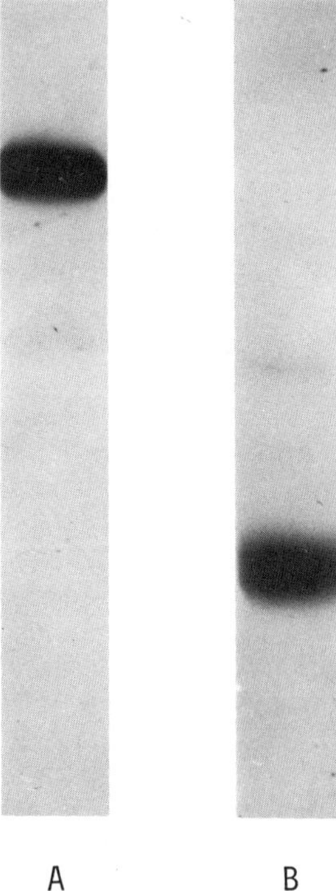


Fig. 3. Acrylamide gel electrophoresis of Cu-chelatin at pH 8.9 (A) and pH 3.2 (B).

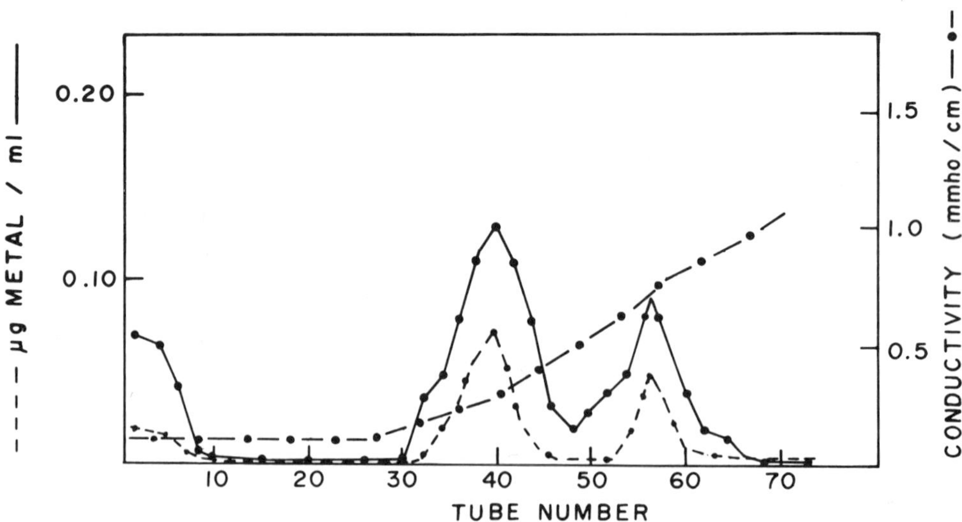


Fig. 4. DEAE-cellulose chromatography of Cu-thionein formed *in vivo*. —— Cu; - - - Cd.

be detected in the sample before or after DEAE-cellulose chromatography. This experiment clearly demonstrated that Cu-chelatin is distinct from thionein, and is induced in rats exposed to copper, but not in animals treated with Cd, Zn, Hg or Ag. The experiment described in Fig. 4 shows that the specificity is at the level of induction and not at the level of binding.

The difference between Cu-chelatin and metallothionein is further corroborated by the amino acid composition shown in Table 4. Notably, the cysteine content of chelatin is about 15 percent, compared to 25 to 30 percent in thionein. Chelatin contains much higher amounts of glutamic acid and the hydrophobic amino acids, while thionein has a very high serine content. Assuming one histidine per molecule of chelatin a molecular weight of 8900 is obtained for this protein. The fact that thionein elutes from sephadex G-75 at a slightly lower elution volume suggests that difference in the calculated molecular weights for the two proteins is probably real. From the copper content of purified chelatin it appears that the protein contains 5 to 6 atoms of copper per molecule. The zinc content is variable, with a maximum of 1.5 atoms per molecule.

Table 4. Comparative Amino Acid Composition of Metallochelatin and Metallothionein

| Residue | Rat Liver Cd,Zn-Thionein Form A | Form B | Rat Liver Cu-Chelatin |
|---|---|---|---|
| | | mole percent | |
| Lysine | 11.6 | 14.1 | 13.1 |
| Histidine | 0 | 0 | 1.1 |
| Arginine | 0 | 0 | 2.7 |
| Aspartic Acid | 8.5 | 8.4 | 9.8 |
| Threonine | 6.2 | 4.0 | 5.1 |
| Serine | 17.5 | 15.7 | 4.2 |
| Glutamic Acid | 3.9 | 7.3 | 11.1 |
| Proline | 3.4 | 3.7 | 3.6 |
| Glycine | 10.3 | 7.2 | 8.0 |
| Alanine | 6.1 | 8.8 | 6.0 |
| Cysteine | 28.8 | 25.4 | 14.6 |
| Valine | 1.7 | 1.4 | 5.2 |
| Methionine | 1.3 | 1.2 | 1.6 |
| Isoleucine | 0 | 0.7 | 4.3 |
| Leucine | 0.6 | 0.7 | 6.3 |
| Tyrosine | 0 | 0 | 1.3 |
| Phenylalaline | 0 | 0 | 2.1 |

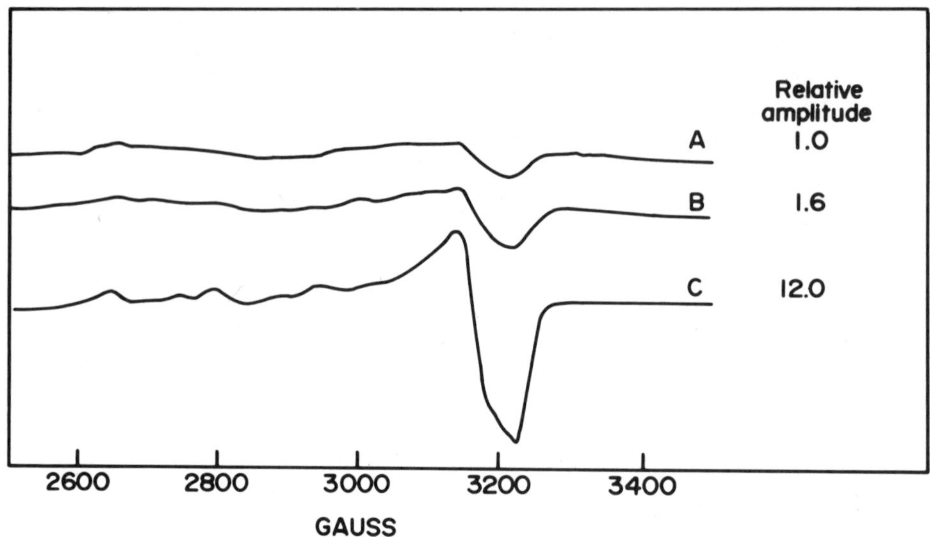


Fig. 5. EPR spectroscopy of Cu-chelatin. See text for details.

## EPR PROPERTIES

Nearly all of the copper in Cu-chelatin was in a diamagnetic form. Addition of 0.1 M EDTA sharpened the broad native EPR signal (Fig. 5,A). When the protein was heated at 100° for 5 min. under anaerobic conditions andthen examined by EPR the signal amplitude increased only slightly (Fig. 5,B). Exposure of the sample to air led to a gradual enhancement of the signal until the intensity corresponded to all of the copper in the sample (Fig. 5,C). In contrast, when the heat treatment was conducted aerobically, maximum signal intensity was observed without a lag period. Heat treatment of Cu-chelatin did not denature the protein but did release the copper as shown by the total separation of protein and metal after passage through Sephadex G-25.

The metal-free chelatin prepared by heating of Cu-chelatin showed very low reactivity with 5,5-dithiobis (2-nitrobenzoic acid) (DTNB), suggesting that the sulfhydryl residues of the protein had been oxidized during the procedure. This was confirmed by the marked enchancement of the rate and extent of reaction with DTNB of the apochelatin after reduction with sodium borohydride.

In view of the possible role of the cysteine moieties in chelatin of copper, the effects of several sulfhydryl-reactive reagents and oxidizing agents on Cu-chelatin were studied. Addition of p-chloromercuribenzene sulfonic acid (PCMBS) to the protein generated the EPR signal of $Cu^{++}$ rapidly even without heating, anaerobically as well as aerobically. The data in Table 5 show that oxidizing agents such as ferricyanide and persulfate and metal ions like $Hg^{++}$ and $Ag^{+}$ also are capable of displacing the copper from its association with the chelatin.

## DISCUSSION

Cu-chelatin, formed in rat liver in response to high levels of copper, has been purified to homogeneity. A variety of studies have shown that Cu-chelatin is distinct from metallothionein. Attention is thus focused on the ability of two different metals, e.g. Cu and Cd, to lead to the specific induction of two different proteins. The molecular basis for this selectivity remains to be established.

The biological function of thionein and chelatin is yet to be defined. While it is clear that large scale accumulation of these proteins in rat tissues occurs only after exposure to high levels of the above-mentioned metals, it is conceivable that these proteins have physiological functions at much lower levels under normal circumstances. Chen _et al_. (22) have also made the observation that Cd injection into rats caused the accumulation of Zn in the thionein fraction. Intriguingly, the Zn levels returned to normal at a rate corresponding to a $t_{\frac{1}{2}}$ of about 5 days whereas the Cd content remained constant at a high level for a long period. Cu-chelatin in rat liver has also been found to have a short half-life (21). The more rapid turnover of Zn and Cu might reflect the fact that they are essential metals in contrast to Cd, and could thus be related to homeostatic mechanisms. The fact that $Cu^{++}$ could be released from Cu-chelatin by oxidizing agents could be used to argue that under

Table 5. Compounds Releasing Copper from Cu-Chelatin

| Reagent | Method | Moles reagent/$Cu^{++}$ |
|---|---|---|
| $Fe(CN)_6^{3-}$ | EPR signal of $Cu^{++}$ | 2.05 |
| $Fe(CN)_6^{3-}$ | Reduction of $Fe(CN)_6^{3-}$ (Spectrophotometric) | 1.95 |
| PCMBS | EPR signal of $Cu^{++}$ | 1.80 |
| $HgCl_2$ | EPR signal of $Cu^{++}$ | 1.8 |
| $AgNO_3$ | EPR signal of $Cu^{++}$ | -[a] |
| DTNB | EPR signal of $Cu^{++}$ | -[a] |
| Persulfate | EPR signal of $Cu^{++}$ | -[a] |

[a] Dash denotes not determined

physiological conditions chelatin serves a storage function and that $Cu^{++}$ is released from it by an oxidative reaction prior to incorporation into other copper proteins. It is hoped that future studies will clarify the biological role of Cu-chelatin as well as metallothionein.

## REFERENCES


1. Margoshes, M., and Vallee, B.L. (1957) J. Amer. Chem. Soc. 79, 4813-4814.
2. Kagi, J.H.R., and Vallee, B.L. (1961) J. Biol. Chem. 236, 2435-2442.
3. Pulido, P., Kagi, J.H.R., and Vallee, B.L. (1966) Biochemistry 5, 1768-1777.
4. Kagi, J.H.R., Himmelhoch, S.R., Whanger, P.D., Bethune, J.L., and Vallee, B.L. (1974) J. Biol. Chem. 249, 3537-3542.
5. Nordberg, G.F., Nordberg, M., Piscator, M., and Vesterberg, O. (1972) Biochem. J., 126, 491-498.
6. Weser, V., Donay, F., and Rupp, H. (1973) FEBS Letters 32, 171-174.
7. Shaikh, Z.A., and Lucis, O.J. (1970) Experentia, 27, 1024-1025.
8. Winge, D.R., and Rajagopalan, K.V. (1972) Arch. Biochem. Biophys. 153, 755-762.
9. Weser, V., Rupp, H., Donay, F., Linnemann, F., and Voelter, W. (1973) Eur. J. Biochem. 39, 127-140.
10. Kimura, M., Otaki, N., Yoshiki, S., Suzuki, M., Horiuchi, N., and Suda, T. (1974) Arch. Biochem. Biophys. 165, 340-348.
11. Cherian, M.G. (1974) Biochem. Biophys. Res. Comm. 61, 920-926.
12. Webb, M. (1972) Biochem. Pharmacol. 21, 2751-2765.
13. Wisniewska, J.M., Trojanouska, B., Piotrowski, J., and Jakubowski, M. (1970) Toxicol. Appl. Pharmacol. 16, 754-763.
14. Yasutomo, S. (1973) Ind. Health 10, 56.
15. Nordberg, M., Trojanowska, B., and Nordberg, G. (1974) Environ. Physiol. Biochem. 4, 149.
16. Bremner, I., and Davies, N.T. (1974) Biochem. Soc. Trans. 2, 425-427.
17. Bloomer, L.C., and Sourkes, T.L. (1973) Biochem. Med. 9, 78-91.
18. Winge, D.R., Premakumar, R., and Rajagopalan, K.V. (1975) Arch. Biochem. Biophys. 170, 242-252.
19. Winge, D.R., Premakumar, R., Wiley, R.D. and Rajagopalan, K.V. (1975) Arch. Biochem. Biophys. 170, 253-266.
20. Premakumar, R., Winge, D.R., Wiley, R.D. and Rajagopalan, K.V. (1975) Arch. Biochem. Biophys. 170, 267-277.
21. Premakumar, R., Winge, D.R., Wiley, R.D., and Rajagopalan, K.V. (1975) Arch. Biochem. Biophys. 170, 278-288.
22. Chen, R.W., Whanger, P.D., and Weswig, P.H. (1975) Biochem. Med. 12, 95-105.

CIRCULAR DICHROISM STUDY OF BOVINE PLASMA AMINE OXIDASE,

A Cu-AMINE OXIDASE

Hiroyuki Ishizaki and Kerry T. Yasunobu
Department of Biochemistry-Biophysics, University of Hawaii
School of Medicine, Honolulu, Hawaii 96822

## INTRODUCTION

Bovine plasma amine oxidase enzyme is a copper amine oxidase which has a molecular weight of about 170,000 (1) and contains 2 gm atoms of copper per mole of protein (2). It is an unusual copper protein since it is pink in color and the apoenzyme is also colored (3). The latter observation has suggested that there is another prosthetic group in the enzyme although the chemical identity of the group has not been established (4). The enzyme catalyzes the general reaction shown in equation 1.

$$RCH_2NH_2 + O_2 + H_2O = RCHO + NH_3 + H_2O_2 \quad (1)$$

The formal mechanism of the oxidative deamination of benzylamine by amine oxidase has been established (5). Recently, the circular dichroism (CD) study of amine oxidase was initiated since the method has been shown to have the potential to act as a sensitive probe of possible conformational changes which occur upon reaction of the substrate and the products of the enzyme reaction with the protein molecule and for the detection of subtle environmental changes which occur upon interaction of strongly absorbing ligands with the protein molecule. In this report, the preliminary CD studies of amine oxidase in the presence of its substrate, some products and copper chelating reagents are presented. The only previous CD study of a Cu-amine oxidase has been the report of Adachi (6) on the *Aspergillus niger* amine oxidase.

## EXPERIMENTAL PROCEDURES

*Preparation of Enzyme.* Bovine plasma amine oxidase was

isolated by the procedure previously reported (7) with some minor modifications. The new steps included a Con A-Sepharose affinity and Bio-gel A-1.5 m chromatography steps after the hydroxylapatite chromatography step of the enzyme purification procedure. The enzyme isolated by this procedure was found to be homogeneous by disc electrophoresis.

Protein Determination. The protein concentration was determined by measuring in duplicate the absorbance at 280 nm using an $E_{1cm}^{1\%}$ value of 20.8, which was obtained by absorbancy and dry weight determinations.

Enzyme Assay. The enzyme activity was determined spectrophotometrically as described by Tabor *et al*. (8) except that it was carried out at 25° rather than 30°. One unit was defined as the amount of enzyme catalyzing a change of 0.001 absorbance per minute at 25° under the standard assay conditions.

CD Spectra. A Cary 61 spectropolarimeter was used to record the CD spectra for the spectral interval from 610 to 190 nm at 25°C. Ellipticity values of the enzyme were measured by cylindrical quartz cells of 0.01, 0.1, and 1.0 cm path length for the spectral intervals of 190 to 250 nm, 250 to 300 nm, and 300 to 610 nm, respectively. The full scale settings were 0.1° for 190 to 250 nm; 0.05° for 250 to 300 nm; and 0.02° for 300 to 610 nm. The data were expressed in terms of mean residues ellipticity, $[\theta]$, for the far- and near-ultraviolet wavelength region and reported as molecule ellipticity, $\theta$, for the visible wavelength region. These ellipticities were defined as $[\theta]$ or $\theta=(\theta^{o})\ (M)/10\ (c)(\ell)$, where $\theta^{o}$ is observed ellipticity in degree at wavelength $\lambda$; M is the mean residue weight, taken as 115 for amine oxidase or the gram molecular weight, which is 170,000; $\ell$ is the path length of cell in centimeter; and c is the protein concentration in grams per $cm^3$. The results for the far- and near-ultraviolet and visible CD spectra of native amine oxidase in the Figures 1-7 represent averages of five spectra. The protein concentrations varied from 6 to 11 mg/ml and 1 to 2 mg/ml for the visible and ultraviolet (far- and near-) spectra, respectively. The specific activities of the enzyme used was 700 to 1000 units/mg. Each spectrum was taken in duplicate.

Preparation of Deoxygenated Enzyme. The enzyme and subsequent transfer of the enzyme to CD cells were carried out in an oxygen-free nitrogen atmosphere by a procedure which will be described in detail in another paper.

## RESULT AND DISCUSSION

Visible CD Spectra of Native and Heat Denatured Amine Oxidase.

visible CD spectrum for amine oxidase shows a broad CD band centered around 610 nm, a negative extremum at 430 nm, and a positive extrema at 370 and 330 nm as illustrated in Fig. 1.

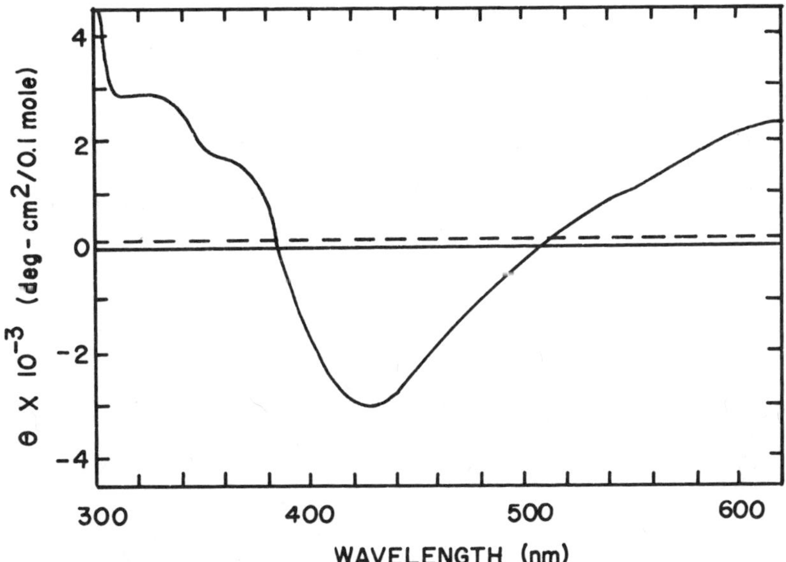


Figure 1. The visible CD spectra of native enzyme (——) and heat denatured enzyme (---) in 0.1 M potassium phosphate buffer, pH 7.0. The enzyme concentration and specific activity were 10.5 mg/ml and 742 units/mg, respectively.

The copper chelate structure in the enzyme is optically active and at least four copper electronic transitions are detected. This multiplicity of CD bands is known to arise from d-d electronic transitions (9). The visible and CD spectral properties of the bovine plasma and *Aspergillus niger* amine oxidases are summarized in Table I.

Table I. Spectral Properties of the Cu-Amine Oxidase

| Form of enzyme | visible spectrum $\lambda max$ (nm) | $\varepsilon_m$ ($M^{-1}cm^{-1}$) | CD spectrum $\lambda max$ (nm) | $\theta$ (deg-cm$^2$/0.1 mole) |
|---|---|---|---|---|
| Aspergillus niger | | | | |
| 1. Oxidized | | | 310? | |
| | 330 | ∿7,745 | 330 | |
| | 410 | | 410 | |
| | 480 | ∿3,511 | 480 | |
| Bovine plasma | | | | |
| 1. Oxidized | 330 | ∿12,698 | 330 | ∿2,950 |
| | | | 370 | ∿1,710 |
| | | | 430 | ∿3,070 |
| | 480 | ∿5,424 | 480? | |
| | | | 610 | ∿2,620 |

From these spectral investigations, the two enzymes show similar but different spectral properties. Another point of interest is the value of the molar absorbancy index of the maxima observed in the visible wavelength region. The $\varepsilon_m$ values of 7,745 and 3,571 at 330 and 480 nm are much higher than was previously calculated by others (10) and approach the values of reported for the "blue" copper proteins (11) in which charge-transfer complexes are reported to exist. All of the CD bands of the enzyme are completely abolished upon heat denaturation of the enzyme. The visible absorption spectrum is also altered; the absorbance in 300 to 400 nm region is increased and a slight decrease in absorbance at 480 nm is observed. However, the enzyme remains pink in color. Heat denaturation thus destroys the optical activity of the copper complex probably by altering the groups in the protein which are liganded to the copper. Adachi has also reported that the CD bands in the visible wavelength region are destroyed by treatment with a 6M guanidine HCl-0.1 M mercaptoethanol solution (6).

Near-Ultraviolet CD Spectra of Native and Heat Denatured Amine Oxidase. The near-ultraviolet CD spectra is characterized by a broad positive band centered at 280 nm (Figure 2).

The origin of this band is probably attributed mainly to an unresolved contribution from several aromatic groups and disulfide bonds. A marked reduction in ellipticity in this region of the spectrum is observed upon heat denaturation of the enzyme at 70$^o$ for 15 min. This represents a change in the local environment of the aromatic amino acid residues in the enzyme which become exposed on heat denaturation of the enzyme.

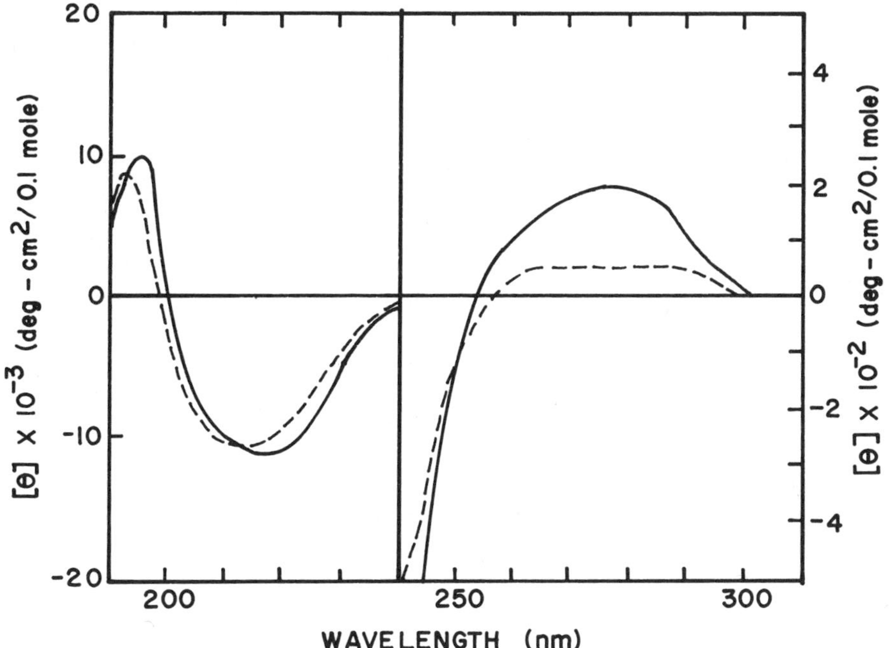


Figure 2. The far- and near-ultraviolet CD spectra of native enzyme (——) and heat denatured enzyme (---) in 0.1 M potassium phosphate buffer, pH 7.0. The protein concentration and specific activity were 1.74 mg/ml and 700 units/mg, respectively.

Far-Ultraviolet CD Spectra of Native and Heat Denatured Amine Oxidase. The far-ultraviolet CD spectra exhibits a negative extremum at 220 nm and positive extremum at 195 nm (Fig. 2). The magnitude of the mean residue ellipticities for the 220 and 195 nm extrema are -11,762 ± 1,346 and 10,643 ± 1,347 deg-cm$^2$/0.1 mole, respectively. The spectrum does not resemble the spectrum of model α-helical polypeptides. The absence of positive 190 nm maximum precludes the presence of a large amount of α-helix (12). Nevertheless, the α-helical content is estimated to be 15% by using the Greenfield-Fasman equation (12). The position of the 220-nm extremum suggests that some β-structure is present in the enzyme (13). Heat treatment of enzyme at 70° for 15 min induces a considerable conformational change in the enzyme as can be seen in Fig. 2. The CD extremum at

220 nm is shifted to 210 nm and a new positive extremum appears at 193 nm. The preliminary indications are that some of the β-structure has been converted to an α-helical structure although other possible conformational changes might also be induced. The estimated α-helical content from the ellipticity value at 208 nm and the use of the Greenfield and Fasman equation indicated that the helical content was 24%. A future investigation on the conformational changes of the enzyme in the presence of α-helix promoting reagents such as trifluoroethanol and sodium dodecyl sulfate may help to confirm that the CD spectral changes in the far-ultraviolet wavelength region are due to an increase in the α-helical content of the enzyme.

Effect of Products, Ammonia and Hydrogen Peroxide on CD Spectra. The visible CD spectra in the presence of ammonium bicarbonate (0.14 M) and hydrogen peroxide (0.15 M) are presented in Figure 3.

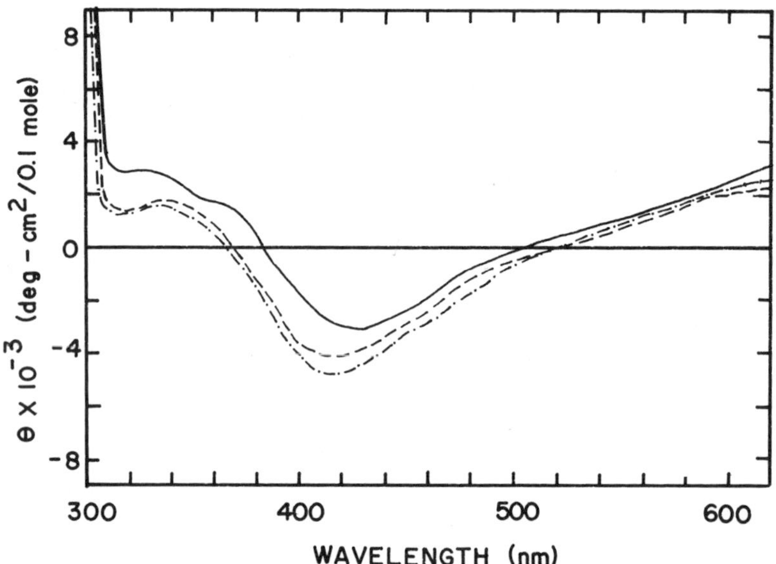


Figure 3. The visible CD spectra of enzyme in 0.1 M potassium phosphate buffer, pH 7.0: (——) enzyme alone; (---) enzyme (protein concentration = 10.26 mg/ml; specific activity = 870 units/mg) to which ammonium bicarbonate (0.14 M final concentration) was added; (-•-) enzyme (protein concentration = 6.48 mg/ml; specific activity = 972 units/mg) to which hydrogen peroxide (0.15 M final concentration) was added.

Since these products are also competitive inhibitors for the oxidative deamination of benzylamine by the enzyme, the CD spectra are useful for probing the nature of the active site and the possible conformational changes which occur upon the interaction of products with enzyme molecule. The binding of these products by the amine oxidase induces similar changes in the visible CD spectrum. The ellipticity between 300 to 400 nm is markly decreased and the negative extremum is shifted from 430 to 410 nm. A concomitant loss in activity is also observed. These findings indicate that when these products react with the benzylamine binding site, the optical activity of the copper complex decreases probably as a result of complex formation with the copper coordination sphere of the enzyme or as a indirect result of gross-protein conformational changes which occur upon binding of the products to the enzyme. The changes observed with $NH_4^+$ ions are reversible but similar studies have not been made with $H_2O_2$ treated enzymes. This is important to investigate since $H_2O_2$ can oxidize the sulfur in thioether linkage present as well as the sulfhydryl groups in the enzyme. In addition, it can oxidize the copper.

The near-ultraviolet CD spectra of the enzyme to which ammonium bicarbonate (0.14 M final concentration) is added is illustrated in Fig. 4. A very small change in the near-ultraviolet region is observed in the presence of 0.14 M ammonium bicarbonate, which suggest that there is virtually no perturbation of the environment of the aromatic amino acid residues in the enzyme.

The analysis of the far-ultraviolet CD spectrum in the presence of ammonia reveals that the position of the CD extrema at 195 and 220 nm are unaltered, while the intensity of CD transitions at these wavelengths are significantly decreased. This demonstrates some sort of conformational change of the enzyme upon product binding. A decrease in ellipticity at 220 nm may be due to a decrease in the β-structure of the enzyme.

Effect of Cyanide and Fluoride on the CD Spectra. The addition of fluoride to the amine oxidase does not cause the position of the visible CD bands to shift but the ellipticity decreases in the 300-400 nm and 500-600 nm region and there is an increase in the negative ellipticity in the 400-500 nm region as shown in Fig. 5. Cyanide has a much more marked effect on the CD spectrum of the enzyme and the position of the CD extrema was altered. Specifically the ellipticity increases in the 300-390 nm region and the 430 nm band shifted to about 460 nm. Since cyanide has been reported to chelate with copper in many copper proteins, these CD spectral changes may result from a replacement of one of the protein ligands in the copper complex by cyanide. Fluoride has also been shown to react with the copper of some copper proteins but it is also a chaotropic reagent and may act by altering the structure of water surrounding the enzyme. However, the latter

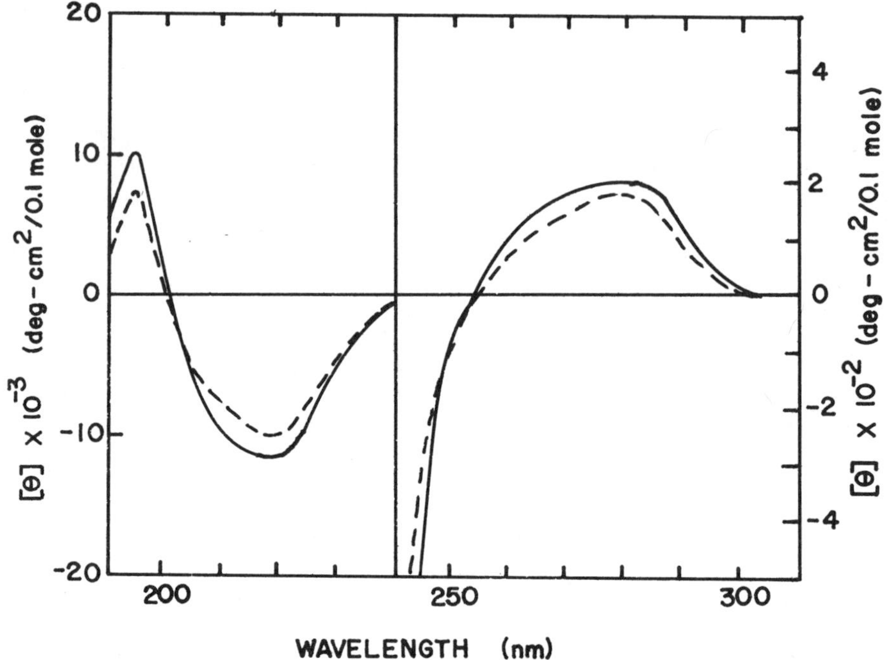


Figure 4. The far- and near-ultraviolet CD spectra of enzyme in 0.1 M potassium phosphate buffer, pH 7.0. The spectra were of enzyme alone (——) and enzyme (protein concentration = 1.17 mg/ml; specific activity = 855 units/mg) to which ammonium bicarbonate (0.14 M final concentration) was added (---).

possibility seems less likely since fluoride had a minimal effect on the near- and far-ultraviolet CD spectra of the enzyme. Thus the differences observed in the visible CD spectrum in the presence of fluoride and cyanide may be due to different modes of binding of cyanide and fluoride with the copper atoms in the enzyme.

The far- and near-ultraviolet CD spectra of the enzyme in the presence of fluoride is shown in Fig. 6. No significant changes are observed, a finding which suggests that fluoride is not acting as a chaotropic reagent but is directly reacting with the copper in

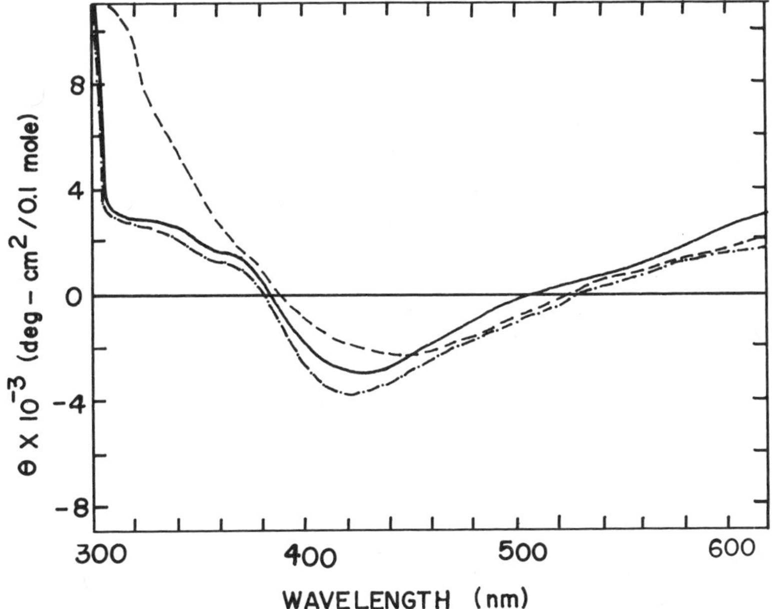


Figure 5. The visible CD spectra of enzyme in 0.1 M potassium phosphate buffer, pH 7.0: (——) enzyme alone; (---) enzyme (protein concentration = 10.86 mg/ml; specific activity = 755 units/mg) to which sodium cyanide (0.01 M final concentration) was added; (-•-) enzyme (protein concentration = 8.09 mg/ml; specific activity = 964 units/mg) to which sodium fluoride (0.1 M final concentration) was added.

the enzyme. On the other hand, in this same wavelength region, cyanide addition to the enzyme significantly reduced the ellipticity in the 210-300 nm region and also in the 190-200 nm region. The negative extremum at around 220 nm was shifted to a lower wavelength (about 5-7 nm). Thus a significant conformational change occurs when cyanide is added to the enzyme. Catsimpoolas and Wood (14) have shown that cyanide can cleave peptide bonds where cystine residues are present but this possibility has not been checked in the present investigation. Since there are also detectable changes in both visible and near-ultraviolet CD spectra when cyanide is present, it is difficult to distinguish whether the observed changes result from a direct interaction of cyanide with the copper chromophore or whether cyanide alters the gross-protein conformation, or if both effects are occurring.

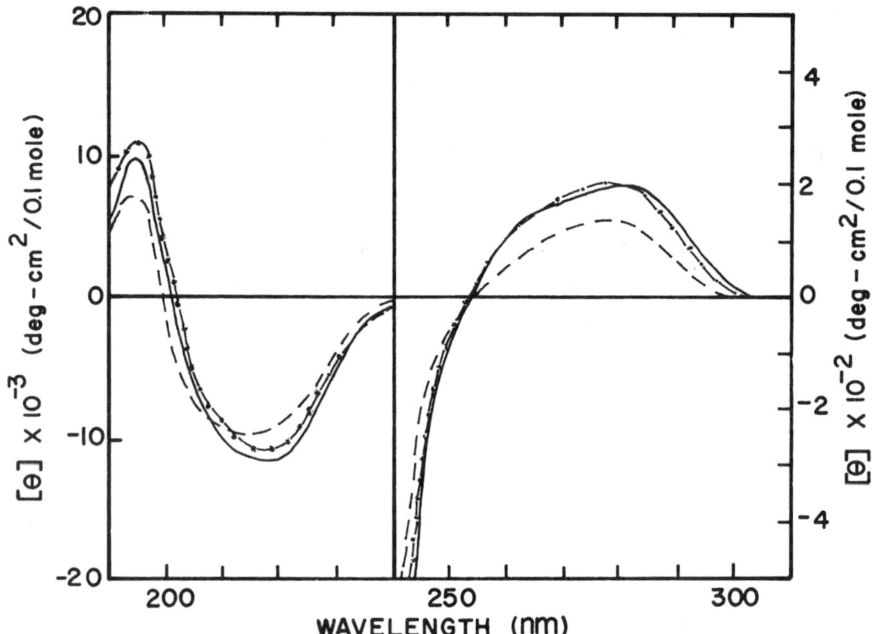


Figure 6. The far- and near-ultraviolet CD spectra of enzyme in 0.1 M potassium phosphate buffer, pH 7.0. The spectra were obtained of enzyme alone (——); enzyme (protein concentration = 1.39 mg/ml; specific activity = 902 units/mg) to which sodium cyanide (0.1 M final concentration) was added (---); enzyme (protein concentration = 1.39 mg/ml; specific activity = 902 units/mg to which sodium fluoride (0.1 M final concentration) was added (-•-).

The Far- and Near-Ultraviolet Spectra of Native and Deoxygenated Amine Oxidase. The deoxygenated enzyme shows virtually no change in both far- and near-ultraviolet spectra except that the ellipticity at 190 nm is increased (Figure 7). Although the origin of this increase at 190 nm is not clear, the protein conformation is relatively unaltered. However, steady state kinetics (5) have indicated that there is a compulsory order of reaction with substrates starting with the amine substrate and then with $O_2$ and therefore no CD changes would be expected with $O_2$ in the absence of the amine substrate.

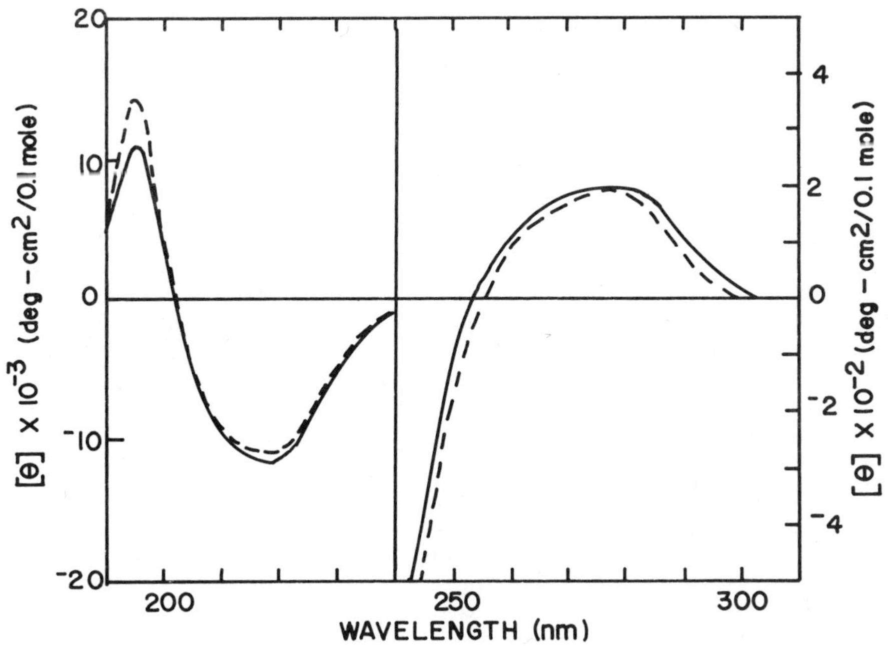


Figure 7. The far- and near-ultraviolet spectra of native enzyme (——) and deoxygenated enzyme (---) in 0.1 M potassium phosphate buffer, pH 7.0. The enzyme concentration and specific activity were 1.57 mg/ml and 750 units/mg, respectively.

## SUMMARY

Among the type II copper proteins, very few or limited CD studies have been performed. Ettinger and Kosman (15,16) have examined the CD spectrum of galactose oxidase and concluded that in the visible wavelength region, possibly 5 CD bands are present although only four are expected on a theoretical basis. In addition, they examined the effects of substrate and products anaerobically and aerobically on circular dichroism spectra. Both the substrate, D-galactose and the product, D-galactohexodialdose had marked effects on the visible and near-ultraviolet CD spectra but not in the far-ultraviolet wavelength region. Very limited CD studies have been performed on the Cu-amine oxidases which are unique copper proteins since they are pink in color. Adachi (6) has examined the visible CD spectrum of the _Aspergillus niger_ amine oxidase and noted that three CD extrema were observed at 330, 410 and 480 nm. These were abolished by treatment of the enzyme with 6 M guanidine HCl-0.1 M mercaptoethanol which showed that the copper complex in the amine oxidase was optically inactive. Possibly five CD extrema were observed at 330, 370, 430, 480 and 610 nm. The CD spectra of the animal and mold Cu-amine oxidases show some similarities but are different.

Also, as mentioned in this report, some preliminary work on the CD spectral properties of the bovine plasma amine oxidase in the presence of its substrate, some products, and copper chelating reagents in the region 190-610 nm have been investigated. The far-ultraviolet CD spectrum shows that there is a relatively small amount of α-helical structure present in the enzyme, and the predominating secondary structure may be a β-sheet structure. There is an indication that the active site of the enzyme may be situated in a segment which has a β-type secondary structure. The presence of aromatic amino acid residues at the active site of the enzyme is feasible but no direct evidence was obtained from this CD study. A conformational change is induced upon heat denaturation of the enzyme. Analysis of the CD spectra suggests the formation of α-helical structure although it is not possible to distinguish whether the transformation occurs from β-structure into α-helices, or from the conversion of the random coil structure portion of the protein into α-helices. The visible CD spectrum indicates the copper complex is optically active. The heat denaturation experiment shows that it is inherently optically inactive and that the Cotton effects are induced upon interaction of the copper with the asymmetric protein. Although the changes in the visible CD spectra of the enzyme in the presence of ammonia and cyanide may be due to gross-protein conformational changes, ammonia and hydrogen peroxide show a relatively similar mode of binding to enzyme, whereas the copper chelating reagents, cyanide and fluoride exhibit a different mode of binding to the copper complex in the enzyme. Such information on the mode of binding of these ligands is valuable for the interpretation of fast kinetics experimental data and elucidation of the reaction mechanism. There are a number of points that must be emphasized at this point. First of all, there is a disagreement whether the amine oxidase contains only copper or whether it also contains, in addition, an aromatic carbonyl containing organic cofactor which is covalently linked to the protein. If an organic cofactor is present, it may contribute to the CD spectrum and complicate the interpretation of the spectrum. On the other hand, if there is no such cofactor in the enzyme, the CD spectral interpretation is straight forward. CD studies of the apoenzyme may show whether an organic cofactor is present in the enzyme. Secondly, there is a need to analyze the CD curves in terms of the theoretical equations developed by Chen _et al_ (17) or by Baker and Eisenberg (18) in order to assess more accurately the secondary structure of the enzyme in the various experiments performed. In addition, $CN^-$, $H_2O_2$, $NH_3$ and $F^-$ all are known to react with copper and other methods such as ESR spectroscopy must be used to look for direct interaction of the copper in the enzyme with these above-mentioned compounds.

## REFERENCES


1. Achee, F.M., Chervenka, C.H., Smith, R.A., and Yasunobu, K.T. (1968) _Biochemistry_ _7_, 4329.
2. Yasunobu, K.T., Ishizaki, H., and Minamiura, N. (1976) _Molecular and Cellular Biochemistry_ in press.
3. Yamada, H., and Yasunobu, K.T. (1962) _J. Biol. Chem._ _237_, 3077.
4. Inamasu, M., and Yasunobu, K.T. (1974) _J. Biol. Chem._ _249_, 5265.
5. Oi, S., Inamasu, M., and Yasunobu, K.T. (1970) _Biochemistry_ _9_, 3378.
6. Adachi, O. (1969) Ph.D. Thesis, Kyoto University
7. Yamada, H., and Yasunobu, K.T. (1962) _J. Biol. Chem._ _237_, 1511
8. Tabor, C.W., Tabor, H., and Rosenthal, S.M. (1954) _J. Biol. Chem._ _208_, 645.
9. Vanngard, T. (1972), in _Biological Applications of Electron Spin Resonance_, Swartz, H.M., Bolton, J.R., and Borg, D.C. ed., New York, N.Y., Wiley Interscience, pp 411.
10. Brill, A., Martin, R., and Williams, R. (1964) in _Electronic Aspects of Biochemistry_, Pullman, B., ed., New York, N.Y. Academic Press, pp. 519.
11. Malkin, R., and Malmstrom, B. (1970). _Advan. Enzymol._, _33_, 177.
12. Greenfield, N., and Fasman, G.D. (1969) _Biochemistry_ _8_, 4108.
13. Timasheff, S.M., Susi, H., Townend, R., Mescanti, L., Gorbunoff, M.J., and Kumosinki, T.F. (1967) in _Conformation of Biopolymer_, Ramachandran, G.N., ed., New York, N.Y., Academic Press, pp. 173.

14. Catsimpoolas, N., and Wood, J.L. (1966) J. Biol. Chem. 241, 1790.
15. Ettinger, M.J. (1974) Biochemistry 13, 1242.
16. Ettinger, M.J., and Kosman, D.J. (1974) Biochemistry 13, 1247.
17. Chen, Y.H., Tang, J.T., and Martinez, H.M. (1972) Biochemistry 11, 4120.
18. Baker, C.C., and Isenberg, I. (1976) Biochemistry 15, 629.

## Acknowledgements

The research project was supported in part by research Grant MH 21539 from the National Institutes of Health, GB 18739 from the National Science Foundation and the Hawaii Heart Association.

SUBJECT INDEX


Acetyl-ferredoxin 6
function of $NH_2$ groups 9
reaction with reductase 9
salt effect on 7
spectrum of 6
Active site of Cu-oxidase 424, 464,501
Adrenodoxin reductase 40,286, 291
Agaricus bispora tyrosinase 466
Allysine 438,460
Amine oxidase
bovine plasma enzyme 575
Cu-amine oxidase 575
CD spectra of 575
Amino acid composition of
azoferredoxin 85
cytochrome P-450 265
protocatechuate 3,4-dioxygenase 131
$P\text{-}450_{11\beta}$ 288
$P\text{-}450_{scc}$ 288
putidaredoxin 265
putidaredoxin reductase 265
Amino acid sequence of
adrenodoxin 266
chloroplast ferredoxins 3,18
C. pasteurianum azoferredoxin 88
cytochrome P-450 268
D. digas rubredoxin 62
D. vulgaris rubredoxin 61
Fe- and Mn-superoxide dismutase 545
protocatechuate 3,4-dioxygenase 132
putidaredoxin 266
Amino tyrosyl adrenodoxin 46
Anacystis nidulans ferredoxin 16
Anaerobic era 561
Arginine modification 9,156
of chloroplast type ferredoxin 9
of submitochonrial particles 156
Ascorbate acid 340,472
Ascorbate oxidase 472
formal mechanism 477
physicochemical significance of 486
previous stopped flow data 478
rate constants of 486
reaction inactivation 482
reversibility of steps 481
step with a pK of 7.9 482
Azoferredoxin component of nitrogen 83
Azurin 242, 425

Bacterial ferredoxin 45
Benzoate 2,3-dioxygenase 118-126
characterization of components 121
from pseudomonas arvilla C-1 119
NADH-cytochrome c reductase in 119
purification of components 119-120
reaction mechanism of 122-125
Binuclear copper clusters 464
active site of oxidases 464
ascorbate oxidase 467,472
Cancer magister hemocyanin 465
ceruoplasmin properties 464,465
criteria for binuclear copper clusters 468,469
dopamine-β hydroxylase 467
mammalian cytochrome oxidase 228,468
mushroom tyrosinase 466
Polyrorus laccase 464,466
type 3 copper protein criteria 464
Bovine adrenodoxin 36
catalytic properties of 52
EPR spectrum of derivatives 48,51
Fe and S determination of 42
molecular and catalytic properties of 39

nitro- and amino-tyrosyl derivatives of 47,49
preparation of 37
reaction with adrenodoxin reductase 52,53
reconstitution of apoprotein 38
superoxide formation 42
tyrosine modification of 46

Bovine adrenal cytochrome $P\text{-}450_{11\beta}$ 281,290
absorption spectrum of 285
amino acid composition 288
effect of Triton X-100 296
molecular weight of 283
products from deoxycorticosterone 287
purification of 282,293
substrate induced spectral changes 286

Bovine adrenal cytochrome $P\text{-}450_{scc}$
absorption spectra, of 284
active form of $P\text{-}450_{scc}$ 307
cleavage of 20S, 22 R-dihydroxy chloresterol 306
molecular weight of 283,309
purification of 281
role in steroid genesis 303
stoichiometry of side chain cleavage 304
substrate induced spectral changes 286

Bovine heart cytochrome oxidase 228
heme a and Cu environment of 228
involvement of Cu in CO-derivative of 235
microenvironment of heme a and Cu in 238
new 575 nm absorbing species in the reduced form 229
new photo- and theochromism in 229

Bovine pineal tryptophan hydroxylase 103-117
activation by sulfhydryl reagents 103,107
activity determination of 104
comparison with brain enzymes 114,115
kinetics of 107
p-chlorophenylalanine as substrate 109,110
product inhibition of 112,113
pterin dependency 103
purification of 105
role in serotonin biosynthesis 103
substrate and cofactor effects 110,112
substrate specificity of 108,109

Brain
tryptophan hydroxylase 103

Carbon-13 NMR of ferredoxin 31

Carboxypeptidase A hydrolysis of ferredoxin 4

Catalase 369,493

Ceruloplasmin 505
ability to oxidize biogenic amines 511
biological role and oxidase activity 517
catalytic activity 506
ferroxidase activity 510
ferroxidase and iron mobilization 517
inhibitors 516
mechanism of oxidation of aromatic amines 512
regulation of plasma or tissue levels of biogenic amines 524
substrate specificity 508
transport of copper 520

Chloroplast ferredoxin 1,16
amino acid sequence of 3,18,19
arginine residues of 9
tryptophan residues of 9

CD spectrum of Cu-amine oxidase 575-588
effect of ammonia and $H_2O_2$ 580
effect of cyanide and fluoride 581
effect of heat 576
effect of $O_2$ 585

Clostridium acidi-urici ferredoxin 2

fluorescence spectrum of 10
iron-sulfur cluster of structure 17
structure-function 1,16
Clostridium ME ferredoxin 2
Clostridium pasteurianum hydrogenase 68
activity, iron and spin content of 74,75
Fe-S centers of 68,73
function of 68
mechanism of action 76,79
physicochemical properties of 70,71
spectral properties of 72-73
Clostridium pasteurianum ferredoxin 2
Clostridium pasteurianum nitrogenase 83
azoferredoxin component 83
partial amino acid sequence of 83
Collagen 438,443,448,460
Copper chelation 565
amino acid composition of 571
biosynthesis of 567
compounds which release Cu from 573
EPR spectra of 572
properties of 567
purification of 569
Copper oxidases 408,424,438,447, 464,472,489,505,575
ascorbate oxidase 424
azurin 424
carbonic anhydrase 426
carboxypeptidase 426
CO-bovine carbonic anhydrase spectra of complexes 432
CO-stellacyanin spectra of complexes 432
Cu(II) carbonic anhydrase optical properties of 426
diamine oxidase 427
EPR spectrum 424
galactose oxidase 427
geometry of complexes 429
laccase 424,428
optical, magnetic and redox properties 425
oxyhemocyanin 426,428
oxytyrosinase 426,428
pentacoordinate complexes 434
stellocyanin
optical properties of 425
type 1, copper 424
type 2, copper 424
optical and magnetic properties of 427
type 3, copper optical properties of 424,428
Copper reactions
EPR studies of 470
ferricyanide and ferrocyanide 470
Cu,Zn-superoxide dismutase 530
chemiluminescence by $H_2O_2$ 534
mechanistic studies 532
molecular weight 531
peroxidase activity of 534
reaction with azide and cyanide 535
reaction with $H_2O_2$ 533
replacement with other metals 532
sequence of 531
structural studies of 531
x-ray studies of 531
Cytochrome $c_3$ 58
Cytochrome $c_{533}$ 58
Cytochrome oxidase
bovine heart type 228
pseudomonas type 240
Cytochrome P-450 266,270,281,290, 303,314,321

Desmosine 447
Desulfovibrio gigas 57
Desulfovibrio vulgaris 57
Dioxygenases 118,126,127-136
Dithiothreitol 103,121

ETF dehydroxygenase 140
Equisetem ferredoxin 2

Ferredoxin 1,16
Ferroxidase 510

Flavodoxin 59

Galactose oxidase 489-504
  activators and inhibitors of 491-492
  Cu(III) in other enzymes 501
  Cu(III) state 499
  effect of ferricyanide 493
  effect of superoxide dismutase 493
  effect of superoxide onion 493
  EPR studies of 493
  kinetic properties 491
  mechanism of 500
  redox potential of 496
  superoxide and Cu(III) 489

Heme
  heme C and D of Pseudomonas cytochrome oxidase 248
  in peroxidase 382
Horseradish peroxidase 382
Hydrogenase 68

Indoleamine 2,3-dioxygenase 335,343,354
  copper content and heme content during purification 356
  effect of methylene blue 377
  Fe and Cu-content during purification 345,354
  from rabbit intestine 343,354
  inhibition by cuprous chelating agents 350
  inhibition of superoxide dismutase 341
  Km for superoxide anion 338
  Km for tryptophan 339
  molecular weight of 346,357
  new assay 335
  purification of rabbit intestinal enzyme 344,355
  reaction sequences 342
  spectrum of 349,360
Iron-sulfur clusters 17,42,51,73, 122,139,141,154,169,182
Iron-sulfur dioxygenase 127-136
  dioxygenase 127
  protocatechuate 3,4-dioxygenase 127
  pyrocatechase 132-134
Iron-sulfur proteins
  clustering of Fe-S centers and flavin 142
  Fe-S centers of the cytochrome $b$-$c_1$ segment 141
  future studies of 146
  in ETF dehydrogenase 139
  in NADH dehydrogenase 138
  in mitochondrial electron transport system 136-148
  in succinate dehydrogenase 140
  oxidation states and redox states of 143
  rates of oxidation-reduction of 144
Iron-superoxide dismutase 540-551
Isodesmosine 438,447

Laccase 408-423
  Chu shi laccase 409
  EPR spectrum of laccase 413
  extinction coefficient at 330 nm and 615 nm 411
  Ken shii laccase 409
  kinetics of copper reduction-oxidation 414
  Mao Ba-III laccase 409
  model of laccase 418
  oxidation-reduction of copper 414
  oxidation reduction potential of 413
  purification of 409
  reductive titration of laccase with absorbate 413
  Rhus succedana laccase 408
  Rhus vernicifera laccase 408
  spectrum of ascorbate oxidase 410
  spectrum of lacase 410
  total copper and valence states of laccase 412

turnover number 417
urshiol 408
Lathyrism 438,460
Liver microsomal mixed function oxidase 270
Liver microsomal P-450
benzphetamine demethylation by 274
butylatedhydroxytoluene effect 318
catalytic mechanism of 270
dimethylaminoazobenzene and diethylnitrosoamine effect 315
effect of cytochrome $b_5$ 278
hydroperoxide substitution 273
induction of synthesis 315
multiple forms of 270,271
P-450 levels during carcinogenesis 316,317
$P\text{-}450_{LM2}$ 271
$P\text{-}450_{LM4}$ 271
phenolorbital induced 270
proposed reaction mechanism 277,278
role of carcinogenesis prevention 314
spectral intermediate of 274
two electron mechanisms 272
Lysyl oxidase 438,447
allysine 438
β-amino proprionitrile 438
chick cartilage enzyme 438
collagen cross-linking 438,439
dehydrohydroxylysinohydroxynorleucine 438,444
dehydrohydroxylysinonorleucine 438,444
effect of aspartic acid on activity 440
fibril formation 440
function of 438
hydroxyallysine 438
lysyl oxidase assay 439,440
purification from cartilage 440
SDS-acrylamide electrophoresis 442
substrate cross-linking 438,439
Lung lysyl oxidase 447
activity of lung 452
activity with age 453
apoenzyme 458
assay of 454
pH optimum 456
physicochemical properties 458
purification of 448
Lysine
in chloroplast ferredoxin 6

Manganese superoxide dismutase 540,551
Melatonin biosynthesis 103
Model P-450 compounds 321
axial ligation in oxidized form 321
b-type cytochrome 322
high spin forms 323
low spin forms 325
physicochemical properties of Fe(III) thiolates 324
protoporphyrin dimethyl ester 323
comparison with chloroperoxidase 328
x-ray structure of 329
Metallothionein
amino acid composition 567
Cd,Zn,Hg and Ag-forms 566
Metals
analysis of Cu 344,355,365,376
Mitochondria 137,138,150,161,182, 228
Monooxygenase 103
Myoglobin 385

NADH dehydrogenase 138
NADH- and NADPH-ubiquinoine reductase 150
butanedione effect on transhydrogenase 156
composition of 152
energy coupling properties of 158
NADH dehydrogenase 153

NADPH oxidation by 154
number of Fe-S centers 154
reactions catalyzed by 151
reductase (complex I) 150,151
reduction of chaotropes 152
reduction of Fe-S centers of 157
subunits of 153
trypsin treatment of 156
Nitrogenase 83
NMR of S. maxima ferredoxin
$^{13}$C-NMR spectrum 32

Oxygen
oxygen carrying proteins 382,465
oxygen content during evolution 561

Peroxidase
comparison with myoglobin 386
distal base of myoglobin and hemoglobin 382
effect of $pK_3$ on Kr and Ko 383,384
heme linked protonation 385
heme linked protonation in reduced forms 382
horseradish isozymes 382
myoglobin 385
Phenylalanine hydroxylase 91-102
hydrate biopterin as intermediate 101
hypothetical intermediate 98
hypothetical mechanism of 96
new intermediate 91-102
sephadex G-50 effect 93
spectrum of new intermediate 96
stimulating protein 91
Plant ferredoxin
structure-function studies 1,16
Protein=metal interactions
x-ray analysis 389,532
Protocatechuate 3,4-dioxygenase 127
amino acid composition of 131
end group analysis 132
subunit structure of 129
Pseudomonas arvilla C-1
benzoate hydroxylase 118-126
Pseudomonas cytochrome oxidase 240
EPR spectra 249
heme c in 241
heme d in 241
reaction with azurin 242
reaction with CO 244
reaction with $O_2$ 246
removal and reconstitution of heme d 248
stopped flow kinetics 242
Pseudomonas camphor monoxygenase system 254
cytochrome P-450 254
equilibrium states and dissociation constants 257,258,259
heme alignment in $P\text{-}450_{cam}$ 258
Mossbauer and magnetic susceptibility 259
N-ethylmaleimide modification of 258
oxygen binding of P-450 259
partial sequence of $P\text{-}450_{cam}$ 263
putidaredoxin reduction by P-450 258
reaction mechanism of 256
sequence of putidaredoxin 266
substrate binding of 258
thermodynamics of 257
x-ray studies of 258
Putidaredoxin 265
Pyridylethyl ferredoxin 4
Pyrocatechase
subunit structure of 133-135

Rubredoxin 57-67
amino acid sequence of 61-64
function of 60
phylogenetic tree from 65

Scenedesmus quadricanda ferredoxin 1
Serotonin biosynthesis 103
Spirogyra ferredoxin 2

Spirulina maxima ferredoxin
  $^{13}C$-NMR spectrum of 31
Spirulina platensis ferredoxin 1
  acetyl-derivative 6
  arginine residues of 9
  carboxypeptidase digestion of 4
  fluorescence spectrum 10
  N- and D- forms 7
  pyridylethyl-derivative 4
  reconstitution of 4
  tryptophan residues of 9
  tyrosine residues in 2
Stellacyanin 411, 428
Succinate dehydrogenase 161,182
  action of carboxins 172
  detection by ESR 184
  discovery of non-heme iron 183
  effect of $CN^-$,TFF,carboxin and Q depletion 164
  effect of preincubated succinate 214
  EPR studies of 168
  EPR of reconstitutively inactive enzyme 201
  Fe-S centers 170
  Fe-S centers in soluble enzyme 193
  ferricyanide reductase activity 175
  g = 1.94 signal 191
  HIPIP and reconstitution activity effect of membrane interactions 175
  labile sulfide in 183
  labile sulfide in soluble enzyme 192
  natural electron acceptor of 218
  problems of electron transfer 211
  reaction with cyanide 204
  reconstitution activity 185
  thiocyanide liberation 208
  turnover number of different preparations 162
  2-theonyltrifluroacetone effect 211
  types and localization of Fe-S centers 214-218
  Q-cycle 222
Sulfate reducing bacteria 7
Sulfhydryl group
  in azoferredoxin 89
  in ferredoxin 12,16
Superoxide dismutase 426,530,540,551
  absence of Cu,Zn-enzyme 546
  antibody to Cu,Zn-enzyme 556
  antibody to Fe-enzyme 556
  bacillus stearothermophilus 542
  chick liver mitochondrial superoxide dismutase 541
  Cu,Zn-type 551
  cyanide sensitivity 551
  distribution in photosynthetic organism 551
  E. coli Fe- and Mn-superoxide dismutase 540
  Fe and Mn-type 540,551
  Fe-enzyme in procaryotes 546
  $H_2O_2$ production 553
  human liver SOD 542
  in angiospermae 557
  in bacteriophyta 560
  in chloroplast 551,552
  in eucaryotes 548
  in eucaryotes algae 551
  in fern and moss 558
  in higher plants 557
  in procaryotes algae 559
  in photobacterium leiognathi 556
  in photosynthetic bacteria 560
  induction by light 554
  lammaellae enzyme 547
  $NH_2$-terminal sequence homologies 545
  photobacterium sepia and leiognathic SOD 542
  phylogenetic distribution of SOD 546
  Plectoneme boryanum 541
  Pleuotus olearsis superoxide dismutase 541
  sacchromyces cerevisiae SOD 542
  spectral properties of the Mn- and Fe-superoxide dismutase 543

streptococcus mutants 542
structure and metal content 541
subcellular distribution of superoxide dismutase 546

Taro ferredoxin
preparation of 21
reductase preparation of 21
Transferrin 400
anion structural requirements 405,406
carbonate site 402
effects of chelating agents 402-404
Fe release 401
Fe site 401,402
functions of 400
mechanisms of Fe release 405
sequence of binding 401
synergestic ions 404
Tropoelastin 447
Tryptophan in chloroplast-type ferredoxin 9
Tryptophan hydroxylase 103
Tryptophan 2,3-dioxygenase
from pseudomonas 363,374
allosteric nature 374
copper content 377
copper content during purification 365,367
$^{64}$Cu-binding 379
Cu-dispute 374,375
effect of copper-specific chelators on enzyme activity 369
EPR spectra 376
ESR spectrum 376
forms of different oxidation states 380
non-involvement of copper 363
Pseudomonas acidovorans enzyme 374
Pseudomonas flourescens enzyme 363,374
rat liver enzyme 374
Tsou effect 211
Two iron ferredoxin 16
hydrolysis by proteases 22-30
Tyrosine hydroxylase 103

Ubiquinoine 146,150,165,187
Urushiol 408

X-ray absorption spectroscopy 389
electron storage rings 389
EXAFS spectra 390
extended fine structure 397
of $FeF_2$ and $FeF_3$ 391
of $KMnF_3$, $KFeF_2$, $KCOF_3$, $KNiF_3$, and $KZnF_3$ 392
of rubredoxin 390
of superoxide dismutase 395
synochroton radiation 389

Zinc
in carbonic anhydrase 426
in superoxide dismutase 531, 551